Supramolecular Chemistry of Biomimetic Systems

Junbai Li
Editor

Supramolecular Chemistry of Biomimetic Systems

Springer

Editor
Junbai Li
Institute of Chemistry, Chinese Academy of Sciences
Beijing
China

ISBN 978-981-10-6058-8 ISBN 978-981-10-6059-5 (eBook)
DOI 10.1007/978-981-10-6059-5

Library of Congress Control Number: 2017949116

Printed on acid-free paper

This Springer imprint is published by Springer Nature
The registered company is Springer Nature Singapore Pte Ltd.
The registered company address is: 152 Beach Road, #21-01/04 Gateway East, Singapore 189721, Singapore

Preface

Molecular biomimetics is an emerging active field and important frontier with interdisciplinary subjects of chemistry, physics, molecular biology, and nanotechnology. Taking lessons from biology, molecular biomimetic is the attempt to construct new highly ordered supramolecular structures through artificial molecule design and assembly at molecular level, which mimic the structures and functions of a unit or organ in biological systems. It is based on the fact that many biomolecules or bioactive molecules like polysaccharides, lipid, peptide, proteins, and dendrimers can self-assemble into well-defined structures and further to a supramolecular architecture while combining with other organic, inorganic, or metal oxide compounds. The molecular biomimetic approach opens up new avenues for the design and utilization of multifunctional molecular systems with a wide range of applications in nanotechnology and biotechnology.

In this book, we present recent achievements in biomimetic systems based on supramolecular chemistry and their potential applications in the biomedical field. These include biomimetic membranes, biomacromolecules supramolecular chemistry, hierarchical organic/inorganic hybrid systems, and molecular assembly of motor proteins. We will show our readers the exciting challenges in this unique research area and we hope to convince them of many new research opportunities.

I fell honored to edit this book. I would like to sincerely acknowledge every contributors who have accepted my invitation and taken much time to write the chapters. Dr. Yi Jia has spent a lot of time to organize every chapters and revise the entire book Figures and Tables. She shall be also deeply acknowledged. I am also grateful to all help from the surrounding students.

Beijing, China
April 2017

Junbai Li

Contents

About the Editor

Junbai Li obtained his Ph.D from the Department of Chemistry, Jilin University in 1992. He was then a postdoctoral fellow at the Interface Department of the Max Planck Institute of Colloids and Interfaces in Germany from 1994–1996. He is currently a professor at the Institute of Chemistry, the Chinese Academy of Sciences. His main research interests are on molecular assemblies of biomimetic systems, self-assembly, biointerfaces, design and synthesis of bioinspired materials with various nanostructures.

Part I
Introduction to Supramolecular Chemistry and Biomimetic Systems

Chapter 1
Molecular Biomimetics and Molecular Assembly

Junbai Li

Abstract Molecular biomimetics is mimicking the structures and functions of biological systems at the molecular scale via molecular assembly of biomolecules or synthetic components. It not only provides experimental models for guiding researches on biological mutation and evolution in organisms, but also opens up new avenues for the design and fabrication of novel functional materials. In this chapter, main contents of individual chapters are briefly introduced to give readers an overview of this book.

Keywords Molecular biomimetics · Layer-by-layer assembly · Self-assembly · Biomimetic membranes · Biomacromolecules · Motor proteins · Dendrimer · Polyoxometalates

Biological systems have evolved over millions of years to adapt to nature and develop highly fascinating structures and outstanding performances. Biomimetics, inspired by biological structures and their functions, focused on emulating or duplicating biological systems using biomolecules or synthetic components via diverse approaches. With the rapid developments of molecular biology and nanotechnology, biomimetics is now entering the molecular scale, that is, molecular biomimetics. The capability of biological molecules to self- and co-assemble into highly ordered nanostructures provides molecular assembly with a novel approach in which well-defined structures and biological functions similar to those existing in nature are produced. Thus, molecular assembly of biomimetic systems opens up new avenues for the design and fabrication of well-defined nanostructures and multifunctional materials with a wide range of applications in nanotechnology and biomedicine. A major advantage of molecular assembly of biomimetic systems is that both biochemical and physical parameters can be controlled precisely.

J. Li (✉)
Beijing National Laboratory for Molecular Sciences, CAS Key Laboratory of Colloid, Interface and Chemical Thermodynamics, Institute of Chemistry, Chinese Academy of Sciences, 100190 Beijing, China
e-mail: jbli@iccas.ac.cn

J. Li (ed.), *Supramolecular Chemistry of Biomimetic Systems*,
DOI 10.1007/978-981-10-6059-5_1

Therefore, it is feasible to utilize biomimetic systems as experimental models for guiding research on biological mutation and evolution in organisms. More importantly, biomimetics is not limited to just copying nature, with the development of modern biology, scientists can directly utilize biological units themselves to construct new types of systems sometimes as hybrid nanostructured materials. As will be illustrated in this book, natural molecular machines such as motor proteins are integrated into the engineering of active biomimetic systems, as well as the construction of artificial molecular machines via molecular assembly of synthetic components.

This book covers practical techniques and potential applications of biomimetic systems based on supramolecular chemistry. It also presents the advantages and an overview of supramolecular chemistry of biomimetic systems that consists of the following four topics covered in individual chapters.

1.1 Nanoarchitectonics of Biomimetic Membranes

Biomembranes are key components in biological systems that are composed of proteins, lipids, and polysaccharides. They form the natural boundary between cells to separate inner components from outer environment. The interactions between biomembranes and various biomolecules play an important role in many biological processes, including molecular recognition, signal transfer and so forth. Investigation of these processes requires constructing artificial model membranes. Fortunately, as one of the most important assembly strategies, molecular biomimetics will be the optimal method, which have great implications in exploring membrane materials and intensifying membrane processes.

In Chap. 3, we primarily focus on nanoarchitectonics of biomimetic membranes and introduce various examples of nanoarchitectonics approaches for bio-related thin films and membranes, including two-dimensional biomimetic membrane (molecular recognition, receptor tuning, and nanomaterial film for life control) and layer-by-layer biomimetic membrane (bioreactor, hierarchic assembly, sensing, and drug delivery).

1.2 Biomacromolecules Based Molecular Assembly

Self-assembly of biological building blocks has attracted increasing attention due to their biocompatibility, biodegradability, functional versatility, easy available for bottom-up fabrication, and widespread applications. Many biomolecules including polysaccharides, peptides, and proteins can interact and self-assemble into highly ordered supramolecular architectures with functionality. The self-assembly process is achieved through synergistic effects of some weak noncovalent interactions such as hydrogen bonding, electrostatic interaction, hydrophobic forces, π–π stacking interactions, chiral dipole–dipole interactions, and so on. Therefore, the precise control of supramolecular architectures and performances could be easily achieved by changing the concentration, solvents or adding functional molecules. These biomolecules-based assemblies have potential applications in both biological and nonbiological areas, including drug delivery, blood substitutes, bioseparation, biosensors, photocatalysis, as well as functional templates for nanofabrication.

In Chaps. 4–6, we mainly describe the molecular assembly of polysaccharides and proteins. Layer-by-layer assembled polysaccharides and proteins multilayers/microcapsules were constructed via different driving forces (electrostatic interaction, hydrogen bonding, covalent bonding, etc.). Layer-by-layer assembly technique provides these multilayers/microcapsules with engineered features including sizes, shape, thickness, composition and permeation. The applications of these biomacromolecules-based multilayers/microcapsules in drug delivery, blood substitutes, biosensors and photoelectrochemical cells were addressed.

In Chap. 7, we present the controlled self-assembly of aromatic dipeptides and the modulation of the self-assembled nanostructures. Besides, the co-assembly of aromatic dipeptides with other functional molecules and materials, including porphyrins, azobenzenes, photosensitizers, polyoxometalates, quantum dots, and glutaraldehyde, were discussed. Finally, we highlight the applications of peptide-based nanomaterials in the areas of biomimetic photosystem, photocatalysis, and biomedicine.

In Chap. 8, we summarized the recent advances in self-assembly based surface modification of different guest substances on the surfaces of cellulose nanofibers of bulk natural cellulose substances for the preparation of functional nanomaterials as well as their related applications. Versatile guest substrates such as metal oxide, nanoparticles, small molecules, polymers, biomacromolecules, and sp2-hybridized carbons were employed for the self-assembly approaches, yielding various nanostructured materials with diverse applications.

1.3 Molecular Assembly of Motor Proteins and Artificial Micro/Nanomotors

Motor proteins or molecular motors are active proteins that abound in biological cells, which could transform chemical energy into mechanical work. They play essential roles in the activities of cells and regulate specific functions through their stimuli–responsive mechanical motions. Molecular motors can be divided into two categories: linear molecular motors (myosin, kinesin, dynein, etc.) and rotary molecular motors (F_oF_1-ATP synthase, bacterial flagella). Linear motors myosin drive muscle contraction and kinesin or dynein transport vesicles from one end of the cell to the other. All of these linear motors are supplied by biological energy through the hydrolysis of the universal "fuel" molecule, adenosine 5′-triphosphate (ATP). The rotary molecular motor F_oF_1-ATP synthase (or F_oF_1-ATPase) is responsible for the catalytic synthesis of ATP molecules in biological organisms. Therefore, the integration of natural molecular machines such as motor proteins into active biomimetic systems is of great importance that has great implication in biological process and has great potential as smart multifunctional micro/nanodevices.

In Chap. 9, we present a general view of myosin family and focus on the structure, function, and regulation of three myosins, i.e., smooth muscle myosin-2 (SmM), myosin-5 (Myo5), and myosin-19 (Myo19).

In Chap. 10, we introduced how biomimetics can be applied to the reconstitution of F_oF_1-ATPase in artificial systems and explored their biomimetic performance in vitro. Several functional components, such as actin filaments, metal nanoparticle/nanorods, dimeric particles and DNA double-strand linked gold nanorod were employed to couple with F_oF_1-ATPase to perform special task or act as biosensors. The reconstitution of F_oF_1-ATPase in liposomes, layer-by-layer assembled protein or polymer-based microcapsules was particularly stressed and their application in the synthesis of ATP was demonstrated. Moreover, the controlled synthesis of ATP could be easily modulated by external fields and these biomimetic systems can serve as a smart container for the storage and release of the synthesized ATP.

Although biological motors are capable of complex and intricate functions, a key disadvantage of their application in vitro arises in their inherent instability and restrictions in their operational conditions. By contrast, wholly synthetic systems can tolerate much more harsh conditions than biological motors and offer considerable advantages in the development of smart micro/nanodevices.

In Chap. 11, we introduced the recent progress on the fabrication and regulation of synthetic micro-/nanomotors based on the controlled molecular assembly (especially layer-by-layer assembly) assisted by several "top-down" technologies. Diverse functional components, such as biomacromolecules, metal nanoparticles (nanoshells or layers) were employed as building blocks and were incorporated into the assemblies. The potential applications of the self-assembled micro-/nanomotors in biomedical fields, such as targeted delivery and remotely controlled release of drugs as well as the photothermal therapy of cancer cells were demonstrated.

1.4 Hierarchical Dendrimer, Polyoxometalates Complexes and Inorganic-Organic Hybrid Systems

Dendrimers, also known as cascade, cauliflower, or starburst polymers are booming in recent years because of their unique structures and properties. They are a class of highly branched, monodispersed, multivalent, synthetic macromolecules with abundant terminal functional groups. They have significant advantages over other polymers due to their well-defined structure, possibility of facile surface modification, and capacity of entrapping diverse molecules or nanoparticles. High generation dendrimers exhibit three-dimensional spherical structure with their sizes and physicochemical properties resembling those of biomolecules, e.g., proteins, therefore they are often referred to as 'artificial proteins'. All these merits enabled them an excellent building block for molecular biomimetic systems.

Poly(amidoamine) (PAMAM) dendrimers are the first complete dendrimer family to be synthesized, characterized and commercialized. In Chap. 12, we review some recent advances made in multifunctional poly(amidoamine) dendrimer-based nanoparticles for gene delivery applications. The surface amine groups of poly (amidoamine) dendrimers were conjugated with functional molecules (hydrophobic moieties, β-cyclodextrin, polyethylene glycol, etc.) and targeting ligands (folic acid, arginine-glycine-aspartic peptide), while the unique interior of dendrimers were employed to entrap metal nanoparticles. These modifications render the dendrimer-based vectors with the ability for targeted and enhanced gene delivery, including pDNA and siRNA delivery.

Polyoxometalates (POMs) are a type of inorganic polyanionic clusters bearing well-defined topologic architecture that are comprised of transition oxo-metalates. The framework structures and compositions of polyoxometalates can be modulated widely through simple chemical synthesis. Due to their negatively charged features, active oxygen atoms, various dimensions, acidity and so forth, POM clusters provide various binding modes with metal ions and organic groups closing to biological nanostructure surfaces through hydrogen bonding, coordination, and electrostatic interaction, etc., which make them vigorous candidates for biological utilizations.

In Chap. 13, we first present a brief introduction about POMs' structures and fundamental properties, and then we discuss the interaction of POMs with amino acids, peptides and proteins, the cocrystallization of proteins with the help of POMs, the inhibitory effect of POMs on enzymes and some diseases, the mimetic enzyme functions of POMs for hydrolysis of peptides and proteins, the antiviral, antibacterial and antitumoral activity of POMs, and their bio-imaging features.

In Chap. 14, we present an overview of supramolecular architectures formed by nanoengineered polyoxometalates, including the layer-by-layer assembled POMs multilayer thin films, self-patterning porous films composed of POMs-based giant vesicles, onion-like structures formed by a fullerene C_{60}-POM hybrid and self-assembled honeycomb films of hydrophobic surfactant-encapsulated POMs at air/water interface, as well as the promising applications of those films.

Chapter 2
Advantages of Self-assembled Supramolecular Polymers Toward Biological Applications

Michal Halperin-Sternfeld, Moumita Ghosh and Lihi Adler-Abramovich

Abstract Supramolecular self-assembly provides a means of achieving "bottom-up" fabrication of nanoscale materials. Their mechanical properties and functionality arise from the assembly of relatively simple molecular building blocks. These materials have selective affinity to different interfaces, high capacity for interfacial adsorption, nanostructure, and spontaneous formation of unique nano-self-assemblies which exhibit remarkable simplicity and biocompatibility. Due to these attractive features, supramolecular nanostructures, particularly peptide-based, have recently been explored as effective nanomaterials in applications ranging from controlled release and drug delivery, nano-fabrication, skin care, biomineralization, sensing, antimicrobial materials, and tissue engineering. This range of applications is facilitated by the diverse primary sequences of the short peptides, which can be either biomimetic or de novo designed. Thus, their self-assembling mechanistic processes and nanostructures also vary enormously. This chapter highlights recent advances in studying self-assembled peptide systems, focusing on the formation of different nanostructures and their applications in diverse fields.

Keywords Self-assembly · Hydrogel · Peptides · Nanostructures · Supramolecular polymers

List of Abbreviations

SAP	Self-assembling peptide
CPTNs	Cyclic peptide nanotubes
CPs	Cyclic peptides
PAs	Peptide amphiphiles
RGD	Arginine–glycine–aspartic acid
CryoTEM	Cryogenic transmission electron microscopy
CMC	Critical micelle concentration

M. Halperin-Sternfeld · M. Ghosh · L. Adler-Abramovich (✉)
Department of Oral Biology, The Goldschleger School of Dental Medicine, Sackler Faculty of Medicine, Tel Aviv University, Tel Aviv 69978, Israel
e-mail: LihiA@tauex.tau.ac.il

J. Li (ed.), *Supramolecular Chemistry of Biomimetic Systems*,
DOI 10.1007/978-981-10-6059-5_2

DOX	Drug doxorubicin
CPP	Cell penetrating peptide
GQDs	Graphene quantum dots
FF	Diphenylalanine
Fmoc	Fluorenylmethoxycarbonyl
Boc-FF	Tert-Butyloxycarbonyl-diphenylalnine

2.1 Introduction

2.1.1 Self-assembly in Nature

Nature endows intricate mysteries regarding biologic processes underlying the formation of a wide variety of complex biological structures. Thus, it has been our greatest "teacher" of all times. Do we ever think of what it takes to turn an egg to an entire animal? Or how a densely packed group of hexagonal cells comprise multiple honeycombs parallel to each other, with a relatively uniform bee space? How does a seed turn into a single tree and the trees to a forest? Or how do sand grains collectively form a sand-dune or a desert? How, depending on different environmental circumstances, can simple water molecules take the form of rain to ocean, ice cube to iceberg, steam to cloud, and super-cooled cloud droplets freezing into magnificent snowflakes?

Nature accomplishes all these phenomena with high level of precision and complexity. It offers "bottom-up" nano-fabrication, constructing materials atom by atom or molecule by molecule to produce beautiful supramolecular self-assembled architectures. Self-assembly is a ubiquitous process in biology where it plays numerous important roles and underlies the formation of a wide variety of complex biological structures. Phospholipid molecules spontaneously arrange themselves in exceptionally ordered structures to form cell membranes and compartmentalization of elements within these membranes, a process which has probably led to the foundation of cells and thus to "LIFE." Protein folding, the formation of DNA double helix, morphogenesis, etc., are some of the most complex, spontaneously formed self-assembled systems. One of the easiest modes of understanding the intricacies of living systems is to explore their components. The basis underlying such processes lies in the tendency of single units to interact with each other to create a unique, complex, but highly organized system.

Over the past two decades, materials scientists have aspired to exploit nature's self-assembly principles to develop biomimicking systems. Toward this goal, both biological and synthetic building blocks have been subject of extensive research. In fact, molecular self-assembly has become increasingly important for the development of biomaterials as it offers a great platform for constructing materials with high level of precision and complexity, while integrating order and dynamics, to achieve functions such as stimuli-responsiveness, adaptation, recognition, transport, and catalysis. The importance of peptide self-assembling building blocks has been recognized in the last years, as they offer a great diversity of biochemical (specificity, intrinsic bioactivity, biodegradability) and physical (small size, conformation) properties to form self-assembled structures with different molecular configurations.

This chapter provides an overview of the design principles of peptide self-assembly and illustrates how these principles have been applied to manipulate self-assembly across the scales. Applications of self-assembling peptides as nanobiomaterials, including carriers for drug delivery, hydrogels for cell culture and tissue repair, are also described.

2.1.2 Self-assembly and Supramolecular Chemistry

Molecular self-assembly is a spontaneous organization of molecular units into ordered structures [1]. The assembly process is facilitated by recognition and association of molecules, controlled by the balance of attractive and repulsive forces within and between them, leading to weak non-covalent intra- and intermolecular interactions including Van der Waals, electrostatic, hydrogen bonding, and π–π stacking interactions [2, 3]. Together, these weak interactions form the basis for fabrication of very stable supramolecular architectures and bio-inspired nanomaterials with chemical complementarity and structural compatibility [4]. The arrangement of the nanostructures depends on molecular composition, assembly kinetics, and variations in the assembly environment (pH, solvent, co-assembling molecules, temperature, and ionic strength) [5, 6].

Self-assembly underlies the generation of many biological nanostructures. DNA double helix formation through hydrogen bonding interactions between nucleotide bases, formation of tertiary or quaternary protein structures through folding of a polypeptide chain, and the formation of cell membranes upon self-assembly of phospholipids are some of the most straight-forward and extensively studied self-assembled structures in a biological system.

The understanding of the spontaneous molecular self-assembly process, along with advances in the design and characterization of such self-organization principals in molecular engineering, have led to the fabrication of new materials using natural building blocks, such as phospholipids, oligosaccharides, oligonucleotides, proteins, and peptides. Amongst the wide variety of molecules often used in this nanotechnology field, several unique properties make peptides, peptide derivatives and proteins particularly attractive building blocks for the construction of supramolecular materials. First, they display an inherent ability to self-organize into well-defined structures. Second, they can deliver potent and selective biological signals to cells, thus enabling to manage varying cellular responses. Third, they can be fabricated by a number of means, including solid- and liquid-phase chemistry, biocatalysis, metabolic engineering, and recombinant expression biotechnologies.

Peptide chemistry is well-established, highly reproducible, and allows easy incorporation of non-peptidic moieties [7, 8]. In other words, peptides are easily accessible not only to chemists, but also to biologists, biochemists, and materials scientists who may not be capable of overcoming the synthetic challenges posed by other non-peptidic small molecules. Peptides thus seem ideal candidates for biological applications.

Self-assembling peptide (SAP) systems involve synthetic scaffolds capable of presenting multiple cell-interactive components in spatially resolved networks via supramolecular self-assembly. In addition, peptides have become known as immensely useful building blocks for the formation of self-assembled nanostructures in medical applications due to their biocompatibility, biodegradability, and versatility. They can be used for drug delivery, tissue engineering, scaffolds for regenerative medicine, matrices for cell cultures, antimicrobial agents, imaging tools, energy storage, biomineralization, and membrane protein stabilization. These applications employ self-assembled peptides with bioinspired nanostructures showing different properties, including nanotubes, nanofibers, nanospheres, nanobelts, and hydrogels. A number of peptide-based building blocks, such as amphiphilic peptides, surfactant-like oligopeptides, and aromatic dipeptides, have been designed and developed for generating supramolecular structures and their possible applications in biology and nanotechnology are explored. SAP systems are also capable of displaying functional amino acid sequences or chemical groups on the surface of their self-assembled fibers and these peptides can also serve as anchors for different ligands.

2.2 Supramolecular Materials

2.2.1 Cyclic Peptide Nanotubes

2.2.1.1 Design of Cyclic Peptide Nanotubes

In recent years, considerable effort has been devoted to the preparation of artificial nanotubular materials. One of the most successful approaches for the construction of non-covalently bonded nanotube entities is the self-assembly of cyclic polypeptides in stacks that are stabilized by hydrogen bonds.

In 1993, cyclic peptide nanotubes (CPTNs) were presented as a new class of organic nanotubes in a pioneering work by Ghadiri et al. CPTNs form hollow tubular nanostructures by self-assembly containing even numbers of alternating D- and L-amino acids [9]. This unique architecture results in flat ring-shaped subunits stacked together into a β-sheet conformation through intermolecular hydrogen bonds. The closed cycle and the alternating D- and L-conformations direct the side chains outward of the ring and the backbone amides approximately perpendicular to the ring's plane. The internal diameter of the nanotubes ranges between 7 and 8 Å and can be controlled by changing the number of the amino acids in the cyclic peptide sequence. By tailoring the chemical structure of the cyclic peptide moieties, the self-assembled supramolecular architectures can be adjusted to meet the requirements of a variety of applications, including antibacterial agents, stimuli-responsive nanomaterials, ion channeling, and ion sensing.

2.2.1.2 Application of Cyclic Peptide Nanotubes

(1) Antimicrobial agents

CPNTs have been of particular interest as antibacterial agents [10, 11]. They act on bacterial membranes and other generalized targets, thus making bacterial resistance unlikely [12]. The mechanism of their antibacterial activity includes binding to the target membrane, in which they transform their structure [13, 14]. At a certain threshold concentration, CPNTs permeabilize the membrane, either by forming a discrete pore or by disrupting the bilayer structure [15]. Membrane selectivity, a major requirement for antimicrobial materials, was assessed by subjecting the cyclic peptide to a hemolysis assay.

The antibacterial activity of CPNTs is governed by several factors including the ring size of the cyclic peptide being either six or eight amino acids, the type of amino acids composing in the peptide, and the type and number of basic amino acids in the peptide sequence [16].

In one early example, Ghadiri's group showed that octameric or hexameric alternating D,L-CPNTs cause permeabilization of the bacterial membrane, thereby showing antimicrobial activity [17]. Selective activity against gram positive (S. aureus, MRSA) or gram negative (*E. coli*) bacteria compared to mammalian cells was demonstrated. They also established that control peptides comprising linear sequences did not have antibacterial activity when compared to their cyclic counterparts [17]. In addition, hexameric cyclic peptides were shown to be less effective antibacterial materials than octameric CPNTs. They also found that increasing the number of basic residues in the cyclic peptide increased the antibacterial activity, while switching the chirality of the basic amino acids had no significant effect on either antibacterial or haemolytic activity. Introduction of histidine as a basic residue in the hexa- or octameric cyclic peptide, however, resulted in a loss of broad-spectrum antibacterial activity, along with an increase in haemolytic activity, and thus loss of membrane selectivity. In a recent study, the cyclic peptide Labaditin was proven as highly efficient in killing S. aureus. With assays of membrane permeability, it was found that Labaditin induced leakage in large unilamellar vesicles, via formation of pores [11].

(2) Ion channels

The ability of cyclic peptides to integrate into lipid bilayers and disrupt membranes makes them suitable for ion channeling [18]. Ghadiri's group elaborated this fact and have constructed artificial ion channel models that exploit the self-assembly of conformationally flat cyclic peptides (CPs) into supramolecular nanotubes [19]. They have also designed a cyclic peptide nanotube based on eight amino acids which was shown to serve as an artificial transmembrane ion channel and display transport activities for potassium and sodium [20]. The peptide incorporated leucine and tryptophan residues to favor its partitioning into lipid bilayers. The pores formed spontaneously upon addition of the peptide to an aqueous liposome dispersion. Single-channel conductance and proton efflux (pH-sensitive dye fluorescence) confirmed the presence of ion channels [20]. Another cyclic peptide composed of cyclo[-Trp-Dap-Leu-D-Ala-Trp-Ser-Val-D-Ala-Trp-Ser-Ile-Gly-] was found to be capable of forming artificial transmembrane ion channels by self-assembly of planar peptide rings, with hydrophilic groups arrayed in the interior of the channel [21].

(3) Ion sensors

The self-assembly of cyclic peptides into tubular channels in organosulfur self-assembled monolayers on gold films enables the construction of diffusion-limited size-selective sensors [22], which have been used to probe redox ion complexes in solution. Through measurements of redox activity, Motesharei and Ghadiri showed that negatively charged $[Fe(CN)_6]^{3-}$ and positively charged $[Ru(NH_3)_6]^{3+}$ complexes were able to traverse the channel lumen. However, upon the addition of the larger $[Mo(CN)_8]^{4-}$ anion, no redox activity was observed. In addition, selectivity was confirmed when a mixture of $[Fe(CN)_6]^{3-}$ and $[Mo(CN)_8]^{4-}$ was added to self-assembled monolayers [22]. In the recent works, a new self-assembly process based on an α,γ-cyclic peptide was utilized to design molecular rotors ion sensors [23]. In another study, a phosphorylated cyclic peptide was used as a fluorescent sensor for the detection of uranyl ions with high selectivity and sensitivity [24].

(4) Gene delivery

CPNTs have been used as dose-dependent inhibitors of adenovirus mediated gene delivery. Horne et al. showed that the designed cyclic D,L-α-peptides can rapidly permeate selected cellular membranes and counteract the development of a low pH environment inside endosomes [25]. The CNPTs could potentially prevent adenovirus infection, thus showing potential supramolecular approach toward the design and discovery of broad-spectrum antiviral agents.

(5) Responsive supramolecular polymers

The transition between intermolecular and intramolecular dimerization could be controlled using N-methylated peptides covalently linked together with an azobenzene linker. Upon irradiation at 366 nm, the azobenzene switches from the E to the Z isomer, leading to a switch from intermolecular dimers, which form a supramolecular polymer, to discrete intramolecular dimers [26]. Steinem et al. reported that the supramolecular polymers undergo reversible structural changes at the air–water interface upon UV or visible light irradiation. Such changes could provide a new route toward the design of novel photoactive materials [27].

2.2.2 *Peptide Amphiphiles*

Peptide amphiphiles (PAs) are composed of oligopeptides consisting of an N-terminal alkyl tail, a β-sheet-forming central segment, and a C-terminal functional segment representing a flexible platform for incorporating a variety of different molecular features [28]. These molecules generally assemble into high aspect ratio rods/cylinders with a hydrophobic core consisting of 12–16 carbon alkyl tails and the peptide presented radially from the core. Formation of rods by PAs is driven by a combination of both the hydrophobic interaction between the alkyl tails and the hydrogen bonding of the peptides, hence providing a β-sheet conformation [2]. PAs have been designed with increasingly complex and bulky functional domains such as fluorophores, branched with two arginine–glycine–aspartic acid (RGD) ligands or one YIGSR and one IKVAV ligand, cyclic RGD ligands, and others [28]. Stability can be provided by modifying the N-terminal alkyl by the inclusion of diacetylene groups, whereas flexibility has been demonstrated by incorporating proteolytically susceptible amino acid sequences in the central portion of the PAs [29]. The rigidity of the rods can also be altered by the addition of a phospholipid. A low percentage of phospholipids allows the rheological properties of the gel to be altered with a slight increase in the mechanical properties, possibly due to a more optimal geometry of interactions between PA molecules for hydrogen bonding [30]. Thus PAs, owing to their predictable self-assembly, their ease of synthesis, and their capacity for incorporating a wide variety of functional components and mechanical properties, have been used for a number of interesting applications.

Stupp and coworkers have developed a series of lipid-peptide molecules comprising a hydrocarbon chain (e.g., palmitoyl) covalently attached to an amphiphilic peptide (e.g., VnAnEn) which is able to form β-sheet-rich supramolecular structures. When dispersed in water, these PAs formed hydrogels at concentrations as low as 1% (w/v) in the presence of calcium ions that triggered gelation through charge screening. Interestingly, these PAs kept their hydrogel-forming capacity even after covalent conjugation of dexamethasone or prodan to the peptide moieties via acid-cleavable hydrazone bond [31, 32]. This feature makes these PAs a versatile system for sustained release of a wide range of medicines. While in most cases, PAs were utilized for drug delivery in hydrogel forms, in a recent study, PA fibers conjugated to a collagen binding peptide showed promising results for targeted delivery to an injured artery via systemic administration following vascular intervention. Importantly, findings indicated that applying the specific targeting ligands in combination with fibrous morphology was crucial to get the optimal binding at the site of interest in the vasculature [33].

Recently, Deshmukh et al. have shown the role of water molecules in the dynamic equilibrium of micelle-fiber formation in self-assembly of PAs [34]. The various stages of self-assembly from micelles, fibers, and bundled fibers take place at a particular time scale (Fig. 2.1a), thus providing extensive insights into early fiber formation and the role of solvent in the process of self-assembly [34]. Moreover, the ability of the fibers to structurally adapt to bind important bioactive targets was recently demonstrated [35]. PA molecules formed nanofibers in water, with a diameter of approximately 7 nm and lengths in the range of micrometers, as observed by cryogenic transmission electron microscopy (CryoTEM; Fig. 2.1b–c). Amphiphiles were fluorescently labeled with Cy3 and Cy5 fluorophores and the distribution of dyes integrated into initially single-color nanofibres was quantified using correlative image analysis. Fluorescently labeled nanofibers revealed similar morphology to that of non-labeled counterparts, as shown by cryoTEM (Fig. 2.1d). Diffraction-limited fluorescence microscopy images of Cy3- (Fig. 2.1e) and Cy5-labeled (Fig. 2.1f) PA nanofibers show similar morphology.

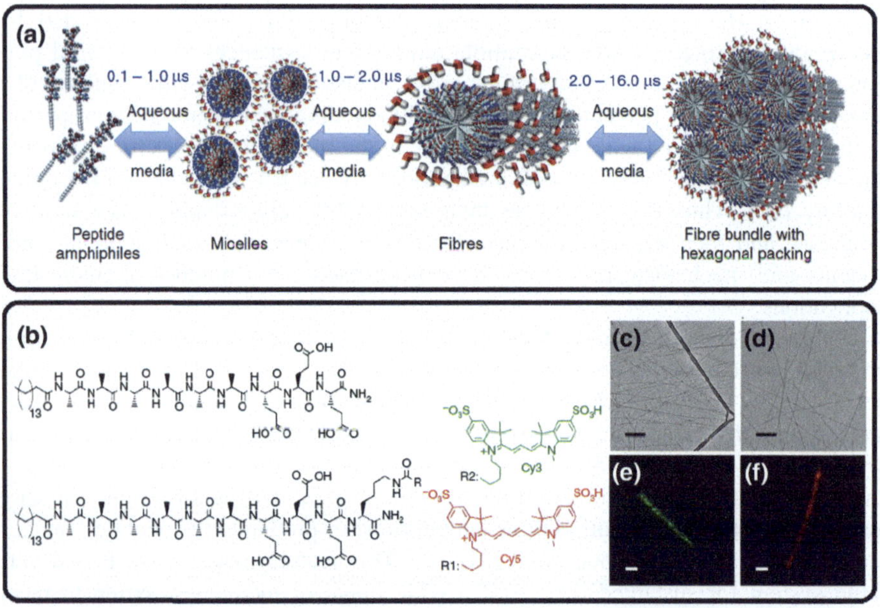

Fig. 2.1 Peptide amphiphiles self-assemble into various morphological structures. **a** Schematic representation of the various stages involved in PA self-assembly with the timescale denoted in *blue* (Reprinted with permission from Ref. [34]. Copyright 2016, Macmillan Publishers Limited, part of Springer Nature. **b** Molecular structure of non-labeled PA and PA molecules labeled with photo-switchable sulfonated cyanine dyes, namely Cy3 (*green*) and Cy5 (*red*). **c–d** CryoTEM images of nanofibres self-assembled in 150 mM NaCl, pH 7.5, **c** from non-labeled PA alone and **d** from a molecular mixture of non-labeled and Cy5-labeled PAs (*scale bar*, 200 nm). **e–f** Diffraction-limited fluorescence microscopy images of Cy3- (**e**) and Cy5- (**f**) labeled PA nanofibres (*scale* bar, 1 mm) (Reprinted with permission from Ref. [35] Copyright 2016, Nature Publishing Group, Macmillan Publishers Limited)

2.2.3 *Surfactant-Like Peptides*

Surfactants are defined as materials that can greatly decrease the surface tension of solvents when used at very low concentrations [36]. A typical surfactant-like peptide molecule consists of two parts: a hydrophobic tail composed of several consecutive hydrophobic amino acids and a hydrophilic head composed of one or two hydrophilic amino acids. Various surfactant-like peptides can be designed by selecting different hydrophobic or hydrophilic amino acids. For example, the hydrophilic head can be designed as positively charged Arg, Lys, and His, or negatively charged Asp and Glu, producing cationic or anionic surfactant-like peptides, respectively. On the other hand, the hydrophobic tail can be designed by choosing different hydrophobic amino acids such as Gly, Ala, Val, Leu, and Ile with different levels of hydrophobicity, so that the overall hydrophobicity of a surfactant-like peptide is controlled. The design of the typical surfactant-like peptide is based not only on the selection of amino acids, but also on the position of the hydrophilic head which can be set at either the C- or N-terminus [37]. The length of the surfactant-like peptide is controlled by the number of hydrophobic amino acids in the tail.

The assembly mechanism of the peptides leads to different nanostructures used for different applications. One self-assembly model of surfactant-like peptides is based on a tail-to-tail alignment to form bilayer structures which further form nanotubes and nanovesicles. In addition to forming a tail-to-tail bilayer structure, traditional surfactants can also form micelles by packing the tails in a hydrophobic core and exposing the hydrophilic heads outside. This kind of self-assembly model has also been observed for surfactant-like peptides, which can form nanofibers instead of nanotubes or nanovesicles [38]. Cationic surfactant-like peptides can form nanofibers, nanorods, and nanospheres with various lengths by packing in the unique orientations as described above.

Surfactants, as their name suggests, have a general tendency of moving to the surface (air/water) or interface (oil/water), and thus they are also known as "surface active agents." The dual characteristics of surfactant molecules show a wide range of properties mainly connected to two key features, namely adsorption at interfaces and self-assembly in bulk solution. Structure–function relationships on the surface and solution properties have been investigated for a wide range of surfactant types with the aim of utilizing them toward relevant applications. Depending on the nature of the hydrophilic moiety ensuring the water affinity of the molecule, surfactant-like peptides can be categorized into anionic, cationic, nonionic and zwitterionic.

2.2.3.1 Anionic Surfactants

Historically, anionic surfactants are the earliest and the most common surfactants. They are surface active substances in which, e.g., one hydrophobic group is connected with one or two hydrophilic groups. In aqueous solution, dissociation occurs into an anion and a cation. The anion is the carrier of the surface active properties. The ionized moiety can be a carboxylate, sulfate, sulfonate, or phosphate. Surfactants can be used as detergents (alkylbenzene sulfonates), soaps (carboxylic acids), foaming agents (lauryl sulfate), wetting agents (di-alkyl sulfosuccinate), and dispersants (lignosulfonates).

2.2.3.2 Cationic Surfactants

Cationic surfactants, which also contain a hydrophobic hydrocarbon group and one or several hydrophilic groups, also dissociate into cation and anion in aqueous medium. Here, however, the cation is the carrier of the surface active properties. Cationic surfactants are characterized by their very high substantivity on various substrates, especially the negatively charged ones, and by the subsequent surface modifications. They are often used as conditioning agents in fabric care and hair products. They are also used as germicidal agents (bactericides and fungicides) and to produce antistatic and hydrophobic effects, and are thus commercially valuable as corrosion inhibitors.

2.2.3.3 Nonionic Surfactants

Nonionic surfactants are surface active substances which do not dissociate into ions in aqueous solutions. The solubility of these substances in water is provided by polar groups such as polyglycol ether groups or polyol groups.

2.2.3.4 Amphoteric or Zwitterionic Surfactants

Amphoteric or zwitterionic surfactants are characterized by one anionic and one cationic functional groups whose dominance is usually regulated by the pH, namely anionic in alkaline pH and cationic in acidic pH. Near the isoelectric point, these surfactants display both charges and are amphoteric in nature with a minimum of interfacial activity and a maximum of water solubility. Amphoteric surfactants, particularly those composed of amino acids, are quite biocompatible and used in pharmaceuticals and cosmetics.

2.2.3.5 Gemini Surfactants

Over the past two decades, dimeric or gemini surfactants attracted considerable interest aiming at developing “next-generation” high-quality surfactants [39]. These surfactants are constructed by linking two monomeric surfactants with a spacer at the level of or in close vicinity to the head groups. They have many unique properties that are superior to those of their corresponding monomers, such as significantly lower critical micelle concentration (CMC), high surface activity, low Krafft temperature, unusual rheological properties, better wetting ability, etc. Their unique aggregate morphologies in solution are also of intense research interest [40]. Gemini surfactants are used as promising surfactants in industrial detergency, as nano-templating agents, as skin care agent, and for catalysis [41].

2.2.3.6 Applications of Surfactant-like Peptides

(1) Stabilization of membrane proteins

Surfactant-like peptides have shown promising potential in the study of membrane proteins. About one-third of all cellular proteins are membrane proteins, which contain at least one transmembrane domain and play vital roles in every aspect of cellular activities, including cell signaling, cell migration and movement, energy transformation, and substance transport [42]. However, due to their unstable nature, stabilization is required for analysis. In the past decades, various traditional surfactants, such as detergents and lipids, were used for stabilization, purification, and crystallization of membrane proteins. However, due to the complexity of membrane protein–detergent–lipid interactions, their efficacy is still lacking. In contrast, based on their molecular structure, surfactant-like peptides can bind to the hydrophobic section of a membrane protein through their hydrophobic tails and sequester it from the water, thus preventing its denaturation [43].

(2) Drug and gene delivery

Since surfactant-like peptide molecules can be easily designed and modified to form various nanostructures, they can be easily tailored for drug or gene delivery. The hydrophobic core within the nanostructure can potentially encapsulate water-insoluble molecules and deliver drugs and other biological molecules. Moreover, the hydrophilic head can be modified with functional groups for cell-targeting.

Surfactant-like peptides have been shown to form vesicular structures, depending on the primary sequence, peptide concentration, pH, and ionic strength of the dispersing media [44–46]. For example, the SA2 peptide (Ac-AAVVLLLWEE) forms discrete nanovesicles with a radius of 60 nm when dispersed in aqueous media at physiological pH. The formed peptide vesicles precipitate out of solution at pH values below the pKa of the glutamic acid side groups, which could be fully reversed when the pH was raised again to 7.4 [45]. SA2 peptide vesicles can be loaded with a photosensitizer with virtually no water solubility. Incubation of different cultured cells (HUVECS, COS-7, and C26) with SA2 peptide vesicles loaded with this photosensitizer resulted in accumulation of the photosensitizer inside the cells. Upon illumination to excite the delivered photosensitizer, concentration-dependent cytotoxicity was observed, in contrast to non-illuminated control cells [46].

Recently, self-assembled polymeric micelles from amphiphilic copolymers and a hydrophobic tail were investigated as nanocarrier systems for the controlled release of various anticancer drugs [47–49]. In an in vitro study, micelles were loaded with ibuprofen and the anticancer drug doxorubicin (DOX) and the sustained release behavior was examined [47]. Due to the incorporation of targeted RGD sequences and the cell penetrating peptide (CPP) residue octa-arginine (R(8)), the micelles could be specifically recognized by cancer cells, as well as be efficiently transported through the cell membrane. The loaded micelles showed high phototoxicity against cancer cells, indicating a powerful potential for effective photodynamic therapy. Furthermore, due to the low cytotoxicity of the peptide against both HeLa and 293T cell lines, the surfactant-like peptide developed in this study may be promising for targeted drug delivery in clinical application [47]. Cui et al. further synthesized reduction-sensitive micelles for enhanced drug delivery effect. These micelles, are stimuli-sensitive and may be triggered by changes in temperature, pH, light, magnetic field, ultrasound, and redox potential, and are thus considered "smart drug carriers" for tumor drug delivery. Upon reaching the target tumor cells, the drug-loaded micelles can be intracellularly localized and subsequently aroused by stimulus to rapidly release the drugs due to the transformation of the chemical structures or the physical properties of the carriers, resulting in aggressive activity inside tumor cells, enhanced therapeutic efficacy, and relatively few side effects. The reduction-sensitive polymeric micelles containing disulfide bonds have been extensively studied for tumor targeting due to the high difference in the redox potential between the mildly oxidizing extracellular milieu and the reducing intracellular fluids. Moreover, tumor tissues are highly reducing and hypoxic compared to normal tissues, with at least fourfold higher concentrations of reducing agents in tumor tissues [50]. The micelles developed by Cui et al. were based on the amphiphilic polymer mPEG-S-S-C16 synthesized by conjugating the hydrophilic mPEG with the hydrophobic alkyl chain by a reduction-responsive disulfide bond followed by self-assembly into micellar aggregates in aqueous solution. These micelles efficiently delivered DOX to the cell nuclei and showed enhanced drug effects when compared to control micelles without a disulfide bond [48].

Sufactant-like peptides have been suggested as DNA delivery systems for gene therapy. Due to the ability of cationic surfactant-like peptides to self-assemble into cationic micelles, the negatively charged DNA binding efficiency was increased [51]. Furthermore, co-delivery of DOX with luciferase reporter gene and p53 gene was demonstrated using micelles [52].

Gemini surfactants have also been proposed as candidates for gene therapy toward mitochondrial diseases. They were shown to successfully deliver a plasmid DNA construct designed by including a codon, which codes for an amino acid only if read by the mitochondrial ribosomes, thus bearing the potential to transform the therapeutic landscape of mitochondrial genetic diseases [53]. Various types of gemini surfactants have been designed for gene therapy for the treatment of various diseases. To this end, they have been utilized as candidates for the formation of non-viral vectors. The transfection efficiency of different types of gemini surfactants has been evaluated in a recent review [54].

(3) Tissue engineering

Surfactant-like peptides can be used as surface modifying molecules for scaffold-free tissue engineering by promoting cell adhesion and growth. A6K, a cationic surfactant-like peptide, was shown to self-assemble on mica surface to form a monolayer, thus turning the surface into a hydrophobic one, which is suitable for cell adhesion and growth. Along with the growth of cultured cells, A6K peptide composed of natural L-amino acids could be gradually biodegraded, and the hydrophilic mica surface could be re-exposed, thereby allowing for an easy release of the cells from the mica surface [55].

Recently, Panda and coworkers have developed a novel surfactant mediated fusion of polylactide particles into scaffold-like structures at room temperature, which enables the fabrication of the desired shape and size. In the presence of ethanol, evenly spread surfactant coated polylactide particles fused immediately into membrane-like structures. These scaffolds supported three-dimensional growth of animal cells in vitro. They were also found to be good wound dressing materials, as well as demonstrated to release a model protein in a sustained manner for a long period of time, making them suitable for various biomedical applications [56].

(4) Template for nano-fabrication

Recently, a surfactant-like AGD peptide, which could undergo self-assembly in nonpolar solvent system, was shown to bear a potential for the fabrication of nanostructures and nanodevices. This peptide was designed to have a shape like an inverted wedge which prevented it from self-assembling in an aqueous solution. However, in a nonpolar mixture of water and tetrahydrofuran, and in the presence of copper ions, the peptide could self-assemble into nanorings by forming reversed micelles [57]. Copper ions were shown to be bound to the negatively charged head of the peptide and embedded in the core of the nanoring. The peptide could thus be applied as a template for fabricating novel metallic nanostructures.

The self-assembly of surfactants has enabled the synthesis of ordered mesoporous silica nanoparticles with high surface area, diverse compositions, variable pore structures, and tunable pore sizes, which are promising materials for various applications, such as drug delivery, catalysis, and sensors [58, 59]. Kim and coworkers have developed a facile, one-step method for achieving systematic control of the surface properties of highly fluorescent graphene quantum dots (GQDs), thus giving them a surfactant-like property. The surface-modified GQDs could effectively stabilize oil-in-water Pickering emulsions and submicron-sized colloidal particles in mini-emulsion polymerization. These newly developed GQD surfactants were also employed in liquid–solid systems, for tailoring the dispersion of graphite in methanol. In addition, their luminescent property makes them potentially applicable as fluorescent sensors and for imaging [60].

2.2.3.7 Amyloid Nanofibrils

Another class of self-assembling peptides is the amyloid nanofibrils, which aggregate into β-sheet-rich supramolecular polymers. They have been reported to play a central role in the pathogenesis of various diseases ranging from neurodegenerative diseases (e.g., Alzheimer's disease and Parkinson's disease) [61] to type II diabetes [62] and cardiovascular diseases [63]. In addition to their pathological role, amyloid fibrils have recently been reported to serve as biocompatible and functional materials. For instance, amyloid fibrils deposited onto an inorganic surface can lead to the formation of a biological thin film that is suitable for bacterial growth [64, 65]. Amyloid fibrils can act as a catalytic scaffold [66] that enhances a biochemical reaction, and as a transporting or storing agent that contains or transmits genetic information [67, 68] and/or hormones [69]. Moreover, amyloid fibrils have recently been employed for developing biomimetic and functional materials whose properties can be controlled. For example, Li et al. developed a biomimetic composite material synthesized by coupling amyloid fibrils and graphene sheets and showed that the properties of such a composite material can be tuned by controlling elements of the chemical environment, such as humidity [70]. Moreover, Tuomas et al. have shown that a thin film made of amyloid protein fibrils was both biologically compatible and highly rigid, with a Young's modulus of up to 5–7 GPa, which is comparable to the highest values for proteinaceous materials found in nature [71].

Recent studies have reported that the mechanical properties of amyloid fibrils are determined by their molecular structures, such as steric zipper pattern or helical pattern, and by their length [72–75]. Furthermore, Lee et al. have recently shown that the structural characteristics (e.g., helical pitch, diameter, and length) of amyloid fibrils can be controlled using microwave-assisted chemistry. The microwave affects the thermodynamics of protein aggregation, which is responsible for the formation of amyloid fibrils [76]. This study may provide insight into the conformational heterogeneity of amyloid fibrils, not only for further understanding the origin of amyloid-driven pathogenesis, which is dependent on the conformational diversity of amyloid fibrils, but also for consolidating a design principle that is applicable for developing biocompatible and biomimetic materials [77].

2.2.4 *Short Aromatic Peptides*

A very interesting class of peptide nanostructures is based on the use of short aromatic peptides which form well-ordered nanostructures. The first peptide described in this group was diphenylalanine (FF), which is the core recognition motif of the β-amyloid polypeptide [78]. Through a systematic reductionist approach, FF was recognized as the smallest sequence to form peptide tubular nanostructures by self-assembly. These biocompatible and water soluble tubes are formed under mild conditions and are easy and inexpensive to manufacture [78]. Since its discovery, the

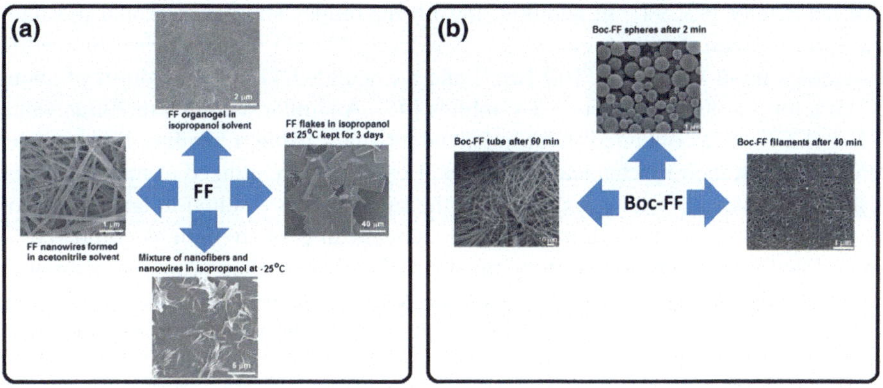

Fig. 2.2 Short aromatic peptide self-assembles into various morphological structures. **a** Diphenylalanine peptide (FF) self-assembles into nanowires in acetonitrile and into organogel, flakes nanofibers and nanowires in isopropanol (Reprinted with permission from Ref [85] Copyright 2014 Royal society of chemistry. **b** Boc-FF peptide exhibits phase transition from spheres to filaments and finally into tubular structures (Reprinted with permission from Ref. [86] Copyright 2014 Macmillan Publishers Limited, part of Springer Nature)

FF motif has gained popularity as a minimalist building block to drive the self-assembly of short peptides and their analogues into different morphologies by subtle chemical changes introduced to the structure of the FF dipeptide or to the solvent [79, 80]. Various FF peptide derivatives were shown to self-assemble, forming ordered nanostructures, including tubes, spheres, plates, and hydrogels [81–84]. In addition, temperature-induced reversible self-assembly of the FF peptide and its structural transition from organogel to crystalline nanowires in acetonitrile and isopropanol solvent have been reported by Huang et al. [85] (Fig. 2.2a). Another interesting example is the protected dipeptide, tert-Butyloxycarbonyl-diphenylalnine (Boc-FF), which can associate into distinct morphologies at the nanoscale as a result of variations in the assembly conditions, such as solvent composition or peptide concentration [80]. Under certain conditions, the process of assembly can undergo several stages, and a clear phase transition is observed. Boc-FF first assembles into spheres, then forms filaments and finally transforms into tubular structures (Fig. 2.2b) [86].

2.2.4.1 The Short Aromatic Peptide Unique Physical Properties

The tubular structures formed by aromatic short peptides have been extensively studied over the past decade, and their unique properties were characterized. The FF nanotubes have been shown to be stable in the presence of organic solvents and to have extraordinary thermal stability properties [87, 88]. In addition, the FF peptide nanotubes (PNTs) exhibit a variety of physical and chemical functionalities, such as optical wave guidance [89, 90], luminescence [91], semi-conductivity [92], and piezoelectricity [93, 94]. In addition, the FF nanotubes show remarkable mechanical rigidity, with direct atomic force microscopy (AFM) measurements indicating an average point stiffness of 160 Nm^{-1} and a calculated Young's modulus of about 20 GPa for nanotubular peptide assemblies [95]. A similar Young's modulus value (27 $\pm$ 4 GPa) was obtained in an independent study using a bending beam model [96]. Based on their mechanical properties, FF PNTs were utilized as nanofillers for composite materials with epoxy resin as the matrix. This resulted in an increase of about 70% in shear strength and a 450% increase in peel strength as compared to unmodified epoxy, while preserving the thermal and elongation properties similar to the original resin. These effects exceed the reinforcement effect of several known inorganic nanofillers, making FF PNTs excellent nanofillers for composite materials [97].

The mechanical properties of the spherical Boc-FF structures were also explored. A direct set of measurements, conducted using an AFM diamond-tip cantilever, demonstrated a remarkable Young's modulus of $\sim$275 GPa, an order of magnitude higher than the tubular FF assemblies [98]. The mechanical properties of the nanostructure can also be tuned using two peptides of differing stereochemistry: the L-form of the FF peptide and the D-form of the dinapthylalanine peptide, which co-assemble to form nanotubular structures. Elevating the portion of the dinapthylalanine peptide in the peptide mixture decreases the nanostructure's stiffness [99]. A density function theory (DFT) study was used to evaluate the basis for the mechanical rigidity of the nanotubes, suggesting that despite the porous nature of the crystal lattice, there is an array of rigid nanotube backbones with interpenetrating "zipper-like" aromatic interlocks that result in stiffness and robustness [100].

2.2.4.2 Low Molecular Weight Hydrogels

Chemical modification of short peptides by a variety of aromatic groups may aid in self-assembly via π–π stacking. For example, it has been shown that the addition of an aromatic group, such as carbobenzyloxy, naphthalene, or fluorenylmethoxycarbonyl (Fmoc), to the N-terminus of some peptides allows them to form stable hydrogels [101]. The short hydrogel-forming aromatic peptide systems are composed of four main components: a synthetic aromatic moiety coupled to the N-terminus of a short sequence peptide, typically a dipeptide or even a single amino acid [2], a linker between the peptide sequence and the N-terminal aromatic moiety which influences the structural orientation of the nanostructure [102], and a C-terminus which can be functionalized [103–105] and is also essential for the balance between protonated and ionized forms. Together, these four segments form a stable self-supporting system, which enables self-assembly of a short peptide sequence, whereas previous systems required a minimum of eight amino acids in each peptide chain [28]. This design is possible due to the combination of π–π stacking interactions and hydrogen bonding of the peptide portion [106].

Hydrogels are of great interest as a class of materials for tissue engineering and regeneration, since they offer 3D scaffolds to support the growth of cultured cells.

The short aromatic peptide-based hydrogels are self-supporting and display rheological behavior that is characteristic of solid like gel materials [101, 107, 108]. In particular, Fmoc-FF hydrogel can support cell growth, release of small molecules in a controllable manner, and exhibits remarkable physical properties. The Fmoc-FF hydrogel storage modulus G' was shown to be higher than 2×10^4 Pa, whereas other peptide hydrogels have a G' value of 50 Pa at a frequency of 1 Hz [108].This study was later expanded by examining additional members of the aromatic dipeptide family, with a new set of Fmoc-peptides, which included both natural and non-natural aromatic amino acids. One product of the peptide building blocks assembly was a hydrogel that presented the cell-adhesive arginine–glycine–aspartate (RGD) motif as a bioactive ligand at the fiber surface and thus mimicked certain essential features of the extracellular matrix [83, 106]. The nanofibrous hydrogel presented by Zhou et al. is a mixture of two short aromatic peptide derivatives, Fmoc-FF and Fmoc-RGD. Cylindrical nanofibres interwoven within the hydrogel causes the presence of RGDs in tunable densities on the fiber surface [106]. These scaffolds may offer an economical approach for fabricating 3D-culture scaffolds with other bioactive ligands for in vitro tissue regeneration.

Moreover, QDs were incorporated into 3D fibrous organogel scaffold based on the FF peptide. To prepare the organogels, FF dipeptide is first dissolved in HFIP and the organic solvent is then added to enable gel formation. The gelation process can be observed exclusively in chloroform or in aromatic solvents, such as toluene or xylene. The FF gelation occurred in the presence of QD solution and the gels displayed photoluminescence from the embedded QDs. The emission maxima of the QDs in fibrous networks were slightly blue-shifted compared to those of free QDs, indicating the attachment of the QDs to the fibrils, but they maintained the

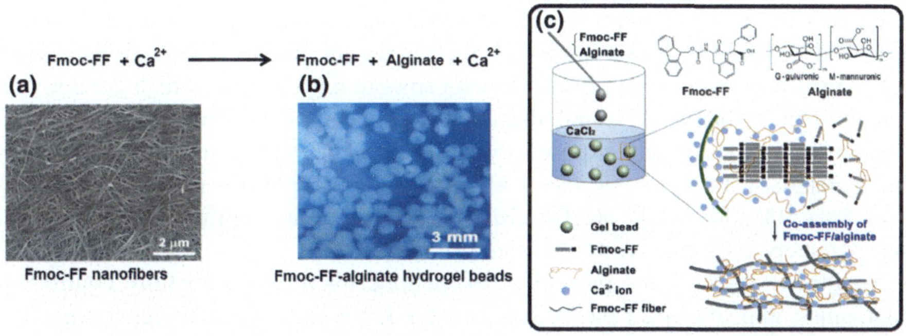

Fig. 2.3 Fmoc protected aromatic peptide self-assembles into various morphological structures in the presence of Ca^{2+} ions. **a** SEM images of Fmoc-FF nanofibers formed in the presence of Ca^{2+} ions. **b** SEM images of Fmoc-FF-alginate hydrogel beads formed in the presence of Ca^{2+} ions. **c** Schematic representation of Ca^{2+} ion-triggered co-assembly of peptide and alginate at an aqueous liquid–liquid interface to synthesize Fmoc-FF–alginate hydrogel beads (Reprinted with permission from Ref. [110] Copyright 2016 Royal society of chemistry)

original photoluminescence colors [109]. A new approach was reported by Xie et al. [110] who used calcium ions to trigger the self-assembly of the Fmoc-FF peptide into nanofibers with diameter of about 30 nm (Fig. 2.3a) while alginate was rapidly crosslinked by the calcium ions, thus forming stable hybrid hydrogel beads (Fig. 2.3b). Figure 2.3c shows a schematic representation of co-assembly of peptide and alginate at an aqueous liquid–liquid interface induced by calcium ions to synthesize Fmoc-FF–alginate hydrogel beads. They further used these hydrogel beads for drug delivery applications in vitro.

2.2.4.3 Applications of Short Aromatic Peptides

The minimal size and the simple synthesis of the short peptide building blocks, as well as the efficient and reproducible assembly procedures and unique physical properties, have attracted many research groups to develop various applications utilizing the aromatic dipeptide system. The short aromatic peptides are used in various applications ranging from biomedical applications, including nanomedicine, drug delivery, tissue engineering and biomaterials, to technological applications, including metal–organic frameworks applications, ultra-sensitive sensors, energy storage devices, and hydrophobic coatings [111–113].

Recently it was shown that self-assembled tetra-peptides which include the FF motif are substrates of cathepsin proteases, which are highly elevated in cancer and other pathologies. The degradation of one of these substrates by cathepsin led to the release of $91.8 \pm 0.3\%$ of the incorporated anti-cancerous drug Doxorubicin from the nanofibers within 8 h. Therefore, such peptides could serve as a platform for targeted drug delivery to pathologies in which protease activity is highly elevated [114]. Furthermore, Bonetti et al. showed that fluorinated β-peptides containing two aryl units can form nanotubes that are stable to protease degradation and to heating up to 120 °C and can locate in the perinuclear region of cells. Cytotoxicity assays were performed on cultured primary human smooth muscle cells, which are derived from human blood vessels, and thus offer a good model to represent the most abundant cell type directly exposed to nano-therapeutics present in the blood system. After 48 h, no cytotoxic effect was noted for peptide concentrations ranging from 1 to 100 μM, while at 200 μM cell viability was significantly reduced to approximately 80% relative to the control [115].

Another example is the growth of FF peptide microrods with fully controlled polarization and improved piezoelectricity for fabricating a power generator. The outputs of the FF peptide-based generators were shown to exceed other bio-inspired generators, and are even better than some inorganic generators. This study makes a significant step toward developing bio-inspired materials for piezoelectric devices to be used to generate renewable and biocompatible energy sources and for biomedical applications, and opens up a portal to the next generation of multifunctional electronics compatible with human tissue [116].

2.2.5 Coiled Coils

The coiled coil is one of the basic folding patterns found in native proteins, consisting of two or more α-helical peptides that are twisted around each other in super helical manner [2]. The primary structure of coiled coil forming proteins is characterized by a heptad repeat pattern $(abcdefg)_n$, where n is the number of repeats. Positions a and d are occupied by hydrophobic amino acids that form the hydrophobic core of the coiled coil, which is characterized by a tight packing of the hydrophobic amino acid side chains in a "knobs-in-holes" fashion, while e and g are charged amino acid functional groups, which can interact to form stabilizing inter-strand salt bridges. The hydrophobic interaction between residues a and d and the electrostatic interaction between residues e and g contribute to the stability of the coiled coil structure [2, 117]. Expansion of the hydrophobic core to include the e and g positions and restriction of ionisable residues to positions b, c and f improved the oligomerization state and thermal stability of the coiled coil [118, 119]. In this stable arrangement, the side chains were positioned along the outside of the coil. The specificity and stability of coiled coils depend on the number of helical strands and the orientation of the helices and are affected by hydrophobic core packing and inter-helical ionic interactions [117].

The highly selective and specific binding properties of coiled coils and their easy manipulation using external stimuli, as well as the biological applications of many coiled coils, have stimulated the interest in the use of these fascinating motifs as building blocks for the development of novel bio(hybrid) materials. Coiled coil based bio(hybrid) materials have been recently used for various applications, including drug delivery, protein labeling, and bio-sensing [120].

Like other supramolecular building blocks, coiled coils can be used to construct a variety of distinct nanostructures of controlled size and shape in some selective solvents. For example, Holowka et al. reported the preparation of charged amphiphilic block copolypeptides that formed vesicles and micelles in water. These coiled coil nanovesicles composed of peptides with hydrophilic residues in one part and hydrophobic residues in another part, allowed encapsulation of dextran [121]. The stability and unfolding of coiled coil motifs depend on the temperature, pH, and ionic strength, which is often used in the design of controlled release delivery systems that respond to a specific stimulus [117]. Furthermore, the distinctive association-dissociation and spatial recognition of coiled coils make them ideal candidates for physical cross-linkers of protein-based supramolecular fibrils or polymer hydrogels [122, 123].

A number of examples of coiled coil α-helical fibrous structures have been presented by the Woolfson group [124–126]. Using a bottom-up design approach, they developed a two-component peptide system for making hydrogels, termed hSAFs (hydrogelating self-assembling fibers) [127]. These dual-peptide systems form gel only on mixing, which allows for tight control over assembly. They have a wide variety of potential applications in biotechnology and medicine, such as the controlled delivery and release of cells, cosmetics and drugs, and as supports for cell growth and tissue engineering. The peptide sequences can be engineered to alter the underlying mechanism of gelation and, consequently, the hydrogel properties. Furthermore, the original two-peptide hSAF system can be supplemented with other components to endow cell-binding functions to the system, hence building up complexity and functionality [124].

Recently, Mondal et al. presented the first-ever self-assembling single heptad repeat module, based on the ability of the non-coded α-aminoisobutyric acid to stabilize very short peptides in helical conformation [128]. A conformationally constrained peptide comprised of aromatic, but not aliphatic, residues at the first and fourth positions formed helical fibrillar assemblies. New analogues of this motif can be accurately designed due to the high-resolution crystal structure of the helical assemblies. Substitution at different positions of the heptad sequence while maintaining the relative arrangements of hydrophobic amino acids and introduction of additional functionalities through side chain modifications can give rise to a vast repertoire of helical assemblies with possible applications in bionanotechnology and biomaterials.

2.3 Conclusions and Future Perspectives

SAPs have been gaining increasing attention as versatile structural building blocks with the ability to generate diverse supramolecular architectures with tunable functionalities. Since various functionalities can be incorporated into the peptide sequence, including self-assembly, cell attachment, or signaling domains, versatile, multifunctional structures can be generated from a single molecular entity. Along with their easy production, these advantages make SAPs attractive for various biomedical applications. However, extensive studies are still required addressing the rational peptide design to understand thoroughly the role of hydrophobicity, electrostatics, and size on pattern formation and to examine the fundamental physical interactions that drive their unique self-assembly. Particularly, for biological applications, a systematic structure–function understanding of self-assembled peptides are required to evaluate the properties of peptide nanostructures upon pH change, high protein and salt concentration, control of the size of peptide nanostructures for developing drug delivery systems and biosensors, as well as to develop nanostructures of equivalent dimension for the fabrication of bio-sensing platforms and determine an optimal balance of hydrophobicity and charge to minimize aggregation through hydrophobic groups. Also, the study of unexplored peptide sequences is required to develop the next generation of self-assembled peptides for various applications.

References

1. Lehn JM (2002) Toward complex matter: supramolecular chemistry and self-organization. Proc Natl Acad Sci USA 99:4763–4768
2. Ulijn RV, Smith AM (2008) Designing peptide based nanomaterials. Chem Soc Rev 37:664–675
3. Zhang S (2003) Fabrication of novel biomaterials through molecular self-assembly. Nat Biotechnol 21:1171–1178
4. Zhang S (2002) Emerging biological materials through molecular self-assembly. Biotechnol Adv 20:321–339
5. Cui H, Webber MJ, Stupp SI (2010) Self-assembly of peptide amphiphiles: from molecules to nanostructures to biomaterials. Biopolymers 94:1–18
6. Mandal D, Nasrolahi Shirazi A, Parang K (2014) Self-assembly of peptides to nanostructures. Org Biomol Chem 12:3544–3561
7. Collier JH, Segura T (2011) Evolving the use of peptides as components of biomaterials. Biomaterials 32:4198–4204
8. Yu CY, Huang W, Li ZP, Lei XY, He DX, Sun L (2016) Progress in self-assembling peptide-based nanomaterials for biomedical applications. Curr Top Med Chem 16:281–290
9. Ghadiri MR, Granja JR, Milligan RA, McRee DE, Khazanovich N (1993) Self-assembling organic nanotubes based on a cyclic peptide architecture. Nature 366:324–327
10. Fjell CD, Hiss JA, Hancock RE, Schneider G (2012) Designing antimicrobial peptides: form follows function. Nat Rev Drug Discov 11:37–51
11. Barbosa SC, Nobre TM, Volpati D, Ciancaglini P, Cilli EM, Lorenzón EN, Oliveira ON Jr (2016) The importance of cyclic structure for Labaditin on its antimicrobial activity against Staphylococcus aureus. Colloids Surf B Biointerfaces 148:453–459
12. Roscia G, Falciani C, Bracci L, Pini A (2013) The development of antimicrobial peptides as new antibacterial drugs. Curr Protein Pept Sci 14:641–649
13. Hancock REW, Rozek A (2002) Role of membranes in the activities of antimicrobial cationic peptides. FEMS Microbiol Lett 206:143–149
14. Brogden KA (2005) Antimicrobial peptides: pore formers or metabolic inhibitors in bacteria? Nat Rev Microbiol 3:238–250
15. Epand RM, Vogel HJ (1999) Diversity of antimicrobial peptides and their mechanisms of action. Biochem Biophys Acta 1462:11–28
16. Chapman R, Danial M, Koh ML, Jolliffe KA, Perrier S (2012) Design and properties of functional nanotubes from the self-assembly of cyclic peptide templates. Chem Soc Rev 41:6023–6041
17. Fernandez-Lopez S, Kim HS, Choi EC, Delgado M, Granja JR, Khasanov A, Kraehenbuehl K, Long G, Weinberger DA, Wilcoxen KM, Ghadiri MR (2001) Antibacterial agents based on the cyclic D, L-alpha-peptide architecture. Nature 412:452–455
18. Rodriguez-Vazquez N, Ozores HL, Guerra A, Gonzalez-Freire E, Fuertes A, Panciera M, Priegue JM, Outeiral J, Montenegro J, Garcia-Fandino R, Amorin M, Granja JR (2014) Membrane-targeted self-assembling cyclic peptide nanotubes. Curr Top Med Chem 14:2647–2661
19. Montenegro J, Ghadiri MR, Granja JR (2013) Ion channel models based on self-assembling cyclic peptide nanotubes. Acc Chem Res 46:2955–2965
20. Ghadiri MR, Granja JR, Buehler LK (1994) Artificial transmembrane ion channels from self-assembling peptide nanotubes. Nature 369:301–304
21. Wang D, Guo L, Zhang J, Jones LR, Chen Z, Pritchard C, Roeske RW (2001) Artificial ion channels formed by a synthetic cyclic peptide. J Pept Res Official J Am Pept Soc 57:301–306
22. Motesharei K, Ghadiri MR (1997) Diffusion-limited size-selective ion sensing based on SAM-supported peptide nanotubes. J Am Chem Soc 119:11306–11312

23. Perez-Alvite MJ, Mosquera M, Castedo L, Granja JR (2011) Toward the rational design of molecular rotors ion sensors based on alpha, gamma-cyclic peptide dimers. Amino Acids 41:621–628
24. Yang CT, Han J, Gu M, Liu J, Li Y, Huang Z, Yu HZ, Hu S, Wang X (2015) Fluorescent recognition of uranyl ions by a phosphorylated cyclic peptide. Chem Commun (Cambridge, England) 51:11769–11772
25. Horne WS, Wiethoff CM, Cui C, Wilcoxen KM, Amorin M, Ghadiri MR, Nemerow GR (2005) Antiviral cyclic d, l-α-peptides: Targeting a general biochemical pathway in virus infections. Bioorg Med Chem 13:5145–5153
26. Vollmer MS, Clark TD, Steinem C, Ghadiri MR (1999) Photoswitchable hydrogen-bonding in self-organized cylindrical peptide systems. Angew Chem Int Ed 38:1598–1601
27. Steinem C, Janshoff A, Vollmer MS, Ghadiri MR (1999) Reversible photoisomerization of self-organized cylindrical peptide assemblies at air–water and solid interfaces. Langmuir 15:3956–3964
28. Jung JP, Gasiorowski JZ, Collier JH (2010) Fibrillar peptide gels in biotechnology and biomedicine. Biopolymers 94:49–59
29. Behanna HA, Donners JJ, Gordon AC, Stupp SI (2005) Coassembly of amphiphiles with opposite peptide polarities into nanofibers. J Am Chem Soc 127:1193–1200
30. Paramonov SE, Jun HW, Hartgerink JD (2006) Self-assembly of peptide-amphiphile nanofibers: the roles of hydrogen bonding and amphiphilic packing. J Am Chem Soc 128:7291–7298
31. Matson JB, Newcomb CJ, Bitton R, Stupp SI (2012) Nanostructure-templated control of drug release from peptide amphiphile nanofiber gels. Soft Matter 8:3586–3595
32. Webber MJ, Matson JB, Tamboli VK, Stupp SI (2012) Controlled release of dexamethasone from peptide nanofiber gels to modulate inflammatory response. Biomaterials 33:6823–6832
33. Moyer TJ, Kassam HA, Bahnson ES, Morgan CE, Tantakitti F, Chew TL, Kibbe MR, Stupp SI (2015) Shape-dependent targeting of injured blood vessels by peptide amphiphile supramolecular nanostructures. Small (Weinheim an der Bergstrasse, Germany) 11:2750–2755
34. Deshmukh SA, Solomon LA, Kamath G, Fry HC, Sankaranarayanan SKRS (2016) Water ordering controls the dynamic equilibrium of micelle–fibre formation in self-assembly of peptide amphiphiles. Nat Commun 7:12367
35. da Silva RMP, van der Zwaag D, Albertazzi L, Lee SS, Meijer EW, Stupp SI (2016) Super-resolution microscopy reveals structural diversity in molecular exchange among peptide amphiphile nanofibres. Nat Commun 7:11561
36. Zhao X (2009) Design of self-assembling surfactant-like peptides and their applications. Curr Opin Colloid Interface Sci 14:340–348
37. von Maltzahn G, Vauthey S, Santoso S, Zhang S (2003) Positively charged surfactant-like peptides self-assemble into nanostructures. Langmuir 19:4332–4337
38. Qiu F, Chen Y, Zhao X (2009) Comparative studies on the self-assembling behaviors of cationic and catanionic surfactant-like peptides. J Colloid Interface Sci 336:477–484
39. Menger FM, Keiper JS (2000) Gemini surfactants. Angew Chem Int Ed Engl 39:1906–1920
40. Johnsson M, Engberts JBFN (2004) Novel sugar-based gemini surfactants: aggregation properties in aqueous solution. J Phys Org Chem 17:934–944
41. Kumar V, Chatterjee A, Kumar N, Ganguly A, Chakraborty I, Banerjee M (2014) d-Glucose derived novel gemini surfactants: synthesis and study of their surface properties, interaction with DNA, and cytotoxicity. Carbohydr Res 397:37–45
42. Kiley P, Zhao X, Vaughn M, Baldo MA, Bruce BD, Zhang S (2005) Self-assembling peptide detergents stabilize isolated photosystem ion a dry surface for an extended time. PLoS Biol 3:e230
43. Zhao X, Nagai Y, Reeves PJ, Kiley P, Khorana HG, Zhang S (2006) Designer short peptide surfactants stabilize G protein-coupled receptor bovine rhodopsin. Proc Natl Acad Sci USA 103:17707–17712

44. Gudlur S, Sukthankar P, Gao J, Avila LA, Hiromasa Y, Chen J, Iwamoto T, Tomich JM (2012) Peptide nanovesicles formed by the self-assembly of branched amphiphilic peptides. PLoS ONE 7:e45374
45. van Hell AJ, Costa CICA, Flesch FM, Sutter M, Jiskoot W, Crommelin DJA, Hennink WE, Mastrobattista E (2007) Self-assembly of recombinant amphiphilic oligopeptides into vesicles. Biomacromolecules 8:2753–2761
46. van Hell AJ, Fretz MM, Crommelin DJ, Hennink WE, Mastrobattista E (2010) Peptide nanocarriers for intracellular delivery of photosensitizers. J Controlled Release Official J Controlled Release Soc 141:347–353
47. Chen JX, Wang HY, Li C, Han K, Zhang XZ, Zhuo RX (2011) Construction of surfactant-like tetra-tail amphiphilic peptide with RGD ligand for encapsulation of porphyrin for photodynamic therapy. Biomaterials 32:1678–1684
48. Cui C, Xue YN, Wu M, Zhang Y, Yu P, Liu L, Zhuo RX, Huang SW (2013) Cellular uptake, intracellular trafficking, and antitumor efficacy of doxorubicin-loaded reduction-sensitive micelles. Biomaterials 34:3858–3869
49. Tao Y, Liu R, Chen M, Yang C, Liu X (2012) Cross-linked micelles of graftlike block copolymer bearing biodegradable ε-caprolactone branches: a novel delivery carrier for paclitaxel. J Mater Chem 22:373–380
50. Kuppusamy P, Afeworki M, Shankar RA, Coffin D, Krishna MC, Hahn SM, Mitchell JB, Zweier JL (1998) In vivo electron paramagnetic resonance imaging of tumor heterogeneity and oxygenation in a murine model. Can Res 58:1562–1568
51. Guo XD, Tandiono F, Wiradharma N, Khor D, Tan CG, Khan M, Qian Y, Yang YY (2008) Cationic micelles self-assembled from cholesterol-conjugated oligopeptides as an efficient gene delivery vector. Biomaterials 29:4838–4846
52. Wiradharma N, Tong YW, Yang YY (2009) Self-assembled oligopeptide nanostructures for co-delivery of drug and gene with synergistic therapeutic effect. Biomaterials 30:3100–3109
53. Cardoso AM, Morais CM, Cruz AR, Cardoso AL, Silva SG, do Vale ML, Marques EF, Pedroso de Lima MC, Jurado AS (2015) Gemini surfactants mediate efficient mitochondrial gene delivery and expression. Mol Pharmaceutics 12:716–730
54. Ahmed T, Kamel AO, Wettig SD (2016) Interactions between DNA and Gemini surfactant: impact on gene therapy: part I. Nanomedicine (London, England) 11:289–306
55. Qiu F, Chen Y, Cheng J, Wang C, Xu H, Zhao X (2010) A simple method for cell sheet fabrication using mica surfaces grafted with peptide detergent A(6)K. Macromol Biosci 10:881–886
56. Rajmohan G, Admane P, Anish C, Panda AK (2014) Fusion and self-assembly of biodegradable polymer particles into scaffoldlike and membranelike structures at room temperature for regenerative medicine. Mol Pharm 11:2190–2202
57. Qiu F, Chen Y, Tang C, Lu Y, Cheng J, Zhao X (2012) Formation of reversed micelle nanoring by a designed surfactant-like peptide. NANO 07:1250024
58. Gérardin C, Reboul J, Bonne M, Lebeau B (2013) Ecodesign of ordered mesoporous silica materials. Chem Soc Rev 42:4217–4255
59. Deng Y, Wei J, Sun Z, Zhao D (2013) Large-pore ordered mesoporous materials templated from non-Pluronic amphiphilic block copolymers. Chem Soc Rev 42:4054–4070
60. Cho H-H, Yang H, Kang DJ, Kim BJ (2015) Surface engineering of graphene quantum dots and their applications as efficient surfactants. ACS Appl Mater Interfaces 7:8615–8621
61. Dobson CM (2003) Protein folding and misfolding. Nature 426:884–890
62. Hoppener JW, Ahren B, Lips CJ (2000) Islet amyloid and type 2 diabetes mellitus. New Engl J Med 343:411–419
63. Rapezzi C, Quarta CC, Riva L, Longhi S, Gallelli I, Lorenzini M, Ciliberti P, Biagini E, Salvi F, Branzi A (2010) Transthyretin-related amyloidoses and the heart: a clinical overview. Nat Rev Cardiol 7:398–408
64. Zhou Y, Smith DR, Hufnagel DA, Chapman MR (2013) Experimental manipulation of the microbial functional amyloid called curli. Methods Mol Biol (Clifton, N.J.) 966:53–75

65. Zhou Y, Smith D, Leong BJ, Brannstrom K, Almqvist F, Chapman MR (2012) Promiscuous cross-seeding between bacterial amyloids promotes interspecies biofilms. J Biol Chem 287:35092–35103
66. Chapman MR, Robinson LS, Pinkner JS, Roth R, Heuser J, Hammar M, Normark S, Hultgren SJ (2002) Role of Escherichia coli curli operons in directing amyloid fiber formation. Science (New York, N.Y.) 295:851–855
67. Fowler DM, Koulov AV, Alory-Jost C, Marks MS, Balch WE, Kelly JW (2006) Functional amyloid formation within mammalian tissue. PLoS Biol 4:e6
68. Shorter J, Lindquist S (2005) Prions as adaptive conduits of memory and inheritance. Nat Rev Genet 6:435–450
69. Maji SK, Perrin MH, Sawaya MR, Jessberger S, Vadodaria K, Rissman RA, Singru PS, Nilsson KP, Simon R, Schubert D, Eisenberg D, Rivier J, Sawchenko P, Vale W, Riek R (2009) Functional amyloids as natural storage of peptide hormones in pituitary secretory granules. Science (New York, N.Y.) 325:328–332
70. Li C, Adamcik J, Mezzenga R (2012) Biodegradable nanocomposites of amyloid fibrils and graphene with shape-memory and enzyme-sensing properties. Nat Nano 7:421–427
71. Knowles TPJ, Oppenheim TW, Buell AK, Chirgadze DY, Welland ME (2010) Nanostructured films from hierarchical self-assembly of amyloidogenic proteins. Nat Nano 5:204–207
72. Sawaya MR, Sambashivan S, Nelson R, Ivanova MI, Sievers SA, Apostol MI, Thompson MJ, Balbirnie M, Wiltzius JJ, McFarlane HT, Madsen AO, Riekel C, Eisenberg D (2007) Atomic structures of amyloid cross-beta spines reveal varied steric zippers. Nature 447:453–457
73. Choi B, Yoon G, Lee SW, Eom K (2015) Mechanical deformation mechanisms and properties of amyloid fibrils. Phys Chem Chem Phys 17:1379–1389
74. Knowles TP, Fitzpatrick AW, Meehan S, Mott HR, Vendruscolo M, Dobson CM, Welland ME (2007) Role of intermolecular forces in defining material properties of protein nanofibrils. Science (New York, N.Y.) 318:1900–1903
75. Yoon G, Kwak J, Kim JI, Na S, Eom K (2011) Mechanical characterization of amyloid fibrils using coarse-grained normal mode analysis. Adv Funct Mater 21:3454–3463
76. Lee G, Lee W, Lee H, Lee CY, Eom K, Kwon T (2015) Self-assembled amyloid fibrils with controllable conformational heterogeneity. Sci Rep 5:16220
77. Cherny I, Gazit E (2008) Amyloids: not only pathological agents but also ordered nanomaterials. Angew Chem Int Ed Engl 47:4062–4069
78. Reches M, Gazit E (2003) Casting metal nanowires within discrete self-assembled peptide nanotubes. Science (New York, N.Y.) 300:625–627
79. Ryu J, Park CB (2008) High-temperature self-assembly of peptides into vertically well-aligned nanowires by aniline vapor. Adv Mater 20:3754–3758
80. Adler-Abramovich L, Gazit E (2008) Controlled patterning of peptide nanotubes and nanospheres using inkjet printing technology. J Pept Sci 14:217–223
81. Reches M, Gazit E (2005) Self-assembly of peptide nanotubes and amyloid-like structures by charged-termini capped diphenylalanine peptide analogues. Isr J Chem 45:363–371
82. Reches M, Gazit E (2006) Designed aromatic homo-dipeptides: formation of ordered nanostructures and potential nanotechnological applications. Phys Biol 3:S10–S19
83. Orbach R, Mironi-Harpaz I, Adler-Abramovich L, Mossou E, Mitchell EP, Forsyth VT, Gazit E, Seliktar D (2012) The rheological and structural properties of Fmoc-peptide-based hydrogels: the effect of aromatic molecular architecture on self-assembly and physical characteristics. Langmuir 28:2015–2022
84. Reches M, Gazit E (2004) Formation of closed-cage nanostructures by self-assembly of aromatic dipeptides. Nano Lett 4:581–585
85. Huang R, Wang Y, Qi W, Su R, He Z (2014) Temperature-induced reversible self-assembly of diphenylalanine peptide and the structural transition from organogel to crystalline nanowires. Nanoscale Res Lett 9:653

86. Levin A, Mason TO, Adler-Abramovich L, Buell AK, Meisl G, Galvagnion C, Bram Y, Stratford SA, Dobson CM, Knowles TP, Gazit E (2014) Ostwald's rule of stages governs structural transitions and morphology of dipeptide supramolecular polymers. Nat Commun 5:5219
87. Adler-Abramovich L, Reches M, Sedman VL, Allen S, Tendler SJ, Gazit E (2006) Thermal and chemical stability of diphenylalanine peptide nanotubes: implications for nanotechnological applications. Langmuir 22:1313–1320
88. Sedman VL, Adler-Abramovich L, Allen S, Gazit E, Tendler SJ (2006) Direct observation of the release of phenylalanine from diphenylalanine nanotubes. J Am Chem Soc 128:6903–6908
89. Li Q, Ma H, Wang A, Jia Y, Dai L, Li J (2015) Self-assembly of cationic dipeptides forming rectangular microtubes and microrods with optical waveguiding properties. Adv Opt Mater 3:194–198
90. Yan X, Li J, Möhwald H (2011) Self-assembly of hexagonal peptide microtubes and their optical waveguiding. Adv Mater 23:2796–2801
91. Amdursky N, Molotskii M, Aronov D, Adler-Abramovich L, Gazit E, Rosenman G (2009) Blue luminescence based on quantum confinement at peptide nanotubes. Nano Lett 9:3111–3115
92. Hauser CAE, Zhang S (2010) Nanotechnology: peptides as biological semiconductors. Nature 468:516–517
93. Bosne ED, Heredia A, Kopyl S, Karpinsky DV, Pinto AG, Kholkin AL (2013) Piezoelectric resonators based on self-assembled diphenylalanine microtubes. Appl Phys Lett 102:073504
94. Handelman A, Beker P, Amdursky N, Rosenman G (2012) Physics and engineering of peptide supramolecular nanostructures. Phys Chem Chem Phys 14:6391–6408
95. Kol N, Adler-Abramovich L, Barlam D, Shneck RZ, Gazit E, Rousso I (2005) Self-assembled peptide nanotubes are uniquely rigid bioinspired supramolecular structures. Nano Lett 5:1343–1346
96. Niu L, Chen X, Allen S, Tendler SJB (2007) Using the bending beam model to estimate the elasticity of diphenylalanine nanotubes. Langmuir 23:7443–7446
97. Even N, Adler-Abramovich L, Buzhansky L, Dodiuk H, Gazit E (2011) Improvement of the mechanical properties of epoxy by peptide nanotube fillers. Small 7:1007–1011
98. Adler-Abramovich L, Kol N, Yanai I, Barlam D, Shneck RZ, Gazit E, Rousso I (2010) Self-assembled organic nanostructures with metallic-like stiffness. Angew Chem Int Ed 49:9939–9942
99. Sedman VL, Chen X, Allen S, Roberts CJ, Korolkov VV, Tendler SJB (2013) Tuning the mechanical properties of self-assembled mixed-peptide tubes. J Microsc 249:165–172
100. Azuri I, Adler-Abramovich L, Gazit E, Hod O, Kronik L (2013) Why Are Diphenylalanine-Based Peptide Nanostructures So Rigid? Insights from first principles calculations. J Am Chem Soc 136:963–969
101. Yang ZM, Gu HW, Zhang Y, Wang L, Xu B (2004) Small molecule hydrogels based on a class of antiinflammatory agents. Chem Commun 2:208–209
102. Fleming S, Debnath S, Frederix PW, Tuttle T, Ulijn RV (2013) Aromatic peptide amphiphiles: significance of the Fmoc moiety. Chemical Commun (Cambridge, England) 49:10587–10589
103. Debnath S, Shome A, Das D, Das PK (2010) Hydrogelation through self-assembly of fmoc-peptide functionalized cationic amphiphiles: potent antibacterial agent. J Phys Chem B 114:4407–4415
104. Ryan DM, Doran TM, Anderson SB, Nilsson BL (2011) Effect of C-terminal modification on the self-assembly and hydrogelation of fluorinated Fmoc-Phe derivatives. Langmuir 27:4029–4039
105. Hughes M, Xu H, Frederix PWJM, Smith AM, Hunt NT, Tuttle T, Kinloch IA, Ulijn RV (2011) Biocatalytic self-assembly of 2D peptide-based nanostructures. Soft Matter 7:10032–10038

106. Zhou M, Smith AM, Das AK, Hodson NW, Collins RF, Ulijn RV, Gough JE (2009) Self-assembled peptide-based hydrogels as scaffolds for anchorage-dependent cells. Biomaterials 30:2523–2530
107. Jayawarna V, Ali M, Jowitt TA, Miller AF, Saiani A, Gough JE, Ulijn RV (2006) Nanostructured hydrogels for three-dimensional cell culture through self-assembly of fluorenylmethoxycarbonyl-dipeptides. Adv Mater 18:611–614
108. Mahler A, Reches M, Rechter M, Cohen S, Gazit E (2006) Rigid, self-assembled hydrogel composed of a modified aromatic dipeptide. Adv Mater 18:1365–1370
109. Yan X, Cui Y, He Q, Wang K, Li Junbai (2008) Organogels based on self-assembly of diphenylalanine peptide and their application to immobilize quantum dots. Chem Mater 20:1522–1526
110. Xie Y, Zhao J, Huang R, Qi W, Wang Y, Su R, He Z (2016) Calcium-ion-triggered co-assembly of peptide and polysaccharide into a hybrid hydrogel for drug delivery. Nanoscale Res Lett 11:184
111. Marchesan S, Vargiu AV, Styan KE (2015) The Phe-Phe motif for peptide self-assembly in nanomedicine. Molecules (Basel, Switzerland) 20:19775–19788
112. Yan X, Zhu P, Li J (2010) Self-assembly and application of diphenylalanine-based nanostructures. Chem Soc Rev 39:1877–1890
113. Adler-Abramovich L, Gazit E (2014) The physical properties of supramolecular peptide assemblies: from building block association to technological applications. Chem Soc Rev 43:6881–6893
114. Ben-Nun Y, Fichman G, Adler-Abramovich L, Turk B, Gazit E, Blum G (2016) Cathepsin nanofiber substrates as potential agents for targeted drug delivery. J Controlled Release Official J Controlled Release Soc
115. Bonetti A, Pellegrino S, Das P, Yuran S, Bucci R, Ferri N, Meneghetti F, Castellano C, Reches M, Gelmi ML (2015) Dipeptide nanotubes containing unnatural fluorine-substituted β2, 3-diarylamino acid and l-alanine as candidates for biomedical applications. Org Lett 17:4468–4471
116. Nguyen V, Zhu R, Jenkins K, Yang R (2016) Self-assembly of diphenylalanine peptide with controlled polarization for power generation. Nat Commun 7:13566
117. Yu YB (2002) Coiled-coils: stability, specificity, and drug delivery potential. Adv Drug Deliv Rev 54:1113–1129
118. Fletcher NL, Lockett CV, Dexter AF (2011) A pH-responsive coiled-coil peptide hydrogel. Soft Matter 7:10210–10218
119. Deng Y, Liu J, Zheng Q, Eliezer D, Kallenbach NR, Lu M (2006) Antiparallel four-stranded coiled coil specified by a 3-3-1 hydrophobic heptad repeat. Structure 14:247–255
120. Apostolovic B, Danial M, Klok H-A (2010) Coiled coils: attractive protein folding motifs for the fabrication of self-assembled, responsive and bioactive materials. Chem Soc Rev 39:3541–3575
121. Holowka EP, Pochan DJ, Deming TJ (2005) Charged polypeptide vesicles with controllable diameter. J Am Chem Soc 127:12423–12428
122. Kumar P, Pillay V, Modi G, Choonara YE, du Toit LC, Naidoo D (2011) Self-assembling peptides: implications for patenting in drug delivery and tissue engineering. Recent Pat Drug Deliv Formul 5:24–51
123. Xu C, Breedveld V, Kopecek J (2005) Reversible hydrogels from self-assembling genetically engineered protein block copolymers. Biomacromolecules 6:1739–1749
124. Mehrban N, Abelardo E, Wasmuth A, Hudson KL, Mullen LM, Thomson AR, Birchall MA, Woolfson DN (2014) Assessing cellular response to functionalized alpha-helical peptide hydrogels. Adv Healthc Mater 3:1387–1391
125. Bromley EH, Sessions RB, Thomson AR, Woolfson DN (2009) Designed alpha-helical tectons for constructing multicomponent synthetic biological systems. J Am Chem Soc 131:928–930

126. Pandya MJ, Cerasoli E, Joseph A, Stoneman RG, Waite E, Woolfson DN (2004) Sequence and structural duality: designing peptides to adopt two stable conformations. J Am Chem Soc 126:17016–17024
127. Banwell EF, Abelardo ES, Adams DJ, Birchall MA, Corrigan A, Donald AM, Kirkland M, Serpell LC, Butler MF, Woolfson DN (2009) Rational design and application of responsive alpha-helical peptide hydrogels. Nat Mater 8:596–600
128. Mondal S, Adler-Abramovich L, Lampel A, Bram Y, Lipstman S, Gazit E (2015) Formation of functional super-helical assemblies by constrained single heptad repeat. Nat Commun 6:8615

Part II
Biomimetic Membranes

Chapter 3
Nanoarchitectonics of Biomimetic Membranes

Katsuhiko Ariga

Abstract Emerging concept, nanoarchitectonics, which is an emerging concept is especially suitable for fabrication of materials systems from organic and biomaterials through self-organization and directed-organization. Nanoarchitectonics tries to control uncontrollable situations through harmonizing various factors and interactions. The most successful example of functional harmony of nanoscale objects can be seen in biological systems and biomimetic systems. In this chapter, we here focus on nanoarchitectonics of biomimetic membranes. This chapter introduces various examples of nanoarchitectonics approaches for bio-related thin films and membranes: (1) two-dimensional biomimetic membrane (molecular recognition, receptor tuning, and nanomaterial film for life control); (2) layer-by-layer biomimetic membrane (bioreactor, hierarchic assembly, and sensing and drug delivery).

Keywords Nanoarchitectonics · Langmuir–Blodgett film · Layer-by-layer assembly · Molecular recognition · Bioreactor · Sensor · Drug delivery

3.1 Introduction: Nanoarchitectonics

Although structures and properties of individual molecules and nanomaterials have certain level of varieties, objects from their assemblies exhibit much more complex, useful, and attractive functions. Therefore, investigation and understanding of individual molecules is not enough to explore functions of materials. It may be especially emphasized in biological and/or biomimetic materials that alter their properties depending on spatial and temporal organization of component molecules. Materials and systems constructed with weak interactions including hydrogen

---

K. Ariga (✉)
World Premier International (WPI) Research Center for Materials Nanoarchitectonics (MANA), National Institute for Materials Science (NIMS), 1-1 Namiki, Tsukuba 305-0044, Japan
e-mail: ARIGA.Katsuhiko@nims.go.jp

J. Li (ed.), *Supramolecular Chemistry of Biomimetic Systems*,
DOI 10.1007/978-981-10-6059-5_3

bonding, hydrophobic interaction, and coordination often exhibit novel functions that cannot be expected simple summation of individual properties.

The above-mentioned organization of component molecules and materials is supposed to be done in microscopic and nanoscopic structural precision. As methodology to control of structures in submicro- and nano-regimes, nanotechnology has been paid much attention. This concept considerably promoted science and technology of materials systems in these small regions. The nanotechnology concept has potentials to fabricate nano-objects into highly functional materials. So far, nanotechnology approach has many successes in controlling systems and materials with hard inorganic and metallic materials. However, application of nanotechnology to organic and biological components is not always successful.

In order to improve the above-mentioned situation, a certain kind of paradigm shift from current nanotechnology to nanoarchitectonics is necessary [1, 2]. This nanoarchitectonics concept was first proposed by Masakazu Aono at an international conference in 2000. After one decade past, nanoarchitectonics gradually became an emerging concept in many research fields from fundamental physicochemical science to practical targets including energy/environmental technologies and biological/biomedical applications [3]. The nanoarchitectonics is especially suitable of fabrication of materials systems from organic and biomaterials through self-organization and directed-organization [4].

For fabrication and organization of soft organic and biological components, keeping their dynamic natures are important for utilization of their advanced functions. If materials systems include features of dynamism in nanoscale phenomena, formation and maintenances of systems suffer influences from uncontrollable disturbances and fluctuation including thermal/statistical fluctuations and even sometimes quantum effects. In such materials systems, external actions such as signal inputs to the systems often induce perturbations to themselves and surroundings, resulting in additional further disturbances. Nanoarchitectonics tries to control such uncontrollable situations through harmonizing various factors and interactions. The most successful example of functional harmony of nanoscale objects can be seen in biological systems. In biological systems, countless numbers of nanoscale components from molecules to biomaterials are working together under unavoidable and substantial thermal fluctuations. Biological systems and biomimetic objects might be regarded as the most important target of nanoarchitectonics.

However, construction of three-dimensional biomimetic systems is rather tough subject. As an initial trial, systems with reduced dimension, two-dimensional systems, might be better target. On the basis of these considerations, we here focus on nanoarchitectonics of biomimetic membranes. This chapter introduces various examples of nanoarchitectonics approaches for bio-related thin films and membranes: (1) two-dimensional biomimetic membrane (molecular recognition, receptor tuning, and nanomaterial film for life control); (2) layer-by-layer biomimetic membrane (bioreactor, hierarchic assembly, and sensing and drug delivery).

3.2 Two-Dimensional Biomimetic Membrane

As seen in self-assembled monolayers (SAM) and Langmuir–Blodgett (LB) films (especially Langmuir monolayers), amphiphiles with various functional groups can be assembled and organized in two-dimensional plane. The formed structures can be regarded as mimics of cell membrane surface.

3.2.1 Molecular Recognition

Molecular recognition is one of the fundamental supramolecular events. Selective recognition of signal molecules at biomolecular receptors can transmit external stimuli into cells. Highly specific reactions by enzymes are supported by precise molecular recognition at the enzyme reaction pockets. Genetic information can be handled through precise base-pairing of DNA and RNA. These molecular recognition events do not occur in molecularly dispersed states in solution. In fact, recognition of biofunctional molecules is always carried out at certain kinds of interfacial media such as cell membrane surfaces, protein surfaces (outside and inside), and surfaces of bi-macromolecules such as DNA. Therefore, two-dimensional biomimetic membranes can supply nice models to investigate molecular recognition of bio-related molecules.

One example of effective molecular recognition of biomolecules was first suggested by Kitano and Ringsdorf with shifts of surface-pressure/molecular area isotherms of an adenine-based amphiphile in the presence of thymine derivative at the air–water interface [5]. This example suggests contribution of hydrogen-bonded base-pair mimics in the presence of water although the corresponding monomeric base-pair formation is not achieved in bulk aqueous phase. Kunitake and co-workers systematically investigated molecular recognition of various biomolecules [6]. They synthesized amphiphilic molecules with molecular recognition sites as hydrophilic head groups that were spread as a monolayer at the air–water interface. At the formed two-dimensional biomimetic membranes (cell surface mimics), recognition of aqueous biomolecules can be investigated using various analytical methods including FT-IR spectroscopy and X-ray photoelectron spectroscopy (XPS). Efficient molecular recognition with enhanced hydrogen bonding capability at the air–water interface was observed for many biomolecules including amino acids, peptides, sugars, nucleic acid bases, and nucleotides (Fig. 3.1).

With fixing molecular recognition pair, molecular recognition efficiencies were investigated at various environments, molecularly dispersed states in water), mesoscopic dynamic interfaces (micellar and bilayers surfaces in water), and macroscopic interface (the air–water interface) (Fig. 3.2) [7]. Biding constants between guanidinium and phosphate significantly changed from 1.4 M^{-1} (in water dispersion) to 10^2–10^4 M^{-1} (aqueous mesoscopic interface) and 10^6–10^7 M^{-1}

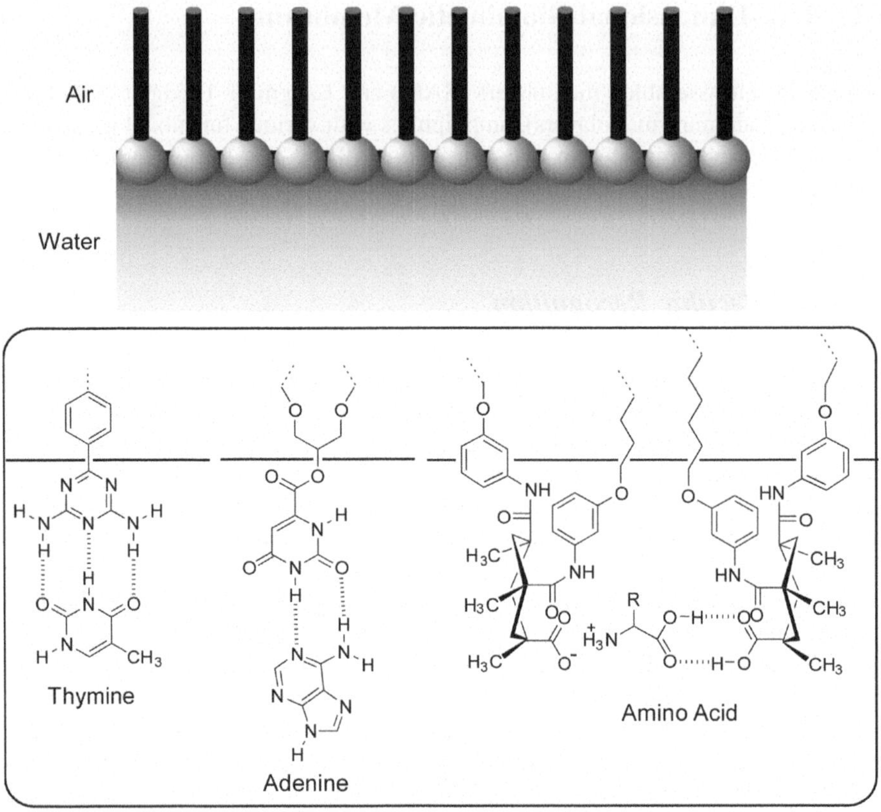

Fig. 3.1 Efficient molecular recognition for biomolecules with enhanced hydrogen bonding capability at the air–water interface

(macroscopic air–water interface). The obtained surprising results indicate that surrounding environments have crucial effects on molecular–molecular interaction, resulting in incredible modulation on molecular recognition efficiency.

Molecular recognition at the air–water interface was also theoretically investigated by Sakurai and co-workers using quantum chemical calculations [8, 9]. As illustrated in Fig. 3.3, a model recognition pair, (hydrophilic) phosphate and (hydrophobic) gaunidinium was put at various positions between high dielectric water phase and low dielectric lipid phase. A energy of the phosphate and guanidinium is plotted as a function of distance between this recognition pair to give its binding energy. The calculated binding constant decreases as the position of the recognition pair shits toward water phase with high dielectric constant, and finally becomes zero completely in water phase. However, the position of the phosphate group exactly at interface line with the phosphate position within aqueous phase can regenerate the binding energy and constant that were obtained in actual experiments. Even though binding site itself stays in water medium, effects from low dielectric phase at

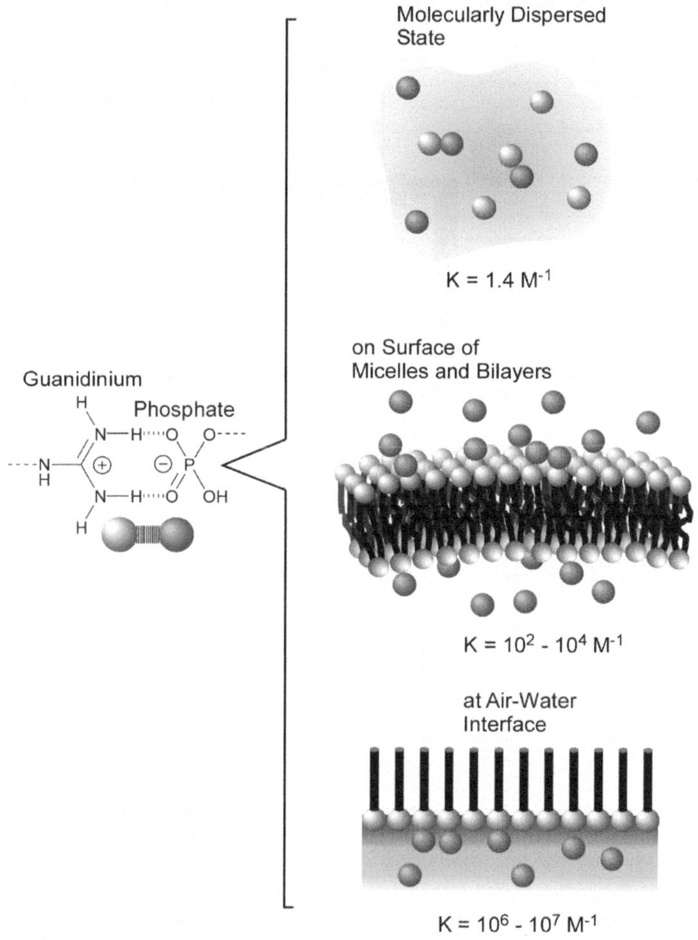

Fig. 3.2 Molecular recognition efficiencies between guanidinium and phosphate functional groups were investigated at various environments, molecularly dispersed states in water), mesoscopic dynamic interfaces (micellar and bilayers surfaces in water), and macroscopic interface (the air–water interface)

vicinity has sufficient contributions in molecular recognition. It is a key mechanism of enhanced efficiency of molecular recognition at interfacial environments.

In addition to promoted molecular interaction, two-dimensional interfacial environments supply media for nanoarchitectonic formation of molecular receptor sites through self-assembly. Because of high advantages in high motional freedom at the air–water interface, multiple kinds of amphiphile host molecules with different recognition sites can be spontaneously assembled into complicated receptor

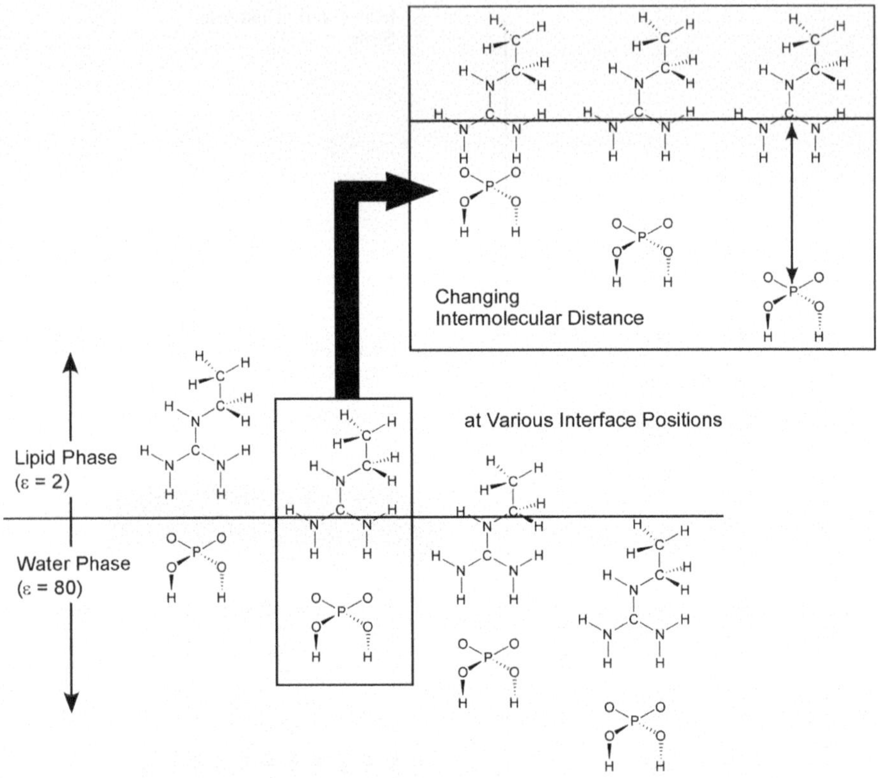

Fig. 3.3 Theoretical models for molecular recognition at the air–water interface using (hydrophilic) phosphate and (hydrophobic) gaunidinium recognition pair

structures. Some examples are shown in Fig. 3.4. An example of binary system of monoalkyl guanidinium and dialkylmelamine can give recognition site for flavin mononucleotide (FMN). Highly efficient recognition site for FMN with a binding constant of more than 10^7 M^{-1} was formed just by mixing two rather simple host molecules [10]. More complicated binding site for flavin adenine dincleotide (FAD) can be created by trinary recognition system at the air–water interface [11]. In the latter case, one FAD molecule is supposed to bind diaminotriazine at isoalloxiazine ring, two guanidinium molecules at phosphate groups, and one orotate molecule at adenine sites. This two-dimensional molecular assembly strategy can be also used for different functional purpose such as nanofabrications not limited to molecular recognition [12]. Array of specific molecular pair of several amphiphile molecules with aqueous guest molecule leads to formation of regular two-dimensional surface pattern structures with nanometer structural precision.

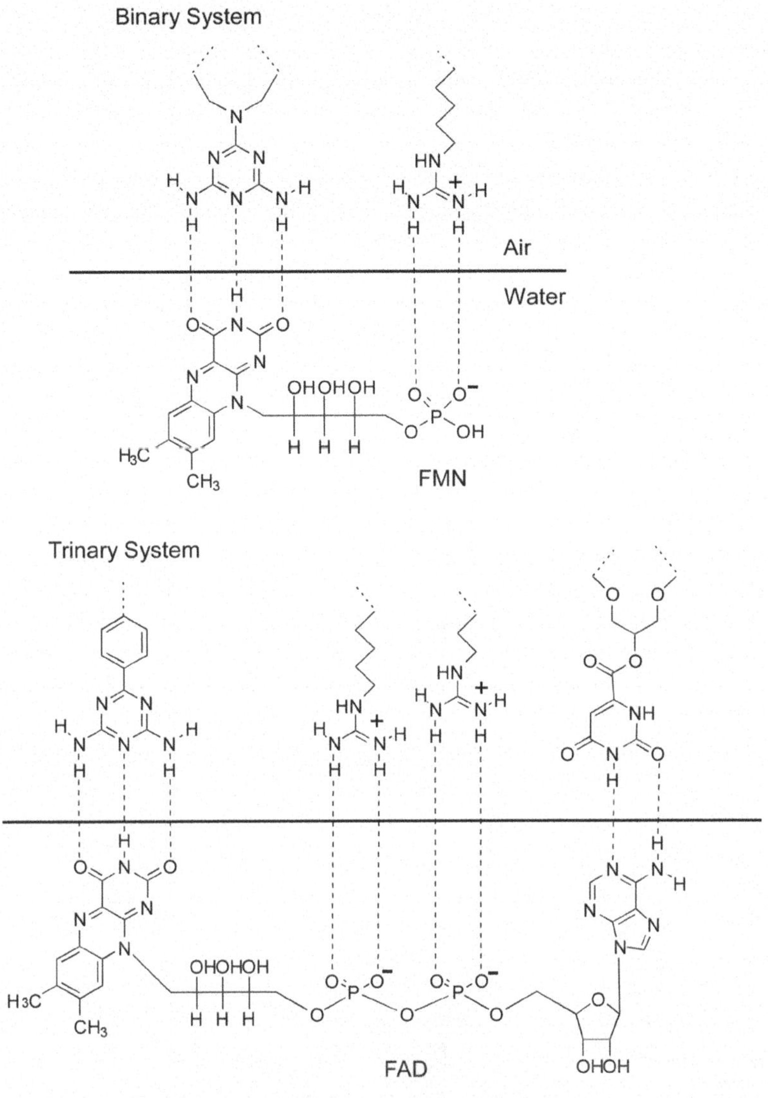

Fig. 3.4 Multiple kinds of amphiphile host molecules with different recognition sites: binary system and trinary

3.2.2 Receptor Tuning

If we consider two-dimensional interfacial environment from lateral direction, another important role of interfaces can be recognized. It is a role for connection between macroscopic mechanical actions and nano-level molecular functions.

At dynamic interfacial media, macroscopic motions including compression, expansion, and bending are applicable to lateral directions of the interface and properties in nanometer-scale dimension will be induced at the same time. Therefore, molecular systems can be controlled at dynamic interfacial media upon application of macroscopic inputs such as mechanical motions. A novel conceptual paradigm, hand-operating nanotechnology, has been proposed on the basis of these characteristics of dynamic interfaces [13, 14]. With this concept, we successfully control functions of molecular machines by hand-motion-like conventional macroscopic mechanical motions. For example, capture and release of guest molecules by molecular machines with capability of reversible cavity formation was demonstrated by hand-motion-like macroscopic mechanical motions [15, 16]. In addition, completely efficient energy conversion from macroscopic motions to molecular-level motion of molecular pliers at the interface has been recently achieved [17].

The above-mentioned methodology is applied for fine tuning of receptor structures, resulting in molecular recognition with molecular tuning concept. As shown in Fig. 3.5, conformational structures of a molecular receptor, armed cyclen (a 1,4,7,10-tetraazacyclododecane core with four cholesteric side arms as *N*-substituents), was finely tuned by application of lateral mechanical force [18]. Because this receptor can be twisted, resulting in modification of surface chiral environments. For example, chiral selectivity in amino acids binding (their L-isomer and D-isomer) can be tuned by applied surface pressures. The most stable conformation of the molecular receptor at the air–water interface can be continuously shifted upon lateral mechanical force, and dramatic changes of chiral selectivity of guest binding

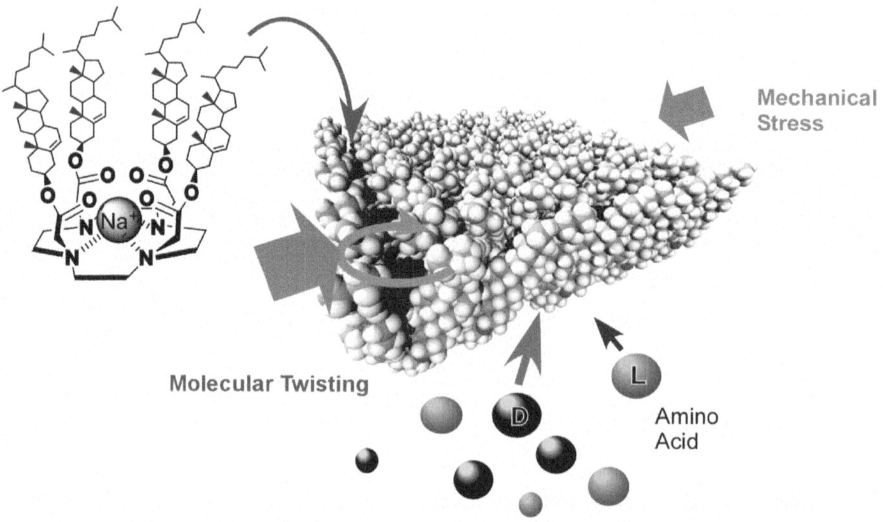

Fig. 3.5 Chiral discrimination of amino acids by application of lateral mechanical force of a molecular receptor, armed cyclen (a 1,4,7,10-tetraazacyclododecane core with four cholesteric side arms as *N*-substituents)

are resulted. Similarly, structural tuning of another receptor, cholesterol-substituted triazacyclononane, resulted in highly efficient discrimination between thymine and uracil derivatives [19] that cannot be distinguished by naturally occurring nucleic acids, DNA and RNA. This molecular tuning strategy may be capable to create molecular recognition sites better than those of standard biological systems.

In Fig. 3.6, major types of molecular recognition mechanisms are historically categorized [20]. Pioneering systems for molecular recognition as seen in host-guest chemistry by crown ethers and cyclodextrins basically uses one energy most stable states. This mechanism actually received Nobel Prize of Chemistry in 1987. One epoch-making innovation was made by Shinkai et al. to create two stable states though photo-isomerization of azobenzene bridge between two crown ether units in order to switch recognition efficiency [21, 22]. This is regarded as switching mechanism in molecular recognition. This switching mechanism is origin of molecular machine operation that is basically corresponding to creation of multiple states and switching between them. Molecular machines have received Nobel Prize of Chemistry in 2017. Unlike these two previously used mechanisms, our molecular tuning mechanism can continuously utilize numerous numbers of molecular states. Among these infinite possibilities, we can select the most desirable state. Therefore, receptor tuning for molecular recognition can be defined as completely new mode in molecular recognition.

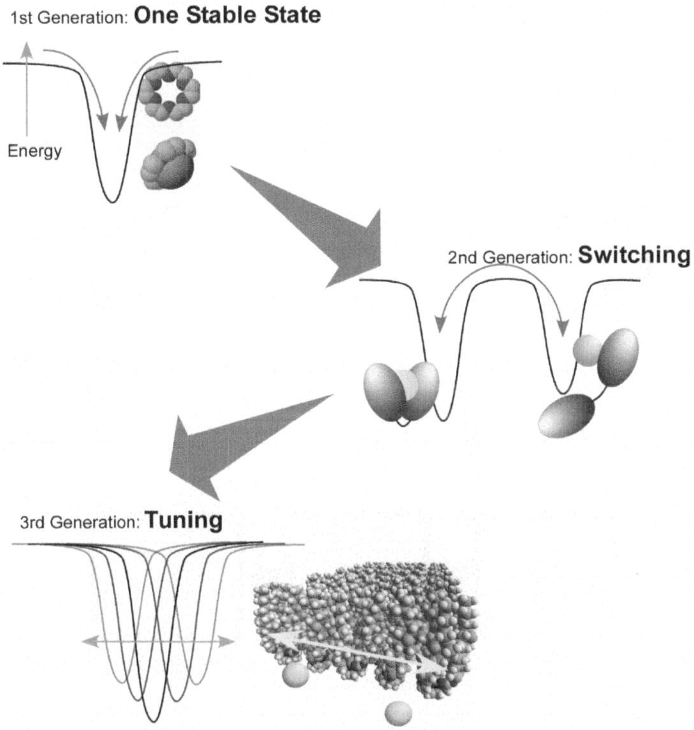

Fig. 3.6 Major types of molecular recognition mechanisms: one stable state; switching; tuning

3.2.3 Nanomaterial Film for Life Control

Fabrication of nanoscale and microscale structures at interfacial media is also effective to control life activity control. For example, aligned array of fullerene whiskers were architected at the air–water interface and were used as a scaffold for cell culture (Fig. 3.7) [23]. Incubation of mouse skeletal myoblast C2C12 cells on the fabricated fullerene whisker arrays induced well-aligned cell growth, and the aligned direction was highly correlated with the direction of the fullerene whisker arrays. Possibilities of cell control were not limited to cell alignment. Even cell differentiation was regulated by the fabricated fullerene whisker arrays. Along directions of cell growth, cell fusion was induced and formation of oriented myotubes progresses. In addition, myogenic differentiation can be promoted in both early and late stages.

Such scaffold films for cell culture was further controlled by external mechanical flow in our new fabrication method, vortex-flow LB technique (vortex-LB) [24]. Mechanical stirring of the medium under the monolayer of nanomaterials activated mechanical flow of water (or solvents) and induce alignment of nanomaterials at the interface. For example, oriented alignments of fullerene whiskers can be achieved along with flow of centrifugal rotation at liquid surface. Straight alignment and curved alignment were both created depending positions from the center of vortex flow. Transfer of the aligned fullerene whiskers array on various types substrates

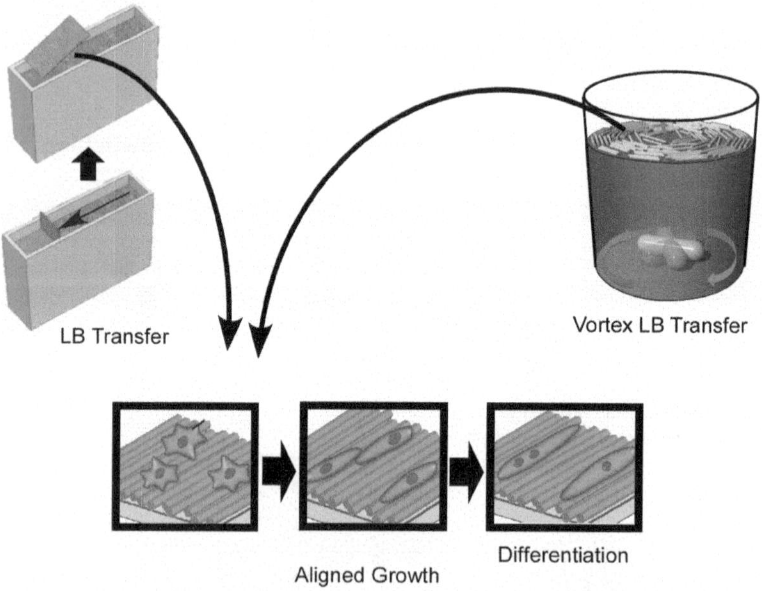

Fig. 3.7 Aligned array of fullerene whiskers architected at the air–water interface for a scaffold for cell culture

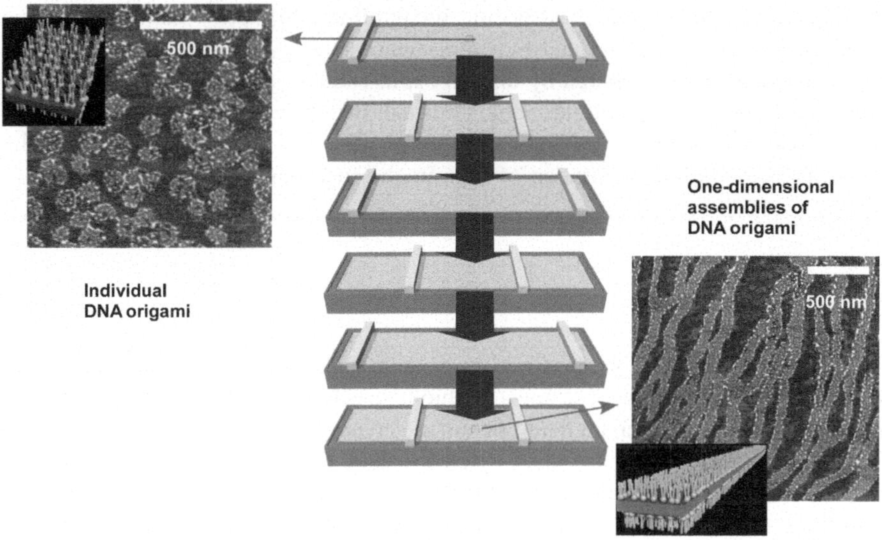

Fig. 3.8 Self-alignment of the DNA origami pieces upon macroscopic mechanical motions at the air–water interface

including paper, mica, glass, gold, and boron nitride. Human osteoblast MG63 cells were adhered preferentially at the surface of aligned fullerene whiskers, and the adhered cells grow along direction of aligned fullerene whiskers. In addition, cell proliferation experiments confirmed low toxic nature of the used fullerene whiskers.

The interfacial process can organize nanoarchitectonics objects of biomaterials, DNA origami [25]. DNA origami sheets with rectangular shape (90 nm $\times$ 65 nm) were first complexed with cation lipid molecules to dissolve the DNA origami in organic solvent. The resulting solution of the DNA origami was spread on water to form Langmuir film. Macroscopic mechanical motion at the air–water interface can induce self-alignment of the DNA origami pieces (Fig. 3.8). Repeated compression–expansion cycles between 3 and 30 mN m^{-1} twice resulted in formation of long belt structures with length up to 4900 nm with keeping its height and width in monomeric size. In this fabrication process, dynamic motions are necessary for formation of unidirectional assembly of the DNA origami pieces. If the DNA origami on water is placed at both low and high pressures, formation of one-dimensional array cannot be achieved.

3.3 Layer-by-Layer Biomimetic Membrane

The alternate layer-by-layer (LbL) assembly is applicable to huge variety of target materials including various polymers, inorganic nanomaterials, and biomaterials with many kinds of driving forces such as electrostatic interaction, hydrogen bonding, coordination, and bio-specific recognition [26]. Only with inexpensive apparatuses such as beakers and tweezers, nanometer level thin films in designed later sequence can be prepared by dipping, spin-coating, and spraying. Not limited to use flat films as morphologies, LbL films can be fabricated into various types of shapes including spheres, capsules, and tubes. The LbL assembly is one of the powerful methods for preparation of biomimetic membranes.

3.3.1 Bioreactor

The LbL assembly is a method highly capable for immobilizing biomaterials in well-organized ways. Because most of biomaterials such as proteins working in aqueous media have certain amount of surface charges, these biomaterials are easily assembled into layered organization by electrostatic LbL assembly. The assembling processes are usually done in bio-friendly their aqueous solutions under mild conditions. In addition, it is proved that various stabilities (thermal, pH variation, and duration) of proteins immobilized within the LbL multilayer films with poly-electrolytes are significantly enhanced.

One of attractive application for protein multilayered films would be sequential enzyme reactor [27]. As shown in Fig. 3.9, the multi-enzyme reactors including layers of glucose oxidase (GOD) and glucoamylase (GA) with polyelectrolyte layers were prepared on an ultrafilter. An aqueous solution of starch substrate was placed on the top of enzyme films and then moderate pressure was applied. Starch was first converted into glucose by GA through hydrolysis of the glycoside bond in

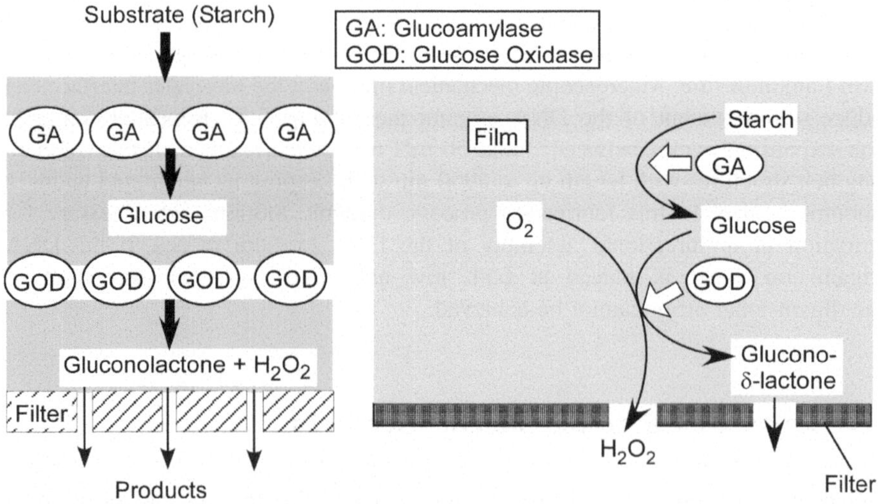

Fig. 3.9 Multi-enzyme reactors including layers of glucose oxidase (GOD) and glucoamylase (GA) with polyelectrolyte layers prepared on an ultrafilter

starch. Glucose was then converted to gluconolactone by GOD with H_2O_2 as a co-product. In addition unreacted starch was not detected in the filtrate at all, because starch with high molecular weight cannot pass through the filter. As a result, separation of substrate and products without additional procedures.

Because the LbL strategy allows free choice of layer sequences of two kinds of enzymes and polyelectrolytes, relation between layer sequences of the enzymes and reaction efficiency. Just by changing the dipping sequences, various film structures with desirable layering structures are easily nanoarchitected. Therefore, optimization of reactor capability through adjusting the number of layers, layering sequence, and layer separation can be achieved. In the cases of GA-GOD multilayer reactors, the enzyme sequence with the top of GA and the bottom of GOD is essential. In addition, separation between the GA layer and GOD layer is also crucial, because the products of reaction by GOD may have effects of inhibition to GA. Design approach for multi-enzyme reactors by the LbL assembly is a versatile method to create novel reactors with free choice of enzymes and their performance optimization.

3.3.2 Hierarchic Assembly

Biological organization themselves basically have hierarchic nature, as seen in biomolecules, organelles, cells, tissues, and organs. The LbL assembly is also effective to architect complicated hierarchic structures not limited to simple layered structures. For preparation of nanostructured materials with high biofunctions, architecting hierarchic structures based on the LbL assembly would be one of the attractive methods.

As artificial cell membranes, spherical capsule objects with a skin of lipid bilayer membranes have been well investigated as seen extensive research of liposomes and vesicles. More advanced mimics of biomembranes would be their hierarchic assemblies. However, these spherical cell mimics often have limited mechanical stability and tend to cause collapse with rapture of their capsule structures upon assembling process. In order to overcome this problem, Katagiri et al. developed lipid bilayer vesicles with covalent linkage of siloxane frameworks, which can be architected into advanced organization as multi-cellular systems. The prepared vesicles possess the siloxane networks covalently attached to the surface of bilayer membrane. These objects were named as cerasome (ceramics + soma) (Fig. 3.10) [28]. Cerasomes are prepared using alkoxysilane-bearing lipids that are dispersed in a slightly acidic aqueous solution. Formation of bilayer vesicle structures and the siloxane network structures spontaneously proceed to results in robust cell mimics.

Upon design of the alkoxysilane-bearing lipids, both cerasomes having positive and negative surface charges can be prepared. These charged cerasomes can be subjected to LbL assembly for formation of advanced hierarchic cell mimic assemblies (Fig. 3.11). For example, anionic cerasome objects were assembled in layer-by-layer fashion with cationic polyelectrolytes, poly(diallyldimethylammonium

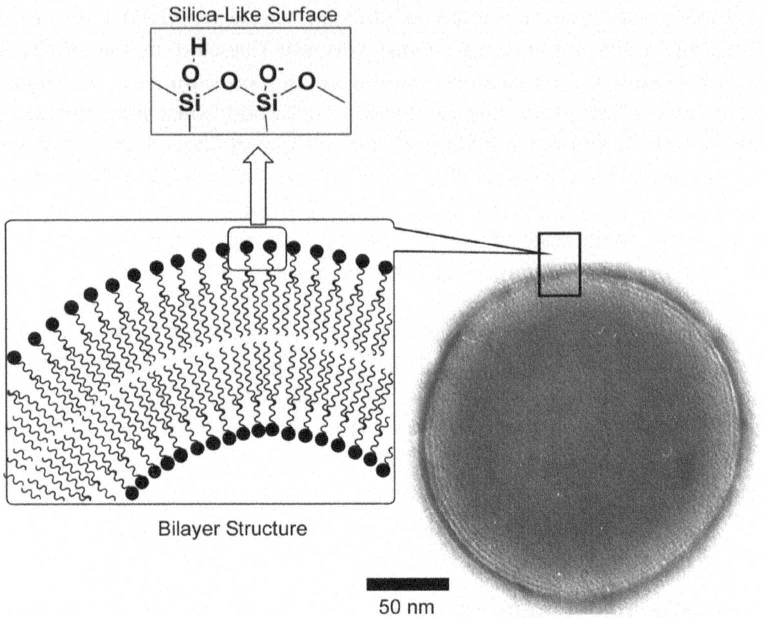

Fig. 3.10 A cerasome of the siloxane networks covalently attached to the surface of bilayer membrane

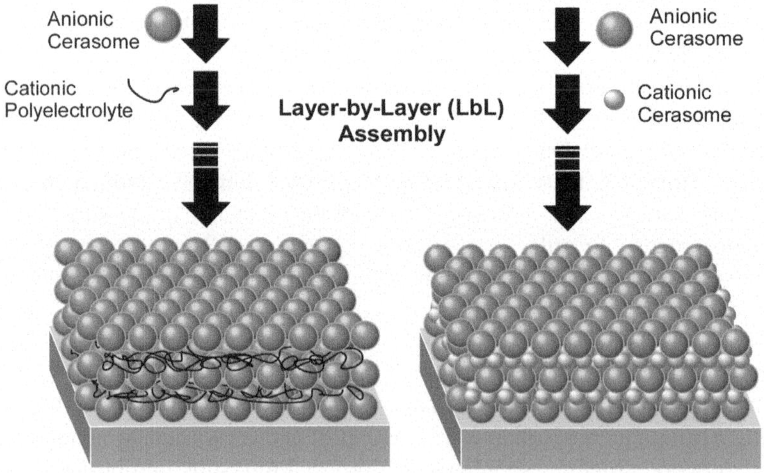

Fig. 3.11 LbL assemblies of cerasomes for formation of advanced hierarchic cell mimic assemblies

chloride) (PDDA) [29]. In addition, direct LbL assemblies between anionic cerasomes and cationic cerasomes without using intermediate polyelectrolytes were also accomplished to form hierarchic assemblies only with cell membrane mimics [30]. Successful processes of these LbL assembly of robust cell mimics, cerasomes, were confirmed by continuous mass changes of the assemblies upon measurements by a quartz crystal microbalance (QCM). Preservation of spherical shapes of individual cell spherical mimics was also confirmed by observation with atomic force microscopy (AFM).

These attempts to architect hierarchic assemblies of cell membrane mimics would promote cell membrane research from individual cell stage to tissue mimic stage. This progress makes the related research fields more applicable to biomedical fields.

3.3.3 Sensing and Drug Delivery

The LbL assemblies can be used for various applications including advanced sensing and drug delivery systems. The LbL assemblies between reduced graphene oxides and ionic liquid can be used as sensor materials for toxic aromatic gasses [31]. Graphene oxide sheet was reduced to reduced graphene oxide sheet in the presence of ionic liquids (imidazolium salts) in water. This process gives formation of charged composites between graphene sheets and ionic liquid. The resultant charged composites were then assembled with anionic polyelectrolytes, poly(sodium styrenesulfonate) (PSS), into layer-by-layer structure (Fig. 3.12). Layer-by-layer architecture of both π-electron rich components, graphene sheets and imidazolium ionic liquids, resulted in formation of highly π-rich nanospace. A QCM sensor coated with these LbL assemblies can be used for highly selective and sensitive detection of aromatic toxic gas substances. Exposure of the QCM sensor with the LbL composite films to various saturated vapors of target gas substances results in an in situ decrease in frequency of QCM. One sensor exhibited considerably higher selectivity for benzene (more than 10 times) over cyclohexane although these substances have similar molecular sizes, molecular weights, and vapor pressures.

For application of hierarchic LbL structures to sensing applications, LbL films of mesoporous carbon capsules with couterionic polyelectrolytes were prepared (Fig. 3.13) [32]. The mesoporous carbon capsule object has homogeneous dimensions ($1000 \times 700 \times 300$ nm^3) with 35-nm-thick mesoporous walls with a uniform pore size distribution centered at 4.3 nm in diameter, and a specific surface area of 918 m^2 g^{-1}. Because the mesoporous carbon capsules have no charges, these capsules were first coated with charged surfactants and subjected to LbL assembly with the aid of appropriate polyelectrolytes onto a QCM sensor plate. Similarly, this sensing system provides high sensitivity to aromatic hydrocarbons such as benzene and toluene as compared with aliphatic hydrocarbons such as cyclohexane. Particular advantage of this sensing system with hierarchic LbL structures is modification of sensing selectivity through doping of the second

Fig. 3.12 Layer-by-layer architecture of both π-electron rich components, graphene sheets and imidazolium ionic liquids, for sensor application

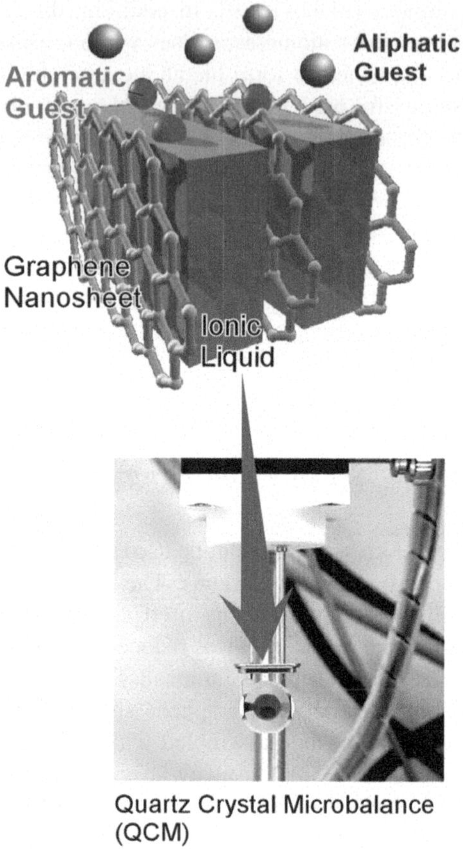

sensing components into the inner cores. For example, sensors with carbon capsule film doped with lauric acid exhibited the greatest affinities for non-aromatic amines. Selectivity of the sensors have strong sensitivity to acetic acid when the mesoporous carbon capsules were doped with dodecylamine.

Sensing performance of hierarchic LbL films in solution phase was also investigated by LbL assembly of mesoporous carbon on a QCM plate (Fig. 3.14) [33]. For successful LbL assembly, negative carboxylate groups was introduced onto surfaces of mesoporous carbon (CMK-3) through partial oxidation and then the charged mesoporous carbon materials were assembled with appropriate polyelectrolytes such as cationic PDDA into layer-by-layer architectures on a QCM sensor. Sensing properties of the LbL-modified QCM in aqueous phase were investigated

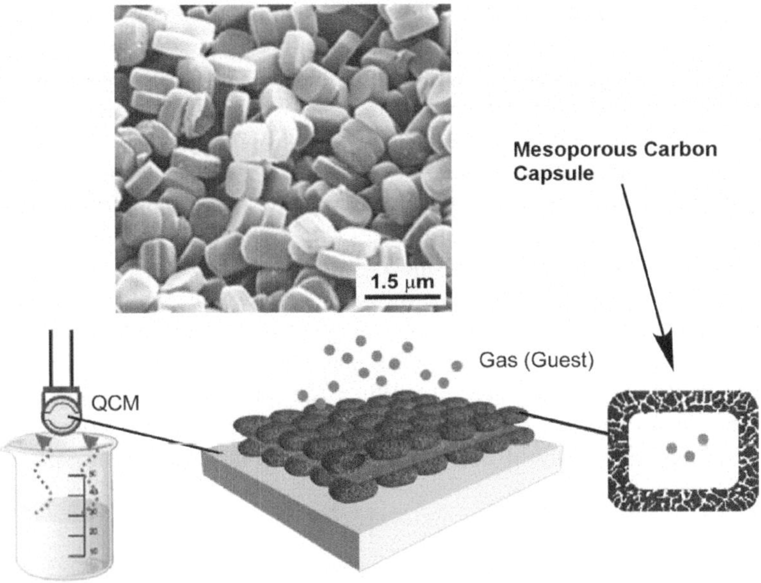

Fig. 3.13 LbL films of mesoporous carbon capsules with couterionic polyelectrolytes for sensor application

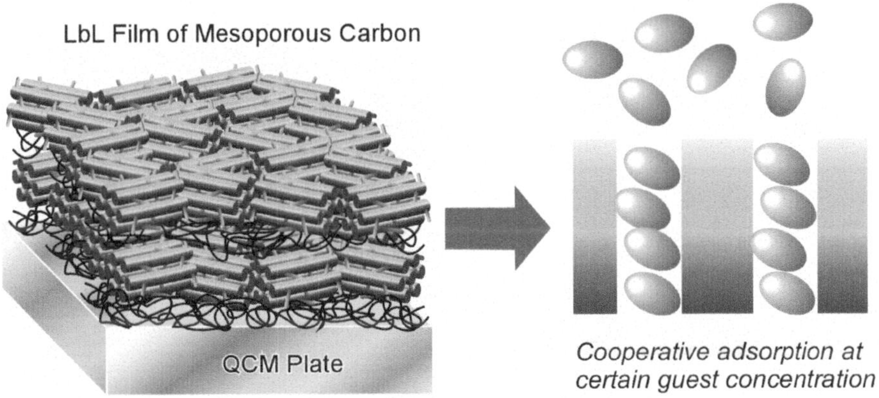

Fig. 3.14 LbL assembly of mesoporous carbon on a QCM plate for sensing application in aqueous phase

after immersion of the sensor plate in water. Injection of tannic acid solution to the water phase immediately caused frequency shifts. The frequency shifts due to adsorption of tannic acid much exceed those for catechin and caffeine. The resulting sensitivity ratios of tannic acid to catechin or caffeine. Effective π–π interactions and hydrophobic effects on the basis of multiple phenyl rings structure of the tannic acid molecule would be more advantageous for facilitate sensing. The adsorption profile of tannic acid to the CMK LbL has a sigmoidal behavior at low concentrations of tannic acid, suggesting highly cooperative nature of adsorption of tannic acids into carbon nanopores. Accumulation of such unusual observation would promote understanding of molecular interactions within nanospaces.

Not limited to sensing application, hierarchic LbL assemblies are used for the other bio-related applications. Hierarchic LbL assemblies between hollow mesoporous silica capsules and PDDA with co-adsorber of silica nanoparticles exhibited unusual release of drug and liquid substances (Fig. 3.15) [34]. Quantitative analyses of water release from the silica capsule LbL films on the basis of QCM measurements exhibited a stepwise behavior of water release even without any external stimuli (automatic on–off periodic release). The similar behaviors were observed for the other liquid drugs, too. Liquid guests entrapped in outer mesopore channels evaporate to the air (release process). When the most of the liquid guest in the mesopore regions released, the release process become almost terminated (stop process). However, supply of the liquid guest from inside reservoir space to the

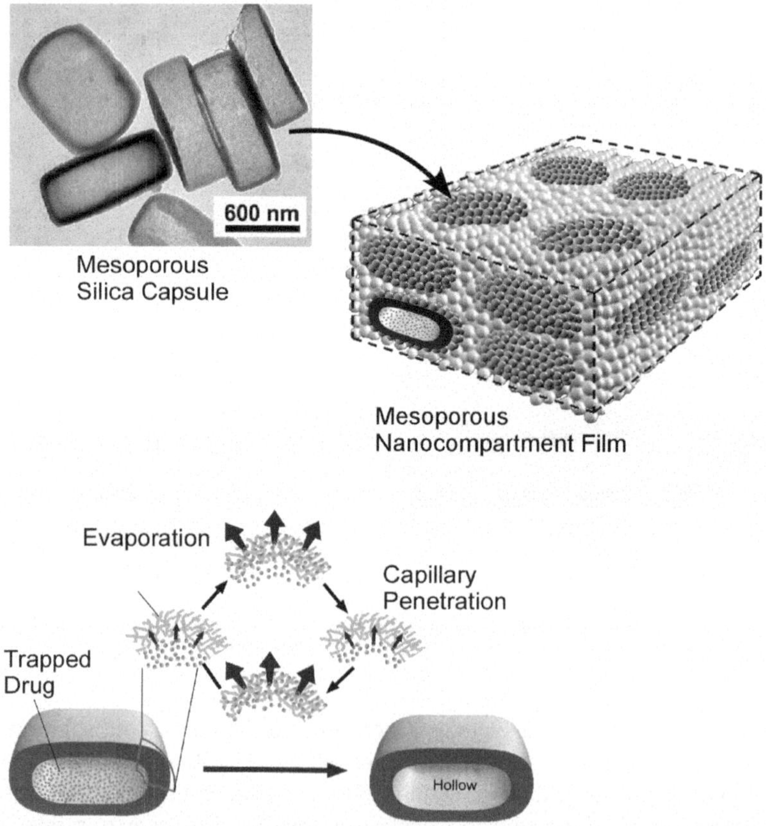

Fig. 3.15 LbL assemblies between hollow mesoporous silica capsules and PDDA with co-adsorber of silica nanoparticles for unusual release of drug and liquid substances

outside mesopore region resume material release process. These on–off materials release process repeated regularly until all the liquid guests were released. This is a rare example of *a stimulus-free controlled release system* of possible drug molecules. The system of great utility for development of energy-less, clean stimulus-free, and patient-friendly controlled drug release applications.

3.4 Summary

This chapter explains various types of biomimetic membranes with some functions. Interfacial environments provide unique properties that are surely useful to promote molecular recognition of biomolecules and to control organization of bio-components. The layer-by-layer assembly is a certain kind of versatile strategy to construct well-designed layered architectures. Especially, construction of hierarchic structures by the LbL assembly would allow us to fabricate advanced biomimetic membranes like tissues. Use of the fabricated structures are demonstrated with their functions in reactors, sensors, and drug delivery. These objects are architected from nanocomponents through various component interactions and self-organization. The latter strategy is supported by emerging concept, nanoarchitectonics. Contributions of nanoarchitectonics to biomimetic membranes would become more and more important in future.

References

1. Aono M, Ariga K (2016) The way to nanoarchitectonics and the way of nanoarchitectonics. Adv Mater 28:922–989
2. Ariga K, Aono M (2016) Nanoarchitectonics. Jpn J Appl Phys 55:1102A6
3. Ariga K, Li J, Fei J, Ji Q, Hill JP (2016) Nanoarchitectonics for dynamic functional materials from atomic/molecular-level manipulation to macroscopic action. Adv Mater 28:1251–1286
4. Ariga K, Kawakami K, Ebara M, Kotsuchibashi Y, Ji Q, Hill JP (2014) Bioinspired nanoarchitectonics as emerging drug delivery systems. New J Chem 38:5149–5163
5. Kitano H, Ringsdorf H (1985) Surface behaviors of nucleic acid base-containing lipids in monolayer and bilayer systems. Bull Chem Soc Jpn 58:2826–2828
6. Ariga K, Kunitake T (1998) Molecular recognition at air-water and related interfaces: complementary hydrogen bonding and multisite interaction. Acc Chem Res 31:371–378
7. Onda M, Yoshihara K, Koyano H, Ariga K, Kunitake T (1996) Molecular recognition of nucleotides by the guanidinium unit at the surface of aqueous micelles and bilayers. A comparison of microscopic and macroscopic Interfaces. J Am Chem Soc 118:8524–8530
8. Sakurai M, Tamagawa H, Inoue Y, Ariga K, Kunitake T (1997) Theoretical study of intermolecular interaction at the lipid-water interface. 1. Quantum chemical analysis using a reaction field theory. J Phys Chem B 101:4810–4816
9. Tamagawa H, Sakurai M, Inoue Y, Ariga K, Kunitake T (1997) Theoretical study of intermolecular interaction at the lipid-water interface. 2. Analysis based on the Poisson-Boltzmann equation. J Phys Chem B 101:4817–4825

10. Ariga K, Kamino A, Koyano H, Kunitake T (1997) Recognition of aqueous flavin mononucleotide on the surface of binary monolayers of guanidinium and melamine amphiphiles. J Mater Chem 7:1155–1161
11. Taguchi T, Ariga K, Kunitake T (1995) Multi-site recognition of flavin adenine dinucleotide by mixed monolayers on water. Chem. Lett. 204:701–702
12. Oishi Y, Torii Y, Kato T, Kuramori M, Suehiro K, Ariga K, Taguchi K, Kamino A, Koyano H, Kunitake T (1997) Molecular patterning of a guanidinium/orotate mixed monolayer through molecular recognition with flavin adenine dinucleotide. Langmuir 13:519–524
13. Ariga K, Mori T, Ishihara S, Kawakami K, Hill JP (2014) Bridging the difference to the billionth-of-a-meter length scale: how to operate nanoscopic machines and nanomaterials by using macroscopic actions. Chem Mater 26:519–532
14. Ariga K, Minami K, Ebara M, Nakanishi J (2016) What are emerging concepts and challenges in NANO? Nanoarchitectonics, hand-operating nanotechnology, and mechanobiology. Polym J 48:371–389
15. Ariga K, Terasaka Y, Sakai D, Tsuji H, Kikuchi J (2000) Piezoluminescence based on molecular recognition by dynamic cavity array of steroid cyclophanes at the air-water interface. J Am Chem Soc 122:7835–7836
16. Ariga K, Nakanishi T, Terasaka Y, Tsuji H, Sakai D, Kikuchi J (2005) Piezoluminescence at the air-water interface through dynamic molecular recognition driven by lateral pressure application. Langmuir 21:976–981
17. Ishikawa D, Mori Y, Yonamine Y, Nakanishi W, Cheung D, Hill JP, Ariga K (2015) Mechanochemical tuning of binaphthyl conformation at the air-water interface Angew. Chem Int Ed 54:8988–8991
18. Michinobu T, Shinoda S, Nakanishi T, Hill JP, Fujii K, Player TN, Tsukube H, Ariga K (2006) Mechanical control of enantioselectivity of amino acid recognition by cholesterol-armed cyclen monolayer at the air-water interface. J Am Chem Soc 128: 14478–14479
19. Mori T, Okamoto K, Endo H, Hill JP, Shinoda S, Matsukura M, Tsukube H, Suzuki Y, Kanekiyo Y, Ariga K (2010) Mechanical tuning of molecular recognition to discriminate the single-methyl-group difference between thymine and uracil. J Am Chem Soc 132: 12868–12870
20. Shirai Y, Minami K, Nakanishi W, Yonamine Y, Joachim C, Ariga K (2016) Driving nanocars and nanomachines at interfaces: from concept of nanoarchitectonics to actual use in world wide race and hand operation. Jpn J Appl Phys 55:1102A2
21. Shinkai S, Ogawa T, Kusano Y, Manabe O (1980) Selective extraction of alkali metal cations by a photoresponsive bis(crown ether). Chem Lett 1980:283–286
22. Shinkai S, Nakaji T, Ogawa T, Shigematsu K, Manabe O (1981) Photoresponsive crown ethers. 2. Photocontrol of ion extraction and ion transport by a bis(crown ether) with a butterfly-like motion. J Am Chem Soc 103:111–115
23. Minami K, Kasuya Y, Yamazaki T, Ji Q, Nakanishi W, Hill JP, Sakai H, Ariga K (2015) Highly ordered one-dimensional fullerene crystals for concurrent control of macroscopic cellular orientation and differentiation towards large-scale tissue engineering. Adv Mater 27:4020–4026
24. Ishihara S, Minami K, Morita H, Yamazaki T, Hanagata N, Miyazawa K, Acharya S, Nakanishi W, Hill JP, Ariga K (2015) Vortex-aligned fullerene nanowhiskers as a scaffold for orienting cell growth. ACS Appl Mater Interfaces 7:15667–15673
25. Yonamine Y, Cervantes-Salguero K, Minami K, Kawamata I, Nakanishi W, Hill JP, Murata S, Ariga K (2016) Supramolecular 1-D polymerization of DNA origami through a dynamic process at the 2-Dimensionally confined air-water interface. Phys Chem Chem Phys 18:12576–12581
26. Ariga K, Yamauchi Y, Rydzek G, Ji Q, Yonamine Y, Wu KCW, Hill IP (2014) Layer-by-layer nanoarchitectonics: invention, innovation, and evolution. Chem Lett 43:36–68

27. Onda M, Lvov Y, Ariga K, Kunitake T (1996) Sequential reaction and product separation on molecular films of glucoamylase and glucose oxidase assembled on an ultrafilter. J Ferment Bioeng 82:502–506
28. Katagiri K, Ariga K, Kikuchi J (1999) Preparation of organic-inorganic hybrid vesicle "Cerasome" derived from artificial lipid with alkoxysilyl head. Chem Lett 1999:661–662
29. Katagiri K, Hamasaki R, Ariga K, Kikuchi J (2002) Layer-by-layer self-assembling of liposomal nanohybrid "cerasome" on substrates. Langmuir 18:6709–6711
30. Katagiri K, Hamasaki R, Ariga K, Kikuchi J (2002) Layered paving of vesicular nanoparticles formed with cerasome as a bioinspired organic-inorganic hybrid. J Am Chem Soc 124: 7892–7893
31. Ji Q, Honma I, Paek SM, Akada M, Hill JP, Vinu A, Ariga K (2010) Layer-by-layer films of graphene and ionic liquid for highly selective gas sensing. Angew Chem Int Ed 49: 9737–9739
32. Ji Q, Yoon SB, Hill JP, Vinu A, Yu JS, Ariga K (2009) Layer-by-layer films of dual-pore carbon capsules with designable selectivity of gas adsorption. J Am Chem Soc 131: 4220–4221
33. Ariga K, Vinu A, Ji Q, Ohmori O, Hill JP, Acharya S, Koike J, Shiratori S (2008) A layered mesoporous carbon sensor based on nanopore-filling cooperative adsorption in the liquid phase. Angew Chem Int Ed 47:7254–7257
34. Ji Q, Miyahara M, Hill JP, Acharya S, Vinu A, Yoon SB, Yu JS, Sakamoto K, Ariga K (2008) Stimuli-free auto-modulated material release from mesoporous nanocompartment films. J Am Chem Soc 130:2376–2377

Part III
Biomolecules Based Molecular Assembly

Chapter 4
Polysaccharides-Based Microcapsules

Yi Jia, Xiyun Feng and Junbai Li

Abstract LbL-assembled microcapsules have gained great interests in diverse fields due to their versatile and engineering properties. In this chapter, we focus on the microcapsules fabricated from biocompatible and biodegradable polysaccharides. Various interactions, including electrostatic interaction, hydrogen bonding, covalent crosslinking, ionic crosslinking, and host–guest interaction are introduced and employed as driving forces to construct polysaccharide microcapsules with specific stimuli-responsivity. The functionalization of polysaccharide microcapsules with bioactive moieties (such as cell surface receptor ligands) is presented and their applications in cancer therapy and blood substitutes are highlighted.

Keywords Polysaccharides · Layer-by-layer assembly · Schiff base · Microcapsule · Cancer therapy · Blood substitute

4.1 Introduction

Multifunctional microcapsule systems have attracted great attention because of their wide potential applications in fields of catalysis, energy storage, cosmetics, especially in drug delivery, and other biomedical applications. In the past few decades, numerous approaches have been developed for the construction of multifunctional microcapsule systems and significant progress has been made. As a powerful molecular assembly technique, layer-by-layer (LbL) assembly has been extensively used for the fabrication of multifunctional microcapsules because it possesses engineered features including size, shape, thickness, composition, permeation, and

Y. Jia · X. Feng · J. Li (✉)
Beijing National Laboratory for Molecular Sciences, CAS Key Laboratory of Colloid, Interface and Chemical Thermodynamics, Institute of Chemistry, Chinese Academy of Sciences, 100190 Beijing, China
e-mail: jbli@iccas.ac.cn

Y. Jia · X. Feng · J. Li
University of Chinese Academy of Sciences, 100049 Beijing, China

J. Li (ed.), *Supramolecular Chemistry of Biomimetic Systems*,
DOI 10.1007/978-981-10-6059-5_4

the capability of incorporating different functional molecules [1–4]. The microcapsules are fabricated by the sequential deposition of polymers onto a sacrificial template followed by the removal of template. Almost all the interactions (such as electrostatic interaction, hydrogen bonding, covalent crosslinking, charge transfer, metal coordination, specific recognition, etc.) could be used as the driving force for the LbL assembly of microcapsules. Therefore, an unprecedented variety of molecules or materials were selected as wall components and a wide variety of LbL-assembled microcapsules were obtained [1, 2, 5–7]. Microcapsules fabricated from biomolecules are of special interest owing to their biocompatibility, biodegradability, specific molecular recognition ability, simple modifiability, and easy availability for bottom-up fabrication. Many biomolecules, such as phospholipids, polysaccharides, peptides, and proteins are employed as building blocks to construct multifunctional microcapsule systems and have successfully applied in drug carriers, bioreactors, biosensors, and other biomedical fields [2, 7–11].

In this chapter, we specifically focus on polysaccharides-based microcapsules and show some of their most significant applications in biomedical field. In details, we first describe the main features of polysaccharides and outline the basic principles for the preparation of polysaccharides-based microcapsules via LbL assembly techniques. We then present the biomedical applications of these polysaccharides-based microcapsules, including cancer therapy and blood substitutes.

4.2 The Structure and Properties of Polysaccharides

Polysaccharides are polymers of monosaccharides linked by glycosidic bonds. These natural polymers are the most abundant renewable resource on the Earth, with an annual formation rate exceeding the world production rate of synthetic polymers by several orders of magnitude. They have various origins including algal origin (e.g., alginate, agar, and carrageenan), microbial origin (e.g., dextran and xanthan gum), plant origin (e.g., cellulose, starch, pectin, and guar gum), and animal origin (e.g., chitosan, chondroitin, heparin, and hyaluronic acid) [8, 12, 13]. The chemical structures and main properties of some conventional polysaccharides are listed in Table 4.1. These polysaccharides are diverse in chemical structures, molecular weight, physicochemical properties as well as biological activities. However, all of the polysaccharides have some favorable characteristics in common such as excellent biocompatibility, biodegradability, stability, hydrophilicity, low toxicity, low immunogenicity, and ease of chemical modification. These outstanding behaviors of polysaccharides are mainly attributed to, (i) originating from extracellular matrices of plant and animal tissues, (ii) bearing numerous terminal hydroxyl groups of glucose units, (iii) existence of a large number of active functional groups (free carboxyl, primary amino, acetamido, etc.). Due to the presence of these active groups on the polysaccharides backbone, chemical functionalization can be easily realized through oxidation, sulfation, amidation,

Table 4.1 The chemical structures and main properties of conventional polysaccharides

Polysaccharides	Main properties
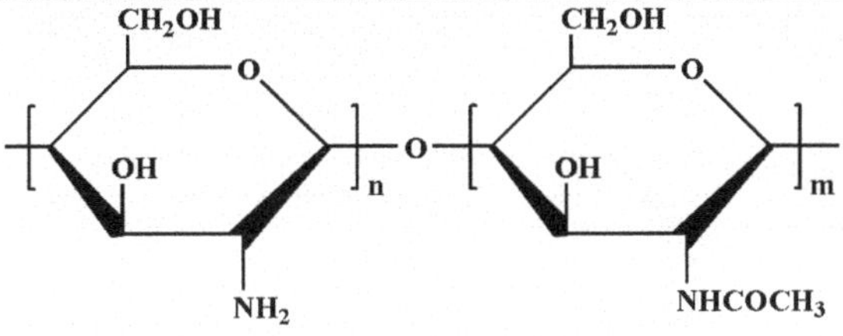 **Chitosan (CHI)**	The only natural polycationic polysaccharide; soluble in acidic solution; facile adherence to mucosal surfaces; wide availability, superior wound-healing, and antibacterial properties
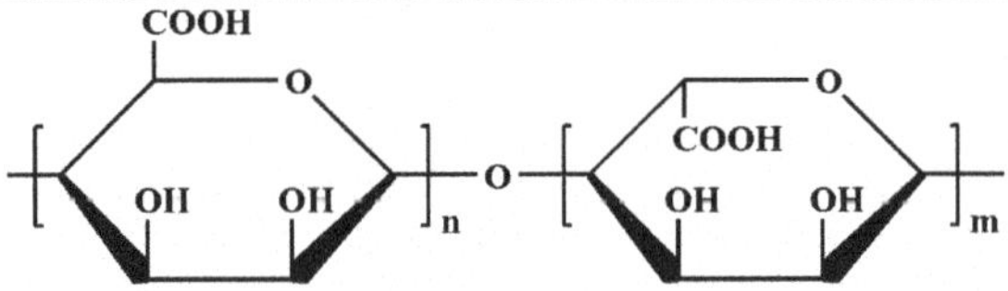 **Alginate (ALG)**	Water soluble at physiological pH and highly mucoadhesive
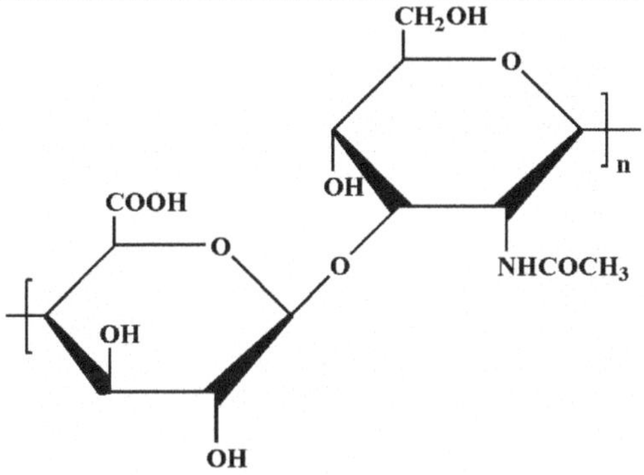 **Hyaluronic Acid (HA)**	Water soluble; non-immunogenic; the only non-sulfated glycosaminoglycans; plays key roles in cell motility, proliferation, and differentiation; can function as a lubricant and shock absorber; active targeting to specific membrane receptors present on most mammalian cells; water retention ability
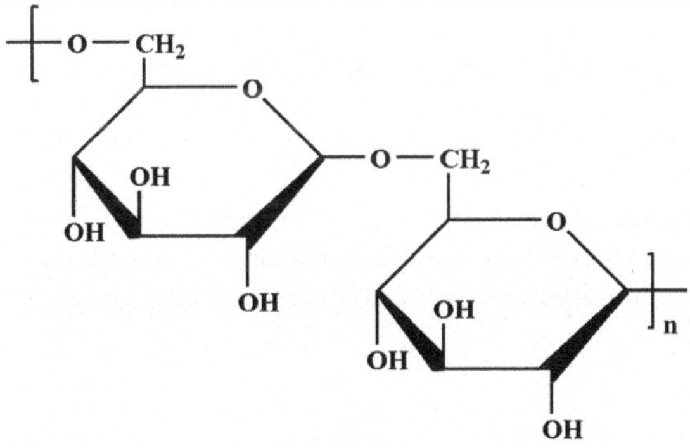 **Dextran**	Highly water soluble, lacks nonspecific cell binding, and resistant to protein adsorption

(continued)

Table 4.1 (continued)

Polysaccharides	Main properties
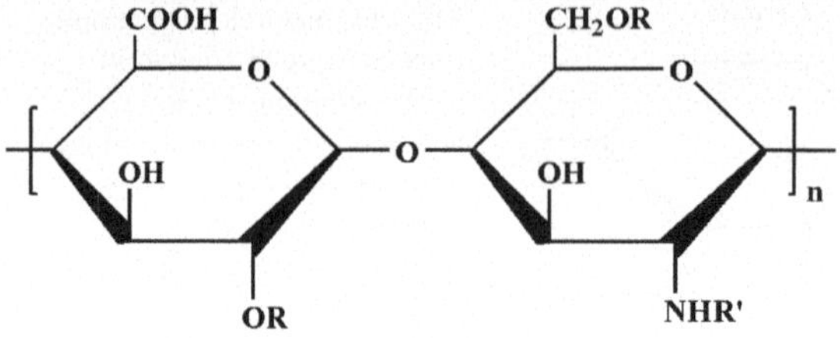 R=H or SO_3H and R'=H, SO_3H or $COCH_3$ **Heparin (HP)**	High negative charge density; anticoagulant activity and antiviral activity; regulate angiogenesis and inhibit complement activation
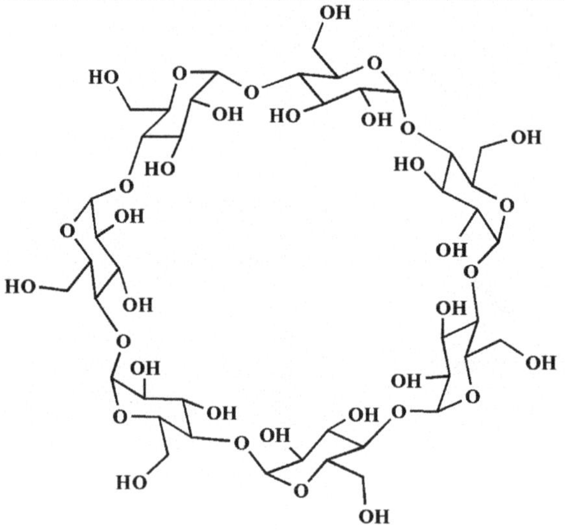 **Cyclodextrin (CD)**	Unique cup-like morphology with a hydrophilic exterior and a hydrophobic cavity that enables them to encapsulate small hydrophobic molecules

esterification, or grafting methods, resulting in many kinds of polysaccharide derivatives [8, 14]. Moreover, these hydrophilic groups could form non-covalent bonds with biological tissues (mainly epithelia and mucous membranes), forming specific bioadhesion on mucosal surfaces (such as colon) and prolonging the drug residence time. All these merits make polysaccharides excellent building blocks for biomedical applications.

4.3 Preparation of Multifunctional Polysaccharide-Based Microcapsules

With the development of biotechnology and nanotechnology, more and more polysaccharide-based microcapsules emerge, which greatly enrich the versatility of polysaccharide-based carriers in terms of category and function. According to structural characteristics, these polysaccharide-based microcapsules are prepared mainly via five mechanisms, namely electrostatic interaction, hydrogen bonding, covalent crosslinking, ionic crosslinking, and host–guest interaction.

4.3.1 Polysaccharide Microcapsules Fabricated via Electrostatic Interaction

Electrostatic interaction is considered as one of the main driving forces exploited for LbL assembly. It is attractive for its simplicity and mild preparation conditions. In a general strategy, oppositely charged components are alternately deposited on templates and hollow capsules are obtained followed by removal of templates. Therefore, the assembly of polysaccharide-based microcapsules can be easily obtained through electrostatic interaction of ionic polysaccharides with oppositely charged polysaccharides or polymers. From the standpoint of polyelectrolyte, polysaccharides can be divided into polycationic polysaccharides (chitosan (CHI)) and polyanionic polysaccharides [alginate (ALG), heparin (HP), chondroitin sulfate (CS), hyaluronic acid (HA), carboxymethyl cellulose (CMC), dextran sulfate (DEXS), etc.]. Among these polysaccharides, CHI is the only natural polycationic polysaccharide derived from partial deacetylation of chitin, which consists of glucosamine and N-acetyl glucosamine units. Owing to its wide availability, superior wound-healing and antibacterial properties, CHI is widely used as polycations in the preparation of polysaccharide-based microcapsules. Any polyanionic polysaccharides or their derivatives could be assembled with CHI through electrostatic interaction, such as CHI/ALG, CHI/HP, CHI/CMC, CHI/DEXS multilayer microcapsules, and so on [15–20]. As alternatives to polysaccharides or their derivatives, other charged polymers like polyelectrolyte, proteins, and polypeptides were also exploited as oppositely charged polymers for the fabrication of polysaccharides-based microcapsules [21–23], which could endow the microcapsules with specific properties or functions. Moreover, these polysaccharides-based microcapsules fabricated through electrostatic interaction are often responsive to pH and ionic strength, making it well suitable for stimulus-responsive drug release.

4.3.2 Hydrogen-Bonded Polysaccharide Microcapsules

Since most polysaccharides have numerous active functional groups (typically, carboxyl, carbonyl, hydroxyl and amine groups, etc.) that can serve as hydrogen bond donors and acceptors, hydrogen bonding was also employed in the LbL assembly of polysaccharides-based microcapsules. Compared to electrostatic interaction, hydrogen bonding allows the fabrication of multilayer microcapsules both in aqueous solution and organic solvents, which increase the diversity of components used in the construction of microcapsules. What is more, hydrogen-bond interaction is sensitive to pH and humidity change, endowing the microcapsules with stimulus-responsive drug release behaviors as well as those fabricated via electrostatic interaction [24]. As an example, poly(N-isopropylacrylamide) (PNIPAAm)/ALG multilayer microcapsules were fabricated through hydrogen bonding between amide carboxyl groups of the PNIPAAm and hydroxyl groups of the ALG [25]. These hydrogen-bonded (PNIPAAm)/ALG microcapsules were demonstrated to be pH-responsive for fluorescent probe molecules FITC-dextran (Mw $\sim$ 2000 kDa). It is permeable for FITC-dextran below pH 5.8, while impermeable above pH 6.8. The distinct permeability of the microcapsules is mainly ascribed to the strength of hydrogen bond and ALG-Mn^{2+} complexation formed at different pH values. In addition, the introduction of PNIPAAm also enabled the polysaccharide microcapsules thermosensitivity. This dual sensitivity may provide new opportunities for these capsules as efficient drug delivery carriers.

In general, hydrogen bonding was not used alone in LbL assembly of polysaccharide-based microcapsules, which is always used in combination with other interactions. In the case of the insulin/ALG microcapsules, both hydrogen bonding and electrostatic interaction contributed to the successful assembly of microcapsules [26].

4.3.3 Covalent-Bonded Polysaccharide Microcapsules

As polysaccharides are often weak polyelectrolytes, the microcapsules fabricated through electrostatic interactions were not strong enough and thus restricted their practical applications. To obtain robust polysaccharide microcapsules, covalent crosslinking was introduced in the assembly process. Most of covalent crosslinked polysaccharide microcapsule systems were prepared using the available $-NH_2$ and $-OH$ groups and crosslinkers that can form a number of linkage chemistries, including amine–carboxylic acid bonding and amine–aldehyde bonding (commonly known as Schiff base bonding) [2, 21]. Here, we specifically focus on polysaccharide microcapsule systems fabricated through Schiff base interaction. First, it is worth to mention the outstanding superiority of Schiff base interaction [2, 27]. (i) Schiff base reaction proceeds under ambient conditions without activation and involves only water as a by-product, avoiding the contamination of the microcapsules by impurities. (ii) Compared to noncovalent bonding, the robustness and stability of covalent Schiff base-bonded polysaccharide microcapsules were significantly improved. (iii) The formation of Schiff base bond enables the microcapsules with autofluorescence, attributing to the $n–\pi^*$ transition of the C=N bonds. (iv) Schiff base bond is a dynamic covalent bond that the stability of these bonds decreases as the pH decreases. All these merits make Schiff base-bonded polysaccharide microcapsules attractive in biomedical application, especially for biological tracing and controlled drug delivery.

Typically, Schiff base-bonded polysaccharide microcapsules can be formed by using small molecule aldehyde or aldehyde polymer as crosslinkers.

4.3.3.1 Glutaraldehyde Crosslinked Polysaccharide Microcapsules

As a familiar crosslinking agent, glutaraldehyde (GA) was frequently used to prepare stable microcapsules. The aldehyde groups of GA can easily react with amino groups of polysaccharides to form Schiff base bonds. After alternate assembly of polysaccharide and GA on decomposable template, hollow single-component or multi-component polysaccharide microcapsules can be obtained [28]. Compared to the microcapsules fabricated through electrostatic interaction, the stability of GA-crosslinked polysaccharide microcapsules against extreme pH treatments was substantially improved. Moreover, other amino-containing compounds such as proteins and peptides could also react with GA to endow the polysaccharide microcapsules with specific recognitions or functions. However, it is noteworthy that GA is relatively toxic to living organisms and it is associated with calcification in certain bioapplications [29]. To reduce the toxicity of GA-crosslinked polysaccharide microcapsules without compromising its stability, lower concentration of GA, or posttreatment with GA after LbL assembly was recommended [15, 30].

4.3.3.2 Aldehyde Polysaccharide-Crosslinked Polysaccharide Microcapsules

In order to avoid the use of toxic GA, researchers tend to pre-functionalized polymer chains of polysaccharides with reactive aldehyde groups. Since most of the polysaccharides have cis-diol groups in their structure, mild periodate oxidation is used to confer the polysaccharide chain with aldehyde functionalities (namely aldehyde polysaccharides) [31], which have highly reactive nature toward amino-containing compounds. Aldehyde polysaccharides are excellent biological crosslinkers due to their low toxicity, biocompatibility, and biodegradability, and they are usually employed both as a crosslinker and as a wall component in covalent assembly of polysaccharide microcapsules [31–33]. Recently, we oxidized ALG to generate alginate dialdehyde (ADA) and prepared CHI/ADA multilayers microcapsules through electrostatic interaction and Schiff base interaction [31], Fig. 4.1. The formation of Schiff base bond enabled the CHI/ADA microcapsules

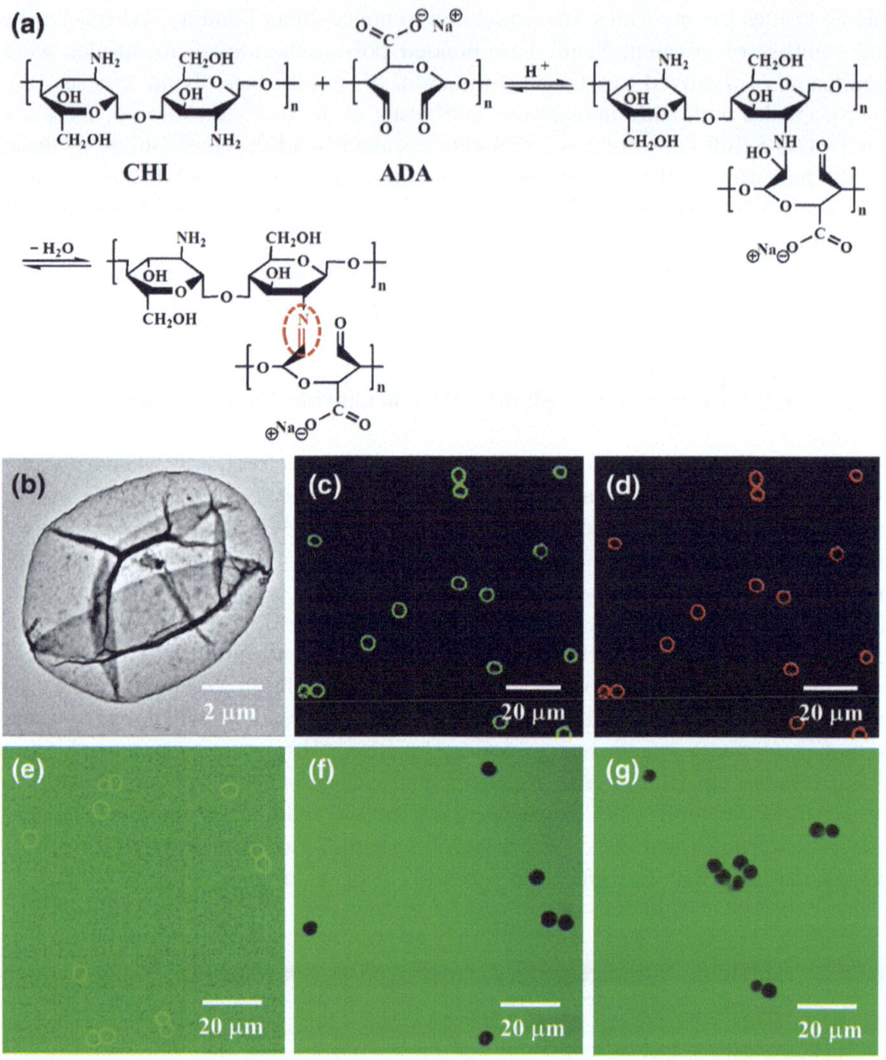

Fig. 4.1 **a** The Schiff base reaction between CHI and ADA, **b** TEM image and **c–d** CLSM images of CHI/ADA multilayer microcapsules, **e–g** CLSM images of CHI/ADA multilayer microcapsules in different pH media with FITC-dextran (20 kDa) as a probe molecule: pH 5, pH 7, and pH 9 (from *left* to *right*). Reproduced from Ref. [31] by permission of The Royal Society of Chemistry

with enhanced stability and pH-dependent permeability. In acid condition, the crosslinked shells become incompact and permit the diffusion of FITC-dextran (20 KDa), while impermeable in neutral and basic solution. This feature is highly preferable for specific pH-triggered drug release. Moreover, the Schiff base-bonded polysaccharide microcapsules were found to be autofluorescent, which ascribe to the n–π* transition of the C=N bonds. The autofluorescence would be beneficial in tracing and monitoring safety and efficacy of microcapsules in organisms, avoiding the use of extra fluorochromes. Therefore, this study provides a simple and promising strategy for making autofluorescent and pH-responsive materials. Based on this, we further used ADA to crosslink with amino-containing small molecule, drugs, and proteins to obtain autofluorescent and pH-responsive polysaccharide microcapsules and confer them with specific or extra functions [33–36]. In addition to ADA, we also synthesized other aldehyde polysaccharides, such as dialdehyde heparin (DHP) and dialdehyde starch (DAS), and then crosslinked them with CHI to prepare polysaccharide microcapsules [31]. The successful preparation of CHI/DHP and CHI/DAS multilayers microcapsules demonstrated that this approach is also applied to other polysaccharides and their derivatives, further confirming the versatility and broad applicability of the method.

4.3.4 Ionic-Crosslinked Polysaccharide Microcapsules

Ionic crosslinking is a specific electrostatic interaction. In ionic crosslinking, polyions or small ionic molecules react with the polyelectrolytes and the formed ionic bond act like bridges along the polymer chains. Compared with other interactions, ionic crosslinking is more simple and mild. Once opposite charged polymers and ions are mixed together, ionic networks are formed immediately. For polycationic CHI and their derivatives, polyanion tripolyphosphate (TPP) is the most widely used ionic crosslinkers [37, 38]. It can form a gel by ionic interaction between positively charged amino groups of CHI and negatively charged counterions of TPP. Compared to other polyphosphates that only bind on the surface of CHI droplets, TPP can diffuse into CHI droplets or films freely to form ionically crosslinked chitosan beads or films [39]. For polyanionic polysaccharides bearing carboxylic groups on the molecular chains, di- and tri-valent ions (Ca^{2+}, Ba^{2+}, Sr^{2+}, Al^{3+}) are suitable ionic crosslinkers. Typically, ALG forms hydrogels by means of Ca^{2+}, which positions in the interstices between G blocks, leading to an ordered conformational structure called "egg-box" array [38]. In our earlier study, we used Ca^{2+} as crosslinkers to alternatively assemble with ALG, nano-hydroxyapatite (nHA) on tube-like template, and 3D hydrogel scaffolds were finally obtained [40]. By simply regulating the Ca^{2+} crosslinking intensity, the elastic modulus, swelling behavior, permeability, and diffusivity of the hydrogel scaffold can be easily tuned. In another study, porous spherical $CaCO_3$ are used as effective 'casting' template to prepare Ca^{2+}-crosslinked ALG microcapsules [41]. Here, $CaCO_3$ microspheres were used not only as a template but also as an ionic crosslinking agent provider. The crosslinking of Ca^{2+} and ALG happened when $CaCO_3$ template was dissolved. Such Ca^{2+}-crosslinked ALG microcapsules or beads are pH-responsive and have been intensively explored for site-specific oral delivery. Generally, Ca^{2+} crosslinked ALG microcapsules showed slow drug release in acidic pH due to the poor swelling. Oppositely, in phosphate buffer and simulated intestinal fluid, the swelling and the release are promoted as phosphate ions extract the Ca^{2+} from ALG hydrogels microcapsules.

4.3.5 Polysaccharide Microcapsules Fabricated via Host–Guest Interaction

Cyclodextrins (CDs) are cyclic oligomers of glucose that have a hydrophilic exterior and a hydrophobic cavity. They can act as hosts to hydrophobic molecules and form water-soluble inclusion complexes with small molecules and portions of large compounds. This unique ability enables CDs-based assemblies widely utilized in the biomedical and pharmaceutical fields to improve bioavailability of poorly soluble or biodegradable drugs and to enhance permeability of biological membranes [42, 43]. Supramolecular microcapsules (SMCs) based on CDs and their derivatives exploited host–guest interactions as the main driving force. Compared to the well-known electrostatic interactions, supramolecular host–guest interactions may offer additional benefits, such as tunable binding affinity, reversibility, incorporation of neutral molecules and/or biomolecules, straightforward control over the growth process and layer thickness by structural and steric design of the building blocks, and so on. Zhang and co-workers reported the preparation of SMCs by alternately depositing carboxymethyl dextran-graft-β-CD (CMD-g-β-CD) and polyaldehyde dextran-graft-adamantane (PAD-g-AD) on $CaCO_3$ particles via host–guest interaction [44]. Simultaneously, adamantine-modified doxorubicin (AD-Dox) was also loaded on the LbL wall via host–guest interaction. Because the AD groups were linked to dextran or Dox via pH-cleavable hydrazone bonds, AD moieties can be removed under the weak acidic condition, leading to destruction of SMCs and release of Dox. Meanwhile, they designed photo-switchable polysaccharide microcapsules based on host–guest interactions between α-cyclodextrin (α-CD) and azobenzene (Azo) [45]. The obtained capsules could be dissociated upon UV irradiation due to the transformation of trans–Azo to cis–Azo, followed by the release of the drug. Analogously, some functional molecules or responsive polymers could be assembled or grafted with CDs to provide the polysaccharide microcapsules with multi-responsiveness or specific recognitions and serve as smart drug reservoir [42, 46].

4.4 Biomedical Applications of the Polysaccharide-Based Microcapsules

4.4.1 *Cancer Therapy*

LbL-assembled polysaccharide microcapsules usually have good biocompatibility, biodegradability, low immunogenicity with tunable size, morphology, surface properties, and permeability. Thus, they have great advantages and important applications in biomedicine. One of the most important applications of polysaccharide microcapsules is used as drug carriers for cancer therapy. Their high loading capacity and tunable properties are favorable to deliver and unload anti-cancer drugs. Effective cancer therapy is achieved by specific targeting to tumor and controlled drug release. Typical examples are summarized in Table 4.2 and introduced.

Table 4.2 LbL-assembled polysaccharide microcapsules and their applications

Assembled pairs	Driving force	Application	References
CHI/ALG	Electrostatic interaction	Folate receptor mediated targeting	[16]
TRAIL/ALG	Electrostatic interaction	Death receptor mediated targeting	[47]
CHI/HP	Electrostatic interaction	Enzyme-responsive drug release	[18]
CHI/DEXS	Electrostatic interaction	Enzyme-responsive drug release	[20]
PNIPAAm/ALG	Hydrogen bond	pH- and temperature-responsive drug release	[25]
Insulin/ALG	Electrostatic interaction and hydrogen bond	Sustained drug release	[26]
CHI/ADA	Electrostatic interaction and covalent Schiff base bond	pH-responsive drug release	[31]
Dox/ADA	Electrostatic interaction and covalent Schiff base bond	pH-responsive drug release	[35]
ADA/CM	Electrostatic interaction and covalent Schiff base bond	pH- and redox-responsive drug release	[34]
PLGA/CHI	Electrostatic interaction and covalent amide bond	pH-responsive drug release	[23]

(continued)

Table 4.2 (continued)

Assembled pairs	Driving force	Application	References
ADA/Ca^{2+}	Ca^{2+} crosslinking	Folate receptor mediated targeting, chemotherapy and PDT	[41]
HA-CD/PLL	Host–guest interaction	CD44 receptor mediated targeting,	[46]
CMD-g-β-CD/PAD-g-AD	Host–guest interaction	pH-responsive drug release	[44]
PAA-C_{12}-Azo/CMD-g-α-CD	Host–guest interaction	Photo-responsive drug release	[45]
ADA/CAT	Covalent Schiff base bond	PDT	[36]
ADA/fLuc	Covalent Schiff base bond	PDT	[33]
CHI/ALG	Electrostatic interaction	PTT	[48]
Hb/DHP	Electrostatic interaction and covalent Schiff base bond	Blood substitutes	[32]

Abbreviations
TRAIL Tumor necrosis factor-related apoptosis-inducing ligands; *ACHI* Adenine-modified chitosan; *HA-CD* HA-modified CD; *CMD-g-α-CD* Carboxymethyl dextran-graft-α-CD; *PAA-C12-Azo* poly(acrylic acid) Naminododecane p-azobenzeneaminosuccinic acid; *PLGA* poly (l-glutamic acid); *fLuc* firefly luciferase

4.4.1.1 Specific Targeting to Tumor

Specific targeting to tumor cells could increase bioavailability of anticancer drugs as well as reduce toxic side effects to normal cells resulting from nonspecific uptake [49]. The strategies for specific targeting polysaccharide microcapsules to tumor cells mainly employ surface functionalization with bioactive moieties such as cell surface receptor ligands (receptor mediated targeting).

Hyaluronic acid (HA) is a natural anionic polysaccharide that represents one of the main constituents of the extracellular matrix (ECM) in mammalian organisms. Moreover, HA can bind to specific receptor cluster determinant 44 (CD44) that overexpress in various cancer cells [50]. When HA is bound to CD44, it is cleaved into lower molecular weight segments as low as 20 kDa by the membrane anchored enzyme Hyal-2 and readily internalized to the cells. Therefore, polysaccharide HA itself is a tumor cell surface receptor ligand and HA-containing microcapsules are suitable for tumor-targeted drug release. Making use of the specific tumor recognition of HA and the unique inclusion complexation of CD, Auzély-Velty and co-workers fabricated polysaccharide microcapsules through LbL deposition of HA-modified CD molecules and poly(L-lysine) (PLL) on calcium carbonate particles [46]. The poorly water-soluble anticancer drug paclitaxel (PTX) molecules were loaded in the LbL assembled wall via selectively complexed by CD. PTX-loaded polysaccharide microcapsules showed greater cytotoxicity than PTX in solution for breast cancer cells that overexpressed HA receptor, suggesting that they were specifically bound to cancer cells and taken up by the cancer cells via HA receptor-mediated endocytosis.

The folate receptor (FR) is a confirmed tumor-associated antigen that binds folate and folate-modified assemblies with very high affinity and shuttles these assemblies inside cells via an endocytic mechanism. A wide variety of drug carriers employed folate as the ligand for FR-positive tumor cells targeting and showed enhanced cytotoxicity to tumors. Our group fabricated CHI/ALG and ALG/Ca^{2+} microcapsules and coated these polysaccharide microcapsules with folate-linked liposomes [16, 41]. In vitro tests show that the introduction of folate on the surfaces of polysaccharide microcapsules leads to selective recognition to cancer cells with increased cell uptakes and enhanced anticancer efficiency. Besides folate, we also exploited TRAIL (tumor necrosis factor-related apoptosis-inducing ligand) as tumor cell surface receptor ligands to modify polysaccharide microcapsules. TRAIL is a cognate ligand of tumor necrosis factor (TNF) receptors. It binds or interacts with cell surface proapoptosis receptors to assemble a cell-death-inducing signaling complex that can induce cell apoptosis. Specially, TRAIL can induce apoptosis selectively in many cancer cell lines, whereas normal cells are relatively resistant [49]. Based on this, we fabricated TRAIL/ALG multilayers on the surfaces of Dox-loaded $CaCO_3$ particles via LbL assembly technique (Fig. 4.2) [47]. Flow-cytometry analysis demonstrated that TRAIL as the outer layer of the system retained its activity and remarkably enhanced the selectivity of polysaccharide nanoparticles to recognize cancer cells. The ratio of TRAIL-functionalized polysaccharide particles uptake by cancer cells reached up to $91.3 \pm 4.1\%$, while nude $CaCO_3$ particles were only $43.5 \pm 3.3\%$. After cell endocytosis, DOX loaded in the polysaccharide particles slowly released in tumor cell. Notable anticancer activity of the nanocomposite was observed due to the synergistic effect of TRAIL targeted extrinsic apoptosis pathway and DOX triggered apoptosis pathway.

4.4.1.2 Controlled Drug Release

Controlled drug delivery systems (DDSs) provide an alternative approach to regulating the bioavailability of therapeutic agents. In DDSs, an active therapeutic is incorporated into a polymeric network structure and the drug is released from the system in a predefined manner. LbL-assembled polysaccharide microcapsules are excellent DDSs that have good biocompatibility, biodegradability, and tunable properties. Controlled drug release from polysaccharide microcapsules can be fine-tuned via endogenous and exogenous stimuli (such as pH, redox, enzyme, glucose, temperature, light illumination, etc.).

Since the tissues around cancer cells have a lower pH than those of normal cells, pH-responsive polysaccharide microcapsules were designed for the controlled release of anticancer drugs. Doxorubicin (Dox) is one of the most common

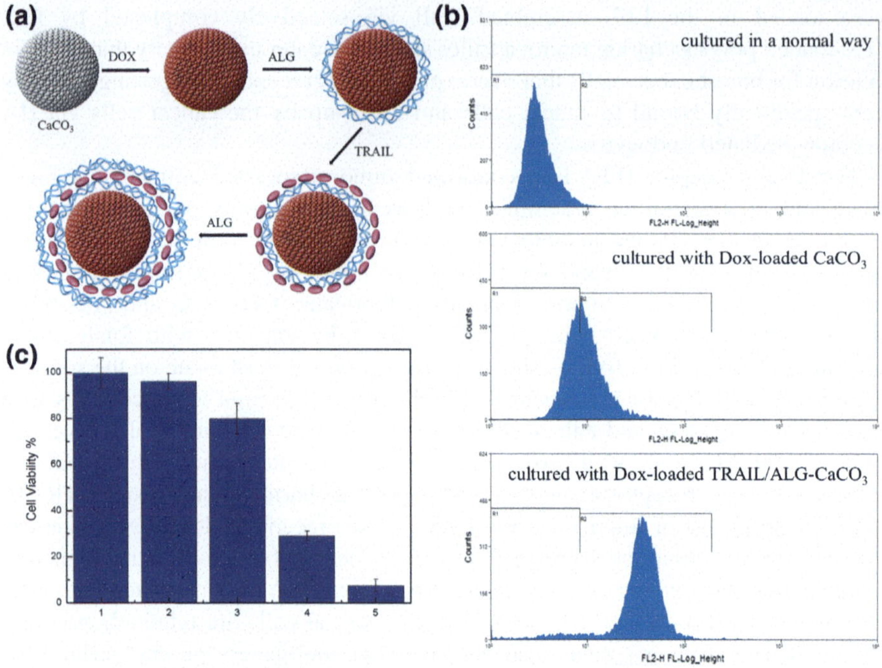

Fig. 4.2 **a** Schematic illustration of the fabrication of Dox-loaded TRAIL/ALG–$CaCO_3$ nanocomposites, **b** Flow-cytometry diagrams of the uptake of hollow shells by HeLa cells cultured in three different way, **c** HeLa cells viability in vitro measured by MTT assay, and the cells were cultured: [1] in a normal way; [2] with nude $CaCO_3$ nanocomposites; [3] with TRAIL/ALG-$CaCO_3$ nanocomposites; [4] with Dox-loaded $CaCO_3$ nanocomposites; [5] with Dox-loaded TRAIL/ALG-$CaCO_3$ nanocomposites. Reproduced from Ref. [47] by permission of The Royal Society of Chemistry

low-molecular chemotherapeutic agents used in the cancer treatment. It could be encapsulated in capsules through preloading in porous template or postloading via physical adsorption. In view of the existence of abundant amino groups on Dox, our group used ADA to crosslink with Dox through covalent Schiff base bond (Fig. 4.3) [35]. As discussed before, the stability of Schiff base bonds decreases as the pH decreases [2]. Thus, controlled release of Dox could be achieved through pH stimulus. As shown in Fig. 4.3b, the release rate of Dox from Dox/ADA capsules was much faster at pH 5.5 compared to that at pH 7.4. Moreover, the Dox/ADA microcapsules induced sustained drug release and exhibited high efficiency against tumor cell proliferation.

Considering the differences between the environment in the bloodstream (neutral pH and low concentration of GSH) and subcellular structures in tumor cells (low pH and high concentration of GSH), we further prepared pH and redox-responsive $(ADA/CM)_5$ microcapsules via covalent Schiff base bond and disulfide bond [34]. The microcapsules were shown to be permeable at low pH due to the hydrolysis of

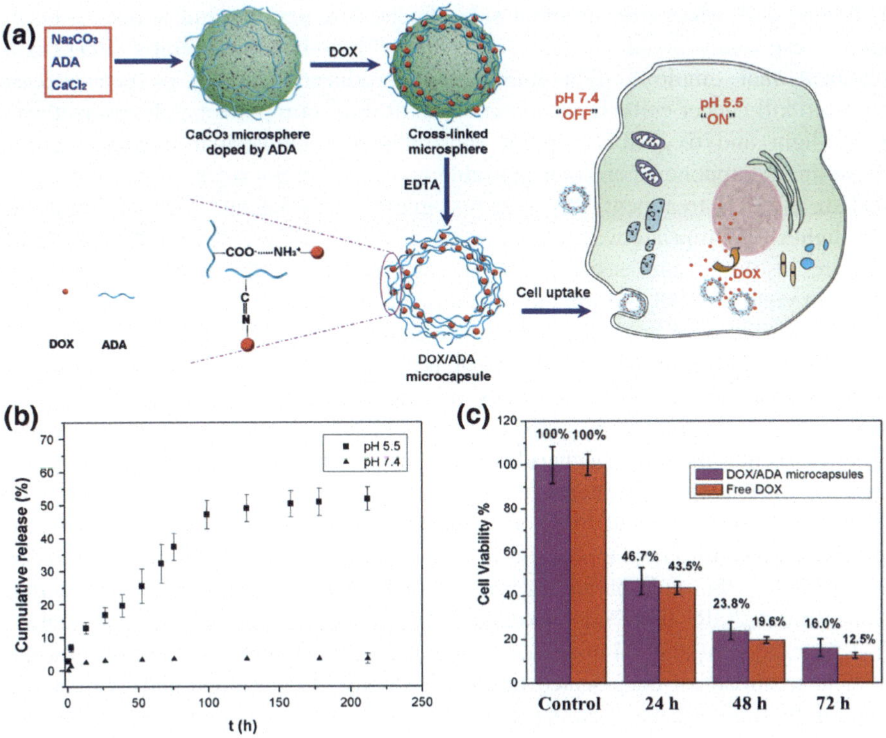

Fig. 4.3 **a** Schematic illustration of the fabrication of DOX/ADA microcapsules against tumor cell proliferation, **b** quantitative analyses of in vitro release of DOX from DOX/ADA microcapsules at pH 5.5 and 7.4 at 37 °C, **c** cytotoxicity results for MCF-7 cells with a drug concentration (14 μg mL^{-1}) incubated with different time stages. Reproduced from Ref. [35] by permission of John Wiley & Sons Ltd

Schiff base bond, or after adding a reducing agent to cleave the disulfide bonds within the formed multilayers. Thus docetaxel-loaded $(ADA/CM)_5$ capsules were readily triggered drug release in tumor cells and prevented unwanted release in bloodstream.

In comparison with endogenous stimuli (pH, redox, enzyme, glucose), exogenous stimuli (light, ultrasound, magnet) are more simple and easy to manipulate. Light is a superior external stimulus for drug release since it can be remotely manipulated to attain both spatial and temporal control with high precision [51]. Photosensitive polysaccharide microcapsules were fabricated by assembling polysaccharides with photosensitive polymers or particles [33, 36, 45]. Drugs can be released from these polysaccharide microcapsules by photocleavage, photoisomerization, or photoactivation of the assembled polymers upon light irradiation.

Photodynamic therapy (PDT) and photothermal therapy (PTT) are two typical and much more promising cancer therapies that have been widely reported [52, 53]. Both therapies possess obviously outstanding advantages such as specific spatiotemporal selectivity, minimal side effects, low toxicity and remote controllability, especially avoiding chemo-resistance. Photodynamic therapy (PDT) is a treatment that employs light illumination produced reactive oxygen species (ROS) to kill cancer cells. It usually consists of three components: photosensitizers (PSs), light and oxygen-containing substrates (e.g., molecular oxygen, water). These three components are indispensable and play crucial roles in the efficacy of PDT. In the PDT treatment, PSs were first injected into patients and then irradiated with light of appropriate wavelength at the diseased tissues. Nevertheless, most of PSs are insoluble and easily aggregate in aqueous media, which reduced the quantum yield of ROS and limited their anticancer efficacy. Therefore, many kinds of PSs' carriers have been developed to overcome this problem. For example, our group employed LbL-assembled ALG/CHI microcapsules as PSs' carriers and loaded hypocrellin B (HB) on shells of microcapsules [54]. These HB-loaded polysaccharide microcapsules showed high cytotoxicity after exposure to visible light. It indicates that polysaccharide microcapsules are suitable as PSs' carriers and can efficiently avoid unwanted aggregation.

PSs are usually excited by short-wavelength UV–Vis light, so its poor tissue penetration has become the obstacle in treating deep-seated tumors below the skin, which hinders the widespread clinical use of PDT [52]. Fortunately, the external near-infrared (NIR) light/X-ray-excited PDT, as well as internal self-illuminating PDT have brought about a novel technique of deep PDT with great promise for the efficient treatment of deep-seated tumors. Recently, we designed and fabricated a kind of bioluminescent microcapsules by covalent LbL assembly of ADA and luciferase (fLuc) on luciferin-coprecipitated $CaCO_3$ microparticles [33]. In the presence of O_2, Mg^{2+}, and ATP, fLuc could catalyze the oxidation of its substrate luciferin to produce light. This light was then used to activate PSs for the production of active oxygen ($^1O_2$), as shown in Fig. 4.4. Cytotoxicity tests confirmed that PSs in the capsules could be activated by the excitation of bioluminescent microcapsules in the dark without external light and effectively prevent the proliferation of tumor cells.

In addition to PSs' aggregation and poor penetration depth of light, PDT is also restricted by the low concentration of molecular oxygen (O_2) in tumor's hypoxic microenvironment [53]. To increase the availability of O_2, an oxygen-generating system was constructed by LbL assembly of CAT/ADA [36]. In this system, catalase (CAT) catalyzes the decomposition of intracellular hydrogen peroxide (H_2O_2) into H_2O and O_2. The produced O_2 can be utilized by PSs to produced $^1O_2$ under light excitation and thus effectively enhanced the anticancer efficiency of PDT. Moreover, the reduced cellular levels of H_2O_2 may also help to inhibit tumor proliferation.

In combination with desirable components, LbL-assembled polysaccharide microcapsules hold great promise for efficient cancer therapy and well tackle the challenges of both chemotherapy and PDT. This also applies for PTT. LbL-assembled CHI/ALG multilayer microcapsules loaded with gold nanorods (GNRs) were constructed and can be used for enhanced PTT (Fig. 4.5) [48]. These polysaccharide microcapsules were served as support matrices and can avoid

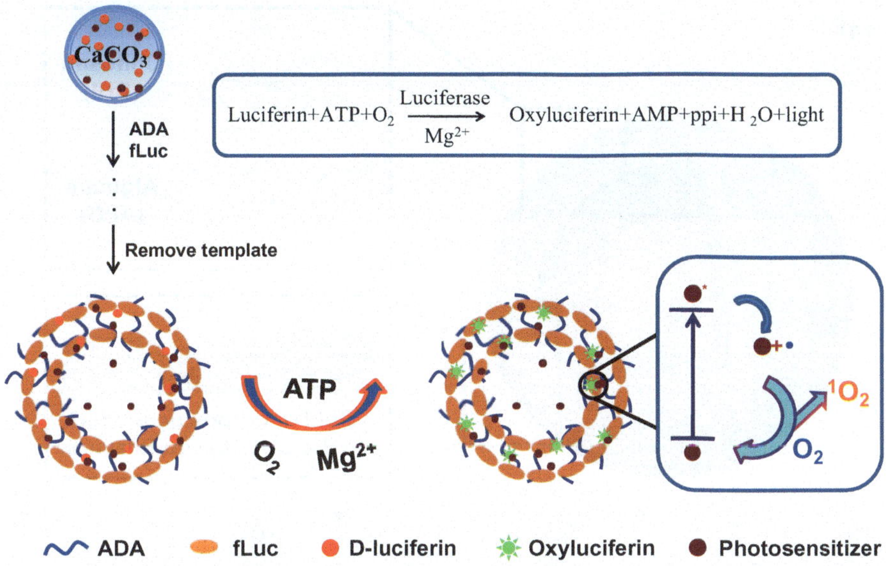

Fig. 4.4 The fabrication and bioluminescent process of ADA/fLuc microcapsules. Reproduced from Ref. [33] by permission of John Wiley & Sons, Ltd

unwanted aggregation of GNRs. GNRs were reported to have a strong absorption in the NIR region and induce faster and higher heating effects compared to other gold materials. Upon irradiation with NIR light, the GNRs absorb the energy and quickly transform it into heat, which may selectively ablate cancerous cells within the irradiated zone and achieve high efficiency to suppress tumor growth. More interestingly, the significant increase of local temperature triggered the autonomous movement of these polysaccharide microcapsules, with the speed ranging from 1.3 to 23.27 μm/s upon the increase of incident laser power. This system is promising as a versatile drug carrier and paves a new way to design self-propelled drug delivery system for diverse biomedical application in future.

4.4.2 *Blood Substitutes*

As essential structural components of both plant and animal cells, polysaccharides play many different roles in vivo. For instance, cellulose and chitin are the major components of cell walls; starch and glycogen act as energy reserves in biological systems; HA and CS are responsible for the unique hydration and mechanical properties of cartilage, synovial fluid, and tendons [55]. Therefore, proper polysaccharides need to be selected for specific bioapplications.

Heparin (HP) is known as an anticoagulant and growth factor-binding agent. It was also found to inhibit cancer cell adhesion, deactivate heparinase, and activate

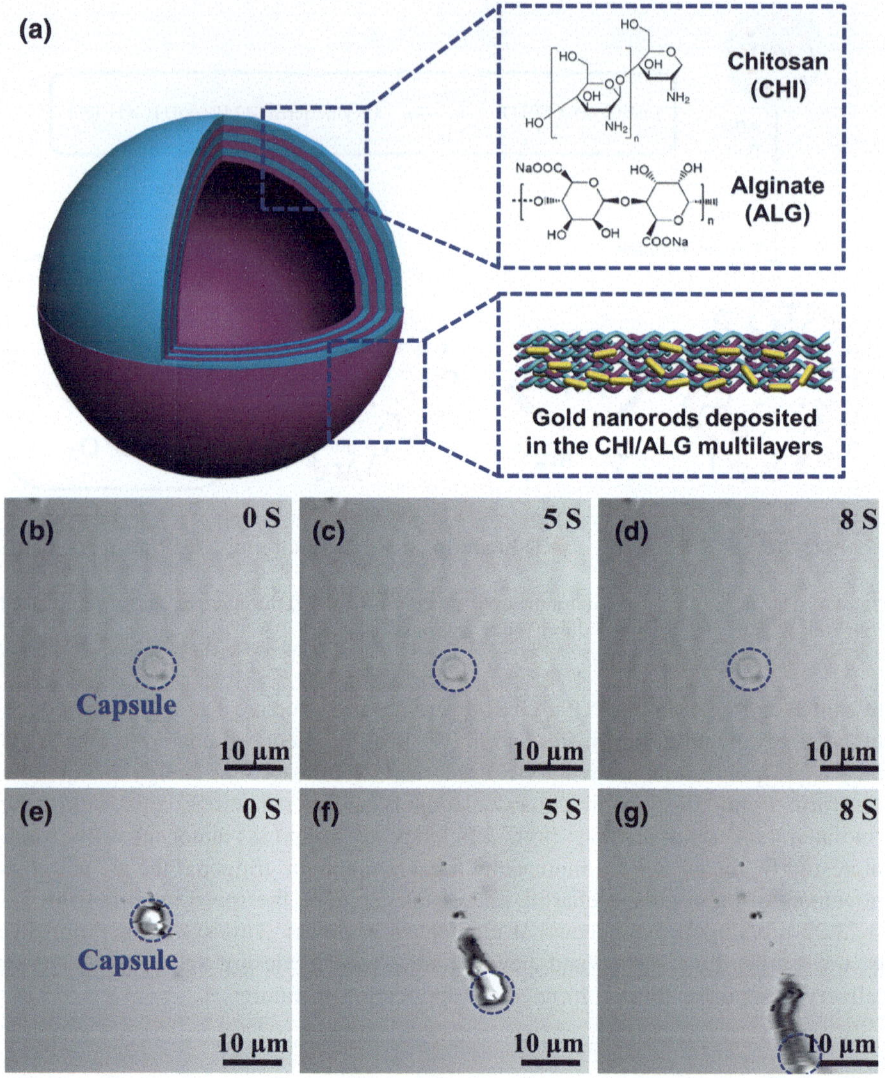

Fig. 4.5 **a** Schematic illustration of CHI/ALG/GNRs multilayer microcapsules, **b–d** video frames of CHI/ALG/GNRs microcapsules without NIR laser illumination; **e–g** video frames of CHI/ALG/GNRs microcapsules under continuous NIR laser illumination (3.23 J/cm^2). Reprinted from Ref. [48], Copyright 2015, with permission from Elsevier

natural killer (NK) cells. Thus, it has been extensively exploited to develop biofunctional multilayer films [17, 56]. However, strong anticoagulant activity limits its use. Dialdehyde heparin (DHP) produced by periodate oxidation of HP, was shown to be a good alternative to HP [32, 56]. It has lower anticoagulant properties than HP, which could prevent severe bleeding complications in biomedicine. Meanwhile, DHP has good biocompatibility, biodegradability, and hemocompatibility as same as HP. In view of its excellent features, DHP was used both as a wall component and as a crosslinker in the fabrication of Hb/DHP multilayer microcapsules to mimic artificial red blood cells (RBCs) (Fig. 4.6) [32]. Hemoglobin (Hb) in RBCs possesses the ability of delivering and releasing oxygen. However, stroma-free Hb could not directly be used as an oxygen carrier since it is liable to dissociate into dimers, leading to severe renal toxicity. The Schiff base bond formed between Hb and DHP in the microcapsules effectively prevent the dissociation of the Hb tetramer into dimers and thus avoiding renal toxicity. Moreover, the Schiff base-bonded Hb/DHP microcapsules were found to be autofluorescent. This is helpful to monitor the microcapsules in humans, avoiding the use of extra fluorochromes. DHP as the outer layer also provides the microcapsules with good "stealth" effect similar to PEGylation, which would be favorable to reduce uptake by macrophages and prolong the blood retention time of microcapsules in vivo. More importantly, DHP-crosslinked Hb maintained its intrinsic bioactivity of reversibly binding and releasing oxygen. All these characteristics make Hb/DHP microcapsules good candidates as blood substitutes.

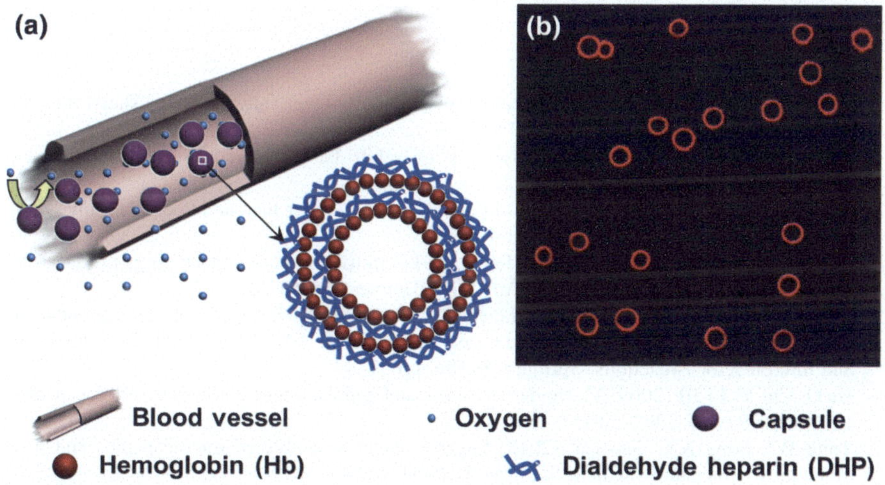

Fig. 4.6 **a** Hb/DHP multilayer microcapsules fabricated through Schiff base interactions for use as blood substitutes, **b** CLSM image of Hb/DHP microcapsules. Reproduced from Ref. [32] by permission of John Wiley & Sons Ltd

4.5 Conclusions

Polysaccharides are the most abundant renewable resources on the Earth that possess excellent biocompatibility, biodegradability, stability, hydrophilicity, low immunogenicity, and ease of chemical modification. They are gaining increasing attention as components to construct well-defined micro/nanometer-sized structures. In this chapter, we highlight recent progress on LbL-assembled polysaccharide microcapsules. Different interactions were used as driving forces to prepare polysaccharide microcapsules and contributed to diverse stimuli-responsivity. In combination with specific bioactive moieties (such as cell surface receptor ligands), the as-prepared functional polysaccharide microcapsules are highly promising as excellent drug carriers for both specific targeting and controlled release in various cancer therapy (chemotherapy, PDT, PTT, combined therapy, etc.). In addition, taking advantages of their unique properties or native functions, polysaccharide microcapsules could also be applied in other biomedical fields, such as blood substitute. With the rapid development of science and nanobiotechnology, LbL-assembled polysaccharide microcapsules are bound to embrace a brighter future and eventually realize important applications in biomedical field.

Acknowledgements The authors acknowledge the financial support from the National Nature Science Foundation of China (Project No. 21433010, 21320102004, 21303221 and 21321063) and National Basic Research Program of China (973 program, 2013CB932800).

References

1. Ariga K, Lvov YM, Kawakami K et al (2011) Layer-by-layer self-assembled shells for drug delivery. Adv Drug Deliver Rev 63:762–771
2. Jia Y, Li JB (2015) Molecular assembly of schiff base interactions: construction and application. Chem Rev 115:1597–1621
3. Ariga K, Yamauchi Y, Rydzek G et al (2014) Layer-by-layer nanoarchitectonics: invention, innovation, and evolution. Chem Lett 43:36–68
4. de Villiers MM, Otto DP, Strydom SJ et al (2011) Introduction to nanocoatings produced by layer-by-layer (LbL) self-assembly. Adv Drug Deliver Rev 63:701–715
5. Ariga K, Ji QM, Hill JP (2010) Enzyme-encapsulated layer-by-layer assemblies: current status and challenges toward ultimate nanodevices. In: Caruso F (ed) Modern techniques for nano- and microreactors/-reactions. Springer, Berlin, pp 51–87
6. He Q, Cui Y, Li JB (2009) Molecular assembly and application of biomimetic microcapsules. Chem Soc Rev 38:2292–2303
7. Tong WJ, Song XX, Gao CY (2012) Layer-by-layer assembly of microcapsules and their biomedical applications. Chem Soc Rev 41:6103–6124
8. Mizrahy S, Peer D (2012) Polysaccharides as building blocks for nanotherapeutics. Chem Soc Rev 41:2623–2640
9. Deshmukh PK, Ramani KP, Singh SS et al (2013) Stimuli-sensitive layer-by-layer (LbL) self-assembly systems: targeting and biosensory applications. J Control Release 166:294–306

10. Lvov Y, Ariga K, Ichinose I et al (1995) Assembly of multicomponent protein films by means of electrostatic layer-by-layer adsorption. J Am Chem Soc 117:6117–6123
11. Feng XY, Du C, Li JB (2016) Molecular assembly of polysaccharide-based microcapsules and their biomedical applications. Chem Rec 16:1991–2004
12. Sinha VR, Kumria R (2001) Polysaccharides in colon-specific drug delivery. Int J Pharm 224:19–38
13. Posocco B, Dreussi E, de Santa J et al (2015) Polysaccharides for the delivery of antitumor drugs. Materials 8:2569–2615
14. Basu A, Kunduru KR, Abtew E et al (2015) Polysaccharide-based conjugates for biomedical applications. Bioconjugate Chem 26:1396–1412
15. Ye SQ, Wang CY, Liu XX et al (2005) Multilayer nanocapsules of polysaccharide chitosan and alginate through layer-by-layer assembly directly on ps nanoparticles for release. J Biomater Sci Polym Ed 16:909–923
16. Qi W, Wang AH, Yang Y et al (2010) The lectin binding and targetable cellular uptake of lipid-coated polysaccharide microcapsules. J Mater Chem 20:2121–2127
17. Thomas MB, Radhakrishnan K, Gnanadhas DP et al (2013) Intracellular delivery of doxorubicin encapsulated in novel ph-responsive chitosan/heparin nanocapsules. Int J Nanomed 8:267–273
18. Qi W, Sun M, Jing ZH (2012) Fabrication of chitosan microcapsules via layer-by-layer assembly for loading and in vitro release of heparin. Supramol Chem 24:350–354
19. Qiu XP, Leporatti S, Donath E et al (2001) Studies on the drug release properties of polysaccharide multilayers encapsulated ibuprofen microparticles. Langmuir 17:5375–5380
20. Itoh Y, Matsusaki M, Kida T et al (2006) Enzyme-responsive release of encapsulated proteins from biodegradable hollow capsules. Biomacromol 7:2715–2718
21. Szarpak A, Pignot-Paintrand I, Nicolas C et al (2008) Multilayer assembly of hyaluronic acid/poly(allylamine): Control of the buildup for the production of hollow capsules. Langmuir 24:9767–9774
22. Balabushevich NG, Tiourina OP, Volodkin DV et al (2003) Loading the multilayer dextran sulfate/protamine microsized capsules with peroxidase. Biomacromolecules 4:1191–1197
23. Yan SF, Rao SQ, Zhu J et al (2012) Nanoporous multilayer poly(l-glutamic acid)/chitosan microcapsules for drug delivery. Int J Pharm 427:443–451
24. Kharlampieva E, Sukhishvili SA (2006) Hydrogen-bonded layer-by-layer polymer films. Polym Rev 46:377–395
25. Wang A, Tao C, Cui Y et al (2009) Assembly of environmental sensitive microcapsules of pnipaam and alginate acid and their application in drug release. J Colloid Interface Sci 332:271–279
26. Zhao J, Cui Y, Wang AH et al (2011) Side effect reduction of encapsulated hydrocortisone crystals by insulin/alginate shells. Langmuir 27:1499–1504
27. Xin Y, Yuan J (2012) Schiff's base as a stimuli-responsive linker in polymer chemistry. Polym Chem 3:3045–3055
28. Manna U, Dhar J, Nayak R et al (2010) Multilayer single-component thin films and microcapsules via covalent bonded layer-by-layer self-assembly. Chem Commun 46:2250–2252
29. Jayakrishnan A, Jameela SR (1996) Glutaraldehyde as a fixative in bioprostheses and drug delivery matrices. Biomaterials 17:471–484
30. Zhao Q, Mao Z, Gao C et al (2006) Assembly of multilayer microcapsules on caco3 particles from biocompatible polysaccharides. J Biomater Sci Polym Ed 17:997–1014
31. Jia Y, Fei J, Cui Y et al (2011) Ph-responsive polysaccharide microcapsules through covalent bonding assembly. Chem Commun 47:1175–1177
32. Jia Y, Cui Y, Fei JB et al (2012) Construction and evaluation of hemoglobin-based capsules as blood substitutes. Adv Funct Mater 22:1446–1453
33. Zhao J, Fei J, Gao L et al (2013) Bioluminescent microcapsules: applications in activating a photosensitizer. Chem Eur J 19:4548–4555

34. Gao L, Fei J, Zhao J et al (2012) Ph- and redox-responsive polysaccharide-based microcapsules with autofluorescence for biomedical applications. Chem Eur J 18:3185–3192
35. Du C, Zhao J, Fei J et al (2013) Assembled microcapsules by doxorubicin and polysaccharide as high effective anticancer drug carriers. Adv Healthc Mater 2:1246–1251
36. Zhao J, Fei J, Du C et al (2013) Assembly of catalase-based bioconjugates for enhanced anticancer efficiency of photodynamic therapy in vitro. Chem Commun 49:10733–10735
37. Jain D, Banerjee R (2008) Comparison of ciprofloxacin hydrochloride-loaded protein, lipid, and chitosan nanoparticles for drug delivery. J Biomed Mater Res Part B 86B:105–112
38. Alvarez-Lorenzo C, Blanco-Fernandez B, Puga AM et al (2013) Crosslinked ionic polysaccharides for stimuli-sensitive drug delivery. Adv Drug Deliver Rev 65:1148–1171
39. Shu XZ, Zhu KJ (2002) The influence of multivalent phosphate structure on the properties of ionically cross-linked chitosan films for controlled drug release. Eur J Pharm Biopharm 54:235–243
40. Du M, Song W, Cui Y et al (2011) Fabrication and biological application of nano-hydroxyapatite (nha)/alginate (alg) hydrogel as scaffolds. J Mater Chem 21:2228–2236
41. Du C, Zhao J, Fei J et al (2013) Alginate-based microcapsules with a molecule recognition linker and photosensitizer for the combined cancer treatment. Chem Asian J 8:736–742
42. Ortiz Mellet C, Garcia Fernandez JM, Benito JM (2011) Cyclodextrin-based gene delivery systems. Chem Soc Rev 40:1586–1608
43. Ma X, Zhao Y (2015) Biomedical applications of supramolecular systems based on host–guest interactions. Chem Rev 115:7794–7839
44. Luo GF, Xu XD, Zhang J et al (2012) Encapsulation of an adamantane-doxorubicin prodrug in ph-responsive polysaccharide capsules for controlled release. ACS Appl Mater Interfaces 4:5317–5324
45. Xiao W, Chen W-H, Zhang J et al (2011) Design of a photoswitchable hollow microcapsular drug delivery system by using a supramolecular drug-loading approach. J Phys Chem B 115:13796–13802
46. Jing J, Szarpak-Jankowska A, Guillot R et al (2013) Cyclodextrin/paclitaxel complex in biodegradable capsules for breast cancer treatment. Chem Mater 25:3867–3873
47. Cui W, Cui Y, Zhao J et al (2013) Fabrication of tumor necrosis factor-related apoptosis inducing ligand (trail)/alg modified $caco_3$ as drug carriers with the function of tumor selective recognition. J Mater Chem B 1:1326–1332
48. Xuan MJ, Shao JX, Lin XK et al (2015) Light-activated janus self-assembled capsule micromotors. Colloid Surf A Physicochem Eng Asp 482:92–97
49. Cui W, Li JB, Decher G (2016) Self-assembled smart nanocarriers for targeted drug delivery. Adv Mater 28:1302–1311
50. Swierczewska M, Han HS, Kim K et al (2016) Polysaccharide-based nanoparticles for theranostic nanomedicine. Adv Drug Deliver Rev 99:70–84
51. Shah S, Sasmal PK, Lee K-B (2014) Photo-triggerable hydrogel-nanoparticle hybrid scaffolds for remotely controlled drug delivery. J Mater Chem B 2:7685–7693
52. Fan WP, Huang P, Chen XY (2016) Overcoming the Achilles' heel of photodynamic therapy. Chem Soc Rev 45:6488–6519
53. Zhou ZJ, Song JB, Nie LM et al (2016) Reactive oxygen species generating systems meeting challenges of photodynamic cancer therapy. Chem Soc Rev 45:6597–6626
54. Wang KW, He Q, Yan XH et al (2007) Encapsulated photosensitive drugs by biodegradable microcapsules to incapacitate cancer cells. J Mater Chem 17:4018–4021
55. Crouzier T, Boudou T, Picart C (2010) Polysaccharide-based polyelectrolyte multilayers. Curr Opin Colloid Interface Sci 15:417–426
56. Cui W, Cui Y, Zhu P et al (2012) An anticoagulant activity system using nanoengineered autofluorescent heparin nanotubes. Chem Asian J 7:127–132

Chapter 5
Hemoglobin-Based Molecular Assembly

Li Duan, Yi Jia and Junbai Li

Abstract Development of protein-based molecular devices is an active area of research due to their broad applications in biotechnology, biorelated chemistry, bioelectronics, and biomedical engineering. Hemoglobin (Hb) is a physiologically important oxygen-transport metalloprotein present in the red blood cells. In this chapter, we present the recent development in fabrication and tailoring of a variety of hemoglobin protein shells via covalent layer-by-layer (LbL) assembly combined with template technique. Also, the developed strategy is effective and flexible, advantageous for avoiding denaturation of proteins. The as-fabricated Hb shells have better applications in drug delivery and controlled release, biosensors, biocatalysis, and bioreactors due to the enhancement of biological availability. In view of the carrying-oxygen function of Hb protein in blood, we particularly focus on the potential applications of hemoglobin-based nanoarchitectonic assemblies as artificial blood substitutes. These novel oxygen carriers exhibit advantages over traditional carriers and will greatly promote research on reliable and feasible artificial blood substitutes.

Keywords Hemoglobin protein · Layer-by-layer assembly · Template technique · Covalent cross-linking · Carrying-oxygen function · Artificial blood substitutes

---

L. Duan (✉)
Northwest Institute of Nuclear Technology, Xi'an, Shaanxi 710024, China
e-mail: duanli@nint.ac.cn

Y. Jia · J. Li (✉)
Beijing National Laboratory for Molecular Sciences, CAS Key Laboratory of Colloid, Interface and Chemical Thermodynamics, Institute of Chemistry, Chinese Academy of Sciences, Beijing 100190, China
e-mail: jbli@iccas.ac.cn

J. Li (ed.), *Supramolecular Chemistry of Biomimetic Systems*,
DOI 10.1007/978-981-10-6059-5_5

5.1 Introduction

Biomimetics has proved very useful in the design and fabrication of new functionally structured materials on the micro- and nanoscale [1]. Biomimetics refers to human-made processes, devices, or systems that mimic or imitate certain aspects of biological systems and have proven useful in providing biological inspiration from natural efficient designs [2–4]. Engineering biomimetic materials encompass a wide variety of research, from nanomaterials to the mechanics of how biological molecules such as proteins and enzymes can function as analogous man-made structures. With the development of modern biology, biomimetics is not limited to just copying nature. Scientists can directly utilize biological units themselves to construct hybrid nanostructured materials. In this chapter, we mainly focus on how molecular biomimetic applied to engineering functional materials, particularly fabricating the hemoglobin protein shells via covalently Layer-by-Layer (LbL) assembly, and its application in biomedical fields. This method to fabricate protein capsules is simple and inexpensive, advantageous for avoiding denaturation of proteins and is applicable to nearly any proteins, which shows a wide range of application.

5.2 Hemoglobin—An Oxygen-Carrying Protein

Hemoglobin (Hb) plays an important role in vital activities. It's the main functional constituent of the red blood cell that causes blood as a red color, and transports oxygen to the cells from the lungs and carbon dioxide away from the cells to the lungs. Red blood cells must contain adequate hemoglobin. A deficiency of hemoglobin would lead to various diseases, such as anemia. In mammals, the Hb protein makes up about 96% of the red blood cells' dry content (by weight), and around 35% of the total content (including water). Each erythrocyte contains 200–300 molecules of hemoglobin. Hemoglobin is also found outside red blood cells and their progenitor lines. Other cells that contain hemoglobin include the A9 dopaminergic neurons in the substantia nigra, macrophages, alveolar cells, and mesangial cells in the kidney. In these tissues, hemoglobin has a non-oxygen-carrying function as an antioxidant and a regulator of iron metabolism [5].

Hemoglobin (Hb) is a conjugated iron–protein compound in which each molecule is a tetramer composed of four monomers held together by weak bonds (Fig. 5.1). It is made up of four globular polypeptide subunits ($\alpha 1$, $\beta 1$, $\alpha 2$, $\beta 2$), noncovalently bound to each other by salt bridges, hydrogen bonds, and the hydrophobic effect. Each globular subunit contains an attached heme molecule composed of iron plus a protoporphyrin molecule. This porphyrin ring consists of four pyrrole molecules cyclically linked together (by methine bridges) with the iron ion bound in the center. The iron ion, which is the site of oxygen binding,

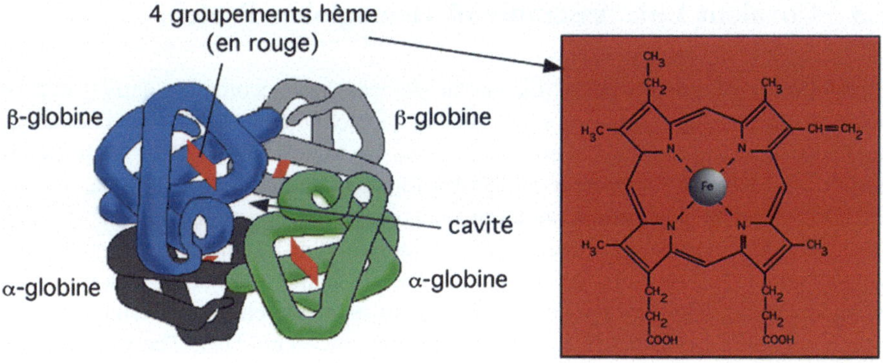

Fig. 5.1 Structure of hemoglobin

coordinates with the four nitrogen atoms in the center of the ring, which all lie in one plane. The iron is bound strongly (covalently) to the globular protein via the N atoms of the imidazole ring of F8 histidine residue (also known as the proximal histidine) below the porphyrin ring. A sixth position can reversibly bind oxygen by a coordinate covalent bond, completing the octahedral group of six ligands. Oxygen binds in an "end-on bent" geometry where one oxygen atom binds to Fe and the other protrudes at an angle. When oxygen is not bound, a very weakly bonded water molecule fills the site, forming a distorted octahedron. The iron ion may be either in the Fe^{2+} or in the Fe^{3+} state, but methemoglobin (Fe^{3+}) cannot bind oxygen [6]. In binding, oxygen temporarily and reversibly oxidizes Fe^{2+} to Fe^{3+} while oxygen temporarily turns into the superoxide ion, thus iron must exist in the +2 oxidation state to bind oxygen. If superoxide ion associated to Fe^{3+} is protonated, the hemoglobin iron will remain oxidized and incapable of binding oxygen. In such cases, the enzyme methemoglobin reductase will be able to eventually reactivate methemoglobin by reducing the iron center.

Several diseases are related to heme and hemoglobin disorders. Gene mutations result in a group of hereditary diseases termed hemoglobinopathies, among which the most common are sickle-cell disease and thalassemia. Decreased levels of hemoglobin and heme synthesis lead to symptoms of anemia, whereas alterations of heme metabolic pathways generate porphyrias syndromes. The decrease of hemoglobin's oxygen-binding capacity affected by molecules such as carbon monoxide (CO) (for example, from tobacco smoking, car exhaust, and incomplete combustion in furnaces) can result in hypoxia and asphyxiation. It is because carbon monoxide competes with oxygen at the same heme binding site on Hb. Hemoglobin binding affinity for CO is 200 times greater than its affinity for oxygen [7], meaning that small amounts of CO dramatically reduce hemoglobin's ability to transport oxygen.

Thus, we have recently devoted to the molecular assembly of hemoglobin shells and its application study to imitate in some sense hemoglobin structural function in the living system and better understand its capacity to select ions.

5.3 Covalent LbL Assembly of Hemoglobin Protein

Development of protein-based molecular devices has become an active area of research due to their broad applications in biotechnology, biorelated chemistry, bioelectronics, and biomedical engineering [8]. Up to now, a variety of methods has been used to construct protein molecular films, including LB technique, sol-gel method, physical adsorption, and covalent cross-linking strategies, etc. [9–11]. Also, the LbL approach has been applied widely in constructing protein multilayer assemblies [12]. The LbL technique was based on the alternating adsorption of charged species onto an oppositely charged substrate via electrostatic interactions, which was introduced firstly by Decher and his coworkers [13]. The procedure is simple and basically applicable to many different kinds of the substrate (1-D, 2-D, or 3-D). Now, different types of films containing charged and uncharged species have been successfully prepared, including polypeptides, polysaccharides, DNA, proteins, viruses, nanoparticles [14–16]. As a molecular assembly technique, LbL assembly has been extensively used in biology, spanning biomimetics, biomedicine, and tissue engineering [17–19].

Since 1996, multi-protein assemblies has been successfully built up on planar solid surface or on 3D colloidal particles through alternated electrostatic adsorption between charged proteins and oppositely charged polymers [20]. And their applications in catalysis and immune sensing have also been demonstrated. However, in all these reports, the deposition of proteins was based on the electrostatic interactions between them and oppositely charged polyelectrolytes. The existence of the other polymers may result in a lack of desirable properties or functionality of such proteins multilayer [8, 21]. In 2004, a novel type of protein nanostructures was constructed with the LbL assembly technique, which was introduced by Martin and his coworkers [22]. They developed a method with glutaraldehyde (GA) as a covalent cross-linker to fabricate hemoglobin (Hb) nanotubes and glucose oxidase (GOX) nanotubes, respectively. The studies showed that the protein nanotubes prepared in this way could remain the electroactivity and catalytic activity. Later, our group extended this strategy to sacrificial 3-D colloidal particles and hollow protein capsules were obtained without the use of any polymer [23–26], different from the assembly of proteins and oppositely charged polyelectrolytes. Here, we mainly introduce the fabrication of this kind of microcapsules and their applications in biomedical engineering and biologically related chemistry, in which they can be flexibly designed to be stimuli-responsive under physiological conditions and allow for the incorporation of uncharged species, which is important for capsule's multifunctionalization.

5.3.1 Hemoglobin Protein Hollow Shells Fabricated Through Covalent LbL Assembly

Our group has recently prepared hemoglobin (Hb) protein microcapsules by using a covalent LbL technique [24]. The use of covalent bonds to assemble LbL microcapsules can provide significant advantages compared to traditional electrostatic assembly [27]. In particular, they have high stability due to the covalent bonds formed, and therefore do not disassemble with changes in pH or ionic strength. In this work, GA is used as a chemical cross-linker because it has less effect on the protein activity. Poly(ethylenimine) (PEI) was firstly adsorbed on template particles to produce an amino-functionalized surface. Then the GA and Hb were alternately adsorbed. Figure 5.2a shows a schematic representation of a covalently cross-linked hemoglobin protein capsules. Direct information on the capsules formation can be obtained from the measurements by TEM and CLSM (Fig. 5.2b). The results show that the wall thickness of capsules can be controlled by the adsorption cycles of alternate GA/Hb. The UV-visible spectra of GA/Hb capsules show the absorption

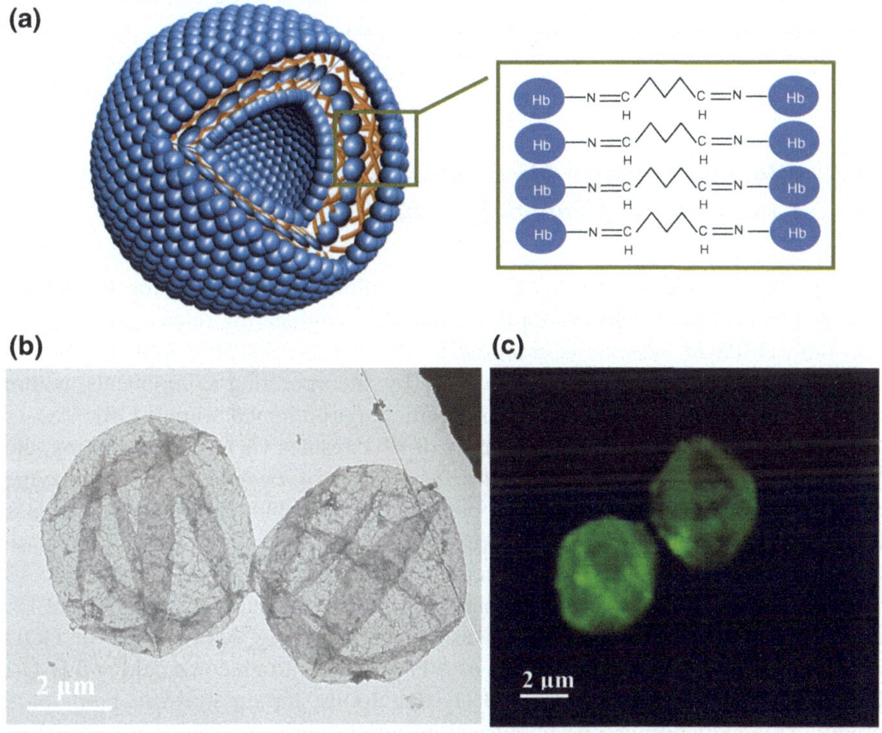

Fig. 5.2 **a** Schematic representation of the assembled hemoglobin protein microcapsules via covalent layer-by-layer assembly. **b** TEM and **c** CLSM images of $(GA/Hb)_5$ capsules at a dried state. Reprinted from Ref. [24], Copyright 2007, with permission from Elsevier

band of heme at 411 nm, indicating that Hb essentially remains in the capsules. Cyclic voltammetry and potential-controlled amperometric measurements confirm that cross-linked Hb capsules keep their heme electroactivity and are not denatured. The typical amperometric response toward the successive additions of H_2O_2 shows the electrocatalytic property of GA/Hb capsules. The permeability of the assembled GA/Hb microcapsules was tested by using fluorescein isothiocyanate (FITC)-dextran with different molecular weight as fluorescent probes. The results show that the $(GA/Hb)_5$ capsules are impermeable to FITC-dextran with molecular weights of 2000 kDa and 500 kDa, while FITC-dextran with molecular weights below 70 kDa can partly, and even completely permeate into the capsules interior. In comparison with traditional $(PAH/PSS)_5$ capsules, the Hb protein shells have a selective permeability. Similarly, the permeability decreases with the increase of wall thickness. Using FITC-dextran with a molecular weight of 20 kDa as a fluorescent probe, the fluorescence recovery in the capsules interior as a function of time is observed at lower excitation intensity. As a consequence, $(GA/Hb)_5$ microcapsules have a permeability of 4×10^{-8} m/s. The improved permeability of Hb capsules is helpful to better tune the storage and release of encapsulated small molecules. The fabrication of Hb capsules could imitate in some sense its structural function in the living system, and may help us to understand its properties and can be expected to have better applications in medicine, catalysis, and cosmetics, as well as biotechnology.

5.3.2 Glucose-Sensitive Microcapsules from Glutaraldehyde Cross-Linked Hemoglobin and Glucose Oxidase

Loading a therapeutic agent into LbL films/capsules is fundamental for developing these systems for drug delivery applications. Also, engineering release of the cargo from the multilayer systems is essential for therapeutic delivery applications. An ideal method to load and release drug is by incorporating components within multilayer films that are responsive to specific environmental stimuli [28].

Encouraged by the successful assembly of Hb capsules via GA cross-linking, the above approach was extended to fabricate a type of comparative, two-enzyme microcapsules including those composed of hemoglobin and glucose oxidase (Hb/GOD capsules), and those composed of catalase and glucose oxidase (CAT/GOD capsules). In this section, we described the fabrication of glucose-sensitive protein multilayer microcapsules by the alternate assembly of Hb and GOD with glutaraldehyde (GA) as cross-linker (Fig. 5.3a) [29, 30]. GOD catalyzes the oxidation and hydrolysis of β-D-glucose into gluconic acid and H_2O_2. Hb can catalyze the reduction of H_2O_2 due to its certain intrinsic peroxidase activity. This reduction can be monitored by nonfluorescent Amplex red, which is oxidized by H_2O_2 into resorufin (a fluorescence dye). Thus, the system involves two enzymatic catalysis reactions, and it offers two advantages. One is that the

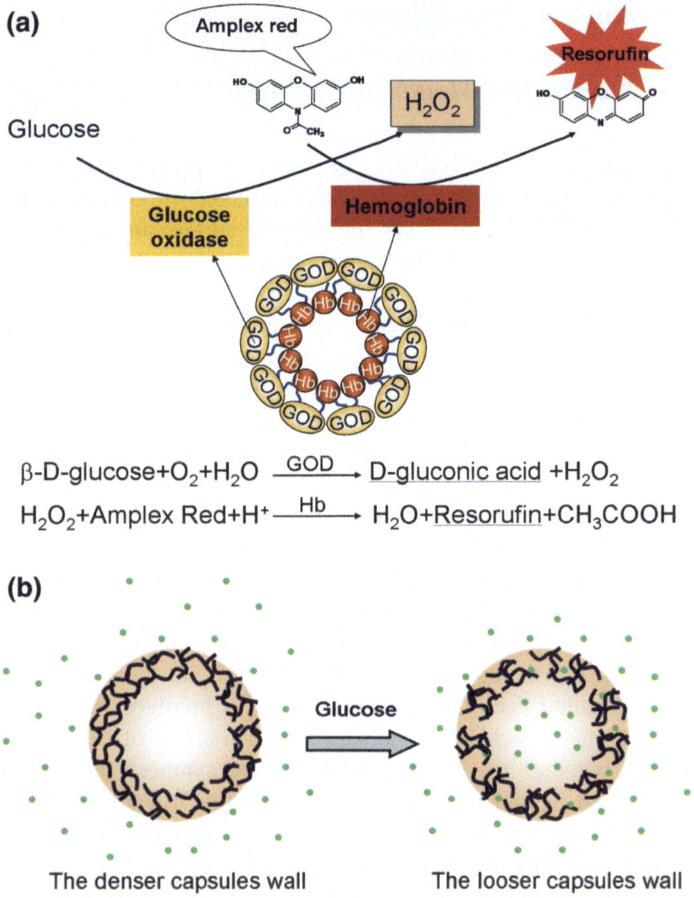

Fig. 5.3 **a** Schematic illustration of coupled enzymatic process based on glucose oxidase (GOx) and hemoglobin (Hb) coimmobilized as LbL microcapsules components; **b** illustration of glucose-stimulated FITC-dextran release (*green circles*). The consumption of glucose enhanced the release of the encapsulated materials. Reprinted with the permission from Ref. [29]. Copyright 2009 American Chemical Society

fluorescence of resorufin makes it convenient to monitor the reactions process, which may facilitate the development of an approach to design new fluorescence sensors for glucose. The other is that the consumption of glucose enhanced the permeability of the capsule wall and increased the release of the encapsulated drugs (Fig. 5.3b), which is highly attractive for the fabrication of glucose-responsive release systems. So, it is anticipated to apply this system in developing a fluorescence sensor to detect glucose qualitatively in vitro. In addition, it is also believed the observed glucose-stimulated enhancement of the wall permeability could play an important role in the development of drug delivery vehicles and may find widespread application in drug loading and release in the coming future.

5.3.3 Assembly of Lipid Bilayers on Covalently LbL-Assembled Hemoglobin Capsules as a Biomimetic Membrane System

In our subsequent work, Hb protein multilayer-supported liposomes or lipid bilayer-coated Hb protein microcapsules were fabricated through the conversion of liposomes into lipid bilayers to cover the capsule surface in analogy to the cell membrane [25]. These lipid-coated protein microcapsules can be considered as an ideally supported biomimetic membrane system to mimic the real cell membrane. It can not only conveniently tune their morphological properties, including the exterior shape, interior space, and shell structure, but also realize their multifunctionalization. Protein multilayer-supported liposome systems should be useful for the understanding of the principles of the interaction between membranes and biopolymers enzyme, and enable the design and application of new biomimetic structured materials.

It is well known that the production of adenosine 5′-triphosphate (ATP) is one of the most important chemical reactions in living organisms. The enzyme primarily responsible for the production of ATP is ATP synthase, which is a rotary motor protein. This enzyme drives the generation of ATP from adenosine 5′-diphosphate (ADP) and inorganic phosphate by utilizing proton gradients. ATP synthase contains two domains, the membrane- embedded F_0 part and the soluble F_1 domain, and is hence also called F_0F_1-ATPase. The F_0 part is involved in proton transport across the membrane, which is accompanied by a rotation of the intramembrane subunits (rotor), whereas the F_1 part catalyzes the synthesis and hydrolysis of ATP. The integrated F_0 and F_1 complex drives protons toward the F_1 side of the membrane, resulting in the synthesis of ATP [31]. Over the last decade, understanding of the functioning of ATPase has advanced dramatically to the point where the enzyme is now being studied by scientists interested in developing nanomachines for information storage and energy interconversion. Several groups have been working on the reconstitution of CF_0F_1 in liposomes to drive the motor proteins to carry out ATP biosynthesis [31–35]. Recently, we have created a route to generate a proton gradient by varying the pH values within and outside polymer microcapsules, with the CF_0F_1 species reconstituted within the walls of the capsules [36]. In this section, we performed the reconstitution of chloroplastic F_0F_1-ATPase (CF_0F_1-ATPase) within lipid bilayer-coated hemoglobin (Hb) protein microcapsules (Fig. 5.4a) to imitate the system in living cells governed by molecular motors [25].

Commonly, the first step of this biomimetic membrane fabrication process consists of preparing the hollow Hb microcapsules by covalent LbL assembly technique, as above-mentioned [36]. Then, CF_0F_1-proteoliposomes were prepared by the incorporation of CF_0F_1-ATP synthase into liposomes according to the method described previously in the literature [37]. Briefly, phosphatidylcholine/phosphatidic acid liposomes have been added into a Triton X-100-solubilized CF_0F_1-ATPase buffer solution, followed by the slow removal of Triton X-100 by

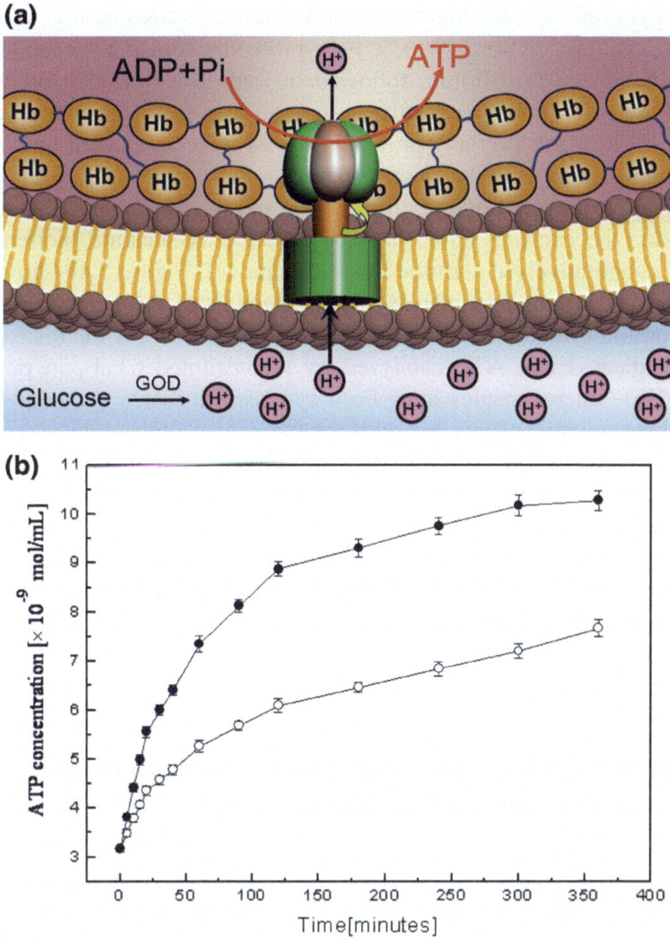

Fig. 5.4 **a** Schematic representation of ATP synthesis catalyzed by CF_0F_1-ATPase reconstituted in lipid-coated Hb microcapsules. **b** ATP synthesis catalyzed by reconstituted CF_0F_1 in proteolipid-coated $(GA/Hb)_5$ microcapsules as a function of the reaction time. Reproduced from Ref. [25] by permission of John Wiley & Sons Ltd

Biobeads SM-2. This leads to the formation of CF_0F_1-proteoliposomes. It is noted that, as phosphatidylcholine lipid is a major component of biological membranes, egg phosphatidylcholine was specially chosen as a lipid model to cover the capsules, but a small fraction of negatively charged lipid, phosphatidic acid, was added in order to promote the adsorption and fusion of vesicles. In other words, the lipid composition on the surface could easily be varied and fine-tuned to specific conditions. The existence and stability of the lipid bilayer were proved by means of CLSM. The fluorescent orb arises from the presence and continuous distribution of lipids on the capsule shells. Following that, a suspension of Hb capsules was first mixed with pyranine by shaking for several minutes, subsequently followed by washing with buffer solution. Next, CF_0F_1-proteoliposomes were added to the mixture and incubated for 30 min, followed by centrifugation and three washing cycles with buffer solution. As a result, the pyranines were encapsulated within the Hb protein capsules, and ATPase–proteoliposomes were assembled onto the outer shell of Hb capsules. It should be noted that only in the case of F_1 subunits of the CF_0F_1 complex extending toward the interior of the assembled capsules, ATP synthesis inside the capsules could become possible. A standard luciferase assay was used to quantitatively determine the ATP amount. Figure 5.4b show that the ATP production continuously increases as a function of the reaction time under driven by a proton gradient between the interior and exterior of the microcapsules. The results indicated that ATP catalyzed by reconstituted CF_0F_1 in proteolipid-coated $(GA/Hb)_5$ microcapsules have been synthesized inside the microcapsules. Such assembled microcapsules are very stable under physiological conditions. Their longer lifetime will enhance the ATP production efficiency. The well-defined microcapsules may serve as containers for the storage of the synthesized ATP as an energy currency. By using this system, it becomes possible to study the function of ATPase in a biomimetic unit in detail. Furthermore, as vital activities need energy, ATP could be released from the assembled capsules to provide energy on demand. Hence, we have built a micrometer-sized energy-storage device suitable to power biological activity.

5.3.4 *Assembled Hemoglobin and Catalase Nanotubes for the Treatment of Oxidative Stress*

Oxidative stress can induce the occurrence of diseases like nonalcoholic steatohepatitis (NASH). The increased oxidative stress in patients will cause the pro-oxidation environment against the antioxidant enzyme activities. In that case, the local H_2O_2 concentration will rise up. The excessive H_2O_2 can be converted to reactive oxygen species (ROS) or released into the extracellular environment. Therefore, how to eliminate the oxidative stress is a big challenge. Hemoglobin (Hb), the main oxygen transporter in erythrocytes, can be expressed in non-hematopoietic organs (in hepatocyte or mesangial cells) to balance the oxidative stress [38]. Oxidative stress upregulates hemoglobin expression, and hemoglobin overexpression suppresses oxidative stress in pathological cells [39, 40]. Thus, the drug constituted mostly by hemoglobin can provide an efficient treatment for oxidative stress induced disease like NASH.

Comparing with other nanostructures, the tubular structure presents several advantages to nanospheres. The multifunctionalities can be introduced on the inner and outer surfaces independently [41]. Second, open-end terminals enable quick loading of target molecules without the structural change. Third, nanotubes can have long circulation persistence in the bloodstream [42, 43]. One of the most efficient methods to prepare the structure-defined nanotubes is using the LbL technique. The flexibility of the LBL assembly allows us to create nanoreactors to scavenge H_2O_2 by an enzymatic reaction.

Hence, our group fabricated biomimetic nanotubes with H_2O_2-scavenging capabilities through LbL assembly of hemoglobin, catalase, and dialdehyde heparin (DHP) (Fig. 5.5a) [44]. Both hemoglobin and catalase are potential candidates to scavenge H_2O_2, and DHP is used as one of the wall components, an auxiliary drug in liver diseases, and also as a cross-linker. With the LbL assembly technique, the

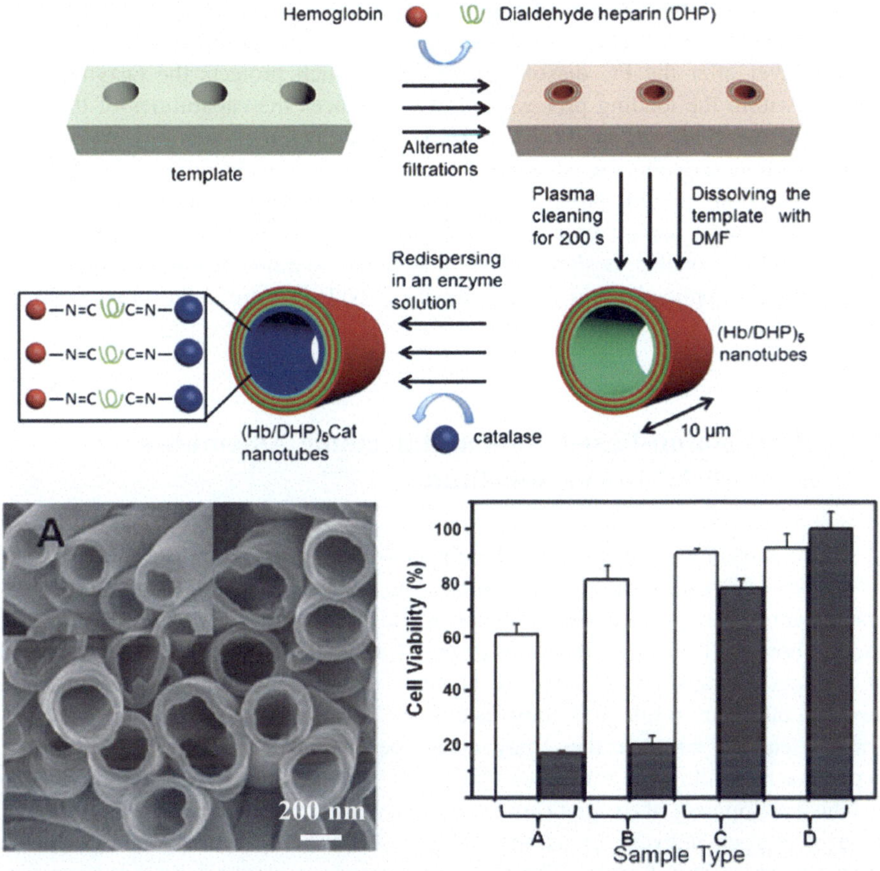

Fig. 5.5 **a** Schematic illustration for the fabrication of $(Hb/DHP)_5$ and $(Hb/DHP)_5$Cat nanotubes. **b** Scanning electron microscopy (SEM) images of the $(Hb/DHP)_5$ nanotubes. **c** $(Hb/DHP)_5$Cat nanotubes improved the HepG-2 cells viability against oxidative stress. HepG-2 viability when exposed to H_2O_2 of (*A*) 1000, (*B*) 500, (*C*) 250, and (D) 0 μM for 24 h as extracellular oxidative stress was evaluated by the CCK-8 assays. The white column represents the experiment group, and the gray column represents the control group. Each error bar represents the mean of at least six measurements (±SD). Reprinted with the permission from Ref. [44]. Copyright 2013 American Chemical Society

size, and behavior of as-prepared nanotubes could be well controlled and adjusted by changing the template inner pore size and the wall components (Fig. 5.5b). All these merits endow the nanotubes with great potential in the treatment of oxidative stress. In this work, in order to assess the H_2O_2-scavenging capacities of the as-assembled nanotubes in vitro, we evaluated cell viability under oxidative stress by the CCK-8 assays. As shown in Fig. 5.5c, when they were exposed to the 1000 and 500 μM H_2O_2, respectively, the improvement of the HepG-2 cells viability by the $(Hb/DHP)_5Cat$ nanotubes is obvious. With no additional H_2O_2, the viability of the experiment group is about 92% of that of the control group, which further proves the biocompatibility of the nanotubes. These results suggested that the assembled $(Hb/DHP)_5Cat$ nanotubes have a positive effect on the HepG-2 cells under oxidative stress. The as-prepared nanotubes were modified with a catalase interior layer after the PC dissolution process, which protects the enzyme from inaction within the loading process and also enhances the capabilities of the nanotubes in the treatment of oxidative stress. With the modification process, the behavior of hemoglobin-based nanotubes is open to a wide range of variations. These variations could provide a basis for maximizing the advantages of hemoglobin-based nanotubes and minimizing its disadvantages by means of systematic and reasonable design, which make the obtained nanotubes promising candidates for applications in the treatment of oxidative stress and other possible biomedical fields.

5.4 Hemoglobin-Based Nanoarchitectonic Assemblies as Artificial Blood Substitutes

Medical interest toward Hb is related to the possibility of its administration as blood substitutes to re-establish oxygen homeostasis in tissues. At present, blood transfusions are mostly applied for this purpose, but the need of the right type of blood and its short shelf life are still serious problems to be overcome. In this case, the use of autologous transfusions prevents the need of cross-matching and virus transmission, although autologous transfusions are considered the safest, they are not always feasible because they may cause perioperative anemia and are more expensive than allogenic transufions [45].

The development of oxygen carriers is particularly indicated in the case of urgent need of oxygen delivery to tissues and to solve the above-mentioned problems related to blood transfusions. The ideal oxygen carrier would deliver oxygen, not transmit disease, not have immunosuppresive effects, would have less strict storage requirements than that for human blood, would not need cross-matching, would be available at reasonable costs, be easy to administer, and able to reach all areas of human body, including ischemic tissues [46].

In the past decades, artificial oxygen carriers have been extensively studied and have gone through a rapid development. Up to now, there are mainly two types of

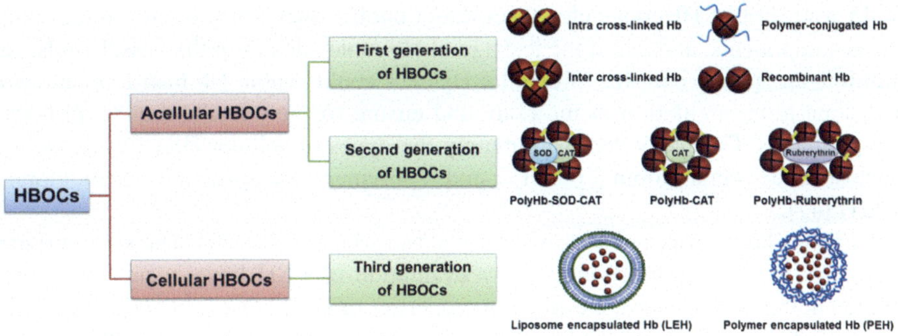

Fig. 5.6 A list of hemoglobin-based oxygen carriers (HBOCs). Reproduced from Ref. [50] by permission of John Wiley & Sons Ltd

oxygen carriers: totally synthetic oxygen carriers (including perfluorochemicals and heme hybrids) [47] and hemoglobin-based oxygen carriers (HBOCs) [48]. Among these oxygen carriers, HBOCs attracted the most attention owing to unique oxygen-transport ability and normal metabolic pathways of Hb [49]. However, stroma free Hb could not directly be used as an oxygen carrier since it dissociated into dimers that have a short circulation time, renal toxicity, high oncotic pressure and high O_2-affinity. To overcome these problems, hemoglobin must be chemically modified or microencapsulated to prevent its dissociation and thus evolved two types of HBOCs: acellular hemoglobin-based oxygen carriers (HBOCs) and cellular hemoglobin-based oxygen carriers, as shown in Fig. 5.6 [50].

Acellular HBOCs refer to chemically modified Hb and can be separated into two generations. The first generation of HBOCs aims to prevent the dissociation of Hb tetramer into a dimer, and mainly includes intra-molecular cross-linked Hb, inter-molecular cross-linked Hb, polymer-conjugated Hb and recombinant Hb [48]. The second generation of HBOCs is on the basis of the first generation that co-assembly of Hb with antioxidant enzymes such as superoxide dismutase (SOD) and catalase (CAT) [51, 52] to further stabilize the assemblies and meanwhile avoid the formation of inactive methemoglobin (metHb). Hb is also intramolecularly cross-linked with ATP, and intermolecularly cross-linked with adenosine, and conjugated with reduced GSH [53] to prevent the dissociation of Hb and the formation of homogeneous polymers, regulate the blood vessel tone, and protect heme from reactive oxygen species. Up to date, some of the above-mentioned Hb-based assemblies have gone through different stages of clinical trials and a few have been approved for routine clinical uses in South Africa or anemia treatment in canines.

Despite acellular HBOCs have made great progress, side effects still occurred in some cases due to the striking different structure in comparison with red blood cells (RBCs). Therefore, the importance of cellular structures like RBCs, that is, the third generation of HBOCs, became apparent. For cellular HBOCs, Hb molecules are encapsulated inside oxygen carriers of various structures. Cellular HBOCs could well protect Hb molecules from direct contact with surrounding tissues and blood components, and avoid the leakage of Hb molecules from the vessel walls. In addition, an appropriate size of cellular HBOCs could enable Hb-based assemblies with long-term circulation in the body and ensure oxygen availability to all body compartments. Generally, there are mainly two types of cellular HBOCs: liposome encapsulated hemoglobin (LEH) and polymer encapsulated hemoglobin (PEH) [47].

LEH has been proposed and investigated as early as the 1970s. The size, surface properties, in vivo efficacy and safety evaluation of LEH have been thoroughly studied in the past decades. Despite the properties of LEH have been well improved through adjusting the components of lipid or surface modification with PEG, some defects still exist. For example, phospholipids are expensive and may induce peroxidation in ischemia reperfusion. In addition, lipid membrane is impermeable to some reducing agents present in the circulating blood that is important for depressing the formation of metHb [54].

Compared to LEH, the researches on PEH have aroused increasingly concerns due to the easy availability, low price, broad variety, and biocompatibility of polymers. Theoretically, PEH is more like a red blood cell than LEH. It has a better physical strength, a better permeability, and ease of adjustment than LEH. Up to date, various strategies were developed to fabricate PEHs and great progress have been made. For instance, a double emulsion-solvent diffusion/evaporation technique was utilized for the encapsulation of Hb in polymers [54]. Chang et al. employed polylactic acid (PLA), poly(lactic acid-co-glycolide) (PLGA) and polyisobutycyanoacrylate as the matrix polymer and obtained Hb-loaded particles or capsules with a diameter of 70–200 nm. As a natural process that spontaneously organizes molecular units into well-ordered structures, molecular self-assembly technique is emerging as a powerful method for functional materials fabrication. It is well documented that amphiphilic block copolymers or grafted copolymers could self-assembly into nano/micro-vesicles or capsules in aqueous solutions. Palmer and coworkers [55] successfully employed amphiphilic diblock copolymers poly (butadiene-*b*-ethylene oxide) (PBD-PEO), poly(ethylene oxide-*b*-lactide) (PEO-PLA) and poly(ethylene oxide-*b*-caprolactone) (PEO-PCL) as building blocks to prepare polymersomes for Hb encapsulation through rehydrate and extrusion method.

LbL assembly technique is a versatile method for fabricating assemblies with a specific shape, size, material composition, and functionality [56]. As mentioned above, our group has successfully fabricated Hb-based hollow microcapsules through covalent LbL assembly technique combined with templating porous particles. It is worth to note that LbL assembly technique enables many conceivable ways to adjust and improve the behaviors of these Hb-based assemblies. First, the size and properties of Hb-based assemblies could be easily controlled by adjusting the size of the template and the assembled composition. Second., some functional particles or reagents like Fe_3O_4 nanoparticles, allosteric effector or methemoglobin reductase could be readily loaded into Hb-based assemblies for further functionalization and optimization of Hb-based oxygen carriers. In our subsequent works

[57–60], the assembly and application of hemoglobin-based microspheres/capsules as artificial blood substitutes were deeply researched. The integration of modern nanobiotechnology and self-assembly technique has greatly facilitated the development of artificial oxygen carriers. The as-prepared Hb spheres may have potential to be developed as useful artificial blood substitutes in future.

5.4.1 Highly Loaded Hemoglobin Spheres as Promising Artificial Oxygen Carriers

In order to enhance the loading content of Hb in a well-defined structure, we explored a facile and controllable avenue to fabricate Hb spheres with a high loading content as promising oxygen carriers [57]. In this work, Hb spheres were prepared by templating decomposable porous $CaCO_3$ particles in combination with covalent LbL assembly. Hb proteins were firstly encapsulated in the porous template particles by the co-precipitation of protein and $CaCO_3$. A biomedical cross-linking agent, glutaraldehyde (GA), was then applied to stabilize the Hb spheres by the formation of covalent bonds between the aldehyde groups of GA and free amino groups of Hb. Covalent cross-linking can stop the rapid decomposition of Hb tetramers into dimers, prolong its vascular retention and eliminate nephrotoxicity [61]. In the following, additional Hb was furtherly assembled on the outer shell of the Hb-loaded $CaCO_3$ particles by covalent LbL assembly with GA to obtain a higher loading of Hb in an individual particle. Finally, the surface of the Hb spheres was further chemically modified by biocompatible polyethylene glycol (PEG) to protect and stabilize the system, prolong the retention time in intravascular circulation and also provide a more stable and semipermeable system, which can allow small molecules such as O_2 and CO_2, the life-sustaining glucose in plasma, reducing agents and metabolic products to diffuse into/out of the system, just more like the active behavior of RBCs.

Figure 5.7a shows a schematic representation of the assembled Hb microspheres with the surface modified by PEG. On the one hand, we obtained a high loading of Hb in an individual particle by a facile way, which is very favorable for application in an urgent need for large amounts of artificial blood, such as natural disasters and battlefield. On the other hand, the surface of the resulting particles can readily be modified chemically and biologically for prolonging the retention time in the body, and introducing new functions such as targeting, antibacterial, antibody function and decreasing cytotoxicity. Compared to the previous reports [62–65], the loading concentration of Hb spheres assembled in the present work would be apparently higher, because a number of micro- and nano-pores of the $CaCO_3$ templates are accessible for Hb molecules (Fig. 5.7b). In addition, the LbL technique can further help the loading of Hb by coating the external surface of $CaCO_3$ with encapsulated Hb inside (Fig. 5.7c). Both factors contribute to the high loading of Hb in the spheres, and also present adjustable permeability [24], which is of great importance

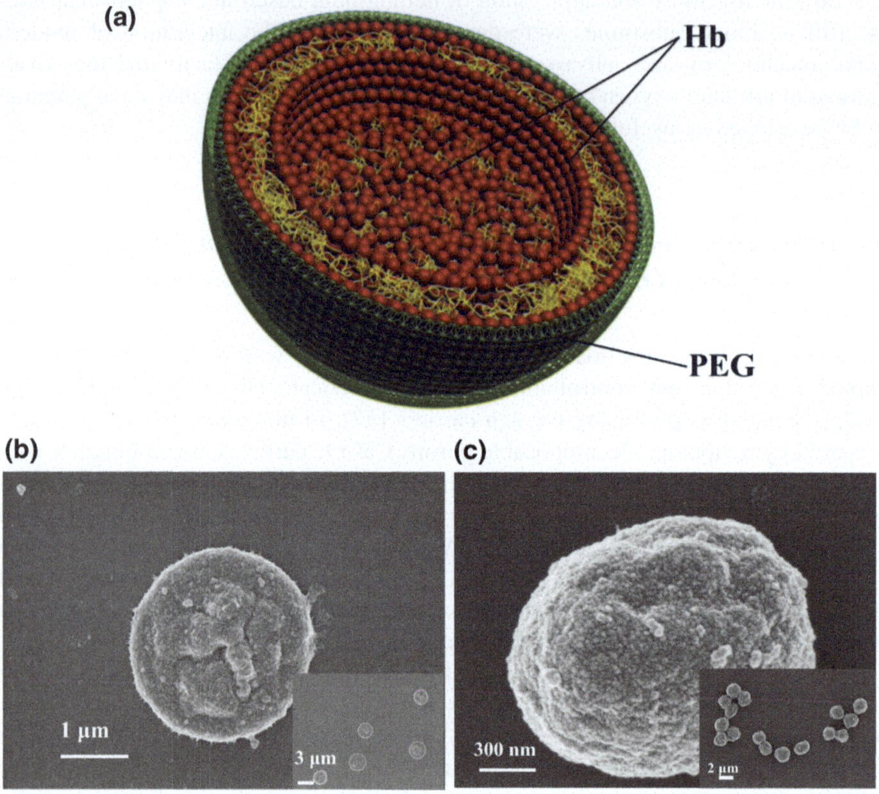

Fig. 5.7 **a** Schematic representation of the assembled Hb microspheres with the surface modified by PEG. **b** SEM images of Hb spheres after removing $CaCO_3$. **c** SEM images of Hb spheres with additional Hbs as the outer shell after removing $CaCO_3$. Reprinted with the permission from Ref. [57]. Copyright 2012 American Chemical Society

for maintaining the normal functionalities of natural RBCs, allowing life-sustaining small molecules such as oxygen, carbon dioxide, glucose, reducing agents and metabolic products to diffuse into/out of the system. More importantly, such Hb protein after covalently cross-linked and modified by PEG retains well its whole structure and especially its oxygen-carrying function. Figure 5.8 shows the results from UV-vis absorption measurement on the carrying and release of oxygen by Hb. It could be obviously seen that the absorption spectrum of the as-prepared Hb spheres shows changes similar to those of deoxygenation and oxygenation of free Hb and cross-linked Hb. All these results demonstrated that the LbL-assembled Hb-based microspheres are much more suitable to function as artificial oxygen carriers, particularly for an urgent need of large amounts.

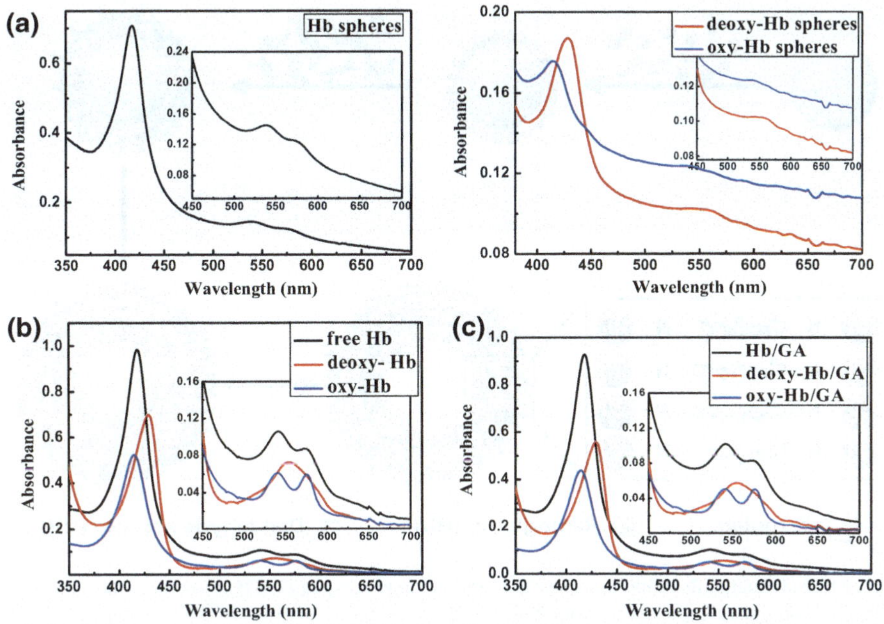

Fig. 5.8 UV-vis absorption spectra of deoxy-Hb and oxy-Hb. **a** Hb spheres; **b** free Hb; **c** cross-linked Hb. Inset image in detail shows the absorption between 450 and 700 nm. Reprinted with the permission from Ref. [57]. Copyright 2012 American Chemical Society

5.4.2 Construction and Evaluation of Hemoglobin-Based Capsules as Blood Substitutes

In the preceding section, we prepared the $(Hb/DHP)_6$ microcapsules to mimic artificial RBCs via covalent LbL assembly technique and template method (Fig. 5.9) [58]. The oxidized heparin (dialdehyde heparin, DHP) was specially used both as one of the wall components and a cross-linker because of its nontoxicity, biodegradability, biocompatibility and hemocompatibility. In addition, the negative charges of DHP on microcapsules' surface would be beneficial to prolong the blood retention time of microcapsules in vivo. The amino groups of Hb react easily with aldehyde groups of DHP through Schiff's base reaction. The formation of Schiff's base bonds among the Hb/DHP multilayers enabled the $(Hb/DHP)_6$ capsules with autofluorescence, which attributed to the n–π* transition of C=N bonds. This property would beneficial in predicting and monitoring the safety and efficacy of $(Hb/DHP)_6$ capsules in humans, while avoiding the use of external fluorochromes for biological tracing. In addition, the cross-linking reaction between amino groups of Hb and aldehyde groups of DHP effectively prevented the dissociation of the Hb tetramer into dimers.

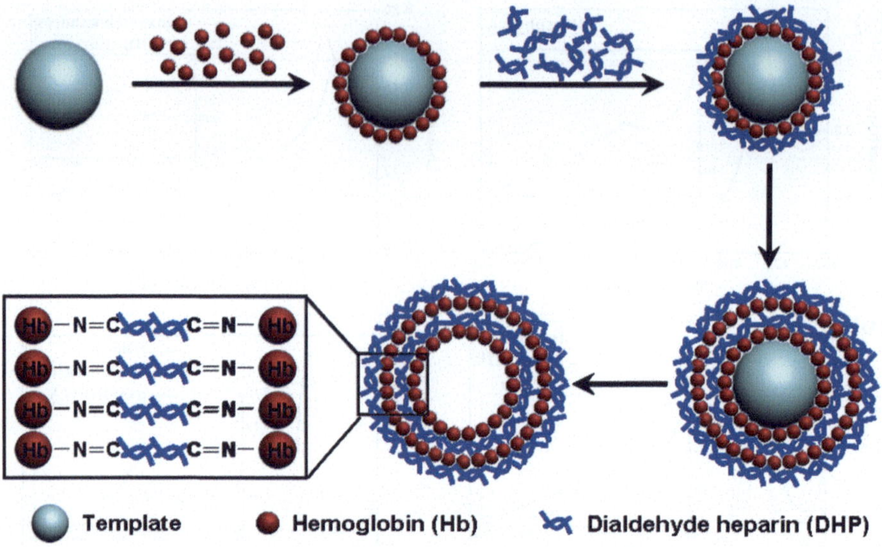

Fig. 5.9 Schematic illustration to show the fabrication process of Hb/DHP microcapsules through Schiff's base bond. Reproduced from Ref. [58] by permission of John Wiley & Sons Ltd

In this work, the biocompatibility of $(Hb/DHP)_6$ microcapsules is evaluated through 3-(4,5-dimethylthiazol-2-yl)-2,5-diphenyltetrazolium bromide) (MTT) assay and cell experiment. Figure 5.10 clearly shows that the cell coexisted with microcapsules without any abnormalities, which confirmed the biocompatibility of the microcapsules. It is also worth to note that microcapsules with red fluorescence are distributed around cells without being uptaken by the cells. It can be attributed to the electrostatic repulsion between negative charges on cell membrane surface and microcapsules surface, which will help to prolong the blood retention time of microcapsules in vivo [66]. To access the degradation behavior, the $(Hb/DHP)_6$ microcapsules were incubated with trypsin solution for 24 h. The results show that no microcapsule was observed after degradation, confirming that $(Hb/DHP)_6$ microcapsules were biodegradable. Hemocompatibility of $(Hb/DHP)_6$ microcapsules is also characterized in terms of prothrombin time (PT), thrombin time (TT), activated partial thromboplastin time (APTT) and hemolysis rate. Oxygen-carrying capacity of the microcapsules is demonstrated by converting the deoxy-Hb state of the microcapsules into oxy-Hb state. All these results demonstrated that the Hb assembled in the $(Hb/DHP)_6$ microcapsules still maintained its own bioactivity and possessed the essential oxygen-transporting function of an oxygen carrier, as well as nontoxicity, biodegradability, biocompatibility, and hemocompatibility. It is indicated that the as-prepared hemoglobin-based capsules have great potentials to function as blood substitutes.

It is worth to note that in view of the advantages of LbL assembly technique, there are many conceivable ways to adjust and improve the behaviors of the

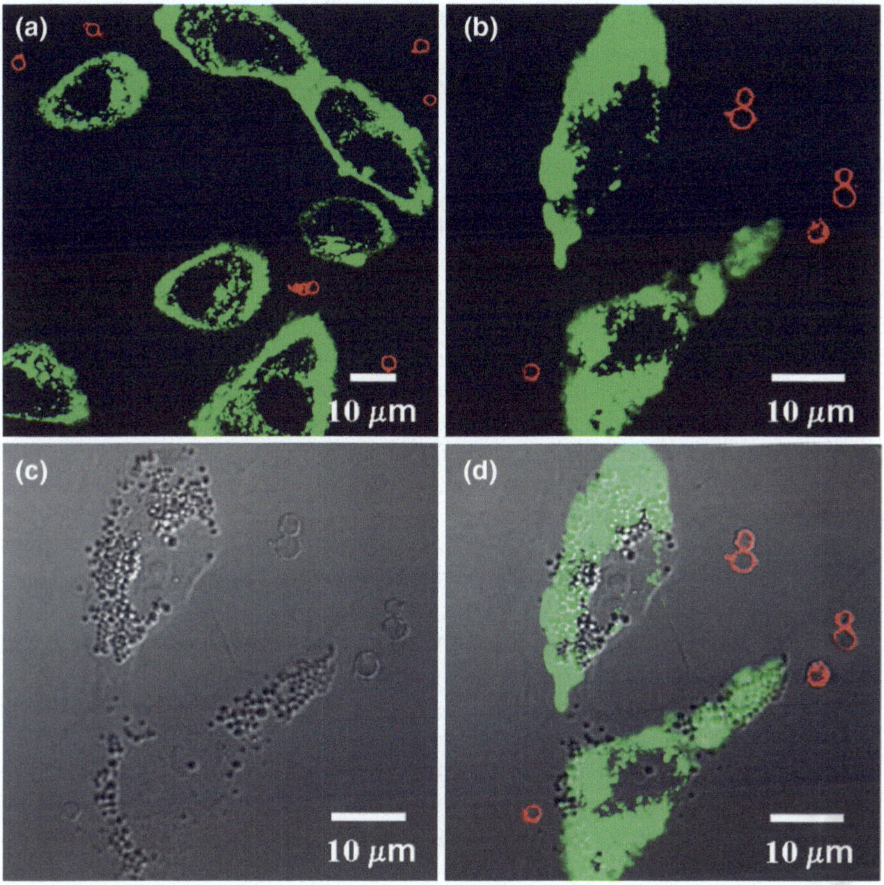

Fig. 5.10 **a**, **b** CLSM images of HUVEC cells stained with Alexafluor 488 incubated with $(Hb/DHP)_6$ microcapsules by excitation at 488 and 559 nm, **c** corresponding bright field image of **b**, **d** the overlapped image of **b** and **c**. Reproduced from Ref. [58] by permission of John Wiley & Sons Ltd

Hb-based capsules that functional groups or reagents (such as allosteric effector, superoxide dismutase, and methemoglobin reductase) may be chemically modified or covalently coupled to hemoglobin, in addition to the possibility of replacing DHP itself by other polysaccharides or their derivatives, such as dialdehyde starch, dialdehyde alginate, etc. [67]. Consequently, the construction and behavior of hemoglobin-based capsules are open to a wide range of variations. These variations could provide a basis for maximizing the advantages of hemoglobin-based capsules and minimizing its disadvantages by means of systematic and reasonable design, which make the obtained microcapsules promising candidates for applications in blood substitutes, oxygen carriers, and other biomedicine fields.

5.4.3 High Impact of Uranyl Ions (UO_2^{2+}) on Carrying-Releasing Oxygen Capability of Hemoglobin-Based Blood Substitutes

Recently, our group investigated the effect of radioactive UO_2^{2+} on the structure and carrying-releasing oxygen capability of hemoglobin-based oxygen carriers in vitro [59]. It is well-known, radioactive metal toxins seriously threaten the human health and environments because of their radiological and chemical toxicity. Radioactive uranyl ion (UO_2^{2+}) is an important contamination discharged into environment mainly from nuclear power reactor effluents. However, it is not so clear how the accumulation of radioactive uranyl ions (UO_2^{2+}) in blood cells can cause immune system damage. It is would be helpful to explore the damage of UO_2^{2+} on the biofunction and physical chemistry mechanism of the blood cells. In this work, the magnetic Hb microspheres fabricated through covalent LbL assembly on iron oxide (Fe_3O_4) loaded porous $CaCO_3$ particles were utilized as blood substitutes for the damage study of UO_2^{2+} on the biofunction of blood cells. The carriers with magnetic properties can help to solve some technical operations [68], which facilitate their trapping from the medium with the help of an external magnet. For the separating process of UO_2^{2+} adsorbed on the Hb microspheres, the magnetically assisted chemical separation (MACS) method is introduced as a rapid and easy avenue based on biocompatible magnetic nanoparticles by applying an external magnetic field. Figure 5.11a, b shows a schematic representation of the fabrication process of the magnetic Hb microspheres and their specific adsorption for UO_2^{2+}. Magnetic nanoparticles of iron oxide (Fe_3O_4) are first encapsulated in porous $CaCO_3$ particles after ultrasonic dispersion. Then, hemoglobin proteins are chemically modified on the surface of the Fe_3O_4-loaded $CaCO_3$ microspheres to form the Hb shells based on covalent LbL assembly. To study its chemical toxicity, the UO_2^{2+} is mixed with the as-assembled magnetic Hb microspheres. The results from UV-vis absorption measurement on UO_2^{2+} before and after being adsorbed on magnetic Hb microspheres (Fig. 5.11c) shows that UO_2^{2+} are highly adsorbed on the as-assembled magnetic Hb microspheres. The damage experiments of radioactive UO_2^{2+} on magnetic Hb microspheres (Fig. 5.11d) obviously proved that the presence of UO_2^{2+} in vivo seriously caused Hb microspheres to be inactivated and denaturalized, and resulted in the loss of the oxygen-transporting capacity. It may be attributed to the highly binding coordination of UO_2^{2+} with an iron atom of Hb, and the tight electrostatic interaction between positive UO_2^{2+} and negative Hb. Our works well confirmed that the accumulation of radioactive uranyl ions (UO_2^{2+}) in blood cells can seriously damage the human health, especially the transporting-oxygen capability of blood cells. In addition, in view of the high adsorption capacity of the as-assembled magnetic Hb microspheres, the assembled magnetic Hb-based microspheres may open a new highly effective way for removal of UO_2^{2+} from the radioactive-contaminated body or from the nuclear power reactor effluents before discharge into the environment. It can also be applied in the removing of other metal toxins from blood due to its biocompatibility.

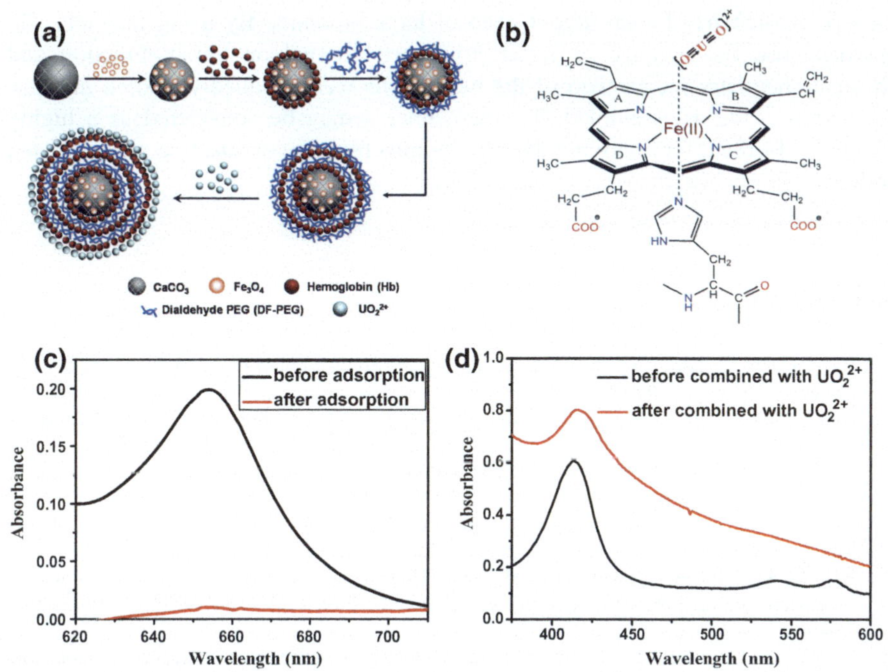

Fig. 5.11 **a** Schematic illustration of the fabrication of magnetic hemoglobin microspheres by the porous template strategy and Layer-by-Layer assembly technique and its adsorption for UO_2^{2+}. **b** The coordinating interaction of UO_2^{2+} with iron ion of heme. **c** UV-vis absorption spectra of UO_2^{2+} in aqueous solution before and after adsorbed on the magnetic Hb microspheres. **d** The sorption percentage of UO_2^{2+} with pure $CaCO_3$ particles and the magnetic Hb microspheres with different adsorption cycles of Hb/PEG as outer shell. Reproduced from Ref. [59] by permission of John Wiley & Sons Ltd

5.5 Conclusions and Perspectives

As an important protein in vital activities, hemoglobin-based molecular assembly has been extensively explored and applied. In this chapter, hemoglobin-based multifunctionalized microcapsules fabricated by covalent layer-by-layer assembly combined with template technique were introduced. The fabrication of Hb capsules could imitate in some sense its structural function in the living system, and may help us to understand its properties and can be expected to have better applications in drug delivery and controlled release, biosensors, biocatalysis and bioreactors due to the enhancement of biological availability. In particular, in view of the carrying-oxygen function of Hb protein in blood, the covalently LbL-assembled Hb microcapsules may have great potential to be developed as useful artificial blood substitutes in future. The integration of modern nanobiotechnology and self-assembly technique has greatly facilitated the development of Hb-based oxygen carriers. The as-prepared Hb spheres with highly loaded content are much more suitable for artificial oxygen carriers, particularly for an urgent need of large amounts. By using magnetic Hb microspheres, the damage of radioactive metal toxins on the biofunction and physical chemistry mechanism of the blood cells were detailedly studied in vitro. Furthermore, the as-assembled Hb microspheres may be considered as a highly effective adsorbent for removing the metal toxins from the radioactive-contaminated body.

References

1. He Q, Cui Y, Li JB (2009) Molecular assembly and application of biomimetic microcapsules. Chem Soc Rev 38:2292–2303
2. Bao G, Suresh S (2003) Cell and molecular mechanics of biological materials. Nat Mater 2:715–721
3. Wendell DW, Patti J, Montemagno CD (2006) Using biological inspiration to engineer functional nanostructured materials. Small 2:1324–1329
4. Lowe CR (2000) Nanobiotechnology: the fabrication and applications of chemical and biological nanostructures. Curr Opin Struct Biol 10:428–432
5. Biagioli M, Pinto M, Cesselli D et al (2009) Unexpected expression of α- and β-globin in mesencephalic dopaminergic neurons and glial cells. Proc Natl Acad Sci USA 106:15454–15459
6. Linberg R, Conover CD, Shum KL et al (1998) Hemoglobin based oxygen carriers: how much methemoglobin is too much? Artif Cells Blood Substit Immobil Biotechnol 26:133–148
7. Guyton AC (2005) Medical physiology, 11th edn, p 509
8. Shutava TG, Kommireddy DS, Lvov YM (2006) Polyelectrolyte/enzyme multilayer as functionalprotective nano/barrier in oxidizing media. J Am Chem Soc 128:9926–9934
9. Uto K, Yamamoto K, Kishimoto N et al (2008) J Mater Chem 18:3876–3884
10. Hwang I, Baek K, Jung M et al (2007) Noncovalent immobilization of proteins on a solid surface by cucurbit[7]uril-ferrocenemethylammonium pair, a potential replacement of Biotin −Avidin Pair. J Am Chem Soc 129:4170–4171
11. Wang Q, Yang Z, Gao Y et al (2008) Enzymatic hydrogelation to immobilize an enzyme for high activity and stability. Soft Matter 4:550–553
12. Onda M, Lvov Y, Ariga K et al (1996) Sequential reactions by glucose oxidase/peroxidase molecular films assembled by layer-by-layer alternate adsorption. Biotechnol Bioeng 51:163–167
13. Decher G (1997) Fuzzy nanoassemblies: toward layered polymeric multicomposites. Science 277:1232–1237
14. Johnston APR, Cortez C, Angelatos AS et al (2006) Layer-by-layer engineered capsules and their applications. Curr Opin Colloid Interface Sci 11:203–209
15. Wang KW, He Q, Cui Y et al (2007) Photosensitive drugs encapsulizated by biodegradable microcapsules applied for photodynamic Therapy. J Mater Chem 17:4018–4021
16. Mercato LLD, Rivera-Gil P, Abbasi AZ et al (2010) LbL multilayer capsules: recent progress and future outlook for their use in life sciences. Nanoscale 2:458–467
17. Tang Z, Wang Y, Podsiadlo P et al (2007) Biomedical application of layer-by-layer assembly: from biomimetics to tissue engineering. Adv Mater 19:906–928
18. De Geest BG, Sanders NN, Sukhorukov GB et al (2007) Release mechanisms for polyelectrolyte capsules. Chem Soc Rev 36:636–649
19. Qi W, Wang AH, Yang Y et al (2010) The lectin binding and targetable cellular uptake of lipid-coated polysaccharide microcapsules. J Mater Chem 20:2121–2127

20. Caruso F, Trau D, Möhwald H et al (2000) Enzyme encapsulation in layer-by-layer engineered polymer multilayer capsules. Langmuir 16:1485–1488
21. Tong WJ, Gao CY, Möhwald H (2006) Single polyelectrolyte microcapsules fabricated by glutaraldehyde-mediated covalent Layer-by-Layer assembly. Macromol Rapid Commun 27:2078–2083
22. Skirtach AG, Déjugnat C, Braun D et al (2005) The role of metal nanoparticles in remote release of encapsulated materials. Nano Lett 5:1371–1377
23. Qi W, Duan L, Li JB (2011) Fabrication of glucose-sensitive protein microcapsules and their applications. Soft Matter 7:1571–1576
24. Duan L, He Q, Yan XH et al (2007) Hemoglobin protein hollow shells fabricated through covalent Layer-by-Layer assembly. Biochem Biophys Res Commun 354:357–362
25. Qi W, Duan L, Wang KW et al (2008) Motor protein CF_0F_1 reconstituted in lipid-Coated hemoglobin microcapsules for ATP synthesis. Adv Mater 20:601–605
26. Duan L, Qi W, Yan XH et al (2009) Proton gradients produced by glucose oxidase microcapsules containing motor F_0F_1-ATPase for continuous ATP biosynthesis. J Phys Chem B 113:395–399
27. He Q, Tian Y, Möhwald H et al (2009) Biointerfacing luminescent nanotubes. Soft Matter 5:300–303
28. Shan E, Han HX, Cosnier S (2007) Self-assembled films of hemoglobin/laponite/chitosan: application for the direct electrochemistry and catalysis to hydrogen peroxide. Biomacromol 8:3041–3046
29. Qi W, Duan L, Yan XH et al (2009) Glucose-sensitive microcapsules from glutaraldehyde cross-linked hemoglobin and glucose oxidase. Biomacromol 10:1212–1216
30. Qi W, Yan XH, Fei JB et al (2009) Triggered release of insulin from glucose-sensitive enzyme multilayer shells. Biomaterials 30:2799–2806
31. Choi HJ, Montemagno CD (2005) Artificial organelle: ATP synthesis from cellular mimetic polymersomes. Nano Lett 5:2538–2542
32. Capaldi RA, Aggeler R (2002) Mechanism of the F_0F_1-ATP synthase, a biological rotary motor. Trends Biochem Sci 27:154–160
33. Fischer S, Etzold C, Turina P et al (1994) ATP synthesis catalyzed by the ATP synthase of *Escherichia coli* reconstituted into liposomes. Eur J Biochem 225:167–172
34. Steinberg-Yfrach G, Rigaud JL, Durantini EN et al (1998) Light-driven production of ATP catalyzed by F_0F_1-ATP synthase in an artificial photosynthetic membrane. Nature 392:479–482
35. Luo TJM, Soong R, Lan E et al (2005) Photo-induced proton gradients and ATP biosynthesis produced by vesicles encapsulated in a silica matrix. Nat Mater 4:220–224
36. Duan L, He Q, Wang KW et al (2007) Adenosine triphosphate biosynthesis catalyzed by F_oF_1 ATP synthase assembled in polymer microcapsules. Angew Chem Int Ed 46:6996–7000
37. Richard P, Rigaud JL, Gräber P (1990) Reconstitution of CF_0F_1 into liposomes using a new reconstitution procedure. Eur J Biochem 193:921–925
38. Buehler PW, D'Agnillo F, Schaer DJ (2010) Hemoglobin-based oxygen carriers: from mechanisms of toxicity and clearance to rational drug design. Trends Mol Med 16:447–457
39. Liu LM, Zeng M, Stamler JS (1999) Hemoglobin induction in mouse macrophages. Proc Natl Acad Sci USA 96:6643–6647
40. Widmer CC, Pereira CP, Gehrig P et al (2010) Hemoglobin can attenuate hydrogen peroxide-induced oxidative stress by acting as an antioxidative peroxidase. Antioxid Redox Signal 12:185–198
41. Kameta N, Minamikawa H, Masuda M (2011) Supramolecular organic nanotubes: how to utilize the inner nanospace and the outer space. Soft Matter 7:4539–4561
42. Komatsu T, Qu X, Ihara H et al (2011) Virus trap in human serum albumin nanotube. J Am Chem Soc 133:3246–3248
43. Geng Y, Dalhaimer P, Cai S et al (2007) Shape effects of filaments versus spherical particles in flow and drug delivery. Nat Nanotechnol 2:249–255

44. Chen B, Jia Y, Zhao J et al (2013) Assembled hemoglobin and catalase nanotubes for the treatment of oxidative stress. J Phys Chem C 117:19751–19758
45. Pape A (2007) Alternatives to allogeneic blood transfusions. Best Pract Res Clin Anaesthesiol 21:221–239
46. Ness PM, Cushing MM (2007) Oxygen therapeutics pursuit of an alternative to the donor red blood cell. Arch Pathol Lab Med 131:734–741
47. Modery-Pawlowski CL, Tian LL, Pan V et al (2013) Synthetic approaches to RBC mimicry and oxygen carrier systems. Biomacromol 14:939–948
48. Chang TMS (2004) Hemoglobin-based red blood cell substitutes. Artif Organs 28:789–794
49. Alayash AI (2014) Blood substitutes why haven't we been more successful. Trends Biotechnol 32:177–186
50. Jia Y, Duan L, Li JB (2016) Hemoglobin-based nanoarchitectonic assemblies as oxygen carriers. Adv Mater 28:1312–1318
51. Nadithe V, Bae YH (2010) Synthesis and characterization of hemoglobin conjugates with antioxidant enzymes via poly(ethylene glycol) cross-linker (Hb-SOD-CAT) for protection from free radical stress. Int J Biol Macromol 47:603–613
52. Hathazi D, Mott AC, Vaida A et al (2014) Oxidative protection of hemoglobin and hemerythrin by cross-linking with a nonheme iron peroxidase: potentially improved oxygen carriers for use in blood substitutes. Biomacromol 15:1920–1927
53. Simoni J, Simoni G, Wesson DE et al (2012) ATP-adenosine-glutathione cross-linked hemoglobin as clinically useful oxygen carrier. Curr Drug Discov Technol 9:173–187
54. Chang TMS (2010) Blood replacement with nanobiotechnologically engineered hemoglobin and hemoglobin nanocapsules. Wiley Interdiscip Rev Nanomed Nanobiotechnol 2:418–430
55. Rameez S, Alosta H, Palmer AF (2008) Biocompatible and biodegradable polymersome encapsulated hemoglobin: a potential oxygen carrier. Bioconjugate Chem 19:1025–1032
56. Ariga K, Yamauchi Y, Rydzek G et al (2014) Layer-by-layer nanoarchitectonics: invention innovation and evolution. Chem Lett 43:36–68
57. Duan L, Yan XH, Wang AH et al (2012) Highly loaded hemoglobin spheres as promising artificial oxygen carriers. ACS Nano 6:6897–6904
58. Jia Y, Cui Y, Fei JB et al (2012) Construction and evaluation of hemoglobin-based capsules as blood substitutes. Adv Funct Mater 22:1446–1453
59. Duan L, Du LL, Jia Y et al (2015) High impact of uranyl ions on carrying-releasing oxygen capability of hemoglobin-based blood substitutes. Chem Eur J 21:520–525
60. Jia Y, Li JB (2015) Molecular assembly of schiff base interactions: construction and application. Chem Rev 115:1597–1621
61. Squires JE (2002) Artificial blood. Science 295:1002–1005
62. Chang TMS, D'Agnillo F, Yu WP et al (2000) Two future generations of blood substitutes based on polyhemoglobin–SOD–catalase and nanoencapsulation. Adv Drug Deliv Rev 40:213–218
63. Dusseault J, Leblond FA, Robitaille R et al (2005) Microencapsulation of living cells in semi-permeable membranes with covalently cross-linked layers. Biomaterials 26:1515–1522
64. Zhao J, Liu CS, Yuan Y et al (2007) Preparation of hemoglobin-loaded nano-sized particles with porous structure as oxygen carriers. Biomaterials 28:1414–1422
65. Sun J, Huang YB, Shi Q et al (2009) Oxygen carrier based on hemoglobin/poly(L-lysine)-block-poly(L-phenylalanine) vesicles. Langmuir 25:13726–13729
66. Komatsu T, Matsukawa Y, Tsuchida E (2001) Reaction of nitric oxide with synthetic hemoprotein, human serum albumin incorporating tetraphenylporphinatoiron(II) derivatives. Bioconjugate Chem 12:71–75
67. Yang Y, Jia Y, Gao L et al (2011) Fabrication of autofluorescent protein coated mesoporous silica nanoparticles for biological application. Chem Commun 47:12167–12169
68. Monier M, Ayad DM, Wei Y et al (2010) Adsorption of Cu(II), Co(II), and Ni(II) ions by modified magnetic chitosan chelating resin. J Hazard Mater 77:962–970

Chapter 6
Photosystem II Based Multilayers

Peng Cai, Guangle Li, Jiao Li, Yi Jia, Zhongfeng Zhang and Junbai Li

Abstract During billions of years of evolution and development, photosynthesis has formed an effective mechanism for solar energy fixation and conversion. The unique property of photosystem II (PSII) to split water in ambient condition makes it the key role in the process of photosynthesis. Assembly of PSII-based multilayers toward the construction of water splitting systems has attracted more and more attention. As a means to study PSII, it might lead to quicker solutions to understand the electron transfer mechanism in such hybrid systems and how activities of PSII can be affected by different physicochemical or environmental factors. Such systems might also provide guidelines for the design and fabrication of artificial photosynthetic energy conversion systems. In this chapter, we concentrate on the design and development of PSII-based water splitting systems, in which photoelectrochemical (PEC) cells utilizing PSII will be discussed in detail.

Keywords Photosystem II · Molecular assembly · Water splitting · Energy conversion · Layer-by-layer assembly · Photoelectrochemical systems · Electron transfer

6.1 Introduction

The development of new technologies for renewable and clean fuels based on solar energy has been greatly accelerated because of global climate and environmental problems caused by continuous consumption of fossil fuels [1–3]. How to capture

P. Cai · Z. Zhang
Tobacco Research Institute of Chinese Academy of Agricultural Sciences, 266101 Qingdao, China

P. Cai · G. Li · J. Li · Y. Jia · J. Li (✉)
Beijing National Laboratory for Molecule Sciences, CAS Key Laboratory of Colloid, Interface and Chemical Thermodynamics, Institute of Chemistry, Chinese Academy of Sciences, 100190 Beijing, China
e-mail: jbli@iccas.ac.cn

J. Li (ed.), *Supramolecular Chemistry of Biomimetic Systems*,
DOI 10.1007/978-981-10-6059-5_6

and transform solar energy efficiently to energy forms that can be directly used for a sustainable future has long been the most important topic in current research [4–6]. Nature provides us the most effective way of harnessing solar energy, that is photosynthesis. Photosynthesis is a sustainable, effective, and complex process that harnesses sunlight for the conversion of CO_2 and water into high energy compounds like carbohydrates and release of O_2 as a by-product. It is possible for mankind to imitate natural photosynthesis and to convert solar energy effectively into clean energy. Among many active proteins involved in photosynthesis, photosystem II (PSII) with unique property of splitting water in ambient condition has been thoroughly investigated [7–10]. The construction of solar energy conversion system based on native proteins, especially PSII, has attracted significant interest [11, 12].

6.1.1 Structure of Photosystem II

PSII is a transmembrane multi-polypeptide complex found in thylakoid membranes of higher plants, algae, and cyanobacteria. It existed as dimeric complex with total molecular weight around 650 kilodalton. Each monomer contains about 20 polypeptide subunits and many cofactors, such as chromophores, plastoquinones, and a manganese–calcium (Mn_4Ca) cluster (Fig. 6.1) [13]. Subunits like CP43 and CP47, which are involved in the transfer of photonic energy, are located in the periphery of PSII. While D1 and D2 subunits are in the center of each monomer, which make up reaction center (RC) complex and are responsible for the transfer of energy in the form of electrons. Other subunits that constitute PSII are distributed among and around the above four subunits. The Mn_4Ca cluster located in close proximity to the P680 is well known as the oxygen evolving complex (OEC) or the water-oxidizing complex (WOC) [10]. Unlike most other enzymes showing variations of divergent and convergent evolution, the organization and operation of PSII's OEC seem to have no difference among all organisms studied thus far [14]. The OEC is comprised of a cubane-like structure with three Mn ions and one Ca ion linked by oxo-bridges and a fourth Mn ion linked to the cubane via one of the oxo groups. Although the detailed mechanism of water oxidation still remains unclear, the coordination of OEC with surrounded amino acid residues certainly plays a key role in the deprotonation of the substrate water molecules. There have been excellent works on the design of artificial manganese clusters for water oxidation [15, 16].

6.1.2 Function of Photosystem II

PSII collects photonic energy through surrounded light-harvesting proteins and transforms it into electrochemical potential within RC complex. With a primary charge separation taking place in a few picoseconds, a chlorophyll center

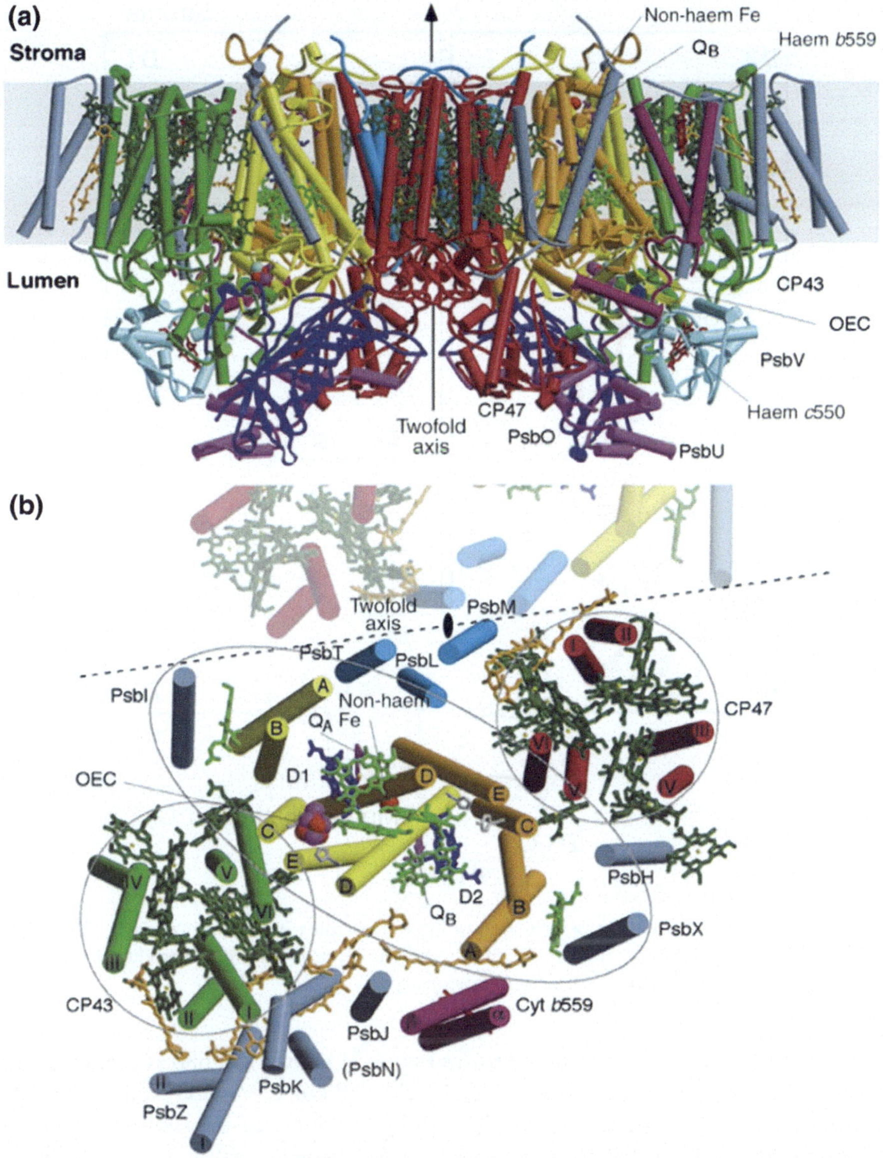

Fig. 6.1 Overall structure of PSII. **a** View of the PSII dimer perpendicular to the membrane normal. **b** View of the PSII monomer along the membrane normal from the lumenal side. Reprinted from Ref. [13], Copyright 2004, with permission from Elsevier

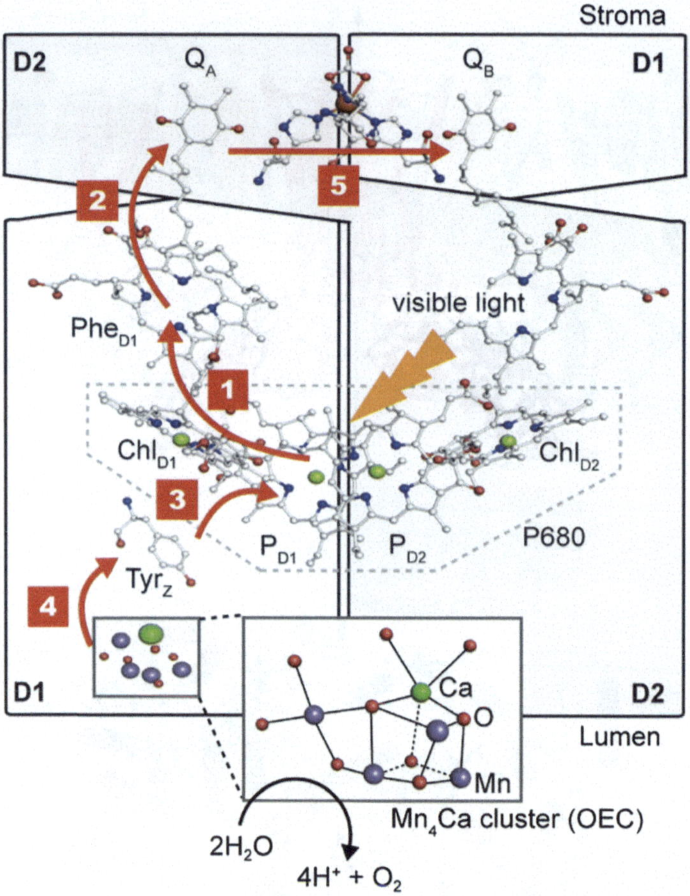

Fig. 6.2 Simplified schematic representation of the RC complex with the D1 (PsbA), D2 (PsbD) subunits and core components involved in photosynthetic electron transfer. Reproduced from Ref. [11] by permission of the Royal Society of Chemistry

(designated P_{D1}, P_{D2}, Chl_{D1}, and Chl_{D2} in Fig. 6.2) known as P680 is oxidized and a pheophytin molecule (chlorophyll molecule without a Mg ion ligated into its tetrapyrrole head group) is reduced. The redox potential of the oxidized primary electron donor in PSII ($P680^+$) is about +1.25 V versus normal hydrogen electrode (NHE), which is high enough to oxidize water in a pH neutral aqueous solution [17]. The OEC is oxidized to reduce $P680^+$, building up oxidizing equivalents to extract electrons from water. On the other side, the redox potential of the anion radical ($Pheo^-$) is about −0.6 V versus NHE [18]. As soon as the primary charge separation takes place, the electron is rapidly transferred by $Pheo^-$ to a tightly bound quinone cofactor, Q_A. Owing to the physiological stability of charge separation [$P680^+$-Q_A^-] for several hundred microseconds [19], the probability for back reaction is considerably reduced. Therefore, the quantum yield for the primary charge separation in RC complex approaches unity. The electron then transfers from Q_A^- to a second quinone molecule, Q_B, which is weakly bound to D1 subunit. After Q_B receives two electrons, it diffuses out into the thylakoid membrane in the form of H_2Q_B to continue the photosynthetic electron transfer chain at the cytochrome b_6f complex. The reducing equivalent is then passed along the electron transfer chain to photosystem I (PSI) and is excited again to give a redox potential of −1 V or more, which drives the fixation of carbon dioxide.

6.1.3 Applications of Photosystem II

At present, investigation of PSII-based systems and their potential applications mainly concentrates in the following two aspects. On one hand, due to its high sensitivity, PSII can be used in biosensing applications to detect herbicides. On the other hand, the robustness against charge recombination and fairly high quantum yield make PSII an ideal building block for the construction of PEC cells.

6.1.3.1 Photosystem II Based Biosensors

The investigation of PSII being utilized for the detection of herbicide has drawn extensive attention since 40% of herbicides could influence the process of photosynthesis by inhibiting PSII activities [20]. Triazine, diazine, phenolic, and urea herbicides could lead to severe damage to PSII. These compounds occupy the Q_B binding site of D1 protein, interacting with different amino acid residues and inhibiting photosynthetic electron transfer process [21]. The signals could be transformed into information such as current, potential, and optical signals, which is directly proportional to the concentration of herbicide being detected. Thus, the concentration of the herbicide could be precisely determined. A biosensor for the detection of triazine and phenylurea herbicides has been reported by Koblizek et al. immobilizing isolated PSII on a graphite-Ag/AgCl screen-printed electrode [22]. The obtained biohybrid electrode can detect diuron, atrazine, and simazine down to nano-molar concentrations.

Heavy metal pollution is one of the most common environmental problems. It works by inhibiting the photosynthetic enzymes, such as protochlorophyllide oxidoreductase, plastocyanin, and PSII. Some heavy metal ions (Hg^{2+}, Cu^{2+}, Cd^{2+}, and so on) could replace the Mg^{2+} ion in the chlorophyll molecule, consequently inhibit photosynthesis. Loranger and Carpentier used thylakoid membranes cross-linked with albumin for phytotoxicity measurements [23]. The immobilized membranes were much more stable and as sensitive as free thylakoids for the detection of inhibitors of photosynthetic electron transfer, including herbicides and heavy metal ions. By measuring the current changes of the electron transport chain, the concentration of Pb^{2+} and Cd^{2+} ions could be detected down to mg L^{-1}.

6.1.3.2 Photosystem II Based Water Splitting Systems

Photosynthesis, an enduring blueprint for photochemical water splitting, harnesses sunlight and transfers it efficiently into chemical energy. Inspired by nature, artificial photosynthesis utilizing synthetic water oxidation catalysts has attracted more and more attention, which carries out light-driven water splitting reaction and releases protons and electrons for the production of clean and renewable energy [6, 24, 25]. One important research approach for artificial photosynthetic system is the development of PEC cells [26]. PEC cells can perform kinds of electrochemical half-reactions. In a simple regenerative cell, the anode and the cathode carry out reversible reactions and the cell generates electricity. As the core enzyme complex in the photosynthetic chain, PSII is responsible for the efficient extraction of electrons from water, evolution of O_2, and establishment of the proton gradient that drives adenosine triphosphate (ATP) synthesis [27]. The investigation of PSII integrated electrodes for the construction of PEC cells will not only help better understand the function and mechanism of PSII and related electron transfer chain, but also provide guidelines for bioinspired artificial photosynthetic systems.

6.2 Photosystem II Based Multilayers for Photoanodes

As basic components of PEC cells, photoanodes must combine good visible light absorption with efficient charge separation and high chemical stability [28]. Unfortunately, no such material has been found to date that possesses all these properties to the desired extent. PSII can absorb light centered at 675 and 420–440 nm. In vivo, the overall quantum yield of charge separation in PSII is reported to be above 85% [29]. Besides, the self-repair machinery of PSII that replaces the photo-damaged D1 protein with a new copy under both low light condition and intense illumination makes it an excellent candidate for the construction of photoanodes [30]. However, the isolated PSII does not yet match the performance of in vivo PSII. The key problem is that PSII is not efficiently wired to the electrode and the electron generated by PSII could not be passed rapidly. At present, a lot of work has been done to the nano-engineering of PEC cells through the assembly of PSII on the electrode surfaces. Various strategies to immobilize the PSII on conductive surfaces and different methodologies to electrically wire it with the electrode are presented. Co-assembly of PSII with other molecules and/or materials for multilayers shows great promise, which will be discussed in the following text.

6.2.1 Integration Photosystem II with a Self-assembled Monolayer

Self-assembly monolayers (SAMs) are a commonly used approach for the surface functionalization of electrode materials. PSII has been integrated with various electrode materials using this approach. For example, PSII with a His-tag on the stromal side of the C-terminus of CP47 or CP43 was linked to a gold electrode modified with a SAM of a nickel-nitrilotriacetic acid (Ni-NTA), as shown in Fig. 6.3 [31]. PSII could be bound tightly to the electrode owing to the strong affinity of Ni-NTA to the His-tag. This method could effectively solve the problem that bare gold surfaces are hydrophobic in nature and do not interact favorably with proteins in direct contact. The Ni-NTA/PSII bilayer modified gold electrode acts as a photoanode and could generate photocurrents upon light irradiation. Furthermore, by the deposition of increasing quantities of gold nanoparticles on the gold surface (Fig. 6.3b), the surface roughness of the electrode was increased. Following the modification of the roughened surfaces with Ni-NTA/PSII bilayer, enhanced photocurrents were achieved (Fig. 6.3c).

Another representative work was reported by Kato et al., which integrated PSII onto SAM-modified electrodes both electrostatically and covalently, Fig. 6.4 [32]. In this study, the PSII was first adsorbed electrostatically onto a nanostructured indium tin oxide (ITO) electrode, which was modified with a SAM of phosphonic acid ITO linkers with a dangling carboxylate moiety (Fig. 6.4a). A higher direct electron transfer (DET) photocurrent density was observed with PSII immobilized onto the negatively charged surface compared to that of positively charged. This phenomenon was attributed to the positive dipole on the electron acceptor side of PSII, which makes the terminal quinones of the PSII in close contact with the electrode surface via Coulomb interactions (Fig. 6.4b). Photocurrent response of mediated electron transfer (MET) of the electrode was also investigated using a diffusional redox mediator, 2,6-dichloro-1,4-benzoquinone (DCBQ). However, no significant difference was observed in MET photocurrent between PSII modified positively and negatively charged surface, which suggests that the electron mediators can, in principle, receive electrons from almost all active PSII immobilized on the electrode independent of the orientation of PSII. Subsequently, treatment of the SAM-CO_2^-/PSII modified electrode with the peptide coupling agents 1-(3-dimethylaminopropyl)-3-ethylcarbodiimide (EDC) and N-hydroxysuccinimide (NHS) was performed to investigate the effect of covalent immobilization (Fig. 6.4c). The observed DET photocurrent of the covalently bonded PSII was almost twofold compared to that of electrostatically adsorbed PSII. Moreover, the photocurrent stability of the photoanode was also increased. This study demonstrates that the linkage between the PSII and the electrode surface is critical for electron transfer efficiencies and enzyme stability in PEC studies.

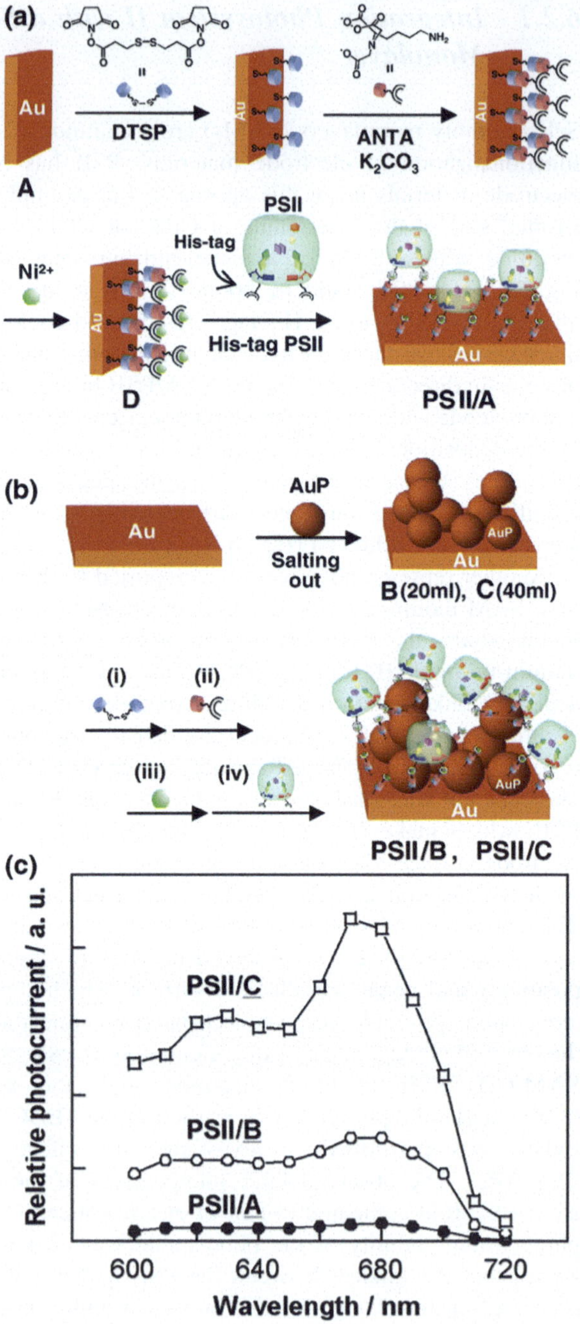

Fig. 6.3 Preparation procedures of the His-tag-PSII immobilized **a** planar (PSII/A) and **b** nanostructured gold electrodes (PSII/B, PSII/C). **c** Action spectra of His-tag-PSII immobilized planar (PSII/A) and nanostructured (PSII/B, PSII/C) gold electrodes. Irradiated light: 680 nm (3.3 mW), electrolyte solution: 0.4 M sucrose, 15 mM $CaCl_2$, 15 mM $MgCl_2$, and 0.1 M NaCl in 40 mM MES-NaOH buffer aqueous solution (pH 6.4). Applied potential: 0.2 V versus Ag/AgCl (sat. KCl). Reprinted from Ref. [31], Copyright 2007, with permission from Elsevier

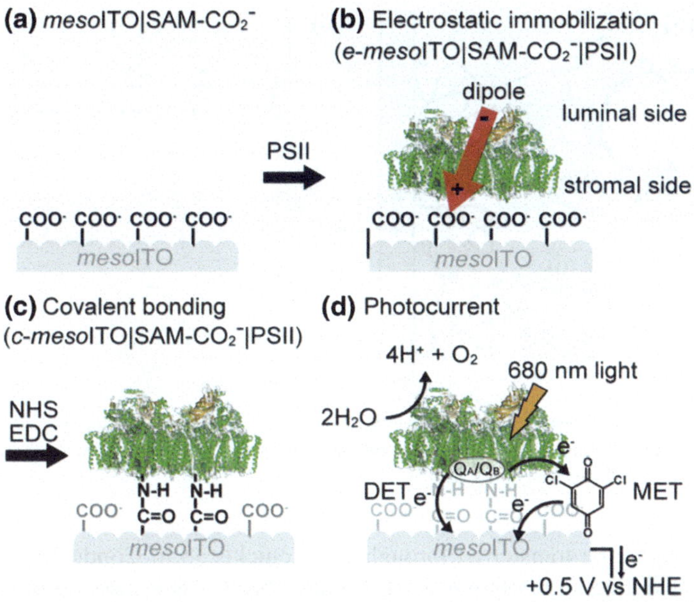

Fig. 6.4 Schematic representation for **a** the assembly of SAM-CO_2^- modified mesoporous ITO electrode. **b** Electrostatic immobilization and **c** Covalent immobilization of PSII on SAM-CO_2^- modified mesoporous ITO electrode. **d** Red-light-driven DET and MET resulting in photocurrents with SAM-CO_2^-/PSII modified electrode. Reprinted with the permission from Ref. [32]. Copyright 2013 American Chemical Society

6.2.2 Integration of Photosystem II with Polymer

Diffusional electron mediators could, in principle, deliver electrons generated from all layers of active PSII to the electrode [11]. However, MET is limited by the mass transport of diffusional redox mediators, which is a relatively slow kinetic process. The use of redox active polymers is an attractive alternative to avoid this problem. In one of the studies, PSII was entrapped in the osmium-containing poly(vinyl) imidazole-based redox polymers and immobilized onto the electrode surfaces, Fig. 6.5 [33]. The redox polymer worked as both immobilization matrix and electron acceptor for PSII. The redox potential of Os^{III}/Os^{II} is +0.39 V versus NHE, which allows for efficient electron transfer between PSII and the electrode. A photocurrent density of 45 μA cm^{-2} for redox polymer/PSII modified electrode was achieved. In the meantime, the system represents a significant improvement of the operational life time with approximately 85% of the initial photocurrent remaining after ~1 h of continuous light irradiation.

Conducting polymers with unique electrical and optical properties as well as electrochemical stability have attracted great interest since they are abundant and cost-effective [34]. Due to conjugated structure and backbone of adjoining sp^2 hybridized orbitals, the delocalized π electrons are formed along their backbone and

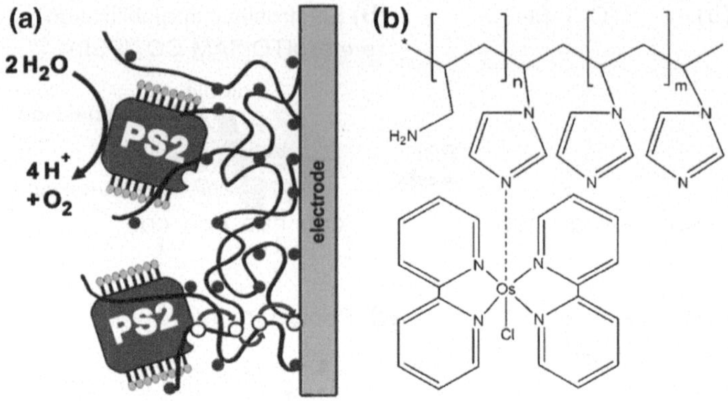

Fig. 6.5 **a** Schematic representation for PSII entrapped within a mediator-modified redox polymer. **b** Structure of poly(vinyl)imidazole Os(bipy)$_2$Cl-polymer. Reproduced from Ref. [33] by permission of John Wiley & Sons Ltd

the electrons can be transferred through them quickly [35]. Conducting polymers such as polyaniline and polypyrrole (PPy) have been widely used in biomedicine, energy storage, and chemical sensors [36–38]. The preparation of electroactive and conducting polymer films can be realized by electropolymerizing relevant monomers onto the electrode [39]. After immobilizing enzymes, the thin films could serve as highly sophisticated nanoreactors for chemical transformations and solar energy conversion [40, 41]. Li et al. developed a novel biohybrid photoanode with enhanced photocurrent and increased stability by assembling PSII-enriched membranes with three-dimensional doped PPy hierarchical nanostructures, Fig. 6.6 [42]. Compared to conventional planar electrode, surface area of the nanostructured electrode is increased to promote the effective loading of PSII (Fig. 6.6a, b). Upon light illumination, benzoquinone (BQ) doped PPy nanowires act as highways to allow electrons from water oxidation by PSII to flow to the ITO electrode. Moreover, redox molecules (BQ) conjugated on the backbone of conducting polymers serve as mediators to permit more efficient electron transfer between PSII and ITO electrode. Indeed, at the same applied bias potential ranging from 0 to 0.5 V versus saturated calomel electrode (SCE), this integrated photoanode always shows a higher photocurrent than PSII modified planar electrode. Notably, the photocurrent is enhanced about 39.0-fold with increased stability at 0.25 V versus SCE, compared to that for ITO/PSII. Such assembled system coupling natural photoactive protein with artificial conducting polymer may have great potential applications in advanced solar energy conversion.

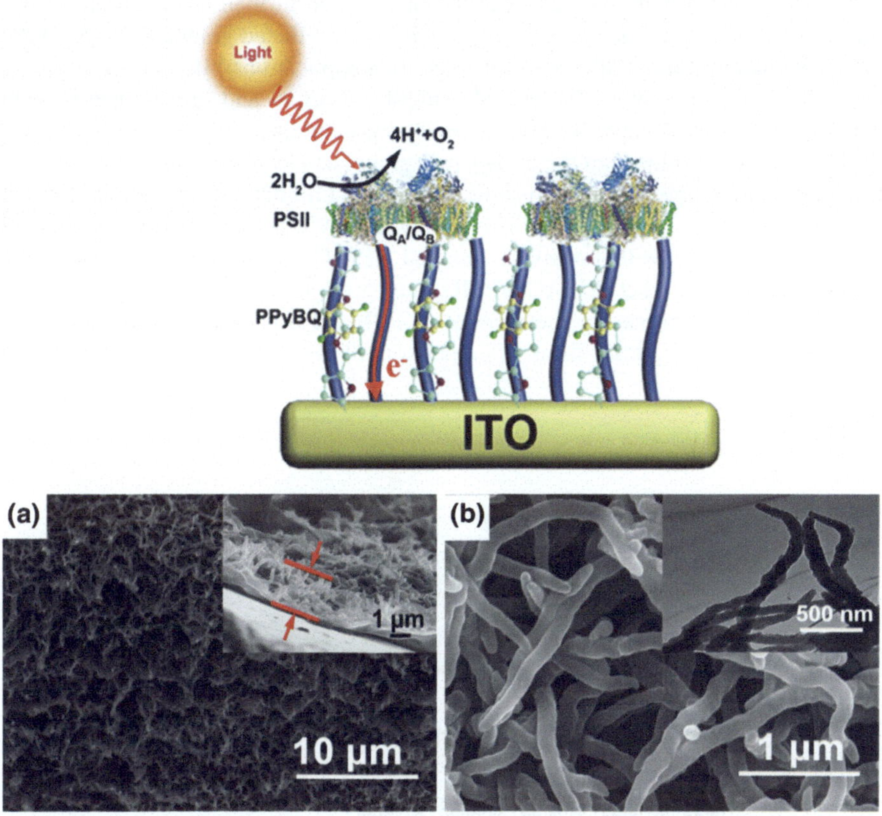

Fig. 6.6 Schematic representation of the biohybrid photoanode integrating PSII with BQ-doped PPy hierarchical nanostructures on the ITO substrate. **a** SEM image of purified PPyBQ nanostructures with a lower magnification. The inset shows their cross section. **b** SEM image of the components with a higher magnification. The inset shows relevant TEM image. Reproduced from Ref. [42] by permission of John Wiley & Sons Ltd

6.2.3 *Layer-by-Layer Assembly of Photosystem II Based Multilayers*

The layer-by-layer (LbL) assembly technique, which was introduced by Decher and colleagues, was initially based on alternating deposition of polyelectrolytes, with opposite charges on a planar substrate [43]. Up to now, the driving forces have been extended to covalent bond, hydrogen bond, base pair interaction, and host–guest interaction. This technique allows the coating of diverse species in various shapes and sizes, with uniform layers and controllable thickness [44–46]. The LbL technique is especially suitable for protein assembly because of the mild assembly condition and convenient assembly process which is beneficial to the maintenance of protein activities.

Graphene-based materials are of great interest due to their extraordinary properties [47, 48]. The solution processing capability of reduced graphene oxide (rGO) makes the material more attractive for large area applications [49]. The transparent nature of graphene enables the use of a highly effective and opaque mediator to amplify photocurrents greatly. For example, by mounting graphene on glass, we can irradiate the cell through the transparent graphene electrode rather than through the solution, enabling the use of opaque mediators at higher concentrations to boost photocurrents [50]. In one study where LbL assembly was used, negatively charged PSII was integrated with positively charged PEI-rGO, Fig. 6.7a [51]. The DET photocurrents of the ITO-(PEI-rGO/PSII)$_n$ increased gradually with the number of bilayers. Notably, increasing the amount of PSII did not lead to enhancement of photocurrent for ITO-PSII films. Comparison of the photocurrents of different

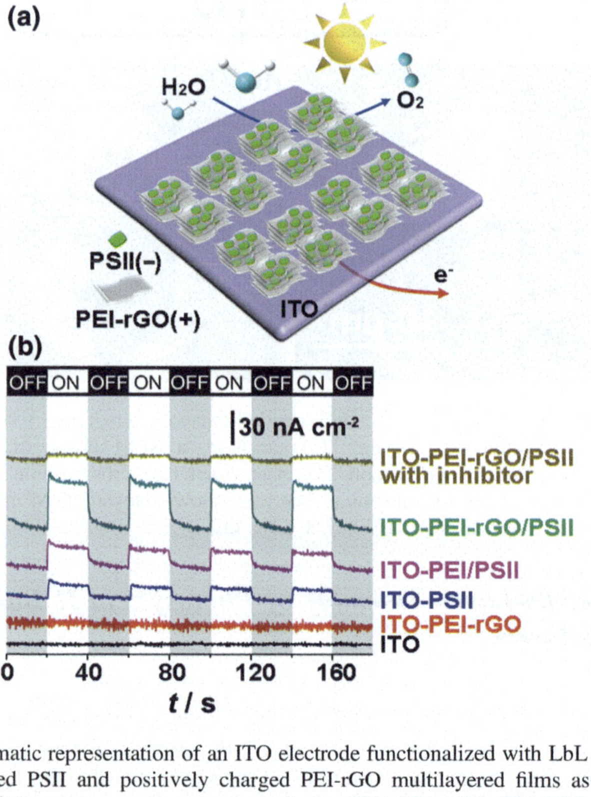

Fig. 6.7 **a** Schematic representation of an ITO electrode functionalized with LbL co-assembly of negatively charged PSII and positively charged PEI-rGO multilayered films as a photoanode. **b** Comparison of the photocurrent responses of pure ITO, ITO-rGO, ITO-PSII, ITO-PEI/PSII, ITO-PEI-rGO/PSII, and ITO-PEI-rGO/PSII with additional 30 μM 3-(3,4-dichlorophenyl)-1,1-dimethylurea (DCMU) as the inhibitor of PSII in an electrolyte buffer solution. The latter four films possess the identical amount of PSII. The above photocurrents were measured with a bias potential of 0.25 V versus SCE in an aqueous electrolyte buffer solution (pH 6.5, 20 mM 4-morpholinoethanesulfonic acid (MES), 50 mM KCl, 5 mM $MgCl_2$ and 3 mM $CaCl_2$) under light illumination (800 nm > λ > 550 nm, P680 nm = 10 mW cm^{-2}) at 25 °C. Reproduced from Ref. [51] by permission of The Royal Society of Chemistry

anodes possessing the identical amount of PSII (Fig. 6.7b) demonstrates that PEI-rGO sheets facilitate the electron transfer between PSII and the electrodes. The co-assembled photoanode with improved stability and ca. twofold enhancement of DET photocurrent confirms that rGO is a promising candidate for the integration with photoactive proteins.

Yehezkeli et al. co-assembled PSII and PSI on ITO electrodes through LbL technique, where two different noble metal-free redox active polymers, poly N,N′-dibenzyl-4,4′-bipyridinium (poly-benzyl viologen, PBV^{2+}) and poly lysine benzoquinone (PBQ), were used as potential electron acceptors and charge carriers, Fig. 6.8a [52]. The photocurrent action spectrum shows that the photocurrent of PBV^{2+}/PSI/PBQ/PSII multilayers is about sixfold enhanced as compared to that of PBV^{2+}/PSI/PBV^{2+}/PSII multilayers (Fig. 6.8b). While the reversed bilayer assembly, consisting of a PSII layer linked to a layer of PSI by PBV^{2+}, generates an inefficient cathodic photocurrent. These results suggest that PBV^{2+} and PBQ are superior charge trapping and electron transport layers for PSI and PSII, respectively. The energy diagram for the coupled electron transfer reactions in the PBV^{2+}/PSI/PBQ/PSII multilayers (Fig. 6.8c) reveals that the appropriate ordering of PSI and PSII layers on the electrode is essential to the efficiency of the resulting photocurrents. The advantage of LbL technique for composite film construction was well manifested.

6.2.4 Integration of Photosystem II with Semiconductor

Semiconductors have many advantages such as superior chemical and optical stability, tunable band structures, and high photocatalytic efficiency [53, 54]. Exploring suitable semiconductor material to construct a new hybrid system with PSII is of great importance. Wang et al. present a hybrid photoanode integrating PSII with a hematite film for water oxidation, Fig. 6.9 [55]. The hematite phase Ti/Fe_2O_3 films exhibit a three-dimensional mesoporous structure with a pore diameter of ca. 200 nm and thickness of 300 nm (Fig. 6.9a, b), which is in favor of PSII adsorption [10]. The fluorescence of the PSII-Ti/Fe_2O_3 films modified electrode suggests that the PSII distributes on the whole surface of the electrode uniformly (Fig. 6.9c). As demonstrated by the author, the electrons generated from PSII multilayers are injected into the valance band of the excited Ti/Fe_2O_3 and significantly enhanced photocurrent is observed.

Among the various semiconductor materials developed to construct PEC cells, TiO_2 is one of the most promising choices due to its low-cost, nontoxic, and so on [56]. A recent study reported a strategy to integrate PSII into a hierarchical nanotubular titania-modified electrode to assemble a hybrid photoanode system (PSII/TiO_2-NT-modified ITO electrode), in which the nanotubular titania materials

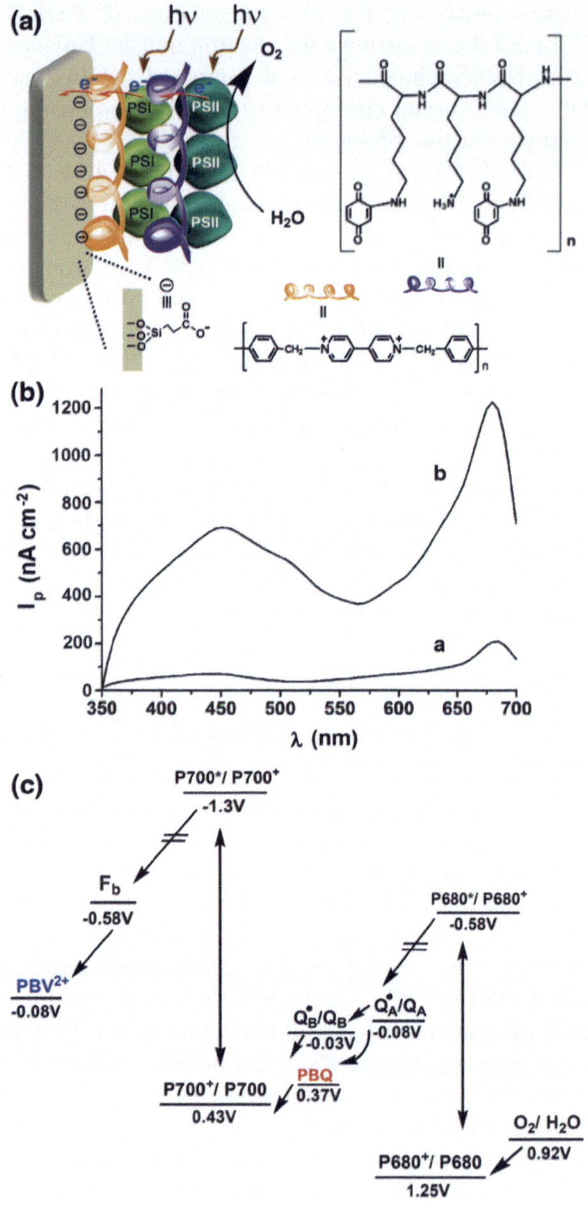

Fig. 6.8 **a** Schematic assembly of PBV^{2+}/PSI/PBQ/PSII multilayers on an ITO electrode. **b** Photocurrent action spectra corresponding to: (curve a) the PBV^{2+}/PSI/PBV^{2+}/PSII multilayer, and (curve b) the PBV^{2+}/PSI/PBQ/PSII multilayers. **c** Energy diagram for the cascaded electron transfer processes in the integrated PSI/PSII multilayers. The effective illumination area was 0.25 cm^2. All measurements were performed in an Ar-deaerated phosphate buffer (0.1 M, pH = 7.2). Reproduced from Ref. [52] by permission of John Wiley & Sons Ltd

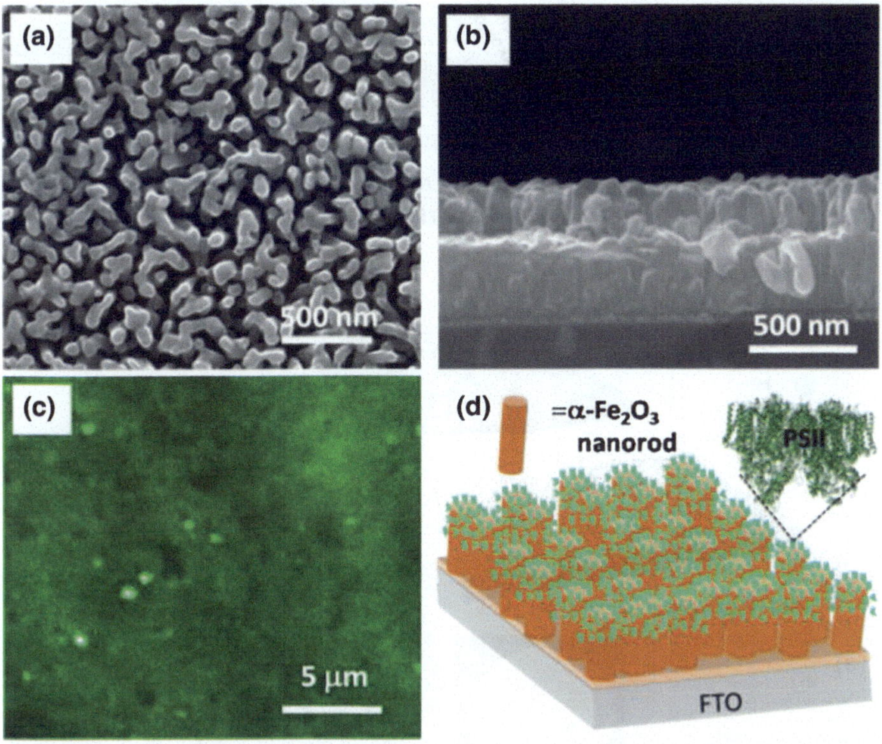

Fig. 6.9 SEM images of the Ti/Fe_2O_3 film: **a** *top view* and **b** *side view*. **c** The fluorescence intensity image of PSII-Ti/Fe_2O_3 hybrid electrode. **d** Proposed schematic configuration of the PSII-Ti/Fe_2O_3 hybrid electrode. Reproduced from Ref. [55] by permission of The Royal Society of Chemistry

(TiO_2-NTs) were achieved by a surface sol-gel method using natural cellulose substances as scaffolds, Fig. 6.10 [57]. The titania materials used to modify the surface of the bare ITO slide show hierarchical porous nanotubular structures replicated from original cellulose (Fig. 6.10a, b). The diameters of individual titania nanotubes range from 50 to 200 nm, and the titania nanoparticles forming the nanotubes are in the pure anatase crystal phase. What's more, the space between each titania nanotube is well suited for PSII loading. PEC measurements under irradiation with red light or white light demonstrate that the PSII/TiO_2-NT-modified ITO electrode reveals superior photocurrent generation ability and an enhanced incident photo-to-current efficiency compared to the PSII and TiO_2-NT-modified ITO electrodes. This semiconductor–protein hybrid photo-bioelectrochemical system enhances DET (1.3 $\mu A\ cm^{-2}$) and MET (10.6 $\mu A\ cm^{-2}$) photocurrents,

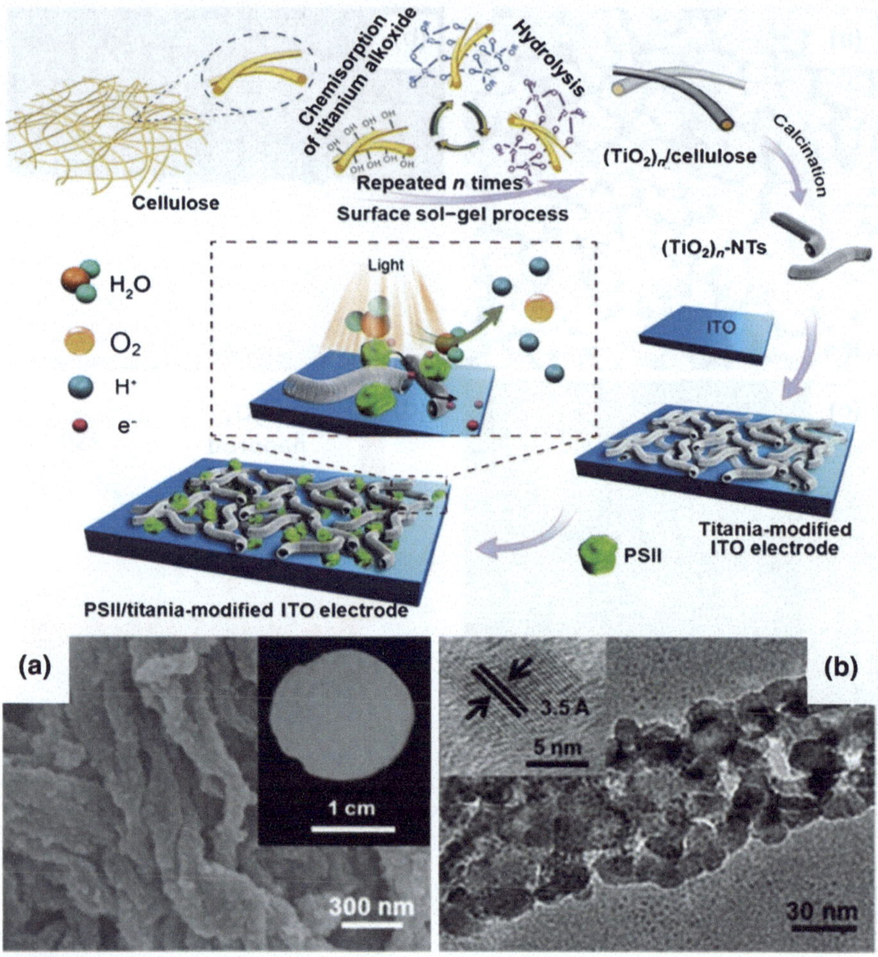

Fig. 6.10 Schematic illustration of a fabrication process of a hybrid PSII/TiO_2-NTs-modified ITO photoanode system. **a** FE-SEM image of the TiO_2-NTs; inset of **a** shows the photograph of this sample. **b** TEM image of an individual titania nanotube; inset of **b** displays the HR-TEM image. Reproduced from Ref. [57] by permission of The Royal Society of Chemistry

benefiting from the well-defined hierarchical nanostructured surface of the electrode for protein loading, and the fine titania nanocrystals facilitating electron transfer from PSII to the electrode.

6.3 Photosystem II for Complete Solar Energy Harvesting Systems

The construction of PSII-based photoanodes has achieved much progress in recent years. However, as electrochemical half-cells, most of these systems included sacrificial electron donors or acceptors for electron transfer and used a potentiostat as a sink for the electrons. Electrically coupling of PSII-based photoanode with a cathode for the construction of complete and autonomous solar energy harvesting systems have attracted more and more attention [58, 59].

Willner group reported the assembly of a photo-biofuel cell that composed of PSII/poly-mercapto benzoquinone (pMBQ) functionalized electrode, acting as a photoanode for the light-induced oxidation of H_2O to O_2, and an electrically wired bilirubin oxidase/carbon nanotubes (BOD/CNTs) modified electrode acting as a cathode for the reduction of O_2 to H_2O, Fig. 6.11a [60]. It was the first photo-bioelectrochemical device (bio-photovoltaic cell) that integrated two inter-connected electrodes. Upon irradiation of the photoanode, electrons generated from water oxidation were transferred from PSII to the pMBQ. The electrons were then injected into the cathode and used for the biocatalytic reduction of O_2. Photocurrent responses upon a cyclic 'ON-OFF' illumination with different light intensities indicate that the current in the cell originates from the photoexcitation of the PSII protein, and the photocurrent density is dependent on the light intensity (Fig. 6.11b). The full-cell performance was further investigated by conducting discharge measurements under variable external resistance, Fig. 6.11c, from which a maximum power output of ca. 18 $\mu W\ cm^{-2}$ was achieved. The main reasons for the relatively low power output are the inappropriate orientation and low surface coverage of PSII on the photoanode, and the relatively low potential difference between the two interconnected electrodes.

Kothe et al. reported the first experimental setup to serially couple PSII-based photoanode and PSI-based photocathode [61]. The resulted photovoltaic cell could operate as a closed system without any sacrificial electron donors or acceptors (Fig. 6.12). The electron transfer communication between PSII and the anode was realized with an imidazole-coordinated bispyridyl osmium complex-based redox hydrogel, which was reported early by Badura et al. [33]. And the PSI was entrapped into a pyridine-coordinated bispyridyl osmium complex-based redox active hydrogel [62]. Upon light irradiation, water molecules are split into electrons and protons. The electrons are transferred to the cathode via the outer circuit, where PSI reduces the electron acceptor (methyl viologen). The methyl viologen radical cation is re-oxidized by molecular oxygen, resulting in water as the final product. The cell could generate a steady-state photocurrent of ca. 1 $\mu A\ cm^{-2}$ with maximal power output of ca. 20 $nW\ cm^{-2}$. By mimicking Z-scheme of natural photosyn-thesis, this work enables the extension of the principle to simultaneous electrical and chemical energy generation.

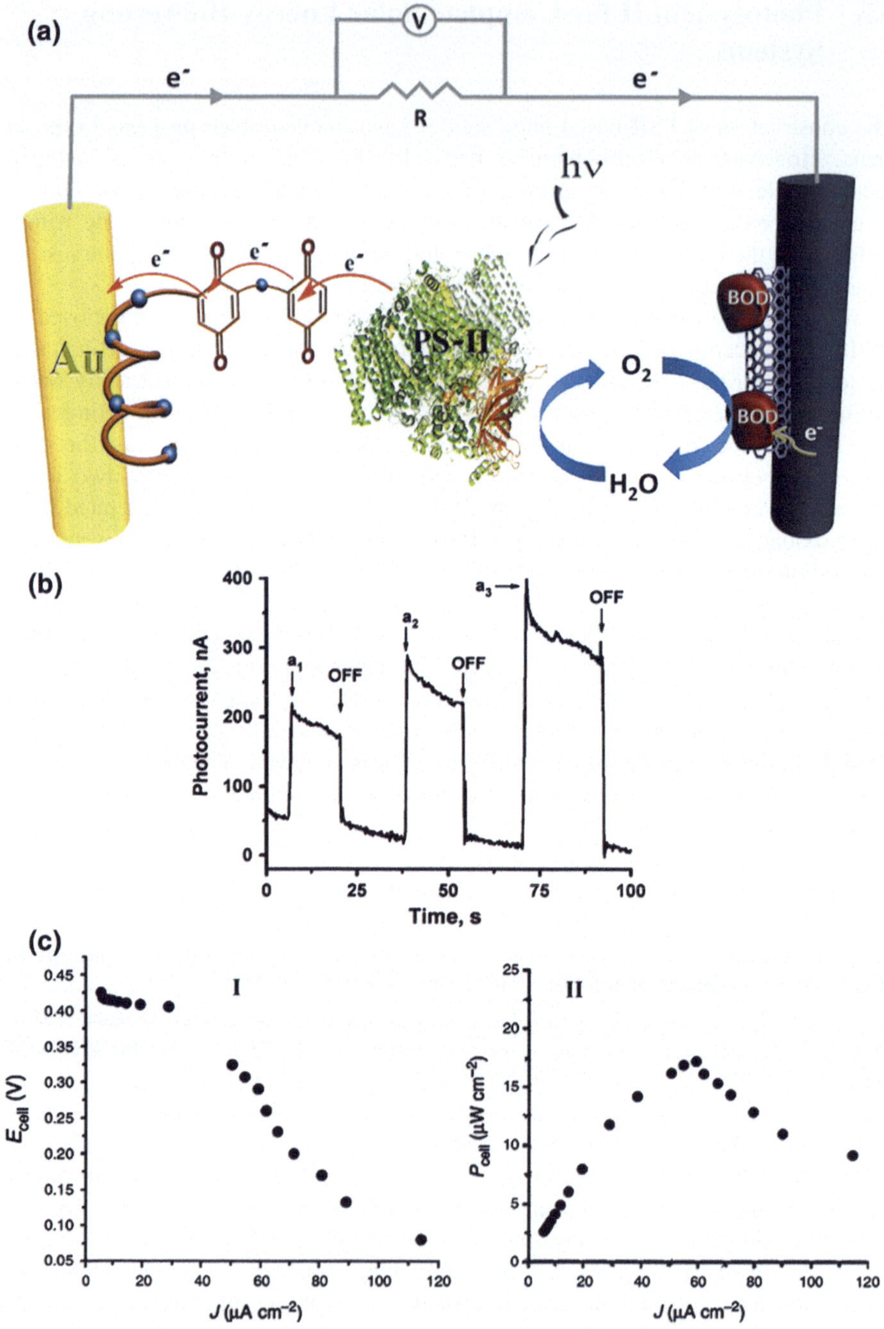
(a)
V
e⁻
R
e⁻
hν
e⁻
e⁻
e⁻
Au
PS-II
O2
H2O
BOD
BOD
e⁻
(b)
Photocurrent, nA
Time, s
a1
OFF
a2
OFF
a3
OFF
(c)
I
Ecell (V)
J (µA cm−2)
II
Pcell (µW cm−2)
J (µA cm−2)

◂**Fig. 6.11 a** Schematic presentation of the pMBQ/PSII//BOD/CNTs photo-biofuel cell. **b** Photocurrent responses of the pMBQ/PSII anode upon the application of intermittent light illumination at variable irradiation intensities: a_1 P = 0.04, a_2 P = 0.07 and a_3 P = 0.10 W. **c** Polarization curve (panel I), and power output (panel II) corresponding to the discharge of the pMBQ/PSII//BOD/CNTs photo-biofuel cell at variable resistances. In all measurements, the electrolyte was a phosphate buffer (pH 7.4, 0.1 M). Reprinted by permission from Macmillan Publishers Ltd: ref. [60], copyright 2012

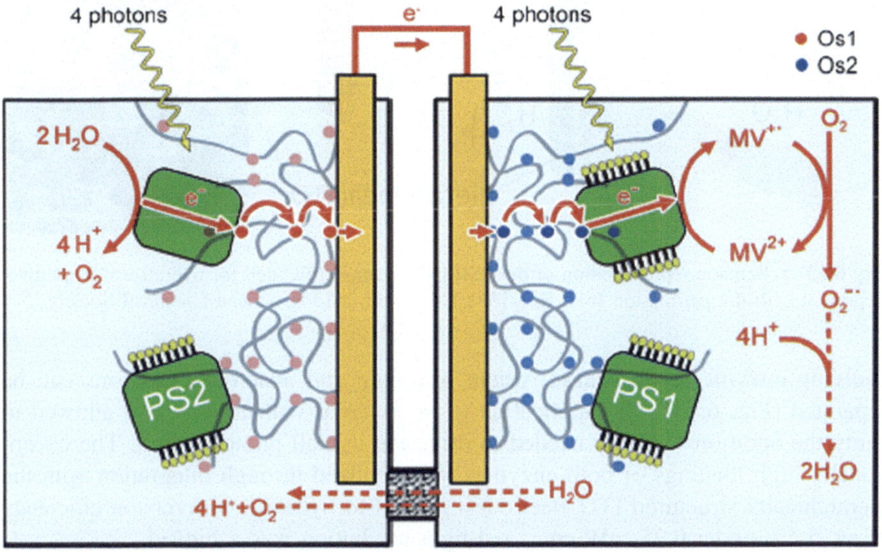

Fig. 6.12 Representation of the proposed biophotovoltaic cell combining a PSII-based photoanode and a PSI-based photocathode. Reproduced from Ref. [61] by permission of John Wiley & Sons Ltd

6.4 Other Photosystem II Based Energy Conversion Systems

It is important to find ways to reduce the cost of water splitting catalysts, improve the conversion efficiency of biomass, and construct the next generation of photovoltaics with better performance for making solar derived fuels competitive with fossil fuels. As pointed out by many scientists, it is equally important to figure it out how to store these solar derived fuels [63, 64]. One important way is to store the energy through chemical bond, like the photosynthesis does. Solar-powered catalytic water splitting process can be exploited for the release of protons and electrons that can be used to produce hydrogen, which possesses remarkably high energy density [65, 66]. PSII-based overall water splitting systems that produce hydrogen at the cathode and oxygen at the anode have drawn enormous attention [67]. Reisner group [68] constructed a PEC cell integrated isolated PSII with the H_2

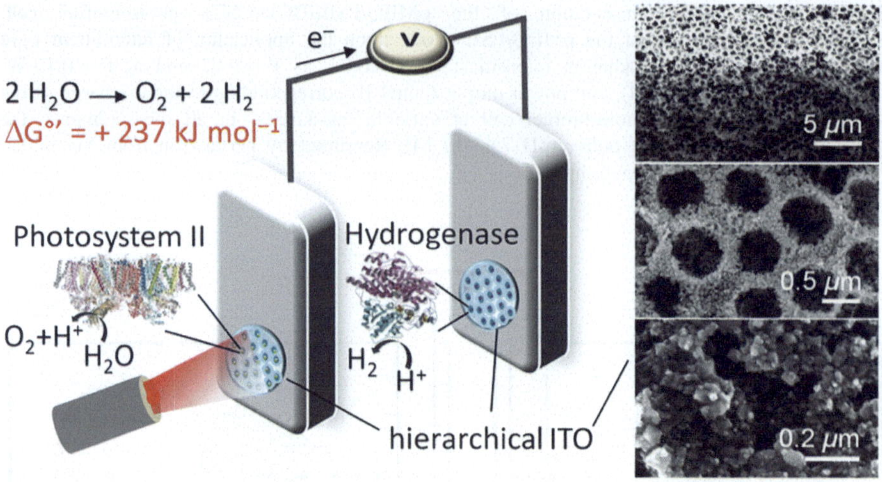

Fig. 6.13 Schematic representation of the PSII-hydrogenase PEC cell for overall water splitting. Reprinted with the permission from Ref. [68]. Copyright 2015 American Chemical Society

evolving enzyme, hydrogenase, where oxygenic and anaerobic reactions can be separated (Fig. 6.13). The application of an external voltage was also allowed to study the additional energy needed to drive the overall photoreaction. The exceptionally high loadings of both enzymes were realized through integration with the hierarchically structured ITO electrodes. A light-to-hydrogen conversion efficiency of ca. 5.4% under 0.25 mW cm^{-2} red light irradiation was achieved.

Besides the construction of PEC systems, the development of PSII-based natural–artificial hybrid photocatalytic systems has also been proposed as an alternative approach to realize overall water splitting. Well water dispersed PSII-inorganic photocatalyst integrations were constructed by Wang et al., Fig. 6.14 [69]. A self-assembly approach driven by hydrophilic interactions was applied for the coupling of the PSII membranes with photocatalyst ($Ru/SrTiO_3:Rh$). An inorganic donor/acceptor pair $Fe(CN)_6^{3-/4-}$ (E^o = 0.358 V vs. NHE) was used as the electron shuttle to transport the electrons from PSII to artificial photocatalysts, because its electrochemical potential lies between that of Q_B and the level of Rh^{3+} doped in $SrTiO_3$:Rh. The photoactive colloidal suspension could carry out overall water splitting reaction and produced H_2 and O_2 without any sacrificial reagent under visible light irradiation. Moreover, no apparent loss of activity was found for at least over 3 h in direct sunlight.

A recent promising application of the isolated PSII component is its integration with ATP synthase (F_oF_1-ATPase) as artificial chloroplast for light-driven ATP synthesis, Fig. 6.15 [70]. The hydrogel-like PSII-based microspheres were prepared through a coprecipitation method, followed by cross-linking with glutaric dialdehyde and subsequent core removing. Isolated F_oF_1-ATPase are reconstituted into liposomes to get proteoliposomes and further coated on the surface of

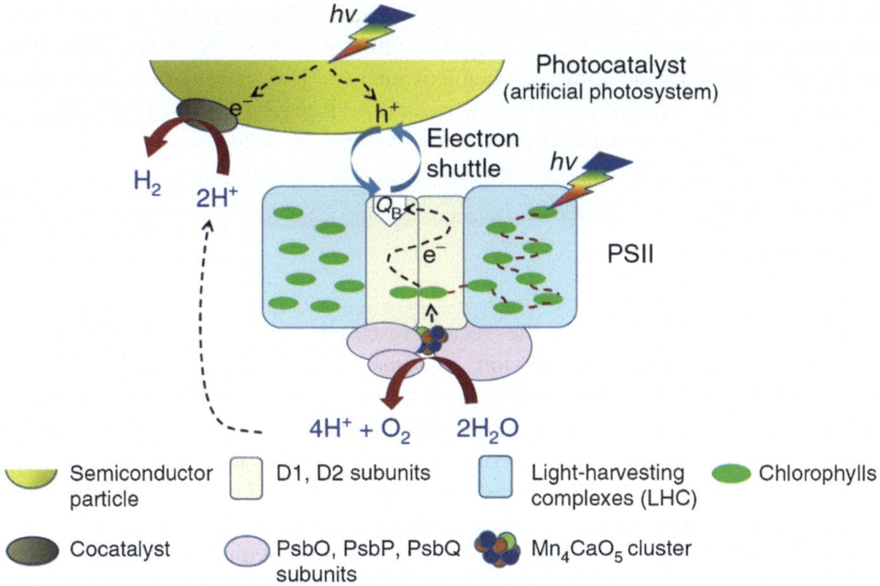

Fig. 6.14 Schematic illustration of the charge separation, electron flow, and water splitting reaction in a PSII-photocatalyst hybrid system for overall water splitting. Reprinted with permission from Macmillan Publishers Ltd: Ref. [69], copyright 2014

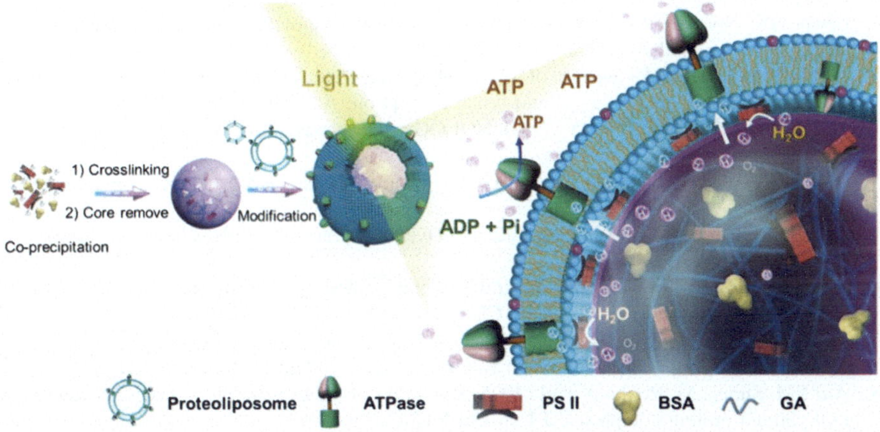

Fig. 6.15 Schematic representation of ATP synthesis catalyzed by co-assembly of CF_oF_1-ATPase and PSII. Reprinted with permission from Ref. [70]. Copyright 2016 American Chemical Society

above-mentioned microspheres. The co-assembled chloroplast-like system could split water and produce ATP under light illumination. Thus, light energy is converted and stored in the form of chemical energy. The system may have great implications in the fabrication of light-triggered ATP-driven micro-/nanodevices or miniature energy harvesting and conversion systems.

6.5 Summary

In this chapter, we reviewed the recent research progress on the design of PSII-based solar energy conversion systems from several research groups. A fundamental topic of high scientific interest is the assembly of PSII-based multilayers for the above purpose as it offers unique advantages over any parallel approach in terms of simplicity, operability, and variety of co-assembly materials. Although the PSII-based solar energy conversion systems are not mature enough for practical applications, continuous advancements in relevant areas could potentially make it an alternative to present day photovoltaics or similar technologies in the future. Such systems might also provide guidelines for the design and fabrication of artificial photosynthetic energy conversion systems.

References

1. Lewis NS, Nocera DG (2006) Powering the planet: Chemical challenges in solar energy utilization. Proc Natl Acad Sci USA 103:15729–15735
2. Li Y (2012) Molecular design of photovoltaic materials for polymer solar cells: toward suitable electronic energy levels and broad absorption. Acc Chem Res 45:723–733
3. Service RF (2014) ENERGY TECHNOLOGY Perovskite solar cells keep on surging. Science 344:458
4. Barber J (2009) Photosynthetic energy conversion: natural and artificial. Chem Soc Rev 38:185–196
5. Hagfeldt A, Boschloo G, Sun L et al (2010) Dye-sensitized solar cells. Chem Rev 110:6595–6663
6. Joya KS, Joya YF, Ocakoglu K et al (2013) Water-splitting catalysis and solar fuel devices: artificial leaves on the move. Angew Chem Int Ed Engl 52:10426–10437
7. Hankamer B, Barber J, Boekema EJ (1997) Structure and membrane organization of photosystem II in green plants. Annu Rev Plant Physiol Plant Mol Biol 48:641–671
8. Aro EM, Suorsa M, Rokka A et al (2005) Dynamics of photosystem II: a proteomic approach to thylakoid protein complexes. J Exp Bot 56:347–356
9. Meyer TJ, Huynh MH, Thorp HH (2007) The possible role of proton-coupled electron transfer (PCET) in water oxidation by photosystem II. Angew Chem Int Ed Engl 46:5284–5304
10. Umena Y, Kawakami K, Shen J-R et al (2011) Crystal structure of oxygen-evolving photosystem II at a resolution of 1.9[thinsp]A. Nature 473:55–60
11. Kato M, Zhang JZ, Paul N et al (2014) Protein film photoelectrochemistry of the water oxidation enzyme photosystem II. Chem Soc Rev 43:6485–6497

12. Yehezkeli O, Tel-Vered R, Michaeli D et al (2014) Photosynthetic reaction center-functionalized electrodes for photo-bioelectrochemical cells. Photosynth Res 120:71–85
13. Iwata S, Barber J (2004) Structure of photosystem II and molecular architecture of the oxygen-evolving centre. Curr Opin Struct Biol 14:447–453
14. Blankenship RE (2010) Early evolution of photosynthesis. Plant Physiol 154:434–438
15. Ruttinger W, Dismukes GC (1997) Synthetic water-oxidation catalysts for artificial photosynthetic water oxidation. Chem Rev 97:1–24
16. Zhang CX, Chen CH, Dong HX et al (2015) A synthetic Mn_4Ca cluster mimicking the oxygen-evolving center of photosynthesis. Science 348:690–693
17. McEvoy JP, Brudvig GW (2006) Water-splitting chemistry of photosystem II. Chem Rev 106:4455–4483
18. Rappaport F, Guergova-Kuras M, Nixon PJ et al (2002) Kinetics and pathways of charge recombination in photosystem II. Biochemistry 41:8518–8527
19. de Wijn R, van Gorkom HJ (2002) The rate of charge recombination in Photosystem II. Biochim Biophys Acta Bioenerg 1553:302–308
20. Varsamis DG, Touloupakis E, Morlacchi P et al (2008) Development of a photosystem II-based optical microfluidic sensor for herbicide detection. Talanta 77:42–47
21. Pfister K, Steinback KE, Gardner G et al (1981) Photoaffinity labeling of an herbicide receptor protein in chloroplast membranes. Proc Natl Acad Sci USA 78:981–985
22. Koblizek M, Maly J, Masojidek J et al (2002) A biosensor for the detection of triazine and phenylurea herbicides designed using photosystem II coupled to a screen-printed electrode. Biotechnol Bioeng 78:110–116
23. Loranger C, Carpentier R (1994) A fast bioassay for phytotoxicity measurements using immobilized photosynthetic membranes. Biotechnol Bioeng 44:178–183
24. Osterloh FE (2008) Inorganic materials as catalysts for photochemical splitting of water. Chem Mater 20:35–54
25. Magnuson A, Anderlund M, Johansson O et al (2009) Biomimetic and microbial approaches to solar fuel generation. Acc Chem Res 42:1899–1909
26. Swierk JR, Mallouk TE (2013) Design and development of photoanodes for water-splitting dye-sensitized photoelectrochemical cells. Chem Soc Rev 42:2357–2387
27. Nelson N, Yocum CF (2006) Structure and function of photosystems I and II. Annu Rev Plant Biol 57:521–565
28. van de Krol R, Liang Y, Schoonman J (2008) Solar hydrogen production with nanostructured metal oxides. J Mater Chem 18:2311–2320
29. Cuni A, Xiong L, Sayre R et al (2004) Modification of the pheophytin midpoint potential in photosystem II: modulation of the quantum yield of charge separation and of charge recombination pathways. Phys Chem Chem Phys 6:4825–4831
30. Tyystjärvi T, Aro E-M, Jansson C et al (1994) Changes of amino acid sequence in PEST-like area and QEEET motif affect degradation rate of D1 polypeptide in photosystem II. Plant Mol Biol 25:517–526
31. Terasaki N, Iwai M, Yamamoto N et al (2008) Photocurrent generation properties of Histag-photosystem II immobilized on nanostructured gold electrode. Thin Solid Films 516:2553–2557
32. Kato M, Cardona T, Rutherford AW et al (2013) Covalent immobilization of oriented photosystem II on a nanostructured electrode for solar water oxidation. J Am Chem Soc 135:10610–10613
33. Badura A, Guschin D, Esper B et al (2008) Photo-induced electron transfer between photosystem 2 via cross-linked redox hydrogels. Electroanalysis 20:1043–1047
34. Li D, Huang J, Kaner RB (2009) Polyaniline nanofibers: a unique polymer nanostructure for versatile applications. Acc Chem Res 42:135–145
35. Abdelhamid ME, O'Mullane AP, Snook GA (2015) Storing energy in plastics: a review on conducting polymers & their role in electrochemical energy storage. RSC Adv 5:11611–11626

36. Janata J, Josowicz M (2003) Conducting polymers in electronic chemical sensors. Nat Mater 2:19–24
37. Fei J, Cui Y, Yan X et al (2009) Controlled fabrication of polyaniline spherical and cubic shells with hierarchical nanostructures. ACS Nano 3:3714–3718
38. Snook GA, Kao P, Best AS (2011) Conducting-polymer-based supercapacitor devices and electrodes. J Power Sources 196:1–12
39. Gu C, Zhang Z, Sun S et al (2012) In situ electrochemical deposition and doping of C-60 films applied to high-performance inverted organic photovoltaics. Adv Mater 24:5727–5731
40. Ariga K, Ji Q, Mori T et al (2013) Enzyme nanoarchitectonics: organization and device application. Chem Soc Rev 42:6322–6345
41. Gizzie EA, Niezgoda JS, Robinson MT et al (2015) Photosystem I-polyaniline/TiO_2 solid-state solar cells: simple devices for biohybrid solar energy conversion. Energ Environ Sci 8:3572–3576
42. Li G, Feng X, Fei J et al (2016) Interfacial Assembly of photosystem II with conducting polymer films toward enhanced photo-bioelectrochemical cells. Adv Mater Interfaces 1600619
43. Decher G (1997) Fuzzy nanoassemblies: toward layered polymeric multicomposites. Science 277:1232–1237
44. Caruso F (2001) Nanoengineering of particle surfaces. Adv Mater 13:11–22
45. Gao C, Yan D (2004) Hyperbranched polymers: from synthesis to applications. Prog Polym Sci 29:183–275
46. Jia Y, Cui Y, Fei J et al (2012) Construction and evaluation of hemoglobin-based capsules as blood substitutes. Adv Funct Mater 22:1446–1453
47. Geim AK, Novoselov KS (2007) The rise of graphene. Nat Mater 6:183–191
48. Nair RR, Blake P, Grigorenko AN et al (2008) Fine structure constant defines visual transparency of graphene. Science 320:1308
49. Li C, Shi G (2012) Three-dimensional graphene architectures. Nanoscale 4:5549–5563
50. Darby E, LeBlanc G, Gizzie EA et al (2014) Photoactive films of photosystem I on transparent reduced graphene oxide electrodes. Langmuir 30:8990–8994
51. Cai P, Feng X, Fei J et al (2015) Co-assembly of photosystem II/reduced graphene oxide multilayered biohybrid films for enhanced photocurrent. Nanoscale 7:10908–10911
52. Yehezkeli O, Tel-Vered R, Michaeli D et al (2013) Photosystem I (PSI)/Photosystem II (PSII)-based photo-bioelectrochemical cells revealing directional generation of photocurrents. Small 9:2970–2978
53. Santato C, Ulmann M, Augustynski J (2001) Photoelectrochemical properties of nanostructured tungsten trioxide films. J Phys Chem B 105:936–940
54. Kudo A, Miseki Y (2009) Heterogeneous photocatalyst materials for water splitting. Chem Soc Rev 38:253–278
55. Wang W, Wang Z, Zhu Q et al (2015) Direct electron transfer from photosystem II to hematite in a hybrid photoelectrochemical cell. Chem Commun 51:16952–16955
56. Burda C, Lou YB, Chen XB et al (2003) Enhanced nitrogen doping in TiO_2 nanoparticles. Nano Lett 3:1049–1051
57. Li J, Feng X, Fei J et al (2016) Integrating photosystem II into a porous TiO_2 nanotube network toward highly efficient photo-bioelectrochemical cells. J Mater Chem A 4:12197–12204
58. Gerster D, Reichert J, Bi H et al (2012) Photocurrent of a single photosynthetic protein. Nat Nanotechnol 7:673–676
59. Mershin A, Matsumoto K, Kaiser L et al (2012) Self-assembled photosystem-I biophotovoltaics on nanostructured TiO_2 and ZnO. Sci Rep 2:234
60. Yehezkeli O, Tel-Vered R, Wasserman J et al (2012) Integrated photosystem II-based photo-bioelectrochemical cells. Nat Commun 3:742
61. Kothe T, Plumere N, Badura A et al (2013) Combination of a photosystem 1-based photocathode and a photosystem 2-based photoanode to a Z-scheme mimic for biophotovoltaic applications. Angew Chem Int Ed Engl 52:14233–14236

62. Guschin DA, Sultanov YM, Sharif-Zade NF et al (2006) Redox polymer-based reagentless horseradish peroxidase biosensors Influence of the molecular structure of the polymer. Electrochim Acta 51:5137–5142
63. Chu S, Majumdar A (2012) Opportunities and challenges for a sustainable energy future. Nature 488:294–303
64. Zhang Q, Uchaker E, Candelaria SL et al (2013) Nanomaterials for energy conversion and storage. Chem Soc Rev 42:3127–3171
65. Protti S, Albini A, Serpone N (2014) Photocatalytic generation of solar fuels from the reduction of H2O and CO2: a look at the patent literature. Phys Chem Chem Phys 16:19790–19827
66. Ahmad H, Kamarudin SK, Minggu LJ et al (2015) Hydrogen from photo-catalytic water splitting process: a review. Renew Sustain Energy Rev 43:599–610
67. Wang W, Wang H, Zhu Q et al (2016) Spatially separated photosystem II and a silicon photoelectrochemical cell for overall water splitting: a natural-artificial photosynthetic hybrid. Angew Chem Int Ed Engl 55:9229–9233
68. Mersch D, Lee CY, Zhang JZ et al (2015) Wiring of photosystem II to hydrogenase for photoelectrochemical water splitting. J Am Chem Soc 137:8541–8549
69. Wang W, Chen J, Li C et al (2014) Achieving solar overall water splitting with hybrid photosystems of photosystem II and artificial photocatalysts. Nat Commun 5:4647
70. Feng X, Jia Y, Cai P et al (2016) Coassembly of photosystem II and ATPase as artificial chloroplast for light-driven ATP synthesis. ACS Nano 10:556–561

Chapter 7
Peptide-Based Supramolecular Chemistry

Qianli Zou, Kai Liu, Manzar Abbas and Xuehai Yan

Abstract Supramolecular chemistry of highly important biomolecules and bioinspired molecules has attracted tremendous interest due to its acknowledged importance in construction of novel functional materials and in revealing the mechanisms of formation and evolution of natural living organisms. As one kind of representative biomolecules, peptides are among the most appealing programmable building blocks for supramolecular self-assembly. In this chapter, we present recent progresses in supramolecular chemistry of self-assembling aromatic dipeptides, including self-assembly of aromatic dipeptides and co-assembly of aromatic dipeptides with various functional molecular motifs, such as porphyrins, azobenzenes, photosensitizers, polyoxometalates, quantum dots, and glutaraldehyde. Particularly, hierarchical self-assembly of peptides and structural transition of the self-assembled peptide architectures are in-depth discussed in controllable fabrication of peptide materials along with revealing the non-covalent interactions that determine the self-assembly and the structure–property relationships of the formed peptide materials. Also, the applications of peptide-based supramolecular materials as optical waveguiding materials, biomimetic energy materials, and biomaterials are highlighted, providing an increased understanding of the role of peptide-based supramolecular chemistry in construction of novel functional materials.

Keywords Peptide · Supramolecular chemistry · Hierarchical self-assembly · Aromatic dipeptide · Optical waveguiding peptide crystal · Biomimetic photocatalysis

---

Q. Zou · K. Liu · M. Abbas · X. Yan (✉)
State Key Laboratory of Biochemical Engineering, Institute of Process Engineering, Chinese Academy of Sciences, 100190 Beijing, China
e-mail: yanxh@ipe.ac.cn
URL: http://www.yan-assembly.org/

Q. Zou · X. Yan
Center for Mesoscience, Institute of Process Engineering, Chinese Academy of Sciences, 100190 Beijing, China

K. Liu · M. Abbas
University of Chinese Academy of Sciences, 100049 Beijing, China

J. Li (ed.), *Supramolecular Chemistry of Biomimetic Systems*,
DOI 10.1007/978-981-10-6059-5_7

7.1 Introduction

Supramolecular self-assembly is a ubiquitous phenomenon in naturally evolving living systems [1–3]. Numerous functional structures and materials in living organisms are formed by self-assembly of biomolecules such as peptides, proteins, lipids, DNAs, and saccharides [4–6], through non-covalent intramolecular and intermolecular interactions [7]. Common non-covalent interactions including hydrogen bonding, π–π stacking, hydrophobic effect, electrostatic interactions, and van der Waals forces [8]. Inspired by naturally occurring self-assembly, supramolecular self-assembly of biomolecules based on non-covalent interactions and biomimetic principles is emerging as an efficient and elegant method to construct functional materials [9–12]. Such materials not only show ordered structures ranging from molecular scale to nanoscale, even to microscale, but also possess fantastic physical, chemical, and biological properties, enabling them as competitive candidates for various applications such as biomaterials, electronic materials, renewable energy, and biomedicine [13–17]. In the bottom-up process of self-assembly, the organization of building blocks can be readily tuned by manipulation of the non-covalent interactions through physical or chemical means [18]. Also, co-assembly of distinctly different building blocks provides an easy but highly efficient way to create nanomaterials with more sophisticated structures and properties [19–23]. Moreover, the self-assembled materials can be endowed with a variety of morphological transitions by incorporation of responsive groups into the building blocks, resulting in control of the structures and properties of the materials on demand in applications [24, 25].

Among natural and nature-derived biomolecules, peptides are appealing building blocks for supramolecular chemistry due to their simple structures, flexibility in sequence manipulation, relatively high stability along with good biodegradability, high biocompatibility, and ease of production on a large scale [26, 27]. The 20 natural amino acids offer enormous sequence space for exploring proper building blocks toward specific self-assembling properties based on their diversity in charge, hydrophobicity, and polarity [28]. Unnatural amino acids have also been incorporated into the peptide sequences to increase the diversity of self-assembling peptides and the complexity of self-assembled materials [29]. Other than the primary structures defined by the sequence of amino acids, peptides as self-assembling building blocks are also characterized by their secondary structures. Secondary structures of peptides can serve as building blocks for higher levels of self-assembly [30, 31]. Hence, peptides are particularly suitable for hierarchical assembly, the most powerful yet challenging tool to nanoengineer functional materials. In the formation of secondary and higher levels of peptide structures, hydrogen bonding, mainly from polar amide, amino, and carboxyl groups in the backbone and side chains of peptides, is recognized as an important and in many situations dominating driving force due to its universality, selectivity, and directionality.

With the advances of synthetic methods for peptides, especially solid-phase peptide synthesis, a lot of peptides have been designed, synthesized, and applied toward supramolecular self-assembly [32, 33]. Successful peptide self-assembly has been demonstrated by using cyclic peptides, dendritic peptides, polypeptides, amphiphilic peptides, and aromatic dipeptides. In many cases, the sequence of peptides is designed by bioinspired approaches and the structures and properties of the self-assembled materials are highly reminiscent of natural biological systems. Due to their biological inspiration, peptide-based supramolecular materials have shown advantages in various biomedical applications, such as drug delivery, tissue engineering, bone regeneration, antibacterium, imaging, and immunotherapy [34–37]. Since self-assembled peptide-based materials are highly ordered at nanoscale, their properties and functions are not only related with the sequence of amino acids but also are related with their supramolecular nanostructures. Thus, the controllable construction and manipulation of peptide nanostructures are the primary goal in the field of peptide-based supramolecular chemistry. By varying self-assembling peptides and the corresponding non-covalent interactions, supramolecular nanostructures, including nanospheres, nanotubes, nanobelts, nanofibers, and nanogels have been obtained in peptide self-assembly [38, 39]. The non-covalent interactions, thermodynamics, and kinetics in formation of the nanostructures and structure–property relationships of the formed nanomaterials are also acquired [40–42]. In return, these results provide information for revealing the non-covalent interactions of natural peptide and protein based systems.

Aromatic dipeptides, mainly derived from L-Phe-L-Phe (FF), are the simplest peptides that exhibit self-assembling characteristics. Self-assembly of FF is inspired by the formation of amyloid plaques by FF-containing polypeptides in the progress of Alzheimer's disease [43]. In recent years, aromatic dipeptides have been demonstrated as versatile building blocks for supramolecular construction of ordered nanostructures [44, 45]. As compared to other self-assembling peptides, aromatic dipeptides have advantages of simpler molecular structures, lower costs of synthesis, and ease of modification and modulation, which are highly valuable for both reveal of inherent self-assembling mechanisms and manipulation of properties toward specific applications.

In this chapter, we focus on supramolecular chemistry of aromatic dipeptides and their analogs. We first present controlled self-assembly of aromatic dipeptides and the modulation of the self-assembled nanostructures with emphasis on self-assembled peptide crystals with optical waveguiding properties. Then, we discuss co-assembly of aromatic dipeptides with other functional molecules and materials, including porphyrins, azobenzenes, photosensitizers, polyoxometalates (POMs), quantum dots (QDs), and glutaraldehyde (GA). Finally, we highlight the applications of peptide-based nanomaterials in the areas of biomimetic photosystem, photocatalysis, and biomedicine.

7.2 Self-assembly of Peptides

Many ordered nanostructures and macrostructures, such as nanotubes, nanowires, nanobelts, macroporous honeycomb scaffolds, crystals, and mesocrystals can be formed through controlled self-assembly of FF and its analogs. Intriguingly, the self-assembled structures show reversible transitions under stimuli, such as dilution and addition of trace solvent. In addition, peptide crystalline structures show attractive optical waveguiding properties.

7.2.1 *Controlled Self-assembly of Peptides*

Self-assembly of FF has attracted a lot of attention in nanotechnology since the formation of nanotubes from FF was observed [43]. Further studies discover that other low dimensional nanostructures, such as nanofibrils and nonowires, can be readily obtained by self-assembly of FF in different conditions [46, 47]. To achieve three-dimensional (3D) self-assembled structures of FF, a simple approach toward hierarchical self-assembly of FF was proposed [48]. In this approach, hierarchical spheres with a diameter of 20 μm were obtained simply by adding tetrahydrofuran (THF) to the solution of FF in 1,1,1,3,3,3-hexafluoro-2-propanol (HFIP). Scanning electron microscopy (SEM) reveals that the peony flower-like spheres consist of hundreds of flakes, which are characterized as mesocrystals. The formation of flakes or flower-like spheres can be induced by other solvents, such as 1,4-Dioxane and pyridine. Detailed analyses show that the structures self-assembled from FF by the solvent exchange method are related with the polarity of solvents: nanotubes are formed in hydrogen bond accepter solvents such as water; flakes are formed in hydrogen bond donor solvents; nanofibrils and gels are formed in solvents that provide no sites for hydrogen bonding, such as toluene. Such solvent-related morphology changes are consistent with the fact that the self-assembly of FF is mainly directed by hydrogen bonding and π–π stacking. It is found that the amount of HFIP, which is usually only regarded as a good solvent for dissolving FF, also has a significant impact on the morphology of the self-assembled materials, probably due to its effect on supersaturation of FF in mixed solvent. In addition, temperature also influences the growth of mesocrystals.

Due to the high roughness and surface area, the flower-like spheres exhibit exciting surface-related properties. For example, when a silica substrate covered by the flower-like spheres was further modified by the vapor of fluoroalkylsilane, the modified surface showed excellent superhydrophobicity as well as oleophobicity, suggesting that the hierarchical self-assembled structures of FF can support the formation of antiwetting surfaces. Peptide-based supramolecular materials can serve as templates for construction of functional metallic nanostructures because peptides can be easily removed by solvents or enzymes. A crust of flower-like spheres was obtained by sputtering Au on the surface of flower-like spheres, followed by removing FF with ethanol [49]. SEM and transmission electron microscopy (TEM) images confirm that the obtained metallic structures consisted of only Au nanoparticles preserve the flower-like morphology. Such hierarchical metallic structures show a remarkable surface enhancement factor for surface-enhanced Raman scattering (SERS) measurements.

Integration of supramolecular self-assembly of FF with the protocol of breath figure is a facile method to generate ordered macrostructures of FF [50]. The formed architectures are honeycomb scaffolds with uniformly distributed and hexagonally organized macropores. The pore sizes and morphology of the scaffolds are readily manipulated by controlling solvent, substrate, air humidity, FF concentration, and air speed. Circular dichroism (CD) and Fourier transform infrared spectroscopy (FTIR) spectra demonstrate that FF in the scaffolds is ordered in the configuration of parallel β-sheet, which is clearly different from the antiparallel configuration of FF organogel formed in toluene. Under ultraviolet (UV) light irradiation, the scaffolds show red-shifted fluorescence, indicating strong π–π stacking and ordered organization of FF in the scaffolds. Significantly, the scaffolds can support the adhesion and growth of human embryo skin fibroblast (ESF) cells, suggesting that such self-assembled macrostructures of FF are promising biomaterials for biomedical applications, such as tissue engineering.

7.2.2 Structural Transition of Self-assembled Peptide Nanostructures

Supramolecular self-assembly of the same peptide can form different nanostructures since self-assembly of peptides is relatively susceptible to changes that influence the non-covalent interactions, providing an opportunity to induce structural transition between self-assembled nanostructures. An extensively investigated example is the reversible transition between cationic dipeptide nanotubes and vesicles (Fig. 7.1a) [51]. Cationic dipeptide H-Phe-Phe-NH_2·HCl (CDP), an analog of FF, self-assembles into nanotubes at high concentration and nanovesicles at low concentration (Fig. 7.1b). Hence, the structural transition is easily manipulated by dilution and concentration. CD spectra reveal that the structural transition occurs along with a change in the second structure of CDP (Fig. 7.1c). CD spectra of assembled CDP nanotubes show characteristic signals of α-helical peptides. Also, the existence of π–π stacking among the aromatic side chains and hydrogen bonding among the main chains is indicated by CD results. No significant signal corresponding to second structure of CDP is observed for the form of nanovesicles, suggesting that the transition from nanotubes to nanovesicles is a disassembly process in which electrostatic repulsion possibly plays a key role other than π–π stacking or hydrogen bonding. Moreover, jointed nanovesicles in the shape of a necklace were observed as intermediate structures in the transition between nanotubes and nanovesicles (Fig. 7.1d). An equation presented to demonstrate the

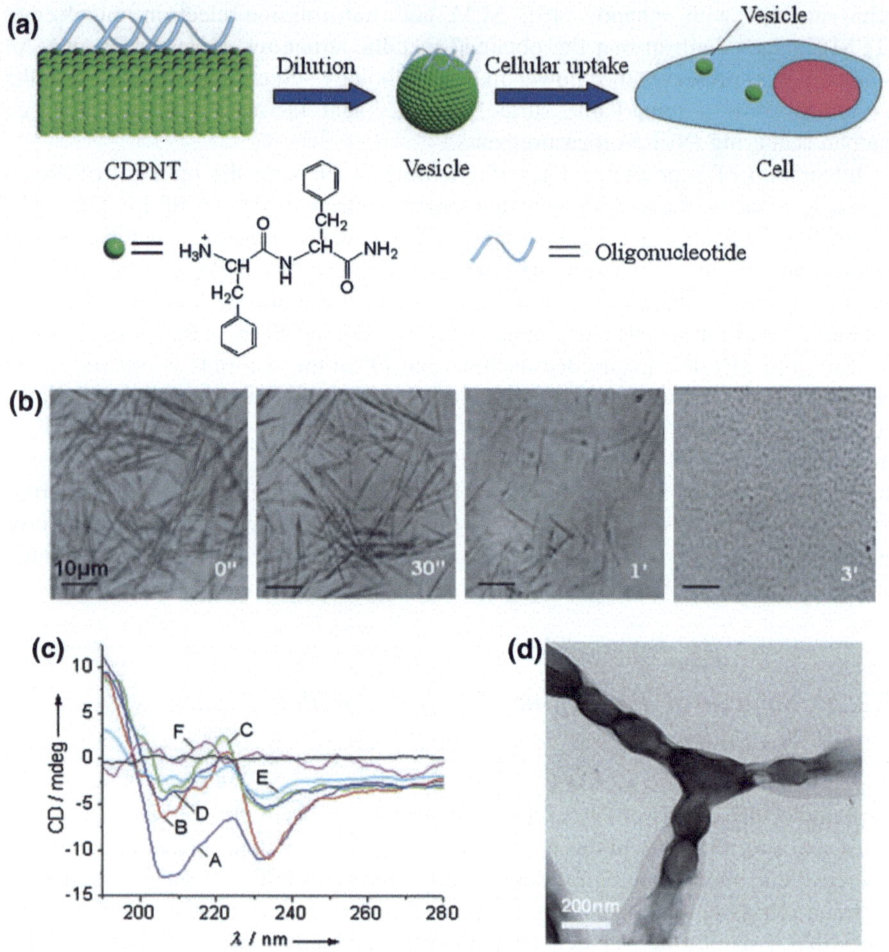

Fig. 7.1 Structural transition of peptide nanostructures. **a** Schematic illustration of structural transition of the CDP nanotubes (CDPNTs) into vehicles for oligonucleotide delivery. **b** Optical images showing the conversion of the CDPNTs into vesicles. **c** CD spectra of the CDPNTs and vesicles at different concentrations (*A* 10, *B* 8, *C* 7, *D* 5, *E* 2, *F* 1 mg mL^{-1}). **d** TEM image of diluted CDPNTs (Reprinted with permission from Ref. [51]. Copyright 2007, Wiley-VCH)

experimental results reveals that the structural transition of peptide-based nanostructures is basically related with the concentration of the building block, the molecular size, and the tension of the solution/aggregate interface [52].

Oligonucleotide delivery has been demonstrated by using CDP nanotubes as carriers for single-stranded DNA (ssDNA) [51]. Negatively charged ssDNA labeled by a fluorescent dye was loaded on the surface of CDP nanotubes through electrostatic interactions. After the nanotubes were changed to nanovesicles under dilution, the fluorescence was found on the obtained nanovesicles, suggesting that ssDNA was retained with CDP nanostructures in the process of disassembly and reassembly of CDP. When HeLa cells were incubated with ssDNA-bounded nanotubes for 24 h, ssDNA was successfully delivered into the cells. The delivery of ssDNA by CDP nanotubes relies structural transition from nanotubes to nanovesicles upon dilution by cell culture media.

Structural transition from nanofibers to nanobelts was observed in an organogel-based system [53]. The system was constructed by in situ encapsulated of a lanthanide ion (Tb^{3+}) and a photosensitizer (salicylic acid) into FF organogels. SEM imaging reveals that the hybrid organogels consist of small fibers. Due to the presence of Tb^{3+} and salicylic acid, the stability and mechanical strength of the hybrid organogels are lower than that of pure FF organogels formed in toluene. However, the stability and mechanical properties of the hybrid organogels were significantly enhanced by the treatments of heat or water. After the treatments, uniform nanobelts were observed as the main components of the gels, suggesting the structural transition from fibers to nanobelts induced by heat and water. FTIR and X-ray diffraction (XRD) analyses show that the encapsulation of Tb^{3+} and salicylic acid and water treatment have no impact on the secondary structure of FF while the heat treatment induces a change from a β-sheet to a β-turn structure in the organogels. Fluorescence spectra demonstrate that the fluorescence emission of Tb^{3+} is enhanced by synergistic energy transfer from FF and salicylic acid.

7.2.3 Solvent-Induced Structural Transition of Peptide Nanostructures

Self-assembly of peptides in solutions not only involves peptide–peptide interactions but also involves peptide–solvent interactions. Hence, solvent can be a predominant factor to tune peptide nanostructures. The solvent-induced structural transition from organogels to flower-like microcrystals was demonstrated in a system containing FF, toluene, and ethanol [54]. Self-assembly of FF in pure toluene generates organogels. SEM images reveal such organogels consist an entangled network of nanofibers and fiber bundles. In changing pure toluene to a mixture of ethanol and toluene, the self-assembled nanostructures of FF vary from organogels to flower-like microcrystals. SEM images show that the flower-like microcrystals are composed with ribbons. FTIR spectra indicate that the nanostructures obtained in pure toluene and in a low amount of ethanol (10%) show the characteristics of antiparallel β-sheet secondary structures, while the flower-like microcrystals obtained in the presence of a high amount of ethanol have parallel β-sheet secondary structures. The impact of solvent on the morphology is due to different properties of ethanol and toluene: toluene molecules participate in the self-assembly of FF though π–π stacking while ethanol interacts with FF through hydrogen bonding.

Hydrogen bonding is an important interaction involved in self-assembly of biomolecules in biological systems. For example, water-related hydrogen bonding is believed to play a key role in the formation of proteins fibrils in the Alzheimer's disease [55, 56]. Systematic investigation of FF self-assembly in dichloromethane in addition of a trace amount of hydrogen bonding solvent provides a direct evidence for the importance of hydrogen bonding in peptide self-assembly (Fig. 7.2a) [57]. In pure dichloromethane, self-assembly of FF leads to the formation of crystalline structures (Fig. 7.2b). FIIR and XRD results show that crystalline structures are similar with the hexagonal structures in nanotubes and single crystals of FF. When a trace amount of hydrogen bonding solvents such as ethanol, dimethylformamide, or acetone, is added to toluene, long fiber or ribbon structures are formed, suggesting that the assembly of FF in one dimension is induced by directional hydrogen bonding (Fig. 7.2c). In contrast, the addition of n-hexane, a solvent that only shows van der Waals forces with FF, cannot promote fiber structures.

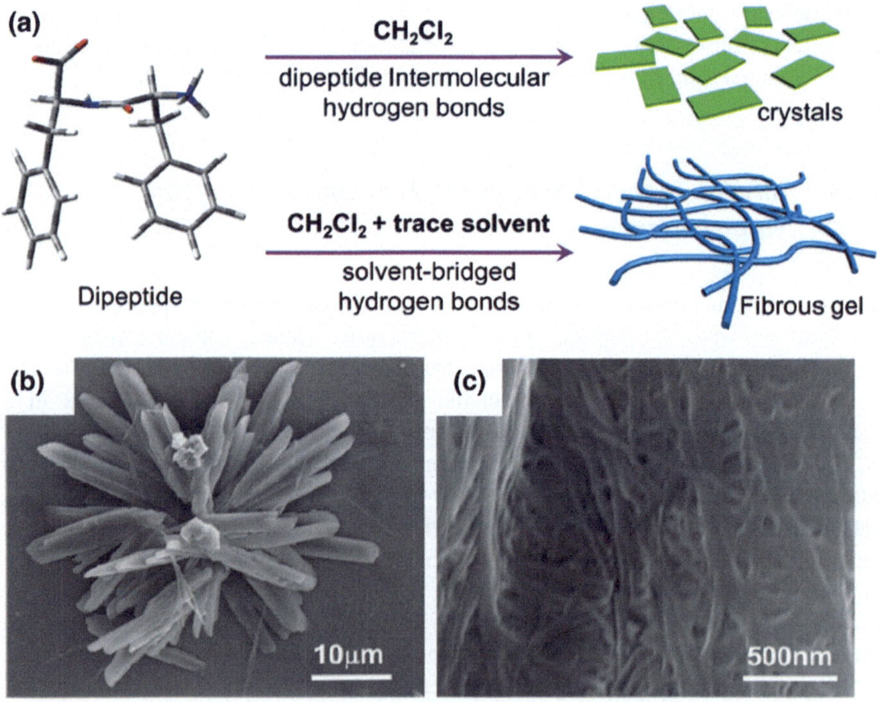

Fig. 7.2 Trace solvent-induced structural transition of FF nanostructures. **a** Schematic illustration of the transition. **b** SEM image of FF/CH_2Cl_2. **c** SEM image of FF/CH_2Cl_2/ethanol (Reprinted with permission from Ref. [57]. Copyright 2016, American Chemical Society)

7.2.4 Self-assembled Peptide Crystals with Optical Waveguiding Properties

The crystalline structures of FF provide platforms for constructing optical waveguiding materials. The waveguide based on self-assembled peptide materials was first demonstrated by aldehyde-induced cyclization and crystallization of FF [58]. To produce the optical waveguiding peptide crystals, organogels were prepared by adding a solution of FF in HFIP into aldehyde-containing toluene. The gels imperceptibly turned into white precipitates in a period of about 1 month. SEM and TEM images demonstrate that the precipitates consist of rectangular platelets, which are composed of anisotropic peptide fiber or ribbon nanostructures. The selected area electron diffraction (SAED) analysis reveals the platelets as single-crystalline peptide platelets. Matrix-assisted laser desorption/ionization time-of-flight (MALDI-TOF) mass spectrometry and ^{1}HNMR spectroscopy show that FF is converted to cyco-Phe-Phe (CPP) in the platelets due to aldehyde-induced cyclization. FTIR and photoluminescence spectra suggest that the formation of platelets is triggered by enhanced hydrogen bonding and π–π stacking interactions. Under excitation at 330–380 nm, the crystalline platelets show blue photoluminescence emission and their ends show remarkably bright spots (Fig. 7.3a, b). When one end of a platelet is excited, bright spots of blue emission can be observed at the other end (Fig. 7.3c, d), suggesting that the light is guided inside the platelet from one end to another. Since the wavelengths of the emission and excitation light are different, the waveguiding based on CPP platelets is a kind of active waveguiding, in which the processes of photoluminescence and waveguiding take place together. Intriguingly, the optical properties of the platelets can be readily tuned by incorporation of other dyes in the process of crystallization. For instance, incorporation of Nile Red into CPP platelets leads to the formation of red light-emitting platelets (Fig. 7.2e, f).

Other than the aldehyde-induced cyclization, CPP platelets can also be prepared by a fast and facile solvothermal approach [59]. It is found that heating the organogels of FF in toluene in sealed vials at a temperature higher than 110 °C generates CPP platelets in 10 min. SEM images reveal that such solvothermally prepared CPP platelets consist of rectangular ultralong nanobelts, similar with those obtained by aldehyde-induced cyclization. Surprisingly, curved crystalline platelets are observed in the precipitates and they show curved optical waveguides upon excitation. Such curved waveguiding along the curved axial of peptide crystalline nanobelts opens a novel way for advanced optical biomaterials.

Self-assembled nanomaterials of FF with crystalline structures also possess optical waveguiding properties. One example is the hexagonal FF microtubes prepared by controlled self-assembly of FF upon solvent thermal annealing [60]. The formation of microtubes from FF is a progress of self-similar-structured growth, which includes the hexagonal packing of FF molecules, ampliative packing

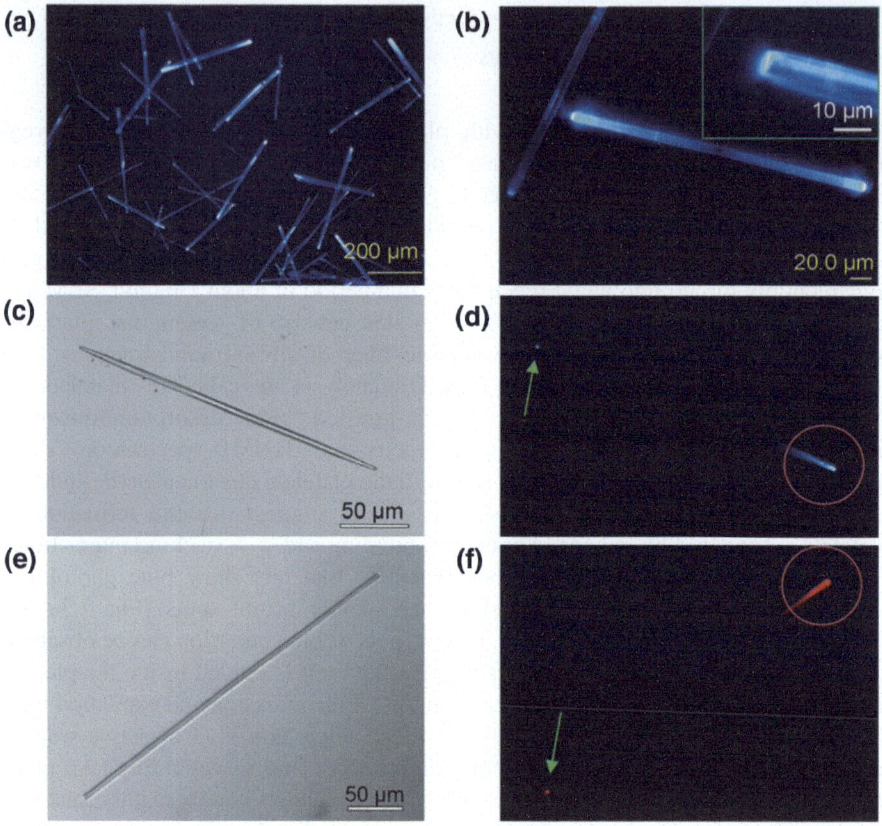

Fig. 7.3 Optical waveguiding of peptide crystals. **a** Photoluminescence image of platelets excited at 330–380 nm. **b** Photoluminescence image of a single platelet. **c** Bright-field image and **d** waveguiding image of a single platelet with local excitation at one end. **e** Bright-field image and **f** waveguiding image of a single platelet incorporating Nile Red dye with local excitation at one end. The *red circle* indicates the excitation area, and the *green arrow* denotes the emission at the other end (Reprinted with permission from Ref. [58]. Copyright 2011, Wiley-VCH)

of the initially formed subunits and the formation of hexagonal nanotubes, and further hierarchical amplification of the nanotubes and the resulting of hexagonal hollow microtubes (Fig. 7.4). Such hierarchically self-assembled FF microtubes show similar XRD patterns with FF single crystals, leading them as promising candidates for active waveguide materials. FF nanofibers, microtubes, and microrods constructed through controlled self-assembly of FF in mixture of HFIP and water with the assistance of ultrasonication also show optical waveguiding characteristics [61]. Moreover, peptide optical waveguide is demonstrated on self-assembled rectangular CDP microtubes and microrods [62].

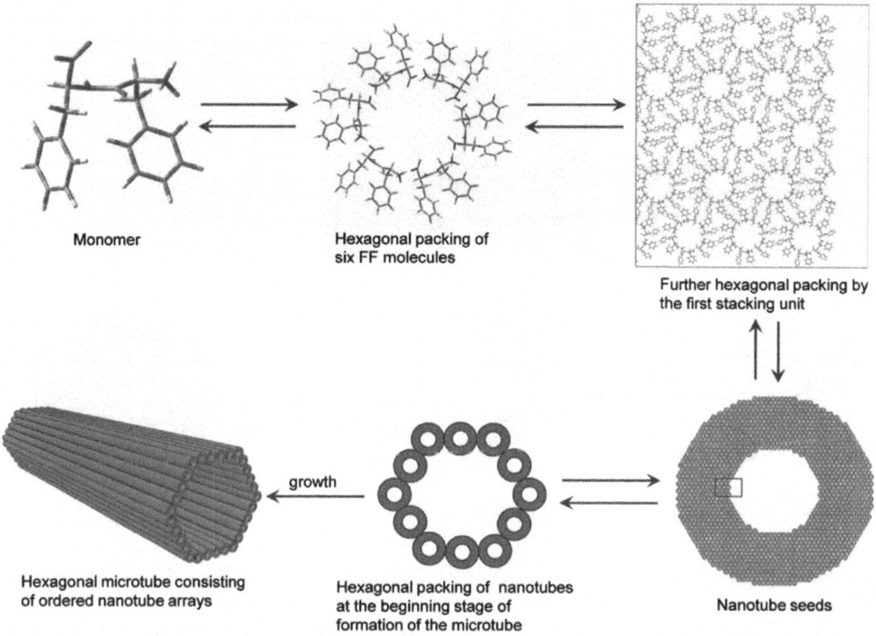

Fig. 7.4 Schematic illustration of the hierarchical self-assembly of FF in the formation of hexagonal FF microtubes through self-similar-structured growth (Reprinted with permission from Ref. [60]. Copyright 2011, Wiley-VCH)

7.3 Peptide-Modulated Self-assembly of Photoactive Molecules

Co-assembly of peptides and other functional molecules provides a simple but efficient strategy for fabrication of functional nanomaterials. Based on such co-assembly strategy, photoactive materials have been constructed through peptide-modulated self-assembly of porphyrins, azobenzenes, and photosensitizers. These photoactive materials show intriguing optical properties due to the ordered arrangement of photoactive molecules inside their nanostructures.

7.3.1 Peptide-Modulated Self-assembly of Porphyrins

Inspired by the natural light-harvesting complexes constructed by chromophores and proteins, a novel and facile strategy has been developed to fabricate light-harvesting architectures via peptide-tuned self-assembly of porphyrins based on the synergy of multiple weak intermolecular interactions [13]. Tetrakis (4-sulfonatophenyl) porphine (TPPS) was chosen as a model light-harvesting

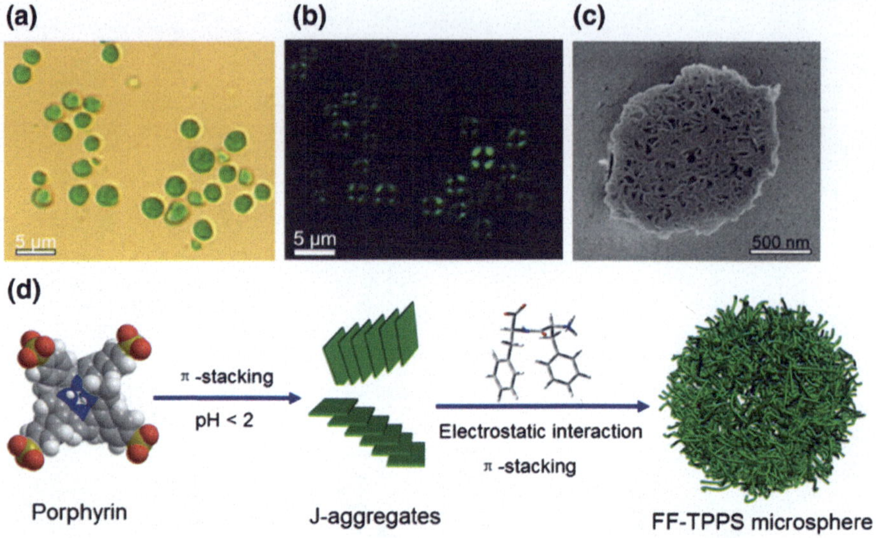

Fig. 7.5 Hierarchical microspheres assembled from FF and TPPS **a** Bright-field image. **b** Polarized image of microspheres when placed between crossed polarizers. **c** SEM image of the cross section of a single microsphere. **d** Proposed hierarchical assembly mechanism (Reprinted with permission from Ref. [63]. Copyright 2014, Wiley-VCH)

porphyrin for its similar photochemical properties to chlorophyll, good water solubility, and capability to self-assemble into higher structures, such as J-aggregation and H-aggregation. Co-assembly of FF and TPPS produces microspheres (Fig. 7.5a) [63]. The microspheres show strong birefringence (Fig. 7.5b), indicating that the TPPS molecules of the microspheres are orderly arranged. SEM images of a single microsphere cross section show that that the microspheres are porous with a multi-compartment interior constructed by many interconnected nanorods (Fig. 7.5c). Uv–vis absorption spectra of the microspheres show a new absorbance peak at 490 nm as compared with that of TPPS, indicating that the TPPS molecules in the microspheres are in J-aggregation. Compared to pure J-aggregated TPPS, this band broadens variously, suggesting that FF perturbs the regular organization of TPPS molecules. The formation of the microspheres is controlled by hierarchical multi-scale self-assembly (Fig. 7.5d). Initially, intermolecular (FF and TPPS) and intramolecular (TPPS) electrostatic interactions combine FF and TPPS molecules together and then co-assemble into nanorods. The nanorods further stack to form microspheres, probably through π–π interactions derived from the benzene ring of FF.

Another light-harvesting model was developed by using hydrophilic L-Lys-L-Lys (KK) to direct the self-assembly of TPPS [64]. In an acid solution, KK and TPPS can co-assemble into long fiber bundles. The fiber bundles are made up of nanorods or nanofibers, which are orderly arranged along the long axis. The fiber bundles show red emission due to the fluorescent of porphyrins. Birefringence

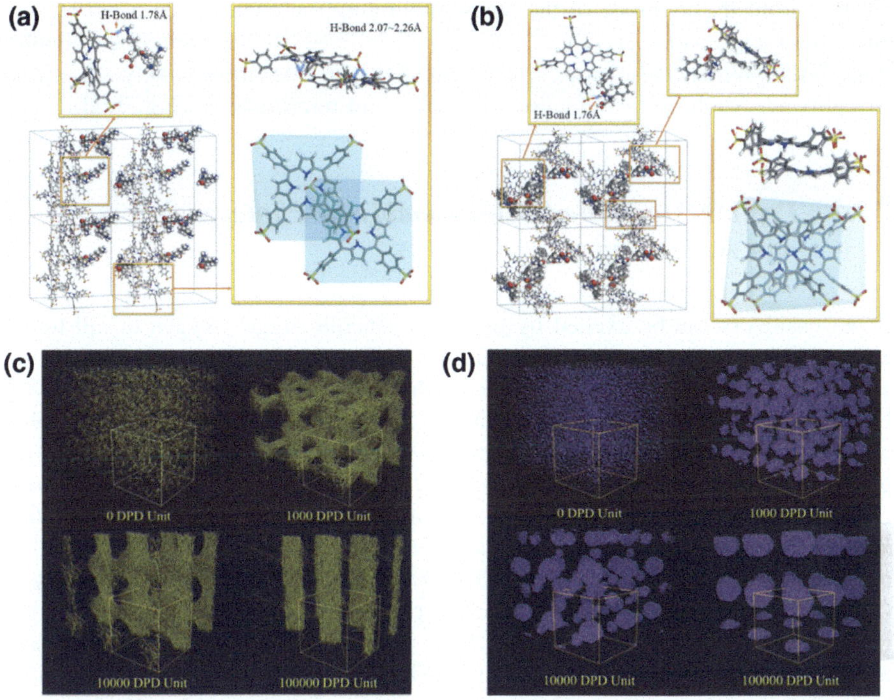

Fig. 7.6 Simulation of peptide-porphyrin hierarchical assembly. Structures of assembled **a** TPPS-KK and **b** TPPS-FF obtained by molecular dynamics simulation. Morphologies of **c** TPPS-KK and **d** TPPS-FF at different simulation times in DPD simulation (Reprinted with permission from Ref. [65]. Copyright 2016, Royal Society of Chemistry)

extinction was observed for two orthogonal fiber bundles at the cross point, suggesting that the nanorods or nanofibers are uniformly distributed on the fiber bundles. Characteristic absorption peak of J-aggregated TPPS was also observed in the fiber bundles, indicating that the TPPS molecules are in J-aggregation. In addition, the chiral signal corresponding to the J-aggregated TPPS increases during the self-assembly process. Similar with FF-promoted self-assembly of TPPS, the KK-promoted self-assembly of TPPS is also a hierarchical organization process. Initially, electrostatic and hydrogen bonding interactions between KK and TPPS promotes the formation of nanorods consisting of J-aggregated porphyrin. Then, the resulting nanorods self-organize at long range to form long fiber bundles.

The self-assembly mechanisms of the dipeptide-porphyrin systems have been further analyzed by multi-scale theoretical study [65]. Molecular dynamics simulation proves that the KK^{3+} molecules are distributed on the surface of H_2TPPS^{2-} nanorods (Fig. 7.6a). The residues of FF^+ (benzyl groups) are more likely to bind with the J-aggregated H_2TPPS^{2-} nanorods via hydrophobic interaction (Fig. 7.6b). Core–shell structure made of H_2TPPS^{2-} and dipeptides has been constructed by dissipative particle dynamics (DPD) simulation. The KK^{3+} molecules tend to be released from the interface of the cores, promoting the coalescence of cores, and thus the J-aggregated H_2TPPS^{2-} are better ordered to generate fiber bundles (Fig. 7.6c). When these rods stack together, FF^+ is more likely to create crosslinking, resulting in formation of equilibrium microspheres (Fig. 7.6d).

7.3.2 Peptide-Modulated Self-assembly of Photosensitizers and Azobenzenes

Photosensitizers can be excited by light and generate single oxygen to kill tumor cells and tissues in photodynamic therapy (PDT) [66–68]. However, most photosensitizers are poorly soluble in water, which results in low bioavailability [69, 70]. Dispersive nanoparticles of photosensitizers have been constructed by co-assembly of photosensitizers with simple dipeptide or amphiphilic amino acid to as photosensitive drug delivery systems [71]. CDP and 9-Fluorenylmethoxy-carbonyl-l-lysine (Fmoc-L-Lys) were used as models of an amphiphilic dipeptide and amino acid, respectively, due to their excellent biocompatibility, nonimmunogenicity, and easy availability. CDP or Fmoc-L-Lys can tune self-assembly of Chlorin e6 (Ce6), a model hydrophobic photosensitive drug, to form uniform nanospheres (Fig. 7.7a, b, c). The Soret band for Ce6 was broadened and red-shifted in the assembled nanodrugs (Fig. 7.7d), indicative of intermolecular hydrophobic and π–π interactions. The nanospheres disassembled at high ionic strength, suggesting the existence of electrostatic forces in the co-assembly process. Therefore, the multiple weak intermolecular interactions, including electrostatic, π–π stacking, and hydrophobic interactions drive the self-assembly.

Tetraphenylethylene (TPE) is a propeller-like molecule with the behavior of aggregation-induced emission (AIE) due to restriction of intramolecular rotations [72]. The emission properties of TPE are usually tailored by using time-consuming organic synthesis. Dynamic regulation of the emission intensity of TPE has been realized by facile peptide-tuned assembly of TPE [73]. After dropping solutions of CDP or Fmoc-L-Lys into TPE-COOH nanoparticle suspensions, the emission of CDP/TPE-COOH nanoparticles has an increase than that of bare TPE-COOH nanoparticles. In contrast, the Fmoc-L-Lys/TPE-COOH nanoparticles show a decreased fluorescence. CDP can enhance the fluorescence presumably due to restriction of intramolecular rotations of TPE molecules via hydrophobic interaction, while Fmoc-L-Lys can weaken the emission of TPE nanoparticles probably due to π–π stack-induced charge transfer.

Photoswitchable sulfonicazobenzene 4-[(4-ethoxy)phenylazo] benzenesulfonic acid (EPABS) was used to manipulate the self-assembly of CDP [74]. The trans-cis conformational change of EPABS induced by light can obviously change the assembled structures of CDP. When EPABS is in the transform, it can co-assemble with CDP into branched structures. After UV irradiation, higher hydrophilic cis-EPABS is formed, leading to disassembly of the structures. The released CDP molecules can form vesicle-like structures via a self-assembly process. Sulfonic

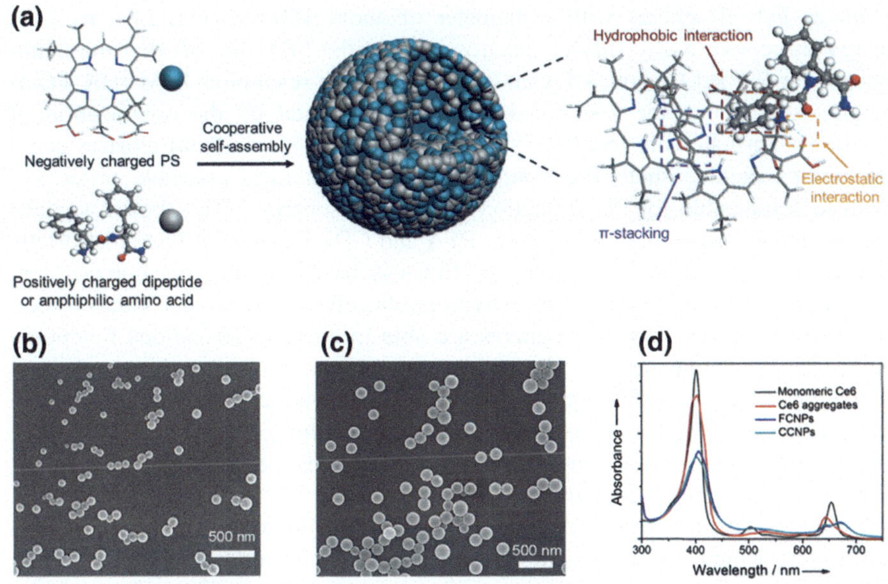

Fig. 7.7 Dipeptide- or amino-acid-tuned self-assembly of photosensitizers (PSs). **a** Schematic illustration. **b** SEM image of assembled nanoparticles using Fmoc-L-Lys and Ce6 as building blocks (FCNPs). **c** SEM image of assembled nanoparticles using CDP and Ce6 as building blocks (CCNPs). **d** Uv–vis absorption spectra of assembled nanoparticles (Reprinted with permission from Ref. [71]. Copyright 2016, Wiley-VCH)

azobenzenes also show a good capability to tune the self-assembly of CDP and diverse assembled structures including urchin-like, flower-like, and plate-like microstructures can be formed [75].

7.4 Self-assembly of Peptide-Inorganic Hybrid Nanomaterials

Self-assembly of peptide-inorganic hybrid nanomaterials is particularly powerful because in such approaches the ease and control offered by the self-assembly of organic components can be combined with the electronic, magnetic or photonic properties of inorganic components.

7.4.1 Peptide-Modulated Self-assembly of Polyoxometalates

POMs are multifunctional nanoclusters of polyatomic ions [76]. Co-assembly of phosphotungstic acid (PTA), a typical POM, with CDP in aqueous solution generates hybrid spheres with a diameter of about 150 nm (Fig. 7.8a, b) [77]. Energy dispersive X-ray (EDX) spectroscopy on the SEM reveals that the nanospheres are formed by both PTA and CDP. The high-resolution TEM (HR-TEM) imaging shows that the assembled spheres are formed by the organization of peptide-encapsulated clusters (PECs), which consist of single POM clusters and a surrounding peptide shell (Fig. 7.8c). Hence, a hierarchical assembly process is believed responsible for the formation of the nanospheres. FTIR spectra suggest that the initial formation of PECs from PTA and CDP is due to strong electrostatic interactions, while further assembly of PECs is based on multiple non-covalent interactions, including, π–π stacking, hydrophobic effect, and van der Waals forces. It is also found that such nanospheres are able to encapsulate various functional molecules and nanomaterials with distinctive properties, such as positively charged Rhodamine 6G, neutral fluorescein isothiocyanate, negatively charged Congo Red, dextran, hypocrellin B nanoparticles, and hydrophilic gold nanoparticles. The adaptive encapsulation of guest materials is easily realized by adding the guest materials to the solution containing PTA before mixing PTA and CDP. Profoundly, the colloid nanospheres of PTA and CDP are responsive to the stimuli of pH and temperature. Moreover, the transition from nanospheres to surface-supported hybrid films under near-infrared (NIR) irradiation is demonstrated [78]. The versatile encapsulation ability of the nanospheres along with their responsiveness to external stimuli suggests such peptide-inorganic hybrid materials are promising for biomedical applications.

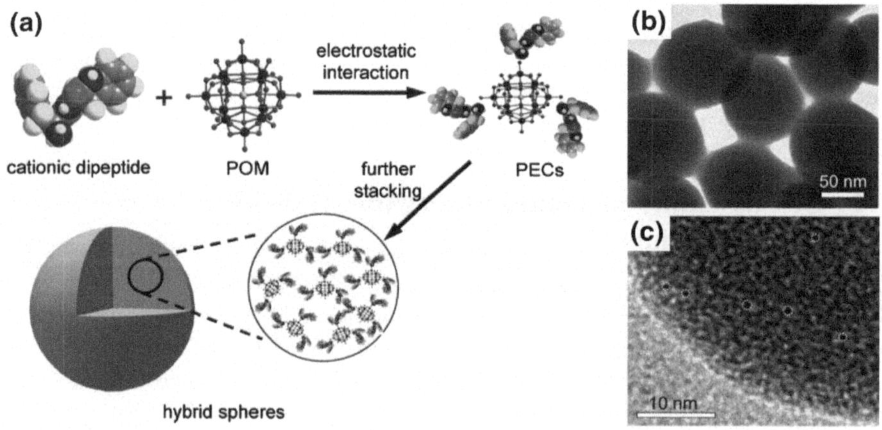

Fig. 7.8 Co-assembly of CDP and POM. **a** Schematic illustration of the formation of the PECs and the hybrid spheres. **b** TEM image of the spheres. **c** HR-TEM image of the spheres showing the PECs (Reprinted with permission from Ref. [77]. Copyright 2010, Wiley-VCH)

7.4.2 Peptide-Modulated Self-assembly of Quantum Dots

Semiconductor QDs are nanoscale particles. Due to their small size, QDs show specific optical and electronic properties [79]. Assembly of such small nanoparticles into higher order architectures while keeping their optical and electronic properties unaffected is of great interest. Supramolecular nanostructures of peptides provide suitable scaffolds to achieve higher order architectures of QDs. The immobilization of QDs by peptides is demonstrated by gelation of a QDs dispersed toluene solution by FF (Fig. 7.9a) [47]. TEM images indicate that most of the QDs are attached to the peptide nanofibers (Fig. 7.9b, c). The gels of the encapsulated QDs show characteristic fluorescence signals of individual QDs though the emission peaks slightly blue-shifted due to the ligand interactions between FF molecules and the QDs (Fig. 7.9c), suggesting that the photophysical properties of the encapsulated QDs remained mainly. Hence, the encapsulation by self-assembled peptide nanostructures provides a suitable method for construction of higher order architectures of QDs.

Since the supramolecular structures of peptides are responsive to external stimuli, the gels of the encapsulated QDs can be manipulated to other architectures [80]. To demonstrate this structural transition for higher order architectures of QDs, gels encapsulated QDs are prepared by gelation of a toluene solution containing QDs by CDP. Then, toluene is removed from the gels by vacuum drying to form the xerogels of encapsulated QDs. By means of adding water to the xerogels and ultrasonic treatment, the xerogels of encapsulated QDs are manipulated to stable nanospheres. In addition, the size of the nanospheres is readily controlled by varying the duration time of ultrasonic treatment and the concentration of CDP. EDX analyses reveal that the colloid nanospheres are composed by both CDP and QDs. The photoluminescence images demonstrate that the assembled QDs in the nanospheres remain their original luminescence properties. Moreover, the colloid nanospheres are stable in a serum-containing cell culture media. When the

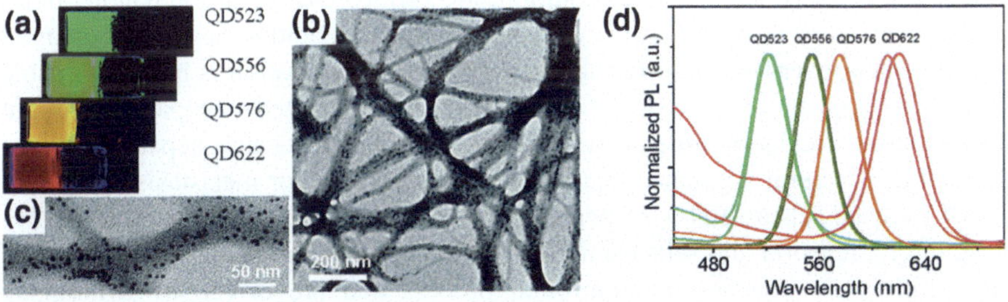

Fig. 7.9 Encapsulation of QDs in FF gel. **a** Photoluminescence photograph of four hybrid gels with different QDs. **b** TEM image of the encapsulated QD523 nanocrystals in the fibril network. **c** Magnified TEM image of the QD523 nanocrystals immobilized to the fibril. **d** Emission spectra of the free QDs in toluene (*solid line*) and the encapsulated QDs in gel (*dash dot*) (Reprinted with permission from Ref. [47]. Copyright 2008, American Chemical Society)

nanopsheres were incubated with HeLa cells, they were successfully internalized by cells and accumulated mainly in the cytoplasm. Cell viability analyses show that the nanospheres are biocompatible at the concentration of 200 μg mL^{-1}. The structural transition of peptide-modulated architectures of QDs along with their high biocompatibility suggest that self-assembling peptides may be widely applicable for construction of biocompatible functional organic–inorganic biomaterials.

7.5 Schiff Base Interaction-Induced Self-assembly of Peptides

In situ generation of peptide-based self-assembling building blocks is realized by Schiff base formation from peptides and GA. It is initially found that CDP and GA assemble into nearly monodisperse nanoparticles with a size of about 450 nm in water gradually when aging the mixture of them at room temperature for about 24 h [81]. The nanoparticles show a positive surface charge, probably due to the presence of positively charged amino groups on their surface. X-ray photoelectron spectroscopy (XPS) spectra reveal that the formed nanoparticles contain carbon–nitrogen double bonds, indicating the formation of Schiff base between the amino group of CDP and the aldehyde group of GA. Spectra of MALDI-TOF mass spectrometry show that the building blocks of the nanoparticles are mainly CDP-2GA-CDP and CDP-3GA-CDP, the products of the reaction of CDP, and oligomeric GA (Fig. 7.10a). Further experiments reveal that the size of the nanoparticles can be controlled by prereaction of CDP and GA in HFIP before the addition of water [82]. When the reaction time for CDP and GA in HFIP is more than 24 h, the size of the obtained nanoparticles drops to 70 nm (Fig. 7.10b, c), suitable for in vivo applications. Mass spectrometry analyses suggest that the variation of building blocks is responsible for the change of the sizes. CDP-2GA-CDP is the main building block for nanoparticles with less prereaction time, while CDP-3GA-CDP is the main building block for nanoparticles with the prereaction of more than 24 h. In addition, the surface change of the nanoparticles can be reversed from positive to negative by surface decorating negatively charged heparin. Such reversion of surface charge by heparin decoration is highly valuable for in vivo applications not only because negatively charged nanoparticles might have a longer circulation time in the bloodstream but also because heparin has been reported as a natural substance that can interfere with tumor metastasis.

GA-induced self-assembly of peptides has also been demonstrated by using FF. Intriguingly, reaction and self-assembly of GA and FF in emulsions of toluene and water follow an interfacially controlled process and are driven by synergistic thermodynamic and kinetic control [83]. At the beginning, the formed nanostructures are crescent-like nanoparticles, which gradually turn into bowl-like nanoparticles along with time and eventually grow into solid interior nanoparticles as the final structures. More sophisticated microstructures can be realized

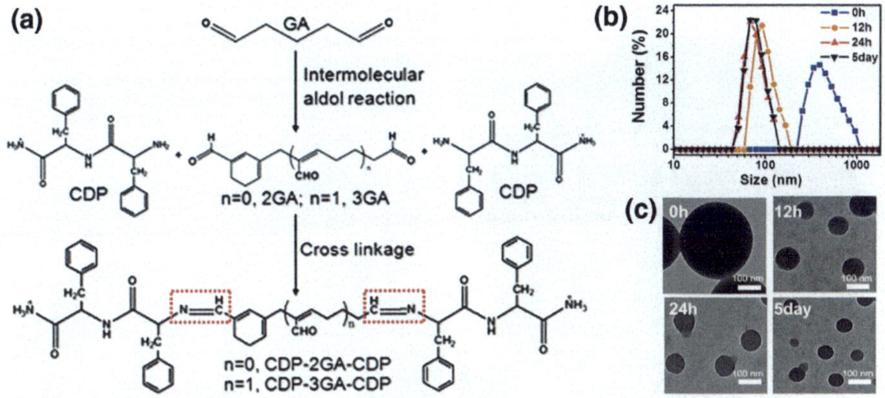

Fig. 7.10 Aldehyde-induced self-assembly of CDP. **a** The formation of CDP-2GA-CDP and CDP-3GA-CDP through reaction between CDP and GA (Reprinted with permission from Ref. [81]. Copyright 2015, Wiley-VCH). **b** Size of the formed nanoparticles under different prereaction times. **c** TEM images of the formed nanoparticles under different prereaction times (Reprinted with permission from Ref. [82]. Copyright 2016, American Chemical Society)

according to the strategy of GA-induced self-assembly of FF by using other emulsions. For instance, multicompartmental hollow spheres have been constructed in emulsions of water/HFIF/hexane [84].

7.6 Applications of Peptide-Based Supramolecular Nanomaterials

Supramolecular peptide nanomaterials have attracted a lot of attentions for biomimetic and biomedical applications due to their advantages of functional diversity and excellent biocompatibility. Nanostructures constructed from peptide-tuned self-assembly of porphyrins have been applied for biomimetic photosystem and photocatalysis. Self-assembled and co-assembled peptide nanoparticles have been evaluated as delivery vehicles for antitumor drugs.

7.6.1 Applications in Biomimetic Photosystem

The FF-TPPS microspheres and KK-TPPS fiber bundles can in situ mineralize Pt nanoparticles via photo-deposition [63, 64]. Porphyrin is excited by light and the resulted anionic porphyrin radical can reduce the K_2PtCl_4 to Pt. After illumination, discrete Pt nanoparticles of 2 nm are formed on the surfaces of the microspheres and Pt nanospheres and Pt nanowires are observed locally on the fiber bundles

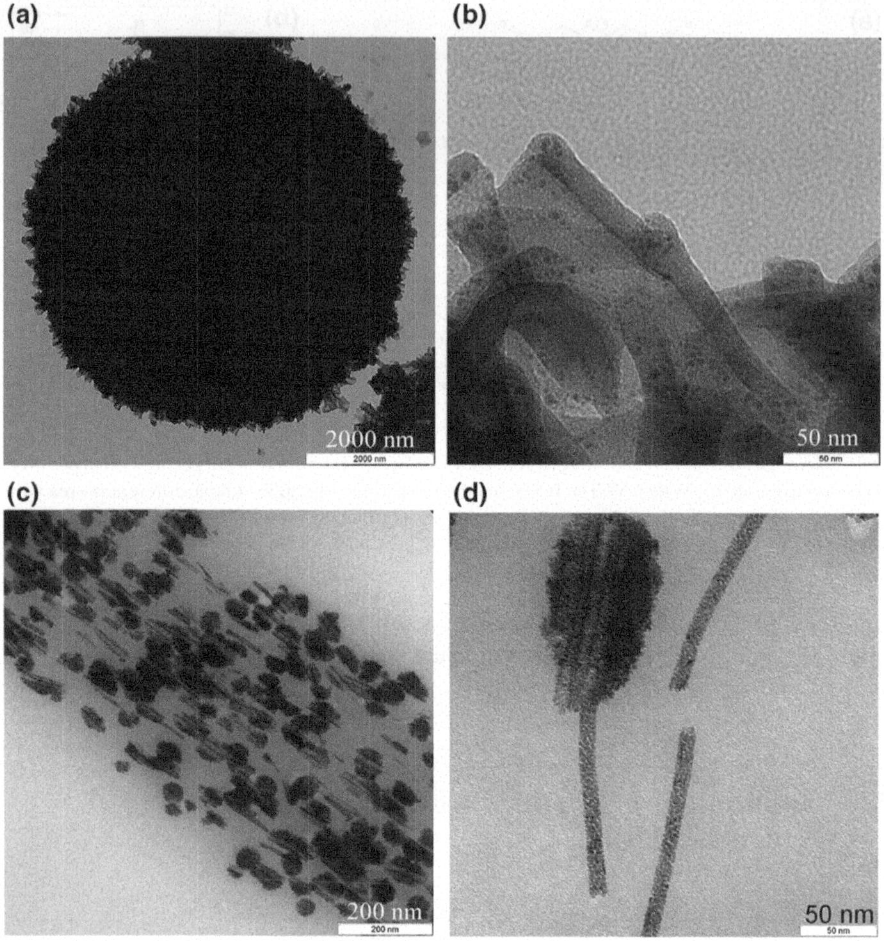

Fig. 7.11 TEM images of the Pt nanomaterials reduced by peptide-porphyrin microspheres or fiber bundles from a photocatalytic pathway. **a** TEM image and **b** higher magnification of Pt incorporated microsphere (Reprinted with permission from Ref. [63]. Copyright 2014, Wiley-VCH). **c** TEM image and **d** higher magnification of Pt nanowires and nanospheres synthesized by fiber bundles (Reprinted with permission from Ref. [64]. Copyright 2015, Wiley-VCH)

(Fig. 7.11). The growth of Pt nanoparticles is oriented along the long axis of the fiber bundles, resulting in the formation of Pt nanowires. In addition, TiO_2 nanoparticles of 5 nm can be mineralized on the surface of the KK-TPPS fibers after incubation with titanium (IV) bis(ammonium lactato)dihydroxide (TiBALDH) in aqueous solution at a high temperature [85]. The HR-TEM analyses show the orientation of the anatase (101) plane. The synthesis of TiO_2 nanoparticles contains several steps. Initially, anionic TiBALDH molecules bind to the amino groups of KK through electrostatic interaction. Subsequently, the mineralization reaction follows the acid–base catalysis of the peptide molecules, leading to the hydrolysis of the TiBALDH complexes, followed by condensation of trihydroxyl species, and ultimately resulting in the formation of TiO_2.

7.6.2 Applications in Photocatalysis

The FF-TPPS microspheres show photocatalytic capability of oxidation of I^- to I_3^- (Fig. 7.12a) [63]. In addition, the microspheres possess enhanced photostability as compared to the pure J-aggregates under illumination for an extended period of time, as proved by the changes of Soret and Q bands. Thus, the yield of iodide is maintained over longer time in the presence of the microspheres (Fig. 7.12b), suggesting that binding of FF to the $[H_4TPPS]^{2-}$ dianions, or the structural stability of the microstructured matrix, may result in the sustained catalytic performance. The microspheres can also catalyze the conversion of organic matter under illumination, such as photo-reduction of 4-nitrophenol (4-NP) to 4-aminophenol (4-AP) (Fig. 7.12c and d).

The co-assembled KK-TPPS fibers can be flexibly self-functionalized of reaction center (TiO_2/Pt) in a volcanic hydrothermal "prebiotic soup": acidic (pH 2), hot (70 °C), and mineral-containing (Na^+, Ti^{4+}, Pt^{2+}, and so forth) water [85]. The porphyrin in the hierarchical fiber bundles has a similar organization to that of chlorosomes. TiO_2/Pt can be regarded as a primitive reaction center model, which uses the energy from light-harvesting porphyrin assemblies to achieve charge separation and catalyze proton reduction for generation of hydrogen. Thus, such system resembles the architectonics of green sulfur bacteria. After illumination by visible light, H_2-evolution occurs on the hybrid fibers in the prebiotic soup (Fig. 7.13a, b). Integration of TiO_2 and Pt simultaneously into the fibers leads to the biggest hydrogen production rate. Interestingly, the H_2 evolution activity of the hybrid fibers turns out to be dramatically increased by 13 times when increasing the NaCl concentration of the prebiotic soup (Fig. 7.13c), probably due to the enhanced exciton delocalization via NaCl-induced coupling of porphyrin monomers or nanotubular subunits. Additionally, the hybrid fibers show sustainable H_2 production in response to light illumination.

7.6.3 Applications in Biomedicine

The assembled photosensitizer nanodrugs (Fmoc-L-Lys/Ce6) have tunable size, high loading efficiency, and responsiveness to the stimuli of pH, surfactants, and enzymes [71]. Therefore, we evaluate their applications in antitumor PDT. The nanodrugs can be uptaken by tumor cells and disassembled in lysosome to release free drugs (Fig. 7.14a). 3-(4,5-dimethylthiazolyl-2)-2,5-diphenyltetrazolium bromide (MTT) assay shows significantly better anticancer efficacy for the nanodrugs

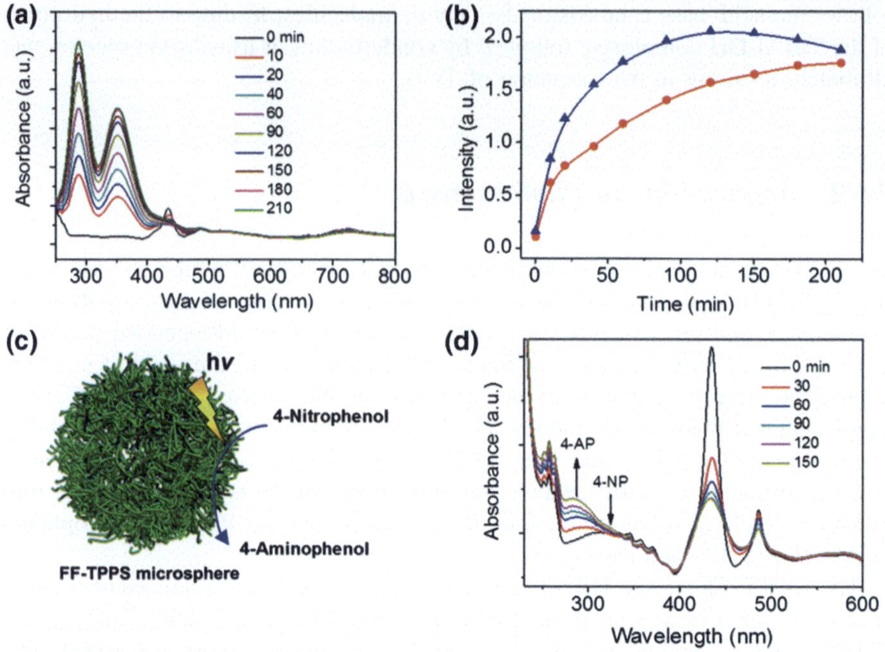

Fig. 7.12 Photocatalysis by FF-TPPS microspheres. **a** UV–vis absorption spectra showing time-dependent changes corresponding to product formation under irradiation of sodium iodide. **b** Plots of intensity at 353 nm versus irradiation time for tri-iodide formation in the presence of a dispersion of the microspheres (*red*) or TPPS J-aggregates (*blue*). **c** Illustration of photocatalytic reduction of 4-nitrophenol (4-NP) into 4-aminophenol (4-AP) in the presence of a dispersion of the microspheres. **d** Time-dependent UV–vis absorption spectra showing the production of 4-AP from 4-NP in the presence of the microspheres (Reprinted with permission from Ref. [63]. Copyright 2014, Wiley-VCH)

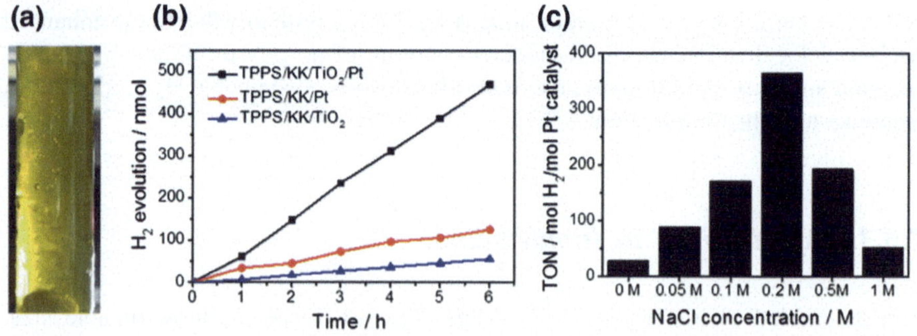

Fig. 7.13 Photocatalytic hydrogen evaluation by TPPS/KK/TiO_2/Pt fibers. **a** Photograph of H_2 bubbles produced from the hybrid fibers in the test tube during illumination. **b** Time dependence of the H_2 production of TPPS/KK/Pt, TPPS/KK/TiO_2 and TPPS/KK/TiO_2/Pt fibers. **c** Effect of NaCl on the hydrogen production of the hybrid fibers (Reprinted with permission from Ref. [85]. Copyright 2016, Wiley-VCH)

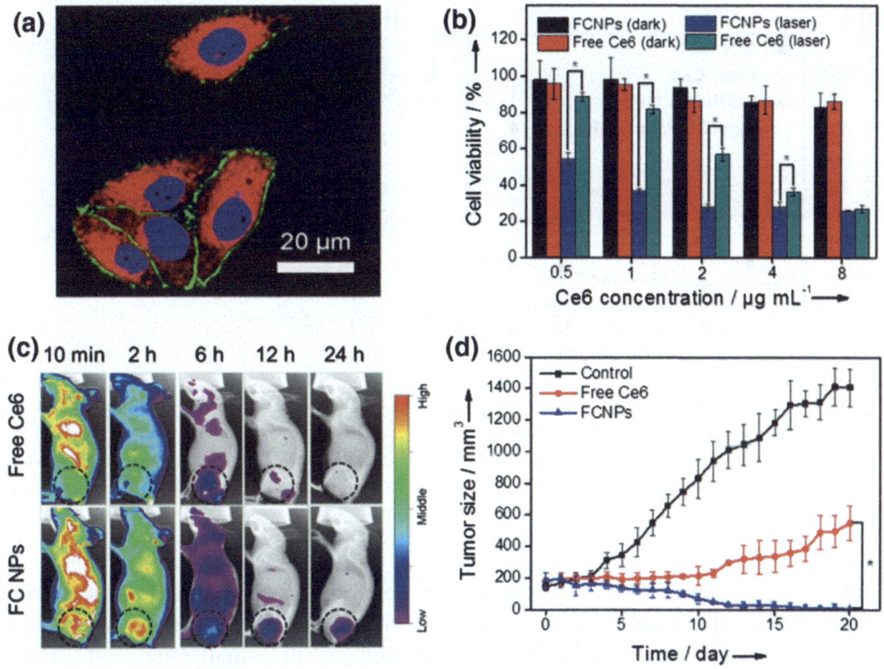

Fig. 7.14 Amphiphilic dipeptide- or amino-acid-tuned self-assembly of photosensitizers for in vitro and in vivo PDT. **a** Internalization of the assembled FCNPs by MCF7 cells. The *red* staining is from the photosensitizer while nuclei are stained by *blue* and cell membrane is stained by *green*. **b** In vitro cytotoxicity and photocytotoxicity of FCNPs. **c** Fluorescence images of tumor-bearing mice showing in vivo distribution of FCNPs and free Ce6. **d** Tumor growth curves of the mice in different groups (Reprinted with permission from Ref. [71]. Copyright 2016, Wiley-VCH)

than free drug (Fig. 7.14b). No cytotoxicity was observed in the dark, confirming the good biocompatibility of the nanodrugs. The nanodrugs also show preferable biodistribution. After intravenous injection, the nanodrugs exhibited sustained and stronger Ce6 fluorescence at the tumor site than free drugs (Fig. 7.14c), mainly due to the enhanced permeability and retention (EPR) effect [86]. These therapeutic features result in good PDT efficacy in vivo, leading to almost complete tumor inhibition after receiving only a single treatment (Fig. 7.14d).

Nanoparticles constructed from GA-induced self-assembly of CDP are versatile drug delivery vehicles for both chemotherapy and PDT. GA/CDP nanoparticles show a loading efficiency of more than 50% in encapsulation of doxorubicin (DOX), a chemotherapy drug [81]. The release rate of DOX from the GA/CDP-DOX nanoparticles is responsive to enzymes. In phosphate-buffered saline solution, only about 12.7 ± 0.3% of the loaded DOX is released within 10 days. In contrast, 75.3 ± 1.4% of the loaded DOX is released in the presence of tyrisin, due to enzyme-induced degradation of GA/CDP nanoparticles.

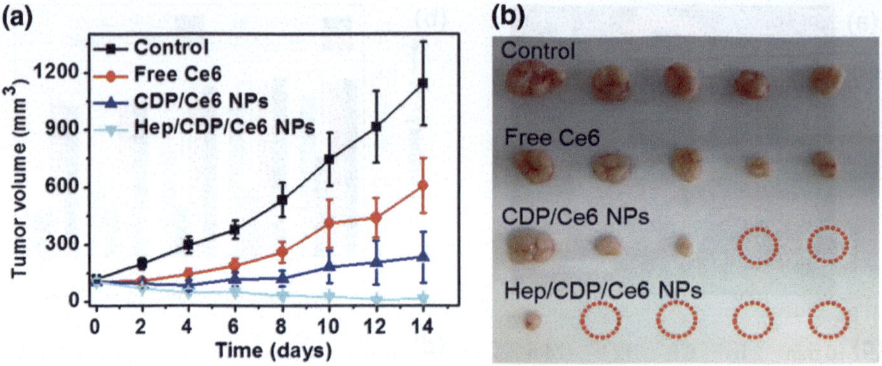

Fig. 7.15 Nanoparticles self-assembled from GA and CDP as drug delivery vehicles of Ce6 toward antitumor PDT. **a** Tumor growth profiles in MCF-7 tumor-bearing mice during PDT. **b** Photographs of the isolated tumors at the end of experiment. The eliminated tumors were expressed in *red circles* (Reprinted with permission from Ref. [82]. Copyright 2016, American Chemical Society)

Significantly, GA/CDP-DOX nanoparticles exhibit a better inhibition capacity toward HeLa cells than the equivalent dose of free DOX, especially at the low concentration of DOX. Also, GA/CDP nanoparticles with a smaller size and surface heparin decoration are demonstrated as efficient drug delivery vehicles for Ce6 [82]. In vivo PDT results show that the drug-loaded nanoparticles (Hep/CDP/Ce6 NPs) are efficient in inhibition of MCF-7 tumors (Fig. 7.15).

7.7 Summary

In summary, peptide-based supramolecular chemistry, including self-assembly of peptides and peptide-modulated self-assembly of functional components have been widely investigated for controllable fabrication of advanced micro- and nanomaterials. Peptide-based materials show tremendous potential for biomimetic and biomedical applications with advantages of tunable micro- and nanostructures and corresponding functions, ease of incorporating organic and inorganic materials, and customized responsiveness to external stimuli. Particularly, hierarchical assembly and structural transition of peptide-based micro- and nanomaterials can be readily realized by manipulation of the non-covalent interactions at different organization levels, providing the formed materials with increasing higher ordered architectures and corresponding functions. As a special kind of peptide-based materials, peptide crystals have shown intriguing optical waveguiding capability promising for optical bio-applications.

Although peptide-based supramolecular chemistry and materials have been significantly promoted by successful examples of both experimental and theoretical studies, there are still formidable challenges in tailored fabrication of peptide-based functional materials for biomimetic and biomedical applications. Even for short peptides, such as dipeptides, the prediction of their self-assembly behavior at various conditions has not been fully realized. Hence, more experiment-based simulations are needed to reveal relationships among environmental parameters, non-covalent interactions, nano- and microstructures, and morphology from the molecular level to material level. With respect to applications in biomimetic energy, integration of more functional units, such as various pigments, charge separators, and reaction centers in the process of peptide-modulated assembly is highly desired in order to mimic the complexity and functions of natural photosystems. For biomedical applications, future clinical translation of peptide-based nanomaterials relies on comprehensive evaluation of their biocompatibility and efficiency in vivo, for example, through the pharmacokinetic and pharmacodynamic analysis. Also, introducing antimicrobial, anticancer, and immunomodulatory sequences into the peptide building blocks is expected to be an efficient method to construct peptide-based functional biomaterials.

Acknowledgements The authors acknowledge financial support from the National Natural Science Foundation of China (Project Nos. 21522307, 21473208, 91434103, and 51403214), the Chinese Academy of Sciences (CAS, Project No. QYZDB-SSW-JSC034), and the Talent Fund of the Recruitment Program of Global Youth Experts.

References

1. Askarieh G, Hedhammar M, Nordling K et al (2010) Self-assembly of spider silk proteins is controlled by a pH-sensitive relay. Nature 465:236–238
2. Ybe JA, Brodsky FM, Hofmann K et al (1999) Clathrin self-assembly is mediated by a tandemly repeated superhelix. Nature 399:371–375
3. Yonekura K, Maki S, Morgan DG et al (2000) The bacterial flagellar cap as the rotary promoter of flagellin self-assembly. Science 290:2148–2152
4. Tanaka S, Kerfeld CA, Sawaya MR et al (2008) Atomic-level models of the bacterial carboxysome shell. Science 319:1083–1086
5. van den Ent F, Amos LA, Lowe J (2001) Prokaryotic origin of the actin cytoskeleton. Nature 413:39–44
6. Luger K, Mader AW, Richmond RK et al (1997) Crystal structure of the nucleosome core particle at 2.8 angstrom resolution. Nature 389:251–260
7. Shih WM (2015) Exploiting weak interactions in DNA self-assembly. Science 347:1417–1418
8. Cerny J, Hobza P (2007) Non-covalent interactions in biomacromolecules. Phys Chem Chem Phys 9:5291–5303
9. Zhang SG (2003) Fabrication of novel biomaterials through molecular self-assembly. Nat Biotechnol 21:1171–1178
10. Hartgerink JD, Beniash E, Stupp SI (2001) Self-assembly and mineralization of peptide-amphiphile nanofibers. Science 294:1684–1688
11. Habibi Y, Lucia LA, Rojas OJ (2010) Cellulose nanocrystals: chemistry, self-assembly, and applications. Chem Rev 110:3479–3500

12. Kuzyk A, Schreiber R, Fan ZY et al (2012) DNA-based self-assembly of chiral plasmonic nanostructures with tailored optical response. Nature 483:311–314
13. Zou QL, Liu K, Abbas M et al (2016) Peptide-modulated self-assembly of chromophores toward biomimetic light-harvesting nanoarchitectonics. Adv Mater 28:1031–1043
14. Stupp SI (2010) Self-Assembly and biomaterials. Nano Lett 10:4783–4786
15. Chen CJ, Liu K, Li JB et al (2015) Functional architectures based on self-assembly of bio-inspired dipeptides: structure modulation and its photoelectronic applications. Adv Colloid Interface Sci 225:177–193
16. Percec V, Dulcey AE, Balagurusamy VSK et al (2004) Self-assembly of amphiphilic dendritic dipeptides into helical pores. Nature 430:764–768
17. Shu SJ, Sun L, Zhang XG et al (2011) Polysaccharides-based polyelectrolyte nanoparticles as protein drugs delivery system. J Nanopart Res 13:3657–3670
18. Rest C, Kandanelli R, Fernandez G (2015) Strategies to create hierarchical self-assembled structures via cooperative non-covalent interactions. Chem Soc Rev 44:2543–2572
19. Mann S (2009) Self-assembly and transformation of hybrid nano-objects and nanostructures under equilibrium and non-equilibrium conditions. Nat Mater 8:781–792
20. Groschel AH, Walther A, Lobling TI et al (2013) Guided hierarchical co-assembly of soft patchy nanoparticles. Nature 503:247–251
21. Xing RR, Jiao TF, Liu YM et al (2016) Co-assembly of graphene oxide and albumin/photosensitizer nanohybrids towards enhanced photodynamic therapy. Polymers 8:181
22. Sharma J, Chhabra R, Cheng A et al (2009) Control of self-assembly of DNA tubules through integration of gold nanoparticles. Science 323:112–116
23. Yan H, Park SH, Finkelstein G et al (2003) DNA-templated self-assembly of protein arrays and highly conductive nanowires. Science 301:1882–1884
24. Ercole F, Davis TP, Evans RA (2010) Photo-responsive systems and biomaterials: photochromic polymers, light-triggered self-assembly, surface modification, fluorescence modulation and beyond. Polym Chem 1:37–54
25. Zimenkov Y, Dublin SN, Ni R et al (2006) Rational design of a reversible pH-responsive switch for peptide self-assembly. J Am Chem Soc 128:6770–6771
26. Zhang SG (2012) Lipid-like self-assembling peptides. Acc Chem Res 45:2142–2150
27. Adler-Abramovich L, Gazit E (2014) The physical properties of supramolecular peptide assemblies: from building block association to technological applications. Chem Soc Rev 43:6881–6893
28. Frederix PWJM, Scott GG, Abul-Haija YM et al (2015) Exploring the sequence space for (tri-)peptide self-assembly to design and discover new hydrogels. Nat Chem 7:30–37
29. Zhou J, Du XW, Yamagata N et al (2016) Enzyme-instructed self-assembly of small D-peptides as a multiple-step process for selectively killing cancer cells. J Am Chem Soc 138:3813–3823
30. Martinek TA, Hetenyi A, Fulop L et al (2006) Secondary structure dependent self-assembly of beta-peptides into nanosized fibrils and membranes. Angew Chem Int Edit 45:2396–2400
31. Aggeli A, Bell M, Boden N et al (1997) Responsive gels formed by the spontaneous self-assembly of peptides into polymeric beta-sheet tapes. Nature 386:259–262
32. Smith KH, Tejeda-Montes E, Poch M et al (2011) Integrating top-down and self-assembly in the fabrication of peptide and protein-based biomedical materials. Chem Soc Rev 40:4563–4577
33. De Santis E, Ryadnov MG (2015) Peptide self-assembly for nanomaterials: the old new kid on the block. Chem Soc Rev 44:8288–8300
34. Cui HG, Webber MJ, Stupp SI (2010) Self-assembly of peptide amphiphiles: from molecules to nanostructures to biomaterials. Biopolymers 94:1–18
35. Gazit E (2007) Self-assembled peptide nanostructures: the design of molecular building blocks and their technological utilization. Chem Soc Rev 36:1263–1269

36. Matson JB, Stupp SI (2012) Self-assembling peptide scaffolds for regenerative medicine. Chem Commun 48:26–33
37. Fernandez-Lopez S, Kim HS, Choi EC et al (2001) Antibacterial agents based on the cyclic D, L-alpha-peptide architecture. Nature 412:452–455
38. Kim S, Kim JH, Lee JS et al (2015) Beta-sheet-forming, self-assembled peptide nanomaterials towards optical, energy, and healthcare applications. Small 11:3623–3640
39. Mandal D, Shirazi AN, Parang K (2014) Self-assembly of peptides to nanostructures. Org Biomol Chem 12:3544–3561
40. Wang J, Liu K, Xing RR et al (2016) Peptide self-assembly: thermodynamics and kinetics. Chem Soc Rev 45:5589–5604
41. Zhang SG, Marini DM, Hwang W et al (2002) Design of nanostructured biological materials through self-assembly of peptides and proteins. Curr Opin Chem Biol 6:865–871
42. Stendahl JC, Rao MS, Guler MO et al (2006) Intermolecular forces in the self-assembly of peptide amphiphile nanofibers. Adv Funct Mater 16:499–508
43. Reches M, Gazit E (2003) Casting metal nanowires within discrete self-assembled peptide nanotubes. Science 300:625–627
44. Yan XH, Zhu PL, Li JB (2010) Self-assembly and application of diphenylalanine-based nanostructures. Chem Soc Rev 39:1877–1890
45. Tao K, Levin A, Adler-Abramovich L et al (2016) Fmoc-modified amino acids and short peptides: simple bio-inspired building blocks for the fabrication of functional materials. Chem Soc Rev 45:3935–3953
46. Amdursky N, Molotskii M, Gazit E et al (2010) Elementary building blocks of self-assembled peptide nanotubes. J Am Chem Soc 132:15632–15636
47. Yan XH, Cui Y, He Q et al (2008) Organogels based on self-assembly of diphenylalanine peptide and their application to immobilize quantum dots. Chem Mater 20:1522–1526
48. Su Y, Yan XH, Wang AH et al (2010) A peony-flower-like hierarchical mesocrystal formed by diphenylalanine. J Mater Chem 20:6734–6740
49. Su Y, He Q, Yan XH et al (2011) Peptide mesocrystals as templates to create an Au surface with stronger surface-enhanced Raman spectroscopic properties. Chem Eur J 17:3370–3375
50. Du MC, Zhu PL, Yan XH et al (2011) Honeycomb self-assembled peptide scaffolds by the breath figure method. Chem Eur J 17:4238–4245
51. Yan XH, He Q, Wang KW et al (2007) Transition of cationic dipeptide nanotubes into vesicles and oligonucleotide delivery. Angew Chem Int Edit 46:2431–2434
52. Yan XH, Cui Y, He Q et al (2008) Reversible transitions between peptide nanotubes and vesicle-like structures including theoretical modeling studies. Chem Eur J 14:5974–5980
53. Liu XC, Zhu PL, Fei JB et al (2015) Synthesis of peptide-based hybrid nanobelts with enhanced color emission by heat treatment or water induction. Chem Eur J 21:9461–9467
54. Zhu PL, Yan XH, Su Y et al (2010) Solvent-induced structural transition of self-assembled dipeptide: from organogels to microcrystals. Chem Eur J 16:3176–3183
55. Knowles TP, Fitzpatrick AW, Meehan S et al (2007) Role of intermolecular forces in defining material properties of protein nanofibrils. Science 318:1900–1903
56. Krone MG, Hua L, Soto P et al (2008) Role of water in mediating the assembly of Alzheimer amyloid-beta a beta 16-22 protofilaments. J Am Chem Soc 130:11066–11072
57. Wang J, Liu K, Yan LY et al (2016) Trace solvent as a predominant factor to tune dipeptide self-assembly. ACS Nano 10:2138–2143
58. Yan XH, Su Y, Li JB et al (2011) Uniaxially oriented peptide crystals for active optical waveguiding. Angew Chem Int Edit 50:11186–11191
59. Li YX, Yan LY, Liu K et al (2016) Solvothermally mediated self-assembly of ultralong peptide nanobelts capable of optical waveguiding. Small 12:2575–2579
60. Yan XH, Li JB, Mowald H (2011) Self-assembly of hexagonal peptide microtubes and their optical waveguiding. Adv Mater 23:2796–2801
61. Li Q, Jia Y, Dai LR et al (2015) Controlled rod nanostructured assembly of diphenylalanine and their optical waveguide properties. ACS Nano 9:2689–2695

62. Li Q, Ma HC, Wang AH et al (2015) Self-assembly of cationic dipeptides forming rectangular microtubes and microrods with optical waveguiding properties. Adv Opt Mater 3:194–198
63. Zou QL, Zhang L, Yan XH et al (2014) Multifunctional porous microspheres based on peptide-porphyrin hierarchical co-assembly. Angew Chem Int Ed 53:2366–2370
64. Liu K, Xing RR, Chen CJ et al (2015) Peptide-induced hierarchical long-range order and photocatalytic activity of porphyrin assemblies. Angew Chem Int Ed 54:500–505
65. Liu K, Kang Y, Ma G et al (2016) Molecular and mesoscale mechanism for hierarchical self-assembly of dipeptide and porphyrin light-harvesting system. Phys Chem Chem Phys 18:16738–16747
66. Yano S, Hirohara S, Obata M et al (2011) Current states and future views in photodynamic therapy. J Photochem Photobiol C 12:46–67
67. Xing RR, Liu K, Jiao TF et al (2016) An injectable self-assembling collagen-gold hybrid hydrogel for combinatorial antitumor photothermal/photodynamic therapy. Adv Mater 28:3669–3676
68. Zhang N, Zhao FF, Zou QL et al (2016) Multitriggered tumor-responsive drug delivery vehicles based on protein and polypeptide coassembly for enhanced photodynamic tumor ablation. Small 12:5936–5943
69. Lucky SS, Soo KC, Zhang Y (2015) Nanoparticles in photodynamic therapy. Chem Rev 115:1990–2042
70. Zhang RY, Xing RR, Jiao TF et al (2016) Carrier-free, chemophotodynamic dual nanodrugs via self-assembly for synergistic antitumor therapy. ACS Appl Mater Interfaces 8:13262–13269
71. Liu K, Xing RR, Zou QL et al (2016) Simple peptide-tuned self-assembly of photosensitizers towards anticancer photodynamic therapy. Angew Chem Int Ed 55:3036–3039
72. Wang J, Mei J, Hu RR et al (2012) Click synthesis, aggregation-induced emission, E/Z isomerization, self-organization, and multiple chromisms of pure stereoisomers of a tetraphenylethene-cored luminogen. J Am Chem Soc 134:9956–9966
73. Liu K, Zhang RR, Li YX et al (2016) Tunable aggregation-induced emission of tetraphenylethylene via short peptide-directed self-assembly. Adv Mater Interfaces 4:1600183
74. Ma HC, Fei JB, Li Q et al (2015) Photo-induced reversible structural transition of cationic diphenylalanine peptide self-assembly. Small 11:1787–1791
75. Ma HC, Fei JB, Cui Y et al (2013) Manipulating assembly of cationic dipeptides using sulfonic azobenzenes. Chem Commun 49:9956–9958
76. Cronin L, Muller A (2012) From serendipity to design of polyoxometalates at the nanoscale, aesthetic beauty and applications. Chem Soc Rev 41:7333–7334
77. Yan XH, Zhu PL, Fei JB et al (2010) Self-assembly of peptide-inorganic hybrid spheres for adaptive encapsulation of guests. Adv Mater 22:1283–1287
78. Xing RR, Jiao TF, Feng L et al (2015) Photothermally-induced molecular self-assembly of macroscopic peptide-inorganic hybrid films. Sci Adv Mater 7:1701–1707
79. Alivisatos AP (1996) Semiconductor clusters, nanocrystals, and quantum dots. Science 271:933–937
80. Yan XH, Cui Y, Qi W et al (2008) Self-assembly of peptide-based colloids containing lipophilic nanocrystals. Small 4:1687–1693
81. Zhang H, Fei JB, Yan XH et al (2015) Enzyme-responsive release of doxorubicin from monodisperse dipeptide-based nanocarriers for highly efficient cancer treatment in vitro. Adv Funct Mater 25:1193–1204
82. Ma K, Xing RR, Jiao TF et al (2016) Injectable self-assembled dipeptide-based nanocarriers for tumor delivery and effective in vivo photodynamic therapy. ACS Appl Mater Interfaces 8:30759–30767
83. Wang J, Shen GZ, Ma K et al (2016) Dipeptide concave nanospheres based on interfacially controlled self-assembly: from crescent to solid. Phys Chem Chem Phys 18:30926–30930
84. Li Q, Ma HC, Jia Y et al (2015) Facile fabrication of diphenylalanine peptide hollow spheres using ultrasound-assisted emulsion templates. Chem Commun 51:7219–7221

85. Liu K, Xing RR, Li YX et al (2016) Mimicking primitive photobacteria: sustainable hydrogen evolution based on peptide–porphyrin co-assemblies with a self-mineralized reaction center. Angew Chem Int Ed 55:12503–12507
86. Petros RA, DeSimone JM (2010) Strategies in the design of nanoparticles for therapeutic applications. Nat Rev Drug Discovery 9:615–627

Chapter 8
Functional Nanomaterials Via Self-assembly Based Modification of Natural Cellulosic Substances

Shun Li, Yuanqing Gu and Jianguo Huang

Abstract Natural cellulose substances possess inherent sophisticated hierarchical structures and morphologies which are impossible to be created by artificial methods at the present time. Precise surface modification of cellulose matters with specific guest substances at the molecular and nanometer scales provides a facile shortcut to combine the unique physical properties of cellulose materials and specifically designed chemical functionalities to give a large variety of new nanomaterials for various practical applications.

Keywords Cellulose · Self-assembly · Nanomaterials · Surface modification · Biomimetic syntheses · Surface sol–gel process

8.1 Introduction

Natural biological organisms exhibit a series of superb properties such as the optical behavior of butterfly wings, superhydrophobicity of water strider legs, and the high porosity of diatoms. Such distinguished functionalities seem to require some complex systems, but these features are intrinsically originated from their naturally produced unique three-dimensional structures actually. Though these natural structures are generally hard to be achieved by artificial approaches, surface modifications of these multilevel structured natural substances with various inorganic or organic guest substrates can readily introduce the amazing structure-related properties of biological species into artificial functional materials [1]. As pioneered by Prof. S. Mann, versatile natural substances, such as silk [2], pollen, insect wings, and so on [3], have been employed as scaffolds for the facile, effective, and environmentally friendly bio-inspired materials syntheses, and a variety of functional native structures were introduced into man-made materials [4].

S. Li · Y. Gu · J. Huang (✉)
Department of Chemistry, Zhejiang University,
Hangzhou 310027, Zhejiang, China
e-mail: jghuang@zju.edu.cn

J. Li (ed.), *Supramolecular Chemistry of Biomimetic Systems*,
DOI 10.1007/978-981-10-6059-5_8

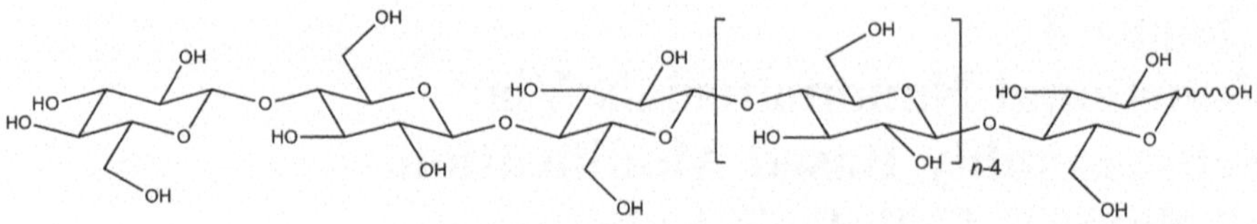

Fig. 8.1 Molecular structure of cellulose (*n* refers to the degree of polymerization). Reprinted with permission from Ref. [6]. Copyright 2005, John Wiley & Sons Ltd.

Among the versatile biological species, cellulose substances are well known as one of the especially important essential materials in both industries and daily lives. In recent research works, it has been revealed that the natural cellulose substances can be used as ideal host matrices for the self-assembly of many kinds of guest substrates to form novel functional materials with specific properties. Natural cellulose is a natural linear polysaccharide of β-(1 $\rightarrow$ 4)-D-glucopyranose (Fig. 8.1). It possesses plenty of surface hydroxyl groups, which form abundant inter- and intramolecular hydrogen bonds. With these hydrogen bonds, the polymer chains assemble into elementary fibrils (diameter 1.5–3.5 nm), bundle to microfibrils (nanofibers, diameter tens to hundreds of nanometers), and further build up randomly cross-linked hierarchical networks of microfibrillar fibers (microfibers, diameter in the micrometer scale or larger), leading to the familiar bulk cellulose substances such as wood, cotton, and moreover the paper-based products after artificial processing [5]. The unique natural fiber-based networks construct a well-defined hierarchical order of supramolecular organization and the partially crystalline fiber domains, which result in the characteristic features of cellulose substances such as flexibility, high porosity, strong mechanical property, hydrophilicity, and significant adsorption as well as swelling behaviors [6].

These low-cost, lightweight, widely available, easily processable, and portable hierarchical-structured natural cellulose substances are ideal choices as platforms for the design and fabrication of advanced functional materials. For example, amorphous domains of natural cellulose fibers of filter paper were etched by the plasma treatment, and the increased surface roughness resulted in enhanced hydrophobicity [7–10]. To apply cellulose substances as substrates, the chemically inert surface of cellulose fibers due to the hydrogen bond networks among the abundant hydroxyl groups is a severe disadvantage for the functional surface modification through self-assembly of determined guest substrates. This drawback has been overcome by ultrathin metal oxide film coating of each cellulose nanofibers at nanometer precisions, which activate the cellulose fiber surfaces and enable self-assembly of versatile guest substrates on them [11]. The sufficient surface hydroxyl groups of the natural cellulose fibers serve as a suitable substrate for thin metal oxide film deposition by the surface sol–gel process. The ultrathin nature of the metal oxide films perfectly reserve the morphological hierarchies and cross-linked structures of natural cellulose substances (such as filter paper and cotton) from macro scales to nanometer level. Self-assemblies with a wide variety of functional guest species on the surface of the as-deposited thin metal oxide films provide a facile and effective route to combine the specific chemical properties of the guest substances and the unique structural features of the natural cellulose substances, targeting the development of a large number of novel nanostructured functional materials. Though self-assembly on cellulose nanofibrils is studied [12] and similar hierarchical structures could be obtained from premodified cellulose pulp, bacterial nanocellulose, or regenerated cellulose dispersion [13], this paper-making approach requires additional process to assemble the modified cellulose fibers into bulk sheets. Obviously, surface modification of the bulk natural cellulose substances is a facile straightforward technique to design and fabricate advanced functional materials [14].

In this chapter, we summarized the recent advances in self-assembly based surface modification of different guest substances on the surfaces of cellulose nanofibers of bulk natural cellulose substances for the preparation of functional nanomaterials as well as their related applications. Versatile guest substrates such as metal oxide, nanoparticles, small molecules, polymers, biomacromolecules, and sp^2-hybridized carbons were employed for the self-assembly approaches, leading to the corresponding nanostructured materials. The natural fiber-based hierarchical networks of the natural cellulose substances are introduced into the artificial materials with tailored functionalities, and the morphologically complex cellulose substances offer effective captivating platforms for surface self-assemblies of varied guest species to nanostructured materials that are superior to the traditional flat two-dimensional matrices.

8.2 Surface Coating of Cellulose Substances with Inorganic Substrates

8.2.1 Deposition of Thin Metal Oxide Films

By applying natural cellulose substances as matrices for surface modification for materials syntheses, the unique natural structure-related functions can be introduced into artificial materials. To fully retain the nature-produced morphological features of cellulose in the corresponding composite products, the key point is to reserve the finest details of the initial cellulose structures. To this aim, the surface sol–gel process-based ultrathin metal oxide film coating of natural cellulose substances was developed, achieving faithful transcription of the macro- to nanoscopic details of cellulose fibers [11]. As illustrated in Fig. 8.2, a dry filter paper washed by ethanol was placed in a suction filtration unit, and titanium *n*-butoxide solution (100 mM in 1:1/v:v toluene/ethanol solution) was first added to allow chemosorption of titanium alkoxide on the hydroxyl-terminated cellulose nanofiber surfaces. Subsequently, ethanol was suction-filtered to remove the excessive solution. Then, pure water was passed through the filter paper to facilitate hydrolysis of the chemisorbed titanium alkoxide, forming an ultrathin titania layer ($\sim$0.5 nm thick) covering each cellulose

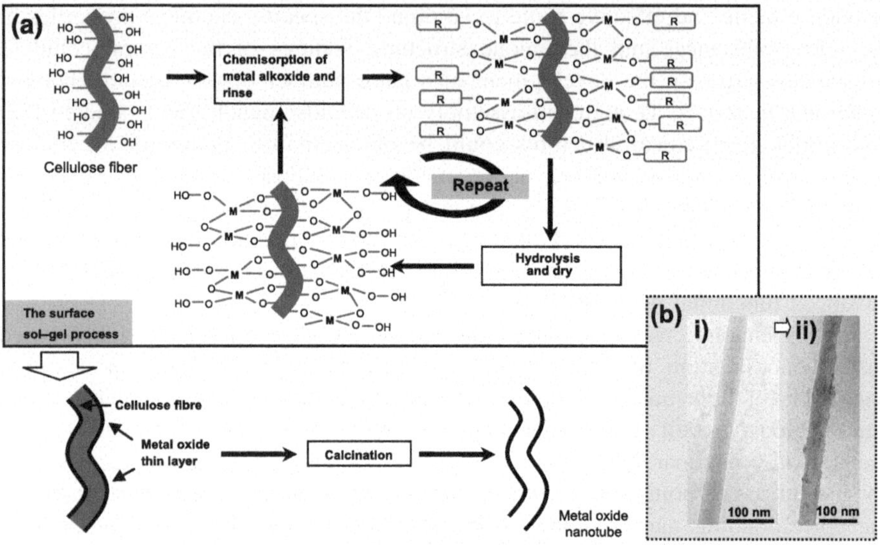

Fig. 8.2 **a** Schematic illustration of the surface modification of natural cellulose fibers with ultrathin metal oxide layers by the surface sol–gel process. **b** Transmission electron micrograph (TEM) images: (i) a virgin cellulose nanofiber of filter paper; (ii) A titania gel film coated cellulose nanofiber; deposition of the ultrathin titania film was repeated 20 times for this sample. Reprinted with permission from Ref. [67]. Copyright 2011, Elsevier Ltd.

nanofiber of the filter paper. By repeating the titania deposition cycle, thin titania films with desired thickness were deposited onto the cellulose fibers. Due to the ultrathin nature of the as-deposited titania gel films, the macroscale as well as nanoscopic details of the cellulose substances were precisely reflected to these thin films. Moreover, the formed new hydroxylated surface enables further modifications with other guest substrates. This facile approach is applicable for the deposition of versatile oxides, and zirconia [11] or silica [15, 16] films were deposited faithfully to coat each cellulose nanofiber of filter paper.

Besides the surface sol–gel process, the ultrathin film coating of titania was also realized by atomic layer deposition (ALD), which is a self-limiting chemical gas-phase methodology for thin film deposition [17]. For the deposition of titania film on each cellulose nanofiber of filter paper, titanium methoxide and water were used as the titanium and oxygen precursors, and the reaction temperatures were controlled at 150 and 250 °C, respectively. Briefly, titanium methoxide vapor was pulsed into the reactor where filter paper was placed; then inert gas was purged to remove the excess titanium precursor; and water vapor was subsequently pulsed into the above reactor to grow a titania (sub) monolayer conformally coating the cellulose nanofibers. The film thickness was precisely regulated by altering the number of the titania deposition cycles (growth rate ~0.03 nm per cycle). Meanwhile, the as-deposited titania films were controlled to be in amorphous or crystalline states according to the deposition temperature. The ultrathin titania films accurately replicated the nanoscale structures of the original filter paper; and the surface area of the titania/cellulose composite was 4 m^2 g^{-1}, which was comparable to that of the original filter paper. The prepared cellulose/anatase titania composite sheet demonstrated effective photocatalytic activity, and the photocatalytic reduction of Ag(I) to Ag by this composite sheet was verified. Similarly, thin alumina film coated cellulose fibers were achieved by ALD process using trimethyl aluminum and water as precursors [18, 19]. In both cases involving ALD processes, additional metal iridium monolayers were deposited onto the surfaces of the metal oxide films precoated cellulose nanofibers to enhance the catalytic performances for photodegradation of methylene blue under ultraviolet (UV) irradiation [18].

Conductive coatings on fibrous networks of bulk cellulose substances are attracting interest for new electronic and other functional systems. Flexible conductive sheets were prepared by depositing thin zinc oxide films on the woven cotton sheet and the paper using diethylzinc and water in an ALD sequence [20]. The thickness of the zinc oxide film was controlled at a growth rate of $\sim$0.21 nm per ALD cycle. According to the 4-probe method, the effective conductivity of the thin zinc oxide film precoated cotton fibers and paper was measured up to 24 S cm^{-1}. Similarly, aluminum oxide films were grown on the surface of the woven cotton fibers at determined growth rates [21, 22].

Natural polysaccharide cellulose is a good source of carbon, and the carbonization of cellulose nanofibers precoated with specific guest species gives a new shortcut to synthesize nanofibrous carbon materials for particular applications. A bulk hierarchical-structured nanofibrous titania-coated carbon material was prepared by applying filter paper as the scaffold and carbon precursor [23]. Each cellulose nanofiber of the filter paper was first coated with ultrathin titania gel films through the surface sol–gel process. Upon the subsequent carbonization at 450 °C under nitrogen gas, the titania surface coating prevented the coalescence effect of the formed carbon species, and the initial cellulose nanofibers were transformed into isolated porous carbon nanofibers with 3–6 nm pores (diameters from tens to hundreds of nanometers). At the same time, the as-deposited amorphous titania gel films were turned into crystalline titania in anatase phase coating the carbon nanofibers (thickness $\sim$12 nm, average anatase titania nanocrystallites size $\sim$4.5 nm). The obtained hybrid material possessed hierarchical network composed of bundled porous carbon nanofibers, which was uniformly coated with ultrathin shells of anatase phase titania (Figs. 8.3a, b). This bulk hybrid sheet was highly porous, and its specific surface area achieved 404 m^2 g^{-1}. With the high porosity, the carbon nanofibers performed as a coadsorbent and significantly enhanced the catalytic efficiency of anatase titania for the photodegradation of various dyes in aqueous solutions. As shown in Fig. 8.3c, only about 50 min was needed to completely decompose the methylene blue aqueous solution (initial concentration 0.03 mM) under high-pressure, fluorescent mercury lamp irradiation.

Titanium carbide is a high-temperature structural material, which shows high chemical stability and significant thermal as well as electrical conductivity. By carbothermal reduction of the paper with each cellulose nanofiber precoated with thin titania films, hierarchical titanium carbide was synthesized [24]. Cellulose

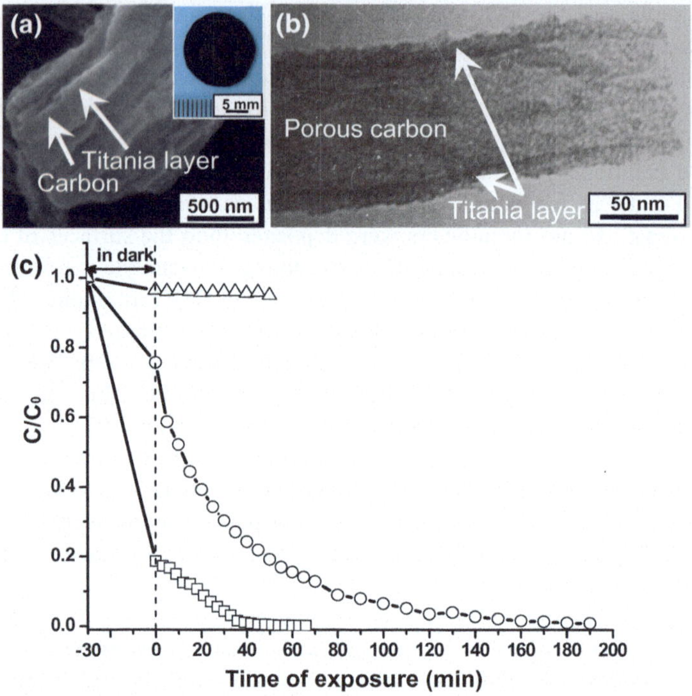

Fig. 8.3 Hierarchical nanofibrous titania-coated carbon material derived from filter paper. **a** Field emission scanning electron microscope (FE-SEM) close-up image of the carbon/titania hybrid fiber obtained by carbonization of filter paper whose nanofibers were precoated with thin titania films, *arrows* indicate the carbon–titania core–shell nanostructure; the inset is a photograph of the black bulk material. **b** TEM micrograph of one individual titania-coated carbon nanofiber. **c** Variations in the concentrations of methylene blue in an aqueous suspension as a function of the irradiation time under high-pressure, fluorescent mercury lamp irradiation for the nanofibrous titania-coated carbon material (*square*), the nanotubular titania material (*circle*), and the carbon material (*triangle*). Reprinted with permission from Ref. [23]. Copyright 2010, John Wiley & Sons Ltd.

substance and aqueous-based Tyzor-LA were employed as the carbon precursor and the titania source, respectively. The as-prepared cellulose/titania composite sheet was carbothermally reduced in argon (temperature above 1300 °C). The final product maintained the original cellulose structures upon the agglomeration of 10–50 nm titanium carbide nanoparticles, which were in cubic phase with high crystallinity.

Alternatively, the pyrolysis of the iodine-adsorbed traditional Japanese washi paper resulted in carbon paper [25]. While the washi paper and ordinary paper both possess natural fiber-based hierarchical cellulose networks, the washi paper consisted of relatively longer microfibers. The washi paper was first exposed to iodine gas at room temperature for several days in a sealed glass vessel, and then the pretreated paper was pyrolyzed at a temperature over 800 °C in nitrogen gas (heating rate 10 °C min^{-1}). The iodine adsorbed on the surfaces of the cellulose fibers in the washi paper inhibited the thermal decomposition of cellulose at the high-temperature carbonization process, which significantly improved the carbon yield and led to the reservation of the fibril structures of the initial washi paper at high quality. The crystalline degree and the electrical conductivity of the obtained carbon paper were improved by the pyrolysis at higher temperatures, and the carbon paper fabricated at 1800 °C exhibited the electrical conductivities of 24–27 S cm^{-1}.

8.2.2 Modification with Nanoparticles

Apart from thin film coating of the cellulose fibers, modification with tiny particles also provides a pathway to obtain bulk functional materials. For example, colloidal titania was fixed onto cellulose fibers of filter paper with silica sol as the binder, retaining improved photocatalytic performance for anionic dyes [26]; zinc oxide crystals (diameter ~5 μm) were adsorbed on filter paper directly, realizing UV light detection through time-dependent current response [27]; zinc oxide nanoparticles (average diameter ~20 nm) were attached onto the fibrous network of filter paper through hydrogen bonding by ultrasonication under base condition, showing antibacterial activity against *Escherichia coli* [28]; the montmorrilonite clay/PAA nanocomposites were coated onto filter paper using a hand coater, achieving superior barrier properties against water and water vapor [29]; redox nanoparticles of cerium oxide (diameter 20 nm) and glucose oxidase were co-immobilized onto the fibers of filter paper through a silanization process, attaining hydrogen peroxide induced colorimetric sensing of glucose with a detection limit of 0.5 mM and a good reversibility [30]; biocidal additives like colloidal silver, silver nitrate, cetyltrimethylammoniumbromid, and octenidine were dispersed in silica sol and coated onto the cellulose fibers' surfaces through the sol–gel process [31–33], demonstrating an excellent long-term inhibition against fungi growth; silver nanoparticles were printed as ink onto paper, performing as radio-frequency identification tags and wireless sensor nodes [34]; manganese ferrite ($MnFe_2O_4$) nanoparticles, CdSe/ZnS quantum dots, silver nanoparticles, poly(ethyl-2-cyanoacrylate), and wax particles were, respectively, modified onto cellulose microfibers through hydroxyl group induced polymerization of ethyl-2-cyanoacrylate as the cross-linker, endowing paper with magnetism, photoluminescence (emission peak at 580 nm), antibacterial effect, and hydrophobicity (water contact angle 140 ± 2°) [35].

Semiconductor nanocrystals show unique features like the relatively pure emission, the high photoluminescence quantum yield, size-depended optical properties, and durability to photobleaching, which promote their extensive studies as functional materials. Yet, such nanocrystals are usually unstable when dispersed in liquid dispersions. For practical applications, it is desirable to stabilize these nanocrystals into certain matrices. Doped zinc sulfide nanocrystals were deposited onto cellulose fibers of bulk paper by direct adsorption of nanocrystals dispersed in aqueous solution [36]. The resulting composite sheet retained the hierarchical structures with zinc sulfide nanocrystals (diameter 30 nm) homogeneously distributed on cellulose fiber surfaces, achieving high tensile strength and flexibility as well as luminescence.

Novel luminescent sheet material was also prepared by the self-assembly of cadmium selenide (CdSe) nanoparticles on natural cellulose substance (common commercial filter paper) as the scaffold [37]. The surface of each cellulose nanofiber of the filter paper was first coated with ultrathin titania film, and then stearic acid was deposited onto these nanofibers covalently. Subsequently, the trioctyl-phosphine-oxide/hexadecane capped CdSe nanoparticles and stearic acid were alternatively self-assembled on the above-mentioned hybrid nanofibers through hydrophobic interactions. The initial morphological hierarchy and the fibrous structure of the natural cellulose scaffold were perfectly preserved in the final product, promising the strong mechanical property and good swelling behaviors. The retained natural structure derived large surface area and high porosity enabled high-density immobilization and stabilization of CdSe nanoparticles without further particle aggregations, resulting in stable green fluorescence of the obtained sheet material for even several months (Fig. 8.4). The luminescent cellulose sheet demonstrated the characteristic UV-Vis absorption band centered at around 500 nm (Fig. 8.4e, dotted line) and a narrow photoluminescent emission peak at near 530 nm (Fig. 8.4e, solid line), which agree with the virginal monodispersed CdSe nanocrystals. This facile precise surface modification approach opens new opportunities for the design and synthesis of cellulose-based stable luminescent materials for practical applications. Alternatively, cellulose was partially regioselective dendritic functionalized as $(HO_2C)_{27}$-dendrimerized-cellulose, and cadmium sulfide (CdS) quantum dots were self-assembled onto the dendrimerized part of the cellulose through hydrogen bonds, forming CdS/cellulose hybrids with corresponding photoluminescence [38].

Functionalization of cellulose substances with magnetic nanoparticles provides chances for important applications such as security tags, magnetic actuators, electromagnetic shielding materials, and magnetographic printing. Recently, magnetic microactuators were prepared through wetting the cellulose fibers of commercial paper with a ferrofluid [39]. The nanofillers impregnated the pores among the fiber-based cellulose networks by forming a magnetic layer. To improve the water resistivity of the composite sheet, polymer layers were further covered on the sheet surface. In another approach, a hydrophobic superparamagnetic nanocomposite with hierarchical cellulose networks was obtained by dip-coating a piece of paper in an ethyl-2-cyanoacrylate monomer solution containing superparamagnetic manganese ferrite colloidal nanoparticles (weight ratio 5/95 of nanparticles/ethyl-2-cyanoacrylate) for 15 s and drying [40]. Upon the moisture initiated polymerization of the reactive monomers on the cellulose fiber surfaces, a thin nanocomposite shell forms around each fiber (100–150 nm thick). Inside the polymer shell, many $MnFe_2O_4$ nanoparticles with average diameter of 9 nm were embedded. The overall physical features of the initial paper were not affected obviously, forming a three-dimensional soft superparamagnetic hydrophobic shell. The nanocomposite coating endowed the cellulose fibers with water repellent (contact angle 120°–140°) and superparamagnetism (saturation magnetization 61 emu g^{-1} at room temperature). The quantity of the nanoparticles entrapped in the composite shells was

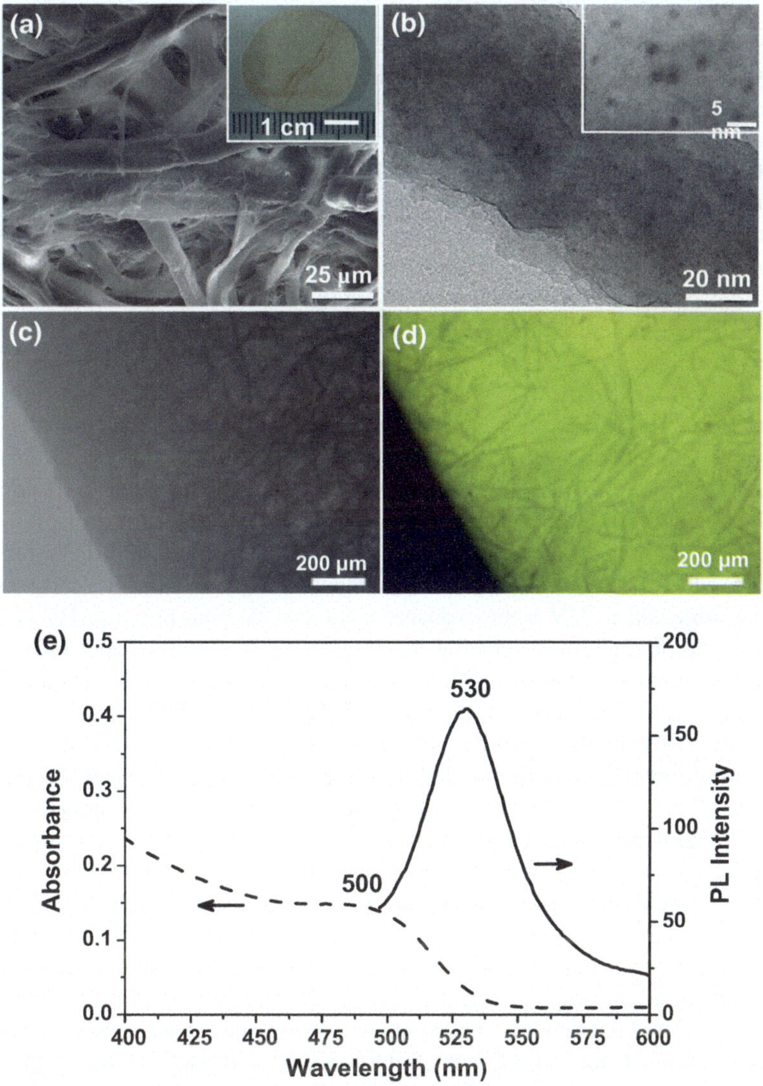

Fig. 8.4 Luminescent cellulosic sheet fabricated through self-assembly of CdSe nanoparticles on cellulose nanofibers. **a** FE-SEM, **b** TEM, **c** bright field micrograph, **d** corresponding fluorescence micrograph of the filter paper/titania/SA/$(CdSe/SA)_5$ sample; the insets of **a** and **b** are the photographs of the luminescent cellulosic sheet and the enlarged TEM image of the surface of the resultant sample, respectively, and **e** UV-Vis (*dotted line*) and photoluminescent (*solid line*) spectra of the luminescent cellulosic sheet. Reprinted with permission from Ref. [37]. Copyright 2011, Royal Society of Chemistry

controlled by their initial concentration in the monomer/nanoparticles solution as well as the density of the overall solution, and further affected the magnetic response of the final product. Such hydrophobic and magnetic material opens up the way for the applications of cellulose sheets in sensors and fluidics uses with huge economic benefits.

Noble metal nanoclusters such as silver nanocrystals (AgNCs) are usually consisted of several to some tens of atoms. Their sizes are typically below 2 nm and are comparable to the Fermi wavelength of electrons, which result in discrete electronic energy levels and molecular-like properties like strong fluorescence. Such fluorescence of the nanoclusters is highly sensitive to the changes in their local microenvironment, which offers a powerful opportunity to develop ultrasensitive nanoscale sensors for the ions such as the copper(II) ion. For practical applications, fluorescent AgNCs were immobilized onto a paper as a sensing platform, forming sensitive test paper for the detection of copper(II) ions [41]. In a typical procedure, azobenzene-modified poly(acrylic acid) (MPAA) was first synthesized by stirring poly(acrylic acid) (PAA), 4-aminoazobenzene, and 1.2 equiv. of dicyclohexylcarbodiimide in *N*-methyl-2-pyrrolidone at 60 °C for 24 h. A freshly prepared silver nitrate solution (19.1 g L^{-1}, 5 mL) was added into the solution of MPAA (5 g L^{-1}, 20 mL). The mixture was incubated in the dark for ~30 min, and then was subjected to UV light exposure by a 250 W high-pressure Hg lamp for 3 min for photoreduction, producing fluorescent AgNCs. Filter paper pieces were then soaked in the solution of AgNCs for a few hours and dried in air under room temperature. AgNCs uniformly adsorbed onto the cellulose fibers of the filter paper pieces, and they kept their fluorescence response toward copper(II) ions. With the presence of copper(II) ions in barreled drinking water and river water samples, the quenching of initial fluorescence was observed by the naked eye under UV irradiation. By applying paper as the platform, a new avenue to the development of robust clusters-based sensing materials is opened up.

Copper is one of the well-known broad spectrum biocides, which shows effective growth inhibition of bacteria, fungus, and algae. Copper-based nanomaterials coating of cellulose fibers of paper based on electrostatic assembly process represents an effective pathway for imparting antimicrobial properties [42]. First of all, carboxymethylated cotton fabric was synthesized by immersing cotton samples in an aqueous mixture of the sodium salt of chloroacetic acid and sodium hydroxide at 60 °C for 15 min. The cleaned fabrics were then immersed in 1–10 mM copper sulfate and incubated at room temperature for 3 h. After washing with pure water, the cotton-M^{n+} chelate was reduced to copper nanoparticles by treating with an aqueous solution of sodium borohydride (50 mM), and dried in air. The cupric ions were grafted onto the chemically transformed anionic cellulose fiber surface through chelating interactions, and zero-valent metal particles formed after the subsequent reduction treatment. A stable and conformal surface coating of sub-5 nm copper nanoparticles were thus deposited onto the cellulose fiber surfaces. The resulting copper–cotton composites demonstrated highly effective antimicrobial activity against a multidrug resistant bacterium *Acinetobacter baumannii* (e.g., 8-log reduction in the target bacterium in 10 min). While the silver-cotton materials identically prepared from silver nitrate as the precursor release silver nanoparticles, no copper nanoparticle was released from the copper–cotton composites, and the latter sample can support mammalian cell growth. These results highlight the nanoparticle–polymer composites with respect to the development of stable, highly efficient antibacterial coatings.

8.3 Self-assembly of Polymers on Cellulose Fibers

8.3.1 Self-assembly of Conjugated Polymers

Conjugated polymers show conductivity when doping with specific electron acceptors or donors. For instance, polypyrrole (PPy), polyaniline (PANI), polythiophene, and their derivatives are an especially important class of polymers. Unfortunately, their potential applications are often faced with limitations because of their inherent intractability, infusible properties, brittleness, and low solubility in most solvents. Thus, the micro-to-nanometer scale morphological control over the conjugated polymer materials is desired. Self-assembly based thin film coating of conjugated polymer on cellulose nanofiber surfaces of natural cellulose substances opens a comparatively direct shortcut to achieve functional composite materials with greater processability, reasonable conductivity, high flexibility, and environmental stability for practical uses.

By the polymerization-induced adsorption process, thin PPy films were deposited coating each cellulose nanofiber of the filter paper, attaining composite materials composed of cellulose/PPy bi-hybrid or cellulose/titania/PPy tri-hybrid nanocables with well-reserved interconnected network structures of the initial cellulose substance (Fig. 8.5) [43]. To obtain the cellulose/PPy composite sheet, a piece of filter paper was immersed into an extremely dilute polymerization solution, in which the pyrrole oligomers were adsorbed onto the cellulose nanofiber surfaces and polymerized into homogeneous ultrathin amorphous PPy films ($\sim$20 nm), forming bi-hybrid nanocable structure. These nanocables were composed of cellulose nanofiber core with seamlessly PPy sheath covered their outside. The thickness of the retained PPy films was precisely controllable by altering the adsorption time. With the retained fiber-based porous network of cellulose, the resultant hierarchical PPy/cellulose composite sheet was provided with meaningful properties like high mechanical strength and large surface area of PPy. Meanwhile, this surface nanocoating approach was also shown to be an effective way to control the surface physical properties of cellulose substances (e.g., the hydrophilicity). This in situ polymerization-based nanocoating of PPy was also applied to cotton fabrics [44]. To achieve the surface nanocoating of PPy, the affinity between pyrrole monomers and/or oligomers for the surface of the matrix substance is crucial. As for the mechanism of surface coating with PPy, it was proposed that PPy oligomers adsorb to the matrices once their molecular length grow too long to be

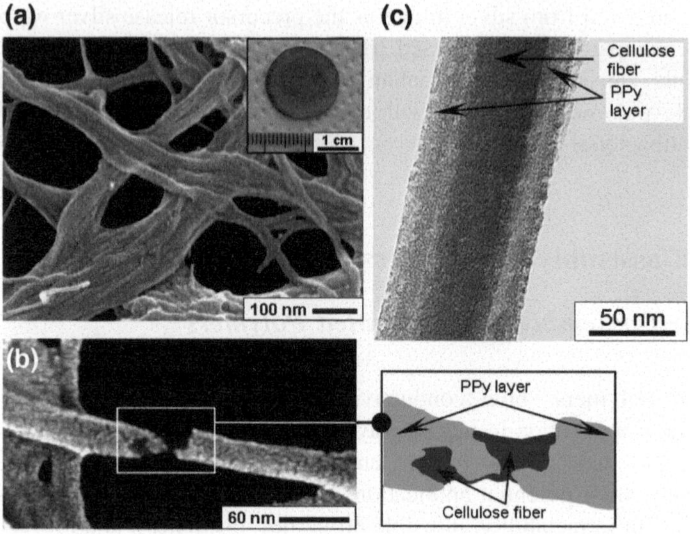

Fig. 8.5 PPy-coated filter paper. **a** FE-SEM image of the sample, envisioning the fibrous assembly; the inset is a photograph of the bulk sheet. **b** FE-SEM image of a PPy-coated cellulose fiber which was broken by extended exposure to the electron beam; morphology of the broken part is schematically illustrated in the right inset. **c** TEM image of a PPy-coated cellulose fiber. Reprinted with permission from Ref. [43]. Copyright 2005, Royal Society of Chemistry

soluble, with subsequent preferential adsorption of oligomers to the PPy nucleation points [45]. When the filter paper is applied as the matrix, the strong chemical interaction between the amine groups in pyrrole and available hydroxyl groups present at the surface of cellulose fibers promote the adsorption of PPy oligomers onto each of the cellulose nanofiber [46]. As revealed by the X-ray photoelectron spectroscopy studies, a shift in the O1 s peak corresponding to C–OH bonds occurred from 533.3 to 530.6 eV for the virginal cellulose and the PANI-coated analogue, which indicates the existence of the chemical bonding between available hydroxyl groups in cellulose and amine groups in the related polymer chains [47]. It is noteworthy the aggressive acidic environment should be avoided for the surface coating of conductive polymer on cellulose fibers, as it would cause degradation of cellulose and thus damage the tensile strength of cellulose composites [48]. This methodology is of sufficient generality and opens new doors for the design and fabrication of nanostructured materials with insoluble conjugated polymers. Instead of copper chloride, iron(III) chloride was applied as oxidative reagent for the in situ polymerization process and cellulose/PPy composite sheet was reported [48].

Alternatively, PPy thin layer was deposited onto the nanofibers of wood cellulose-based commercial filter paper and the *Cladophora* cellulose paper by a dip-coating polymerization process [49]. First, the paper sheets were soaked in the 3.3 M pyrrole solution. The soaked sheets were further immersed in the 3.0 M iron(III) chloride solution for 10 min, and then the composites were washed with water and dried in air. Both composite sheets retained the initial morphological hierarchy as well as the fibrous bundling structures, and the PPy components modified as spherical nanoparticles on the cellulose fibers. By repeating the dip-coating process, the quantity of the PPy nanoparticles was increased. The charge capacity of the dip-coated filter paper ranged from 50 to 80 C g^{-1}. Dip-coating of poly[2-methoxy-5-(2-ethylhexyloxy)-1,4-phenylenevinylene] on pure cotton fibers was also performed [50], and a photoconductive composite with the excellent white-light emission consisted of blue-, green-, and red-light-emitting bands.

Due to the high surface area of the cellulose scaffold, the as-deposited cellulose/PPy composite sheet showed a total anion exchange capacity of *ca.* 1.1 mol kg^{-1}. By changing the electronic charge through the galvanostatic oxidation and reduction of the deposited conformal PPy layer coating on each cellulose nanofiber, the cellulose/PPy composite sheet was used for the extraction and the subsequent release of the negatively charged fluorophore tagged DNA oligomers [51]. The tested $(dT)_6$, $(dT)_{20}$, and $(dT)_{40}$ DNA oligomers were all extracted by the cellulose/PPy composite sheet, and the oligomers were preferentially released during the reduction of the PPy layer. The electrochemically controlled extraction and release behaviors of the cellulose/PPy composite sheet provide an inexpensive and efficient shortcut for the batchwise extraction of a range of different biomolecules. Besides, the application of PPy surface-coated cloth as the heating pad under a fixed voltage was reported [52], and the heat generated was up to 1000 W m^{-2} depending on the percentage of the modified pyrrole.

Compared with other conjugated polymers, PANI possesses unique reversible doping/dedoping chemistry accompanied with color changes. The PANI backbone is protonated by hydrochloride and transforms green colored emeraldine salt type of PANI, and these polymer chains are deprotonated and form blue emeraldine base one with the presence of ammonia. This obvious color change behavior was utilized as an acid/base indicator, and filter paper was coated with PANI film through the in situ polymerization process for the colorimetric detection of gaseous and aqueous ammonia [53]. Originating from the large surface area of cellulose network, the obtained material realized low detection limits for ammonia vapor (45 ppm) and ammonia solution (14 ppm) by the naked eye.

The polymerization behavior of conductive polymer PANI is electrically responsible, and thus PANI coating on cellulose nanofiber of paper was accomplished by electrochemical polymerization of the surface adsorbed aniline monomers on cellulose fibers [54]. Briefly, the paper was first successively printed with silver paste and carbon paste via forcing the respective ink. After drying, the printed paper was immersed in the solution containing 0.3 M aniline in 1 M hydrochloride, and the solution was exposed to a potential cycling ($n = 9$) between −200 and +1200 mV versus the reference electrode at a scan rate of 100 mV s^{-1} employing cyclic voltammetry to promote polymerization. Thus, the stable and uniform conducting PANI film was coated onto the filter paper. With the 1-ethyl-3-[3-dimethylaminopropyl]carbodiimide hydrochloride and *N*-hydroxysuccinimide as the crosslinkers, the monoclonal antibodies of the human cardiac Troponin I were further immobilized onto the fibers of the as-prepared composite sheet, realizing the troponin (a cardiovascular biomarker) detection in a physiological range of 1–100 ng mL^{-1} with a sensitivity of 5.5 μA ng^{-1} mL^{-1} cm^{-2}. This approach shows potential applications for the development of low-cost and high-sensitive point-of-care diagnosis. As the conducting polymer PPy and PANI possess redox properties, further deposition of silver nanoparticles on the polymer precoated cellulose fiber surfaces are allowed to impart antimicrobial activity to it [55].

Poly(3,4-ethylenedioxythiophene) (PEDOT) is a sort of highly conductive polymer with high thermostability and light resistance, which afforded as coating substrates for bulk cellulose substances [56]. To fabricate the conductive PEDOT-coated paper, 3,4-ethylenedioxythiophene and Baytron oxidizing agent (ferric chloride and *p*-toluenesulfonate in methanol) were mixed and spread on the filter paper with a glass rod. In the ambient atmosphere, the mixture underwent polymerization quickly, and the color of the filter paper gradually changed from white to black. Through in situ polymerization of 3,4-ethylenedioxythiophene, a PEDOT layer was uniformly formed on the cellulose fibers of filter paper, forming a hierarchical structured conductive composite paper. Due to the PEDOT coating, the composite paper exhibited the electrical conductivity of $\sim$1.8 S cm^{-1} and drastically increased tensile strength, which are promising for practical applications.

Alternative to the chemical modification of cellulose substances, the printing methodologies with the inks containing certain nanomaterials were also used to deposit guest substrates on paper substances [57]. PANI–dodecylbenzenesulfonic acid patterns were produced on the paper by the roll-to-roll printing method with the gravure printer, achieving the surface resistivity of 10 kΩ sq^{-1} [58]; PEDOT/poly(4-styrene sulfonate) was coated onto paper using a labcoater/dryer, retaining a surface resistivity of 5.5 Ω sq^{-1} [59]. It should be noted that the inkjet printing technique on cellulose fibers is not applicable to involve the direct chemical polymerization of conductive polymers, because the most commonly used oxidants like iron(III) chloride, ammonium persulfate, etc., require low pH (typically pH $\sim$ 2) and the corrosive nature of these oxidants would deteriorate the printing devices [60].

By surface coating of conductive polymers on natural cellulose substances, the unique structural properties of cellulose were introduced into these polymers, achieving flexible, mechanically strong, and low-cost conductive sheets. Though the energy density of this sort of cellulosic materials would be relatively lower than those of many metal-based systems, these conductive composite sheets could be applied to the fields where other devices cannot be used due to the demands for flexibility, conformability, and environmental friendliness, for example.

8.3.2 Modification with Other Polymers

2,7-Poly(9,9-dialkylfluoreneco-fluorenone) (PFFO) is one of the photoluminescent polymers. It possesses fast hole mobility and high photoluminescent quantum yield within emission wavelengths in the blue spectral region, and they blew up increasing interest for potential applications in electro-optics and optoelectronics. PFFO has a good thermal stability in both solutions and thin films, which allows their usage as building blocks to construct many kinds of functional structures. By electrostatic interactions mediated direct adsorption of cationic PFFO colloidal polymer particles on the cellulose fibers of the bulk paper, homogeneous photoluminescent sheet was fabricated [61]. To deposit these luminescent colloidal polymer particles onto cellulose fibers, the fibers in the paper were first oxidized to enhance the electronegativity, and then the polymeric colloidal spheres were self-assembled onto the surfaces of the oxidized cellulose fibers through direct electrostatic interactions, forming composite paper sheets showing efficient yellowish photoluminescence.

Layer-by-layer (LBL) self-assembly has been proven to be a facile and effective approach to modify solid matrix surface through forming thin multilayers with designated composition and properties [62]. With the LBL self-assembly technique, various combinations of oppositely charged polyelectrolytes such as (polydimethyldiallylammonium chloride/poly(styrene sulfonate), polyallylamine (PAH)/PAA, polyethylene oxide/PAA [63, 64], and (2-(dimethylamino)ethyl methacrylate)/(2,2,2-trifluoroethyl methacrylate) were deposited on to the fibers of cellulose substances [65], and the contract angle of the cellulose fibers was increased to 107°–160° depending on the hydrophobicity of the modified multilayer films.

Atom transfer radical polymerization (ATRP) is a technique to form carbon–carbon bond using transition metal catalysts [66]. This method was utilized to graft specific polymers onto cellulose nanofibers, the initially hydrophilic cellulose fibers were turned into hydrophobic ones, resulting in superhydrophobic cellulose-based materials [67]. First, the cellulose fiber was modified via the reaction of the hydroxyl groups with 2-bromoisobutyryl bromide as the initiator; then, with the mediation of tris(2-(demethylamino)ethyl)amine and copper(I) bromide, methyl acrylate was grafted onto the initiator at ambient temperature, and further polymerized employing ethyl acetate. Because the degree of polymerization depended on the ratio between the free initiator and the monomer, ethyl 2-bromoisobutyrate was applied as a sacrificial initiator to control the polymer length of the poly(methyl acrylate)-grafts. With the grafted polymer, the fabricated composite paper became hydrophobic, and the water contact angle reached 133°.

Various cationic polymers have been widely investigated for applications in medical diagnostics and molecular biology due to their unique binding behaviors and availability to interact with biological molecules in sophisticated biological environments. Through ATRP of the initiator bound on the cellulose fiber surface of the paper, linear cationic polymer brushes were grafted onto cellulose paper [68]. To grow the polymer brush on the paper surface, the initiator was first immobilized onto the cellulose fiber surface by immersing the paper in a mixture solution containing 25 mM 2-bromoisobutyryl bromide (BIB), 27.5 mM triethylamine, and a catalytic amount of 2-dimethyl aminopryridine in 40 mL tetrahydrofuran. As the BIB component was less hydrophilic, the BIB grafted paper turned to slightly hydrophobic. By immersing the BIB-modified paper in a reaction mixture of 40 mmol 2-(dimethylamino)ethyl methacrylate, 1 mmol BIB, 0.125 mmol *N*,*N*,*N*′,

N′,*N″*-pentamethyldiethylenetriamine, 0.125 mmol copper chloride, 0.0034 g L-ascorbic acid, and 30 mL tetrahydrofuran, the mixture was left for 30 min to 2.5 h to guarantee polymer further grafted on the BIB-modified cellulose fibers. The cationic polymer grafted paper was able to interact with PicoGreen, demonstrating specific identification of nucleic acids at the nanomolar level in the homogeneous solution and in a biological serum. ATRP with hydroxyl groups of cellulose as the initiator, the surface modifications of cellulose fibers of filter paper were accomplished with versatile guest substances such as PMA [69], poly(2-hydroxyethyl methacrylate) [70], poly(ε-caprolactone), and poly(L-lactic acid) [71]. Instead of directly polymerizing on cellulose surface, polymerization of polystyrene-*b*-poly (ethylene oxide)-*b*-poly(styrene) block copolymer was first proceeded with poly (ethylene oxide)chloro telechelic as macroinitiator and copper chloride/2,2′-bipyridine as the catalyst, and then printed onto paper using a bar coater [72]. The ratio of hydrophilic and hydrophobic blocks in the adsorbed block copolymers served to control the water absorption behavior of paper, and the synthesized block copolymer containing 10 wt% polystyrene retarded water absorption considerably.

Using reversible addition–fragmentation chain transfer polymerization, 2-(dimethylamino)ethyl methacrylate was grafted on to the cellulose fibers of filter paper, forming antibacterial cellulose sheet [73]. After grafting poly(2-(dimethylamino)ethyl methacrylate) chains on cellulose fiber surfaces, the tertiary amino groups of these poly(2-(dimethylamino)ethyl methacrylate) chains were then quaternized using alkyl bromides with varied chain lengths, leading to cellulose modified by quaternary ammonium groups. In the resultant composite sheet samples, the hydrophobicity and the antibacterial activity of the samples were decided by the alkyl chain lengths of the quaternary ammonium groups and the degree of quaternization, respectively. With this polymerization approach, controlled grafting of polystyrene [74], poly(methyl methacrylate), and PMA was accomplished on cellulose fibers [75].

Alternatively, the superhydrophobic cellulose-based materials were fabricated by the means of polymethylsiloxane coating through the chemical vapor deposition (CVD) technique [76] and the solution-immersion process [77]. For the CVD approach, trichloromethylsilane molecules were adsorbed onto cotton fibers and reacted with the halide at the hydroxyl group, followed by hydrolysis in pyridine aqueous solution. After the subsequent polymerization, porous composite sheet with each cotton fiber was coated with polymethylsiloxane. For the second case, cotton fabric or filter paper was applied as the scaffold. The scaffold sheet was immersed in potassium methyl siliconate aqueous solution, and then the mixture was bubbled with carbon dioxide gas to adjust the pH value within a range of 7.5–8.5. The methyl siliconate gradually formed silanol and slowly condensed, giving oligomeric and polymeric siloxane coating on the surfaces of the cellulose fibers. In both cases, the initial porous cellulose networks were well preserved. With the natural cellulose structures and grafted polymers, the initially hydrophilic cellulose surfaces successfully become superhydrophobic ones with satisfying durability.

To improve the surface hydrophobicity of cellulose substance, another practical approach is to increase the surface roughness by attaching polymeric nanoparticles (diameter 90–105 nm) onto the hierarchical fibrous network of paper [78, 79]. Nanoparticles were synthesized by partial imidization of styrene maleic anhydride in the presence of ammonium hydroxide, and the aqueous dispersions of these poly (styrene maleic anhydride) nanoparticles were spread onto paper using a laboratory bar coater with determined bar speed. Through hydrogen bonding, the nanoparticles were adsorbed onto the cellulose nanofibers uncontinuously. The unique rough structure achieved a stable water contact angle of 120 °C, and the wettability of as-prepared paper was controlled by changing the imide content and the average surface roughness.

Plasma-assisted coatings have a significant application potential as nanometer-thick films which can be tailored to impart some desirable surface property to cellulose platform. Coatings are achieved by energizing suitable vapor-phase monomers in a plasma reactor and the activated species are formed as a coating on the surface of cellulose fibers. In this way, perfluoromethylcyclohexane [80], pentafluoroethane, octafluorocyclobutane [81], pentadecafluorooctylacrylate, perfluorododecene, tridecafluorooctene, perfluoromethylcyclohexane [82], methane [83], hexamethyldisilazane, *n*-hexane, and tetraethyl orthosilicate [84] were used as initiate monomers for the surface modification of cellulose fibers, forming hydrophobic sheets.

8.4 Molecular Monolayer Nanocoating on Cellulose Fibers

8.4.1 Self-assembly of Long Alkyl Chain Molecules

The ultrathin titania film deposited coating each cellulose nanofiber of cellulose substances by the surface sol–gel process not only reserves the initial morphology of cellulose fibers, but also activates the chemically inert cellulose surface, allowing further deposition of molecular monolayers with expected functions on the cellulose platform [11]. By immersing natural cellulose substances like filter paper and cotton with each nanofiber precoated by ultrathin titania film (thickness ~2.5 nm), a long alkyl chain siloxane (octyltrimethoxysilane, OTMS) monolayer was deposited on the nanofiber surfaces, giving bulk cellulosic materials with significant superhydrophobicity (water contact angle > 150°) (Fig. 8.6) [85]. The formed superhydrophobic cellulosic materials exhibited remarkable self-cleaning performances and long-term chemical stabilities. The superhydrophobicity remained stable within a wide pH range from 3.0 to 11.0 (Fig. 8.6b). The contact angle decreased appreciably only with strong acidic (pH 1.0) and strong basic (pH 13.0) droplets due to the instability of titania under these conditions. The combination of titania nanocoating and subsequent self-assembly of alkylsilane monolayer opens a facile and effective door to alter extremely hydrophilic natural cellulose into superhydrophobic materials. Various siloxane precursors such as tetraethoxysilane, methyl triethoxysilane, dimethyl diethoxysilane, and trimethyl monoethoxysilane were employed to achieve hydrophobic coating on cellulose fibers of paper through sol–gel dipping process [86]. It was revealed that the hydrophobic behavior of the

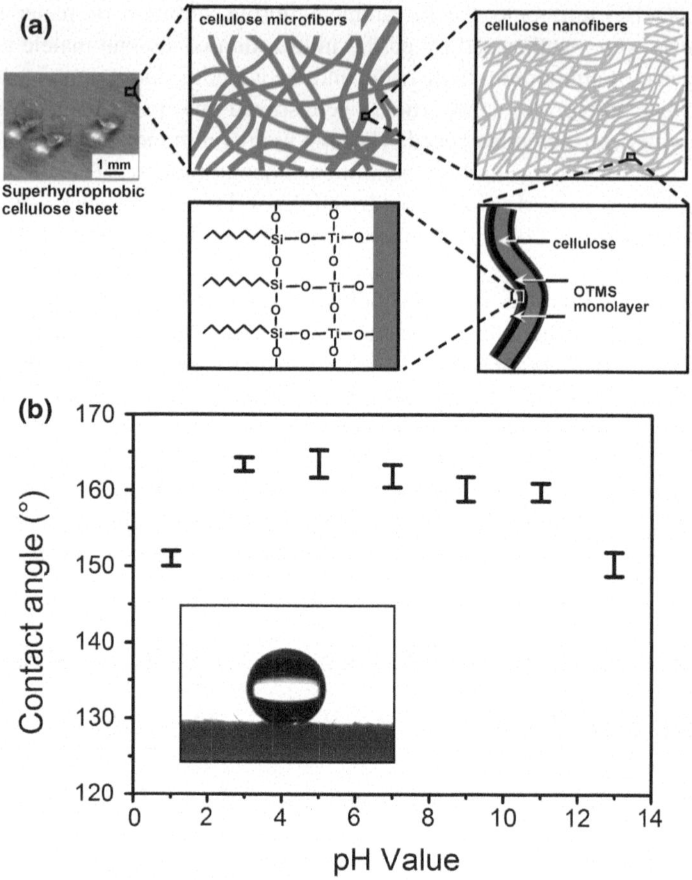

Fig. 8.6 Superhydrophobic cellulosic material fabricated by nanocoating of siloxane monolayer on cellulose fibers. **a** Schematic illustration of the superhydrophobic cellulosic materials fabricated by nanocoating with ultrathin titania/OTMS films. **b** Contact angle measurement of droplets with different pH values on the surface of the superhydrophobic filter paper; the inset shows the photograph of water droplet placed on the surface. Reprinted with permission from Ref. [85]. Copyright 2010, the Chemical Society of Japan

prepared samples is proportional to the methyl number of the siloxane precursor. Increasing the thickness of the coating films simply improved the mechanical and thermal properties of the obtained samples, but it does not affect the hydrophobicity of the samples. Instead of liquid-phase self-assembly, vapor-phase esterification of cellulose was preceded by exposing filter paper to mixed vapor of trifluoroacetic anhydride and acetic acid or trifluoroacetic anhydride and acetic anhydride at room temperatures, and the hydrophobicity was improved to water contact angles with ~98° [87].

While the changes in wettability is increased by surface roughness, the surface modification of the cellulose nanofiber of filter paper with photosensitive azobenzene derivative (CF3AZO) achieved smart materials, which was capable of reversible change in hydrophilic state with external UV irradiation [88]. Each cellulose nanofiber of filter paper was first coated with five-layer thin titania film ($\sim$2.5 nm thick) through the surface sol–gel process to activate the fiber surface, and then the as-deposited filter paper was immersed in CF3AZO solution (10 mM in ethanol) in the dark at room temperature for 24 h to promote self-assembly of CF3AZO molecules onto the titania precoated cellulose nanofibers coordinatively. The resultant yellow cellulose/oxide/CF3AZO composite sheet retained the original micro- to nanoscale rough morphology of the filter paper, and it showed a high hydrophobicity with water contact angle of $145.2 \pm 1.6°$. When the composite sheet was irradiated with UV irradiation for 13 h, the water contact angle was reduced to $24.2 \pm 4.1°$. This hydrophilicity was recovered to the initial antiwetting state after dark storage. The reversible change in water contact angles of the fabricated composite sheet originated from the photoisomerization of the CF3AZO molecules in the surface monolayer. As schematically illustrated in Fig. 8.7a, the conformational transformation of the CF3AZO molecules accompanied by dipole moment was induced, and chain order changes as well as surface hydroxyl groups on the titania gel layer surface were generated under UV irradiation, resulting in the sharp decrease of the contact angle on the surface of the composite cellulose sheet. With similar process, surface modification of CF3AZO was also performed on the thin silica gel film precoated cellulose nanofibers of filter paper. However, no obvious change in the hydrophobic state occurred for the filter paper deposited with thin silica/CF3AZO composite films after UV irradiation. Because the CF3AZO molecules were bonded to the silica gel layer surface through esterification, which caused more disorder of the CF3AZO molecular chains in the monolayers. The bacterial cellulose membrane with UV light controlled hydrophobicity was reported by hydrogen bond mediated adsorption of 1′,3′,3′-trimethyl-6-nitrospiro(2H-1-benzopyran-2,2′-indoline), which shows photochromic behavior accompanied with the chemical structure change of the modified guest substances under UV irradiation [89]. The cellulosic materials with switchable wettability possess great potential applications in various fields such as sensors, intelligent membranes, self-cleaning surfaces, and so on.

As described above, the surface roughness and the surface energy are the main factors adjusting the surface wettability. It is effective to coat hierarchical structured cellulose with low surface energy compounds for superhydrophobic surfaces. However, most of the attained superhydrophobic surfaces are oleophilic, and they have difficulty in repelling liquids with surface tensions which are lower than water (72.8 mN m^{-1}) such as hexadecane (27.5 mN m^{-1}). To improve the surface oleophobicity of the cellulosic materials, commercial laboratory filter paper was first etched using 17.5% sodium hydroxide aqueous solution and then successively surface modified with ultrathin titania films and 1H,1H,2H,2H-perfluorooctyl trimethoxysilane (PFOTMS) monolayers through the surface sol–gel process (Fig. 8.8) [90]. The surface-etched cellulose fibers of filter paper were endowed with enhanced surface roughness, and the thin titania gel films deposited coating each cellulose nanofiber of the filter paper enable further deposition of the PFOTMS monolayer thereon with low surface energy. The obtained composite

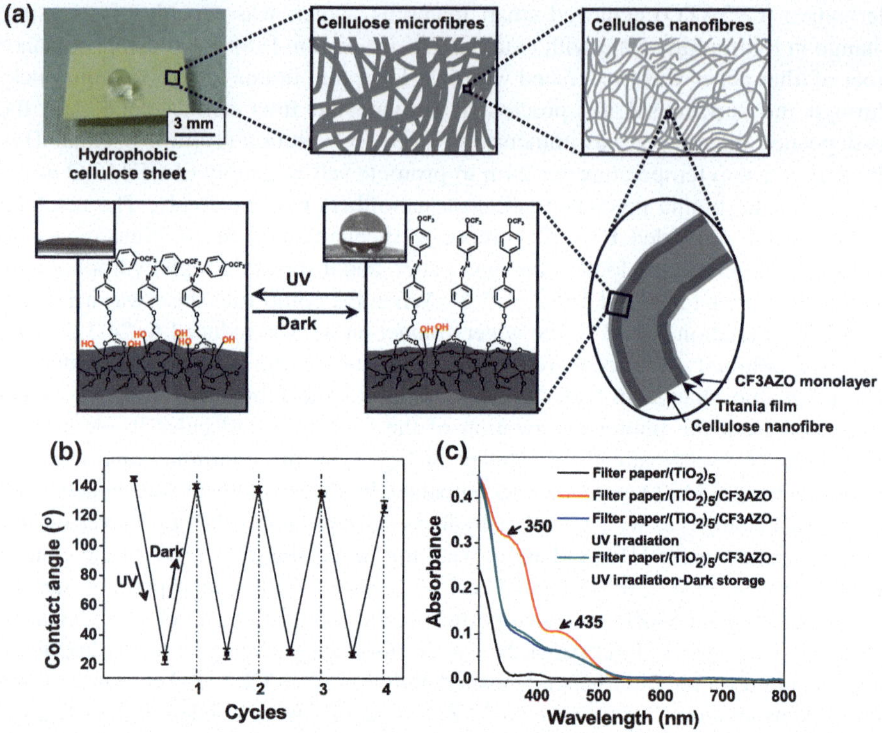

Fig. 8.7 Cellulose substance with reversible photoresponsive wettability. **a** Schematic illustration of the reversible wettability on the surface of the filter paper with each nanofiber surface-coated by the ultrathin titania/CF3AZO film. **b** Reversible contact angle changes on the composite sheet surface during cyclic UV irradiation and dark storage. **c** Solid UV-Vis spectra of the five-layer titania gel film precoated filter paper, and the ultrathin $(titania)_5$/CF3AZO film-modified filter paper before and after UV irradiation, and after subsequent dark storage. Reprinted with permission from Ref. [88]. Copyright 2011, Royal Society of Chemistry

exhibited excellent antiwetting effects both to water and hexadecane droplets. The contact angle for water and hexadecane achieved 158° and 146.5°, respectively. Moreover, as a result of the amphiphobic behavior, the fabricated composite cellulose sheet effectively suppressed the adhesion of bacteria such as lysogenic *Escherichia coli*. The amphiphobic cellulosic materials hold great potentials in versatile fields including self-cleaning surfaces, antifouling surfaces, intelligent membranes, and so on. The superhydrophobic is able to be patterned by printing with high surface energy black ink. The shape and size of the ink islands show different adhesion forces, which enable the manipulation of liquid drops on the paper substrates [91]. Paper was also patterned into hydrophilic and hydrophobic channels, which direct the fluid analyte into different detection zones and sense multiple analytes simultaneously [92, 93]. The patterned paper pieces were used for the colorimetric detection of glucose and protein in artificial urine [94, 95].

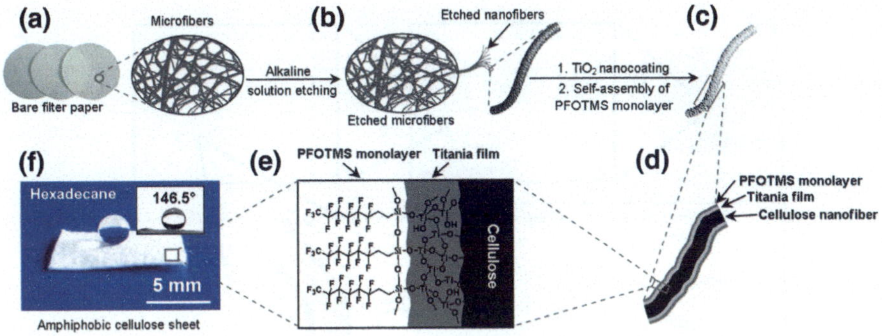

Fig. 8.8 Schematic illustration of the fabrication process of the amphiphobic cellulosic sheet. **a** Cellulose microfibers of the filter paper; **b** rougher cellulose fibers of the filter paper caused by the alkaline solution etching; **c** cellulose nanofiber of the etched filter paper deposited with ultrathin titania/PFOTMS composite films; **d** and **e** structure of the ultrathin titania/PFOTMS film-modified cellulose nanofiber, PFOTMS molecules are covalently bonded onto the titania film surface forming a monolayer; and **f** the resultant amphiphobic cellulosic sheet shows a contact angle of 146.5° for hexadecane. Reprinted with permission from Ref. [90]. Copyright 2012, Royal Society of Chemistry

8.4.2 Self-assembly of Functional Molecules for Colorimetric Sensors

The well-defined natural hierarchical networks of cellulose nanofibers endow natural cellulose substances with high porosity as well as large specific surface area. When such bulk cellulose substances are used as scaffolds for specific surface modification, the unique natural structural features allow assembling of a large number of sensor molecules and enhance the sensing ability of the resulting sensor materials. In earlier studies, paper-based test strip was prepared for the determination of chromium in water samples [96, 97]. By immobilization of di-tetrabutylammonium cis-bis (isothiocyanato)bis(2,2′-bipyridyl-4,4′-dicarboxylato)-ruthenium(II) (N719) monolayer onto each titania ultrathin film precoated cellulose nanofiber of commercial filter paper, a highly selective, sensitive, and reversible cellulose-based chemosensor material was achieved for the colorimetric detection of mercury ions in water [15]. First of all, a titania film of ~7.5 nm thick was deposited onto the individual nanofibers of the filter paper by the surface sol–gel process. Subsequently, N719 molecules were self-assembled onto the as-deposited titania film surface, resulting in an N719 monolayer coating each titania gel film precoated cellulose nanofiber (Fig. 8.9a). The obtained composite cellulosic sheet demonstrated extraordinary high sensitivity as well as selectivity and favorable detection reversibility for the colorimetric sensing of mercury ions in water media. With the presence of mercury ions in aqueous solutions, the mercury ions selectively coordinated with the thiocyanate groups of N719 dye molecules, and the initially purple color of the composite sheet immediately changed into obvious orange. The high surface area of filter

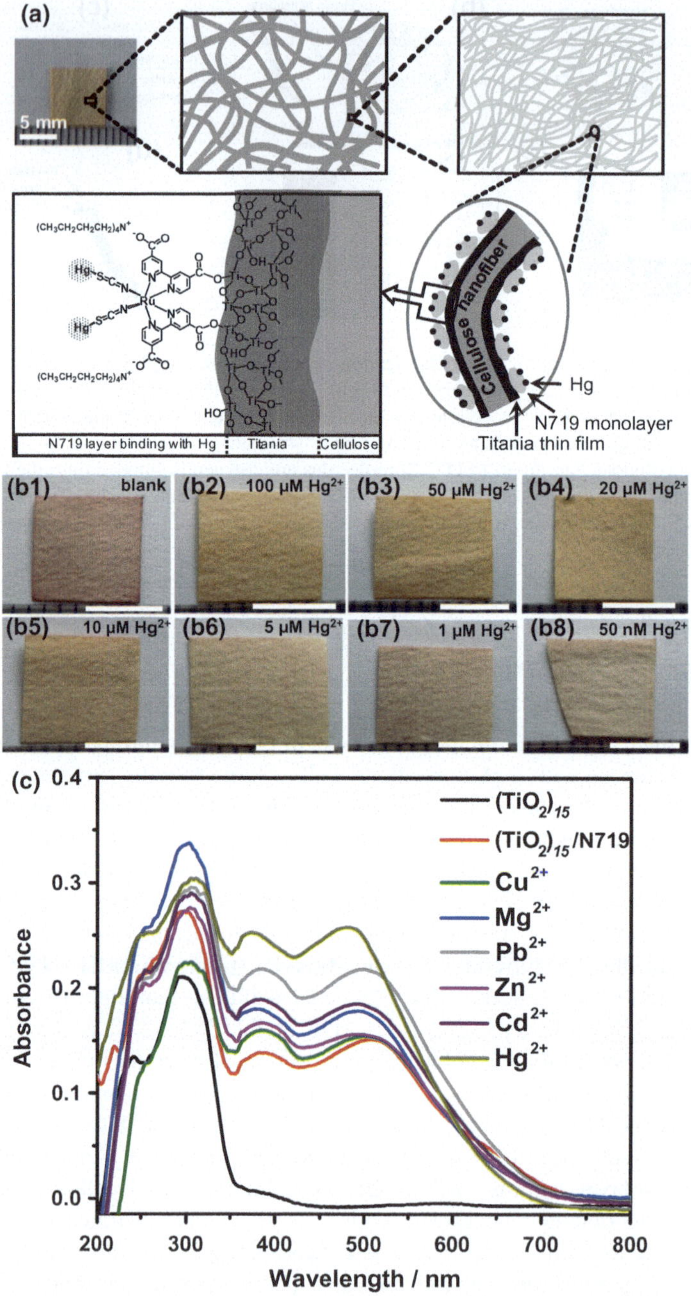

Fig. 8.9 Colorimetric sensing materials yielded by immobilization of ruthenium dye N719 onto titania precoated cellulose nanofibers of filter paper. **a** Schematic representation of mercury ion sensing using titania/N719 multilayer coated cellulose fibers of filter paper. (b1–b8) Color change of the $(\text{titania})_{15}$/N719-modified filter papers upon exposure to mercury nitrate aqueous solutions of different concentrations, scale bars: 5 mm. **c** Solid UV-Vis spectra of the $(\text{titania})_{15}$/N719 multilayer-modified filter papers after dipping in aqueous solutions of various metal ions. Reprinted with permission from Ref. [15]. Copyright 2010, Royal Society of Chemistry

paper scaffold largely increased the sensitivity of the resultant composite sheet, the detection limit of mercury ions reached ~10 ppb (50 nM) by naked eyes (Fig. 8.9b). Even in mixed metal ions solutions with the concentration of other metal ions higher by a factor of 100, the selective detection of mercury ions was achieved (Fig. 8.9c). The specific surface modification of filter paper was not limited to sensitive colorimetric mercury ion detection, it was also applied to the design and fabrication of ion-trapping materials for the adsorption of other heavy metal ions from diluted solutions. By surface modification with (3-mercaptopropyl) trimethoxysilane or *N*-[3-(trimethoxysilyl)-propyl)]ethylenediamine monolayers onto the thin silica film precoated filter paper, effective adsorbent materials for mercury or copper ion were fabricated. Due to the large surface area and the unique swelling behavior of filter paper, these materials were able to adsorb the heavy metal ions from aqueous solutions readily via the filtrations of the solution through the composite sheets.

Mercury in thymine–Hg^{2+}–thymine complex has stronger affinity to the free sulfydryl of cysteine (Cys). On the other hand, through the competitive coordination between the sulfur atoms of di-tetrabutylammonium *cis*-bis(isothiocyanato)bis-(2,2′-bipyridyl-4,4′-dicarboxylato) ruthenium(II) (N719) and mercury ions, the purple ruthenium dye N719 conjugates with mercury ions and form orange N719–Hg^{2+} complex. Taking advantage of these two features, the N719–Hg^{2+} complex immobilized cellulose nanofibers of filter paper yielded sensitive, selective, and reusable solid-phase colorimetric Cys chemosensor [98]. Each cellulose nanofiber of filter paper was first covered with a uniform 10-layer zirconia gel film by the surface sol–gel process, and the filter paper was immersed in a solution of 0.1 M N719–Hg^{2+} complex to promote the immobilization of N719–Hg^{2+} complex onto the surface activated nanofibers of the filter paper. As a result of the specific mercury displacement by Cys from the modified N719–Hg^{2+} complex, the obtained composite sheet generated apparent color change from orange to purple upon contact with aqueous solution containing Cys (Fig. 8.10a). The detection limit of this naked-eye assay achieved 20 μM (Fig. 8.10b), and the selectivity for Cys against disturbance of the other 19 natural amino acids and their mixture was fairly splendid (Fig. 8.10c). Thus, it is demonstrated that for surface decoration of cellulose substances with varied probe molecules, a wide variety of functionalized cellulosic material with engineered function can be applied to detection or analysis of many important target analytes such as biomolecules, organic compounds, inorganic ions, and so on.

Titanium(IV) oxo complex is intrinsically colorless, but it turns to bright yellow upon complexation with hydrogen peroxide through the exclusively selective titanium(IV)-peroxide bond. This unique feature of titanium(IV) oxo species was used to fabricate paper-based colorimetric sensor for hydrogen peroxide [99]. A water solution of titanyl oxalate was drop-casted onto the cellulose fibers of a paper towel, resulting in a homogeneous distribution of titanyl salt among the cellulose networks of the paper towel. By regulating the initial concentration of titanyl oxalate, the loading amount within the paper matrix was controlled. The initially white sensor materials became bright yellow (UV absorption at around 400 nm) upon exposure

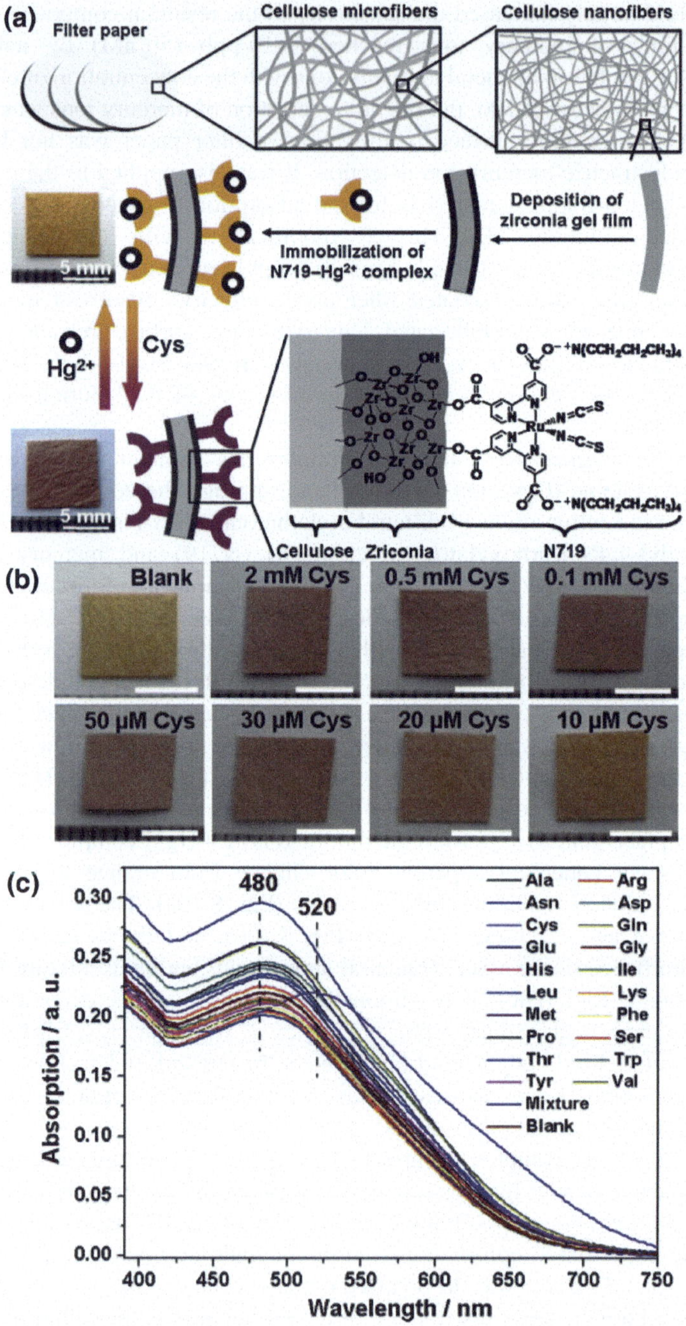
(a)
Filter paper
Cellulose microfibers
Cellulose nanofibers
Deposition of zirconia gel film
Immobilization of N719–Hg²⁺ complex
5 mm
Hg²⁺
Cys
5 mm
Cellulose
Zriconia
N719
(b)
Blank
2 mM Cys
0.5 mM Cys
0.1 mM Cys
50 μM Cys
30 μM Cys
20 μM Cys
10 μM Cys
(c)
480
520
Ala
Arg
Asn
Asp
Cys
Gln
Glu
Gly
His
Ile
Leu
Lys
Met
Phe
Pro
Ser
Thr
Trp
Tyr
Val
Mixture
Blank
Absorption / a. u.
Wavelength / nm

◂**Fig. 8.10** Colorimetric Cys chemosensor for fabricated by N719–Hg complex onto zirconia precoated cellulose nanofibers of filter paper. **a** Color transition of the cellulose/(zirconia)$_{10}$/N719·Hg papers upon exposure to aqueous Cys solutions with varied concentrations, scale bars: 5 mm. The sample denoted as "Blank" is the initial cellulose/(zirconia)$_{10}$/N719·Hg paper. **b** Solid UV-Vis spectra of the cellulose/(zirconia)$_{10}$/N719·Hg papers after being dipped into aqueous solutions of various amino acids and aqueous solution of the mixture of all the other 19 natural amino acids. Reprinted with permission from Ref. [98]. Copyright 2012, Elsevier Ltd.

to the vapor of hydrogen peroxide, whereas no response was found to other common liquids or solid chemicals. The detection limit of the sensor materials achieved 0.4 ppm.

In medical diagnosis of diabetics, frequent daily blood glucose test is usually required. "Dip-and-read" type of test strips are convenient and less costly for end users, as they do not need electronic devices such as glucometers to conduct blood glucose detection. To prepare such test strips, the biodegradable and widely available cellulose paper is an ideal choice as carrier matrix to realize the colorimetric chemistry-based quantitative analysis. For the paper-based glucose test strips, there are three key components: chromogen, enzymes, and gelatin. They were prepared as a mixture and then applied to a paper by soaking/dipping [100]. Briefly, 4.0 cm 2.5 cm filter paper strips were soaked in a mixture solution containing gelatin, glucose oxidase, peroxidase, citric acid buffer solution, 2,4,6-tribromo-3-hydroxy benzoic acid (TBHBA), and 4-aminoantipyrine (APP) for 5 min and dried at room temperature. The positively charged chromogen agent TBHBA/APP was firmly fixed onto the electronegative cellulose fibers of the paper through charge interactions. Glucose oxidase and peroxidase were used to catalyze the reaction between chromogen agent and the target glucose molecules. Meanwhile, gelatin was employed to stabilize and improve the catalytic activity of the above enzymes, and to control the diffusion rate of sample solution to meet the overall color-forming process. With the enzymes possessing increased catalytic activity and durability, the strip sample shows obvious red color (absorption at 510 nm) when glucose is present. The exhibited color intensity was correlated to the glucose concentration in the solutions in the range of 0.18–9.91 mg ml^{-1}. This sensitive paper strips have potential applications in monitoring the blood glucose concentration.

Oligodeoxyfluoroside (ODF) dyes are a category of short DNA-like oligomers with the initial DNA bases replaced by fluorophores. Such dyes are water-soluble, and possess fluorescence properties which are responsive to the changes in their immediate environment. As described above, guest substrates showing good affinity with cellulose can be applied as ink to be printed onto the paper sheets, and tetrameric ODFs of varied distinct sequences were printed on cotton paper by a commercial inkjet printer [101]. 5% w/v polyethyleneglycol (average molecular weight 8000) was added to the aqueous solutions of the synthesiszed ODF (50 mM), and the solution was charged to commercial black inkjet cartridges. Then the ink was printed on a designed square of 100% cotton paper (2 cm^2) six times over the identical position by a commercial thermal inkjet printer to ensure evenness of deposition, and the printed paper was dried in air. The ODFs printed paper was qualified for food spoilage monitoring, which is important to human health. The as-prepared paper exhibited fluorescence intensity and/or color changes during exposure to spoiled food volatiles such as meat (ground beef and shrimp), dairy (milk and cheese), fruit (orange juice), grain (bread), and vegetable (lettuce) sources, and each of the shown color was distinguishable by the relative differences among RGB channels. This approach benefits for the preparation of convenient, effective, and economical chemosensors for a disposable one-time measure of food spoilage.

8.4.3 Modification with Other Small Molecules

Carbohydrates play essential roles in a great majority of living things, and thus glyco-modification of solid cellulose surfaces is attractive for the potential applications. In natural world, condensations of carbohydrates are often accomplished by enzymes. With the assistance of 1,4-β-D-glucan 4-glucanohydrolase from *Trichoderma viride*, surface modification of a cotton fiber filter paper with lactose was successfully achieved [102]. To activate the chemically inert cellulose surface, filter paper was immersed in lithium chloride/dimethylacetamide mixture. As this mixture would inhibit the enzymatic activity, the enzyme molecules were protected by forming water-in-oil emulsion using dioleyl-*N*-D-glucona-L-glutamate. The gluco-modified filter paper showed increased cell adhesion performance, which is attractive for further practical applications. With similar concept, the chemoenzymatic modification of a cellulose fiber surface of filter paper with xyloglucan was realized by employing the combination of lipase and endotransglycosylase for the activation of solid cellulose surfaces [103, 104].

8.5 Biomacromolecule Assembly in Cellulose Substances

For bioactive functional surfaces, immobilization of protein molecules on two-dimensional matrices like gold, silicon, and glass has been carried out. Compared with such conventional flat plates, protein immobilization on the morphologically sophisticated surfaces of cellulose substances affords vast superiority, because the fabricated materials retain the native hierarchical structures as well as the related functions of proteins. Immobilization of protein on filter paper as the matrix forms a new type of bio-nanomaterial [105]. The chemically active surface of the nanometer-thick titania films coated cellulose nanofibers is biocompatible, which is capable of successive modification with biotin and protein (streptavidin) molecules through the high-affinity biospecific biotin–streptavidin interaction. The streptavidin molecules anchored on the three-dimensional cellulose nanofiber surfaces reserved their biological activities. Due to the hierarchical porous network and large surface area of the original cellulose substance, the assembled protein molecules demonstrate high sensitivity with the detection limit of 10^{-9} M

(Fig. 8.11). The exhibited functions were better than those on planar ones, and the resulting cellulose/protein composite sheet showed potential application as biosensors. Cellulose substances with each nanofiber coated with precise titania films at nanoprecision are proved to be promising platforms for construction of biomolecular architectures, which provides a critical pathway to combine the physical features of cellulose substances and the unique properties of biomacromolecules. The streptavidin immobilized cellulose fibers of filter paper allow further modification with biotinylated titania particles (100 nm large) through the biotin–

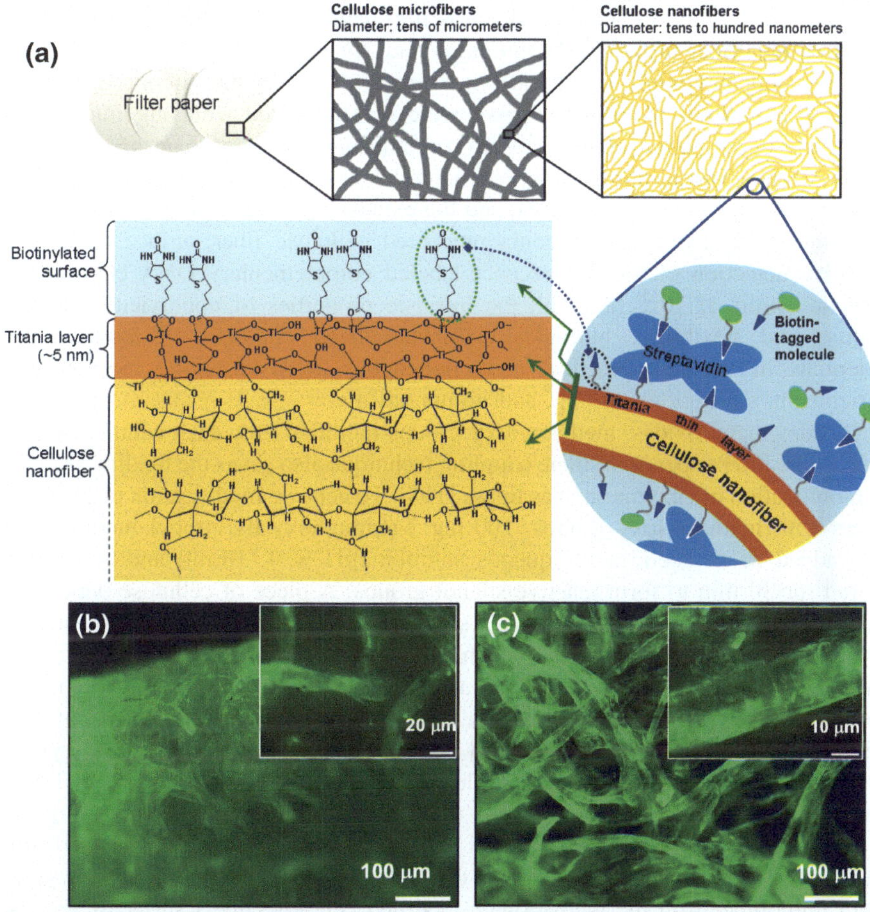

Fig. 8.11 **a** Schematic representation of immobilization of protein molecules (streptavidin) on each cellulose nanofibers and their successive binding to fluorescence-labeled biotin (not to scale). *Left panel* illustrates the formation of biotinylated surface on cellulose nanofiber; and *right panel* represents subsequent binding of streptavidin and biotin-tagged species. The thickness of titania layer is ~5 nm. **b** Fluorescence micrographs of filter paper in which the cellulose fibers were decorated with Alexa 488-labeled streptavidin. **c** Fluorescence micrograph of native streptavidin-modified cellulose fibers upon binding with biotin-4-fluorescein. Reprinted with permission from Ref. [105]. Copyright 2006, John Wiley & Sons Ltd.

streptavidin interaction, which demonstrated efficient catalytic decolorization for UV sensitive dyes [106].

DNA detection has considerable applications in wide areas such as biomedical diagnostics, gene expression, disease prevention, reaction discovery, forensic determination, and so on. For detection of DNA, the immobilization of a DNA probe on a solid substrate has been demonstrated to be an effective approach, which realizes hybridization assay with a target DNA in a homogeneous solution. By immobilization of oligonucleotides on the morphologically complex cellulose substance, a sensitive and duplicated fluorescence-based recognition for corresponding complementary target DNA was achieved at a nanomolar level [107]. A uniform ultrathin zirconia gel film was first deposited onto the surface of each cellulose nanofiber in the common commercial filter paper by the surface sol–gel process using zirconium *n*-butoxide as a precursor. Then the zirconia precoated cellulose nanofibers of the filter paper were functionalized with oligonucleotides through a coordination effect between zirconia and the phosphate group of the probe DNAs. The faithfully reserved natural hierarchical structures of the filter paper provided a large surface area, and the extensive concentration of probe DNA was immobilized onto the zirconia precoated cellulose fiber surface, achieving sensitive detection of the fluorescence-labeled complementary DNA chains with detection limit of 10^{-9} M. With the intrinsic properties of nontoxicity, degradability, and flexibility, biologically functionalized cellulosic materials show potentials in versatile practical and challenging fields including biomolecular enrichment, separation, sensing, and immunoassay.

Besides metal oxides, methacryloxy groups modified onto each cellulose nanofiber of the filter paper by a silane coupling technique also act as the binding sites for immobilization of enzyme such as lipase [108]. To achieve modification of cellulose fibers with methacryloxy groups, 100 mg 3-(trimethoxysilyl)propyl methacrylate was added to an acetic acid aqueous solution (pH $\sim$ 4, 10 ml) and vigorously stirred for 30 min to form reactive silanol groups. A piece of cellulose paper was subsequently immersed in the resulting solution for 2 h, and the acetic acid aqueous solution was evaporated at 40 °C for 3 h under reduced pressure. Then, the obtained paper was heated at 110 °C for 3 h, washed with ethanol, and dried at room temperature. Through the condensation reaction between the Si–OH of the organofunctional silane coupling agent and the C–OH of cellulose, methacryloxy groups were modified onto each cellulose fiber surface. Immobilization of lipase onto the methacrylate-modified paper was then accomplished by immersing the as-modified paper in the lipase solution (pH $\sim$ 6.86, ionic strength: 0.1 M, 10 ml) and stirring at 23 °C for 12 h. Through the hydrophobic interaction, lipase enzymes were immobilized on the methacrylate-modified cellulose fibers surfaces, and the insoluble components adsorbed onto the paper was washed away via stirring the as-prepared sample in a phosphate buffer solution (pH $\sim$ 6.86, ionic strength: 0.1 M, 10 ml) at a stirring rate of 150 rpm for 3 h. The fabricated enzyme immobilized paper exhibited improvement in hydrophobicity (111.3 $\pm$ 1.9°) as well as physical strengths in both dry and wet states. The Young's modulus and tensile strength were enhanced for 0.19 GPa and 3.39 MPa (dry), 0.13 GPa and 5.02 MPa (wet), respectively. The interconnected porous structure of the paper offered preferable flow paths for the reactant solution, resulting in high catalytic activity, selectivity, and reusability for the nonaqueous transesterification of 1-phenylethanol and vinyl acetate to form 1-phenylethylacetate. Hence, the enzyme immobilization of paper provides a green catalytic material for the effective production of useful chemicals. The silane coupling technique is a facile and versatile pathway to introduce functional groups into cellulose fibers, and modification of cellulose paper with amino groups was similarly accomplished [109].

Another practical approach to immobilize biomacromolecules onto cellulose fibers is the electrostatic force driven layer-by-layer self-assembly. By alternative assembly of anionic cibacron blue F3GA dye and cationic lipase that was obtained from *Candida rugosa* on acetylated cellulose template, biocatalytic membrane was fabricated [110]. With the maintained hierarchical cellulose network and the protein bioactivity of lipase, the prepared membrane exhibited an efficient catalytic property for olive oil, which was kept to be about 45% of that of free lipase.

Rapid and accurate determination of human blood type is crucial in many medical procedures. Taking advantage of the strong adsorption performance and high porosity of paper sheets, the rapid, flexible, and reproducible blood typing devices become available [111]. Typically, a 10 mm $\times$ 10 mm filter paper piece was dipped with undiluted commercial antibody solutions and left to dry for 10 min. The affinity between the antibody molecules and cellulose components allows the antibody molecules to strongly adsorb onto the cellulose fibers' surfaces. When blood samples were introduced to the antibody treated paper, the RBCs in the blood were immobilized onto the adsorbed antibody molecules through the specific antibody–antigen interaction. Meanwhile, a substantial fraction of the antibody molecules (<40%) was desorbed from the cellulose fiber surface, and the released antibody molecules facilitate the bridging between the RBCs dispersed in the blood sample and the RBCs immobilized on the cellulose fibers surfaces. Moreover, these desorbed antibody molecules also formed agglutinated RBC lumps, which were too large to pass through the pores of the paper sheet. The unique behaviors of the antibody molecules on cellulose fibers significantly enhanced the blood typing effect and demonstrated obvious purple color after staining, which was far sensitive and swift than the traditional blood typing assays. The paper scaffolds create new chances for swift, sensitive, and easy-to-use diagnostic sensors in clinics, and a series of paper-based blood typing devices have been developed [112–117].

8.6 Sp^2-Hybridized Carbon Assembly in Cellulose Substances

The studies on sp^2-hybridized carbon such as carbon nanotubes (CNTs) and graphene are now proceeding at an astonishing pace because of their attractive properties and great potentials in wide applications like electricity, sensor, and so on. By conformal single-walled CNTs coating on the cellulose fibers of paper through simple solution processes, commercially available paper is made highly conductive [118, 119]. Printable solution processing is a low-cost technique to deposit versatile nanomaterials such as nanocrystals, nanowires, fullerene, and CNTs as ink onto flat substrates like silicon wafers, metallic films, glass, and plastics for large-scale applications. Commercial CNTs were first dispersed in aqueous solution with sodium dodecylbenzenesulfonate as the dispersant, and then the mixture ink was dipped onto paper. The three-dimensional hierarchical porous structure of paper led large capillary force for the ink and promoted strong adhesion for the nanomaterials dispersed in ink without any additives required to support the ink adhesion on the cellulose fibers. After drying out the solvent, the conformal coating of the flexible CNTs was formed on the fibers of paper without damaging the original fiber-based morphological hierarchy of the paper. The CNTs were tightly attached with the cellulose fibers with a high contacting surface area between them. The porous nature and the large surface area of the formed conductive paper guaranteed more efficient accessibility of ions, resulting in a low sheet resistance $\sim 1\ \Omega\ \text{sq}^{-1}$. By applying this approach with the ink of silver nanowires, conductive paper whose cellulose nanofibers coated with dense silver nanowires were obtained. The sheet resistances were affected by the effective film thicknesses of the as-deposited CNTs and silver nanowires. This CNT-conductive paper was qualified for the application as supercapacitors, and the material showed a specific capacitance of 200 F g^{-1}, a specific energy of 30–47 Wh kg^{-1}, a specific power of 200,000 W kg^{-1}, and a stable cycling life of more than 40,000 cycles. In the fabricated conductive sheet, the flexibility of each CNTs, the strong bind of the CNTs with the cellulose fibers, and the natural hierarchical structure of the paper worked together, bringing up the relaxation of the bending strain, high flexibility, and increased mechanical strength, which allow the application of the conductive paper for varied rolled up devices. Meanwhile, the resultant conductive paper was mechanically robust, and no crack or breakage was observed from it in either air or liquids. If the resulting conductive paper is used instead of the heavy metallic current collectors, the battery weight will be decreased for up to 20% with acceptable internal resistance. The concept of solution processing of nanomaterials on hierarchical cellulose paper as a platform provides a simplified route for advanced applications in energy storage and conversion, and the obtained materials are promising for all-integrated paper and textile electronics [120, 121].

With this solution-based process, sensors electrodes were drawn onto a piece of filter paper using graphite conductive ink, giving a sensitive photoconductive infrared (IR) detector [122]. To increase the conductivity to the detector, 1 drop ($\sim$ 35 mL) of an electrolyte solution of potassium bromide was dropped on to the paper surface. Because potassium bromide is transparent to a wide spectrum of light radiation, the origin of the optical absorption of the detector was isolated. The paper detector was dried at 120 °C after the drop deposition to evaporate the water. The hygroscopic cellulose promotes the formation of thin films on its surface, and the C–O bonds in cellulose absorb mid-range IR radiation ($\sim 1000\ \text{cm}^{-1}$) and turn into heat through the radiative heating effect, which contributes in a detectable electric current variation. Thus, the IR radiation sensing was achieved by monitoring the current change. The conductivity of the detectors improved with the addition of glycerol and electrolyte salts. Also, the electrical conductivity of these IR detectors depended largely on humidity and temperature, originating from mobility change in the ionic current involving electrolyte salts dissolved in the thin liquid layer dispersed over the paper. This kind of IR radiation sensor could be applied to recognize IR signatures of specific molecules, and also detect human, animals, and any other type of target with a temperature above the background. Besides, the paper was printed with conductive carbon ink and Ag/AgCl ink, and they were applied as electrodes for electrochemical detection of heavy metal ions [123]. The paper containing the analytes inhibited the convective movement of the solutions after complete wetting of the paper channel with fluids. Therefore, the accuracy of time-dependent chronoamperometry and chronopotentiometry was increased.

Instead of drawing and spreading inks with a rod, a graphene nanosheets suspension was vacuum filtered through a piece of filter paper to obtain flexible energy storage materials [124]. Graphene was first exfoliated from natural flake graphite powders through successive oxidation, thermal expansion, and hydrogen reduction, and was then dispersed in *N*-methylpyrolidone and sonicated to form a graphene nanosheets suspension (average size ~2 μm, concentration ~0.2 mg mL^{-1}). Subsequently, the graphene nanosheets suspension was vacuum filtered through both sides of the filter paper in turn until the initially black suspension became colorless. After vacuum drying at 100 °C for 8 h, *N*-methylpyrolidone component was removed, and a cellulose/graphene membrane was obtained. The porous structure of the filter paper allowed the graphene nanosheets to be filled in the pores. The penetrated graphene nanosheets were strongly bound to the cellulose fibers through the abundant functional groups on these cellulose fibers, forming a continuous conductive interwoven network around the cellulose fibers of the filter paper. The graphene contents of these cellulose/graphene membranes were readily changed by controlling the volume and/or the concentration of the pristine graphene suspension. The resultant membrane combined the macroporous network and high strength of the filter paper as well as the good electrical conductivity and electroactivity of graphene nanosheets. As a result, distinguished mechanical flexibility, practical cyclic stability, excellent specific capacitance, and power performance were achieved. Their electrical conductivity reduced for only 6% after being bent 1000 times; the capacitance per geometric area was 81 mF cm^{-2}; the capacitance just reduced <1% over 5000 cycles of bending. These membranes were employed to prepare freestanding and binder-free electrodes for flexible supercapacitors, and the supercapacitors exhibited a high capacitance per geometric area of 46 mF cm^{-2} for the complete devices under highly flexible conditions. When CNTs were used as fillers instead of graphene, the prepared sheet showed conductivity of 3×10^{-2} S cm^{-1} [125]. The flexible supercapacitators based on the cellulose/graphene membrane have considerable potential applications in flexible and portable micropower devices.

In the above solution-based method, a stable ink with the right rheological properties, a hydrophilic platform, and thermal treatment as well as cleaning for the removal of the surfactants in ink after deposition are required, which pose challenges for a wide range of applications of this methodology. Inspired by the fact that graphite is an outstanding electrode material for electrochemical double layer formation and pencil trace on paper is highly conductive, supercapacitors were fabricated by directly drawing on cellulose paper with a graphite rod or pencils [126]. The solvent-free process is similar to directly drawing with regular pencil on paper. The drawing was repeated accurately with ruler guidance in orthogonal directions for three times to obtain a stripe of uniform coating. The first round of the drawing was carried out in the direction of the fiber orientation in order to prevent fraying of the paper surface, which would cause a significant reduction in electrical conductivity of the paper electrodes. The rough surface of the paper allowed exfoliation and adhesion of the graphitic materials, which resulted in continuous conducting paths on the surface and offered significant adsorption area for double layer formation at the same time. The graphite trace was very stable, and it remained conductive even on bending down to a radius of 2 mm. The obtained sheet resistance was $\sim 2.23\ \Omega\ m^{-1}$. To form the supercapacitor, a virgin paper was sandwiched between two as-prepared conductive paper electrodes. The highly porous paper substrate served as the separator for the supercapacitor, which prevent occasional penetration of the deposited graphites and guaranteed rapid diffusion of ionic species [127]. The resultant supercapacitors demonstrated a high areal capacitance of 2.3 mF cm^{-2} and long-term cycling performance. This solvent-free method is highly scalable, low cost, which can be potentially applied for the fabrication of integrated paper-based energy devices.

To control the load content of CNTs on paper, silica was employed as the carrier [128]. The CNTs were dispersed in the dilute sol and were airbrush sprayed onto paper surface together. Through the sol–gel process, CNTs were fixed in a thin silica film continuously coating the cellulose fiber surfaces. The high degree of connectivity of the stranded structure of the CNTs in the silica films reached a specific electrical conductivity of $3 \times 10^3\ \Omega^{-1}\ m^{-1}$.

8.7 Conclusions

In summary, the surface modification based on the self-assembly of a great variety of guest substrates on the fibers in natural cellulose substances sheds a considerable light on new opportunities for the development of novel nanostructured materials with tailored functionalities. Versatile guest species such as metal oxides, nanoparticles, polymers, small molecules, biomacromolecules, and sp^2-hybridized carbons were successfully modified into fibers of the bulk cellulose substances. The synergetic effects of the natural structural features of the well-maintained unique hierarchical cellulose networks and the specific properties of the guest substrates achieve various high functionalities, yielding unique functional materials. Biomimetic synthetic approaches provide ideal pathways to the fabrication of functional materials [129–131], where the self-assembly technique has been widely adopted for this purpose [132–135]. Self-assembly based surface modification in cellulose substances demonstrates an effective pathway to combine the physical features of biological matters and the chemical properties of the guest substrates, which provides a pathway for the advancement of bio-inspired artificial materials.

Acknowledgements This work was supported by the National Key Project on Basic Research of China (2009CB930104). J. Huang sincerely acknowledges the guidance and support of Prof. Toyoki Kunitake.

References

1. Sanchezl C, Arribart H, Guille MMG (2005) Biomimetism and bioinspiration as tools for the design of innovative materials and systems. Nat Mater 4:277–288
2. Omenetto FG, Kaplan DL (2010) New opportunities for an ancient material. Science 329:528–531
3. Caruso RA (2004) Micrometer-to-nanometer replication of hierarchical structures by using a surface sol–gel process. Angew Chem Int Ed 43:2746–2748
4. Mann S (2009) Self-assembly and transformation of hybrid nano-objects and nanostructures under equilibrium and non-equilibrium conditions. Nat Mater 8:781–792
5. Alava M, Niskanen K (2006) The physics of paper. Rep Prog Phys 69:669–723
6. Klemm D, Heublein B, Fink HP et al (2005) Cellulose: fascinating biopolymer and sustainable raw material. Angew Chem Int Ed 44:3358–3393
7. Balu B, Breedveld V, Hess DW (2008) Fabrication of "roll-off" and "sticky" superhydrophobic cellulose surfaces via plasma processing. Langmuir 24:4785–4790
8. Cunha AG, Gandini A (2010) Turning polysaccharides into hydrophobicmaterials: a critical review. Part 1. Cellulose. Cellulose 17:875–889
9. Vesel A, Mozetic M, Hladnik A et al (2007) Modification of ink-jet paper by oxygen-plasma treatment. J Phys D 40:3689–3696
10. Balu B, Kim JS, Breedveld V et al (2009) Tunability of the adhesion of water droplets on a superhydrophobic paper surface via selective plasma etching. J Adhes Sci Technol 23: 361–380
11. Huang J, Kunitake T (2003) Nano-precision replication of natural cellulosic substances by metal oxides. J Am Chem Soc 125:11834–11835
12. Moon RJ, Martini A, Nairn J et al (2011) Cellulose nanomaterials review: structure, properties and nanocomposites. Chem Soc Rev 40:3941–3994
13. Jia B, Mei Y, Cheng L et al (2012) Preparation of copper nanoparticles coated cellulose films with antibacterial properties through one-step reduction. ACS Appl Mater Interfaces 4: 2897–2902
14. Gandinia A, Pasquini D (2012) The impact of cellulose fibre surface modification on some physico-chemical properties of the ensuing papers. Ind Crop Prod 35:15–21
15. Zhang X, Huang J (2010) Functional surface modification of natural cellulose substances for colorimetric detection and adsorption of Hg^{2+} in aqueous media. Chem Commun 46: 6042–6044
16. Gu Y, Liu X, Niu T et al (2010) Superparamagnetic hierarchical material fabricated by protein molecule assembly on natural cellulose nanofibers. Chem Commun 46:6096–6098
17. Kemell M, Pore V, Ritala M et al (2005) Atomic layer deposition in nanometer-level replication of cellulosic substances and preparation of photocatalytic TiO_2/cellulose composites. J Am Chem Soc 127:14178–14179
18. Kemell M, Pore V, Ritala M et al (2006) Ir/oxide/cellulose composites for catalytic purposes prepared by atomic layer deposition. Chem Vap Deposition 12:419–422

19. Kemell M, Ritala M, Leskelä M et al (2008) Coating of highly porous fber matrices by atomic layer deposition. Chem Vap Deposition 14:347–352
20. Jur JS, Sweet WJ III, Oldham CJ et al (2011) Atomic layer deposition of conductive coatings on cotton, paper, and synthetic fibers: conductivity analysis and functional chemical sensing using “all-fiber” capacitors. Adv Funct Mater 21:1993–2002
21. Hyde GK, Scarel G, Spagnola JC et al (2009) Atomic layer deposition and abrupt wetting transitions on nonwoven polypropylene and woven cotton fabrics. Langmuir 26:2550–2558
22. Hyde GK, Park KJ, Stewart SM et al (2007) Atomic layer deposition of conformal inorganic nanoscale coatings on three-dimensional natural fiber systems: effect of surface topology on film growth characteristics. Langmuir 23:9844–9849
23. Liu X, Gu Y, Huang J (2010) Hierarchical, titania-coated, carbon nanofibrous material derived from a natural cellulosic substance. Chem Eur J 16:7730–7740
24. Shin Y, Li XS, Wang C et al (2004) Synthesis of hierarchical titanium carbide from titania-coated cellulose paper. Adv Mater 16:1212–1215
25. Kyotani M, Matsushita S, Kimura S et al (2012) Efficient preparation of carbon papers by pyrolysis of iodine-treated Japanese paper. J Anal Appl Pyrolysis 95:14–20
26. Pelton R, Geng XL, Brook M (2006) Photocatalytic paper from colloidal TiO_2–fact or fantasy. Adv Colloid Interface Sci 127:43–53
27. Gimenez AJ, Yãnez-Limon JM, Seminario JM (2011) ZnO–paper based photoconductive UV sensor. J Phys Chem C 115:282–297
28. Ghule K, Ghule AV, Chen BJ et al (2006) Preparation and characterization of ZnO nanoparticles coated paper and its antibacterial activity study. Green Chem 8:1034–1041
29. Sun Q, Schork FJ, Deng Y (2007) Water-based polymer/clay nanocomposite suspension for improving water and moisture barrier in coating. Compos Sci Technol 67:1823–1829
30. Ornatska M, Sharpe E, Andreescu D et al (2011) Paper bioassay based on ceria nanoparticles as colorimetric probes. Anal Chem 83:4273–4280
31. Mahltig B, Fiedler D, Böttcher H (2004) Antimicrobial sol–gel coatings. J Sol-Gel Sci Tech 32:219–222
32. Hou A, Shi Y, Yu Y (2009) Preparation of the cellulose/silica hybrid containing cationic group by sol–gel crosslinking process and its dyeing properties. Carbohydr Polym 77: 201–205
33. Xie K, Yu Y, Shi Y (2009) Synthesis and characterization of cellulose/silica hybrid materials with chemical crosslinking. Carbohydr Polym 78:799–805
34. Rida A, Yang L, Vyas R et al (2009) Conductive inkjet-printed antennas on flexible low-cost paper-based substrates for RFID and WSN applications. IEEE Antenn Propag M 51:13–23
35. Bayer IS, Fragouli D, Attanasio A et al (2011) Water-repellent cellulose fiber networks with multifunctional properties. ACS Appl Mater Interfaces 3:4024–4031
36. Small AC, Johnston JH (2008) Novel hybrid materials of cellulose fibres and doped ZnS nanocrystals. Curr Appl Phys 8:512–515
37. Niu T, Gu Y, Huang J (2011) Luminescent cellulose sheet fabricated by facile self-assembly of cadmium selenide nanoparticles on cellulose nanofibres. J Mater Chem 21:651–656
38. Hwang SH, Moorefield CN, Wang P et al (2006) Construction of CdS quantum dots via a regioselective dendritic functionalized cellulose template. Chem Commun 3495–3497
39. Ding Z, Wei P, Chitnis G et al (2011) Ferrofluid-impregnated paper actuators. J Microelectromech Syst 20:59–64
40. Fragouli D, Bayer IS, Corato RD et al (2012) Superparamagnetic cellulose fiber networks via nanocomposite functionalization. J Mater Chem 22:1662–1666
41. Liu X, Zong C, Lu L (2012) Fluorescent silver nanoclusters for user-friendly detection of Cu^{2+} on a paper platform. Analyst 137:2406–2414
42. Cady NC, Behnke JL, Strickland AD (2011) Copper-based nanostructured coatings on natural cellulose: nanocomposites exhibiting rapid and efficient inhibition of a multi-drug resistant wound pathogen, *A. baumannii*, and mammalian cell biocompatibility in vitro. Adv Funct Mater 21:2506–2514

43. Huang J, Ichinose I, Kunitake T (2005) Nanocoating of natural cellulose fibers with conjugated polymer: hierarchical polypyrrole composite materials. Chem Commun 1717–1719
44. Varesano A, Aluigi A, Florio L et al (2009) Multifunctional cotton fabrics. Synth Met 159:1082–1089
45. Vernitskaya TV, Efimov ON (1997) Polypyrrole: a conducting polymer; its synthesis, properties and applications. Russ Chem Rev 66:443–457
46. Johnston JH, Kelly FM, Moraes J et al (2006) Conducting polymer composites with cellulose and protein fibres. Curr Appl Phys 6:587–590
47. Kelly FM, Johnston JH, Borrmann T et al (2007) Functionalised hybrid materials of conducting polymers with individual fibres of cellulose. Eur J Inorg Chem 5571–5577
48. Beneventi D, Alila S, Boufi S et al (2006) Polymerization of pyrrole on cellulose fibres using a $FeCl_3$ impregnation-pyrrole polymerization sequence. Cellulose 13:725–734
49. Olsson H, Carlsson DO, Nyström G et al (2012) Influence of the cellulose substrate on the electrochemical properties of paper-based polypyrrole electrode materials. J Mater Sci 47:5317–5325
50. Zakirov AS, Yuldashev SU (2012) Functional hybrid materials derived from natural cellulose. J Korean Phys Soc 60:1526–1530
51. Razaq A, Nyström G, Strømme M et al (2011) High-capacity conductive nanocellulose paper sheets for electrochemically controlled extraction of DNA oligomers. PLoS ONE 6: e29243
52. Bhat NV, Seshadri DT, Nate MM et al (2006) Development of conductive cotton fabrics for heating devices. J Appl Polym Sci 102:4690–4695
53. Dutta D, Sarma TK, Chowdhury D et al (2005) A polyaniline-containing filter paper that acts as a sensor, acid, base, and endpoint indicator and also filters acids and bases. J Colloid Interface Sci 283:153–159
54. Jagadeesan KK, Kumar S, Sumana G et al (2012) Application of conducting paper for selective detection of troponin. Electrochem Commun 20:71–74
55. Johnston JH, Kelly FM, Burridge KA et al (2009) Hybrid materials of conducting polymers with natural fibers and silicates. Int J Nanotechnol 6:312–328
56. Kawashima H, Shinotsuka M, Nakano M et al (2012) Fabrication of conductive paper coated with PEDOT: preparation and characterization. J Coat Technol Res 9:467–474
57. Ichinose I, Kunitake T (1999) Polymerization-induced adsorption: a preparative method of ultrathin polymer films. Adv Mater 11:413–415
58. Makela T, Jussila S, Vilkman M et al (2003) Roll-to-roll method for producing polyaniline patterns on paper. Synth Met 135–136:41–42
59. Zhou J, Fukawa T, Shirai H et al (2010) Anisotropic motion of electroactive papers coated with PEDOT/PSS. Macromol Mater Eng 295:671–675
60. Winther-Jensen B, Clark N, Subramanian P et al (2007) Application of polypyrrole to flexible substrates. J Appl Polym Sci 104:3938–3947
61. Sarrazin P, Valecce L, Beneventi D et al (2007) Photoluminescent paper based on poly (fluorene-co-fluorenone) particles adsorption on modified cellulose fibers. Adv Mater 19:3291–3294
62. Zhang X, Shi F, Yu X et al (2004) Polyelectrolyte multilayer as matrix for electrochemical deposition of gold clusters: toward super-hydrophobic surface. J Am Chem Soc 126: 3064–3065
63. Lingström R, Wagberg L, Larsson P et al (2006) Formation of polyelectrolyte multilayers on fibers: influence on wettability and fibre/fibre interaction. J Colloid Interface Sci 296: 396–408
64. Lingström R, Notley SM, Wagberg L et al (2007) Wettability changes in the formation of polymeric multilayers on cellulose fibers and their influence on wet adhesion. J Colloid Interface Sci 314:1–9

65. Nurmi L, Kontturi K, Houbenov N et al (2010) Modification of surface wettability through adsorption of partly fluorinated statistical and block polyelectrolytes from aqueous medium. Langmuir 26:15325–15332
66. Nyström D, Lindqvist J, Östmark E et al (2006) Superhydrophobic bio-fiber surfaces via tailored grafting architecture. Chem Commun 3594–3596
67. Huang J, Gu Y (2011) Self-assembly of various guest substrates in natural cellulose substances to functional nanostructured materials. Curr Opin Colloid Interface Sci 16: 470–481
68. Aied A, Zheng Y, Pandit A et al (2012) DNA immobilization and detection on cellulose paper using a surface grown cationic polymer via ATRP. ACS Appl Mater Interfaces 4: 826–831
69. Carlmark A, Malmström EE (2012) Atom transfer radical polymerization from cellulose fibers at ambient temperature. J Am Chem Soc 124:900–901
70. Carlmark A, Malmström EE (2003) ATRP grafting from cellulose fibers to create block-copolymer grafts. Biomacromol 4:1740–1745
71. Lönnberg H, Zhou Q, Brumer H et al (2006) Grafting of cellulose fibers with poly (epsilon-caprolactone) and poly(L-latic acid) via ring-opening polymerization. Biomacromol 7:2178–2185
72. Ibrahim K, Salminen A, Holappa S et al (2006) Preparation and characterization of PS-PEO amphiphilic block copolymers via atom transfer radical polymerization. J Appl Polym Sci 102:4304–4313
73. Roy D, Knapp JS, Guthrie JT et al (2008) Antibacterial cellulose fiber via RAFT surface graft polymerization. Biomacromol 9:91–99
74. Roy D, Guthrie JT, Perrier S (2005) Graft polymerization: grafting poly(styrene) from cellulose via reversible addition-fragmentation chain transfer (RAFT) polymerization. Macromolecules 38:10363–10372
75. Perrier S, Takolpuckdee P, Westwood J et al (2004) Versatile chain transfer agents for reversible addition fragmentation chain transfer (RAFT) polymerization to synthesize functional polymeric architectures. Macromolecules 37:2709–2717
76. Li S, Xie H, Zhang S et al (2007) Facile transformation of hydrophilic cellulose into superhydrophobic cellulose. Chem Commun 4857–4859
77. Li S, Zhang S, Wang X (2008) Fabrication of superhydrophobic cellulose-based materials through a solution-immersion process. Langmuir 24:5585–5590
78. Samyn P, Schoukens G, Van den Abbeele H et al (2011) Application of polymer nanoparticle coating for tuning the hydrophobicity of cellulosic substrates. J Coat Technol Res 8:363–373
79. Stanssens D, Van den Abbeele H, Vonck L et al (2011) Creating water-repellent and super-hydrophobic cellulose substrates by deposition of organic nanoparticles. Mater Lett 65:1781–1784
80. Mukhopadhyay SM, Joshi P, Datta S et al (2002) Plasma assisted hydrophobic coating on porous materials: influence of plasma parameters. J Phys D 35:1927–1933
81. Vaswani S, Koskinen J, Hess DW (2005) Surface modification of paper and cellulose by plasma-assisted deposition of fluorocarbon films. Surf Coat Technol 195:121–129
82. Mukhopadhyay SM, Joshi P, Datta S et al (2002) Plasma assisted surface coating of porous solids. Appl Surf Sci 201:219–226
83. Kim JH, Liu G, Kim SH (2006) Deposition of stable hydrophobic coatings with in-line CH_4 atmospheric rf plasma. J Mater Chem 16:977–981
84. Tan IH, da Silva MLP, Demarquette NR (2001) Paper surface modification by plasma deposition of double layers of organic silicon compounds. J Mater Chem 11:1019–1025
85. Li S, Wei Y, Huang J (2010) Facile fabrication of superhydrophobic cellulose materials by a nanocoating approach. Chem Lett 39:20–21
86. Cappelletto E, Callone E, Campostrini R et al (2012) Hydrophobic siloxane paper coatings: the effect of increasing methyl substitution. J Sol-Gel Sci Technol 62:441–452

87. Yuan H, Nishiyama Y, Kuga S (2005) Surface esterification of cellulose by vapor-phase treatment with trifluoroacetic anhydride. Cellulose 12:543–549
88. Jin C, Yan R, Huang J (2011) Cellulose substance with reversible photo-responsive wettability by surface modification. J Mater Chem 21:17519–17525
89. Hu W, Liu S, Chen S et al (2011) Preparation and properties of photochromic bacterial cellulose nanofibrous membranes. Cellulose 18:655–661
90. Jin C, Jiang T, Niu T et al (2012) Cellulose-based material with amphiphobicity to inhibit bacterial adhesion by surface modification. J Mater Chem 22:12562–12567
91. Balu B, Berry AD, Hess DW et al (2009) Patterning of superhydrophobic paper to control the mobility of micro-liter drops for two-dimensional lab-on-paper applications. Lab Chip 9:3066–3075
92. Martinez AW, Phillips ST, Butte MJ et al (2007) Patterned paper as a platform for inexpensive, low-volume, portable bioassays. Angew Chem Int Ed 46:1318–1320
93. Martinez AW, Phillips ST, Carrilho E et al (2008) Simple telemedicine for developing regions: camera phones and paper-based microfluidic devices for real-time, off-site diagnosis. Anal Chem 80:3699–3707
94. Martinez AW, Phillips ST, Whitesides GM (2008) From the cover: three-dimensional microfluidic devices fabricated in layered paper and tape. Proc Natl Acad Sci U S A 105:19606–19611
95. Martinez AW, Phillips ST, Wiley BJ et al (2008) FLASH: a rapid method for prototyping paper-based microfluidic devices. Lab Chip 8:2146–2150
96. Kong F, Ni Y (2009) Determination of Cr(VI) concentration in diluted samples based on the paper test strip method. Water Sci Technol 60:3083–3089
97. Kong F, Ni Y (2009) Development of cellulosic paper-based test strips for Cr(Vi) determination. BioResources 4:1088–1097
98. Xiao W, Hu H, Huang J (2012) Colorimetric detection of cysteine by surface functionalization of natural cellulose substance. Sens Actuators, B 171–172:878–885
99. Xu M, Bunes BR, Zang L (2011) Paper-based vapor detection of hydrogen peroxide: colorimetric sensing with tunable interface. ACS Appl Mater Interfaces 3:642–647
100. Cha R, Wang D, He Z (2012) Development of cellulose paper testing strips for quick measurement of glucose. Carbohydr Polym 88:1414–1419
101. Kwon H, Samain F, Kool ET (2012) Fluorescent DNAs printed on paper: sensing food spoilage and ripening in the vapor phase. Chem Sci 3:2542–2549
102. Egusa S, Yokota S, Tanaka K et al (2009) Surface modification of a solid-state cellulose matrix with lactose by a surfactant-enveloped enzyme in a nonaqueous medium. J Mater Chem 19:1836–1842
103. Brumer H, Zhou Q, Baumann MJ et al (2004) Activation of crystalline cellulose surfaces through the chemoenzymatic modification of xyloglucan. J Am Chem Soc 126:5715–5721
104. Gustavsson MT, Persson PV, Iversen T et al (2005) Modification of cellulose fiber surfaces by use of a lipase and a xyloglucan endotransglycosylase. Biomacromol 6:196–203
105. Huang J, Ichinose I, Kunitake T (2006) Biomolecular modification of hierarchical cellulose fibers through titania nanocoating. Angew Chem Int Ed 45:2883–2886
106. Ye L, Filipe CDM, Kavoosi M et al (2009) Immobilization of TiO_2 nanoparticles onto paper modification through bioconjugation. J Mater Chem 19:2189–2198
107. Xiao W, Huang J (2011) Immobilization of oligonucleotides onto zirconia-modified filter paper and specific molecular recognition. Langmuir 27:12284–12288
108. Koga H, Kitaoka T, Isogai A (2012) Paper-immobilized enzyme as a green microstructured catalyst. J Mater Chem 22:11591–11597
109. Koga H, Kitaoka T, Isogai A (2011) In situ modification of cellulose paper with amino groups for catalytic applications. J Mater Chem 21:9356–9361
110. Lu P, Hsieh YL (2010) Layer-by-layer self-assembly of Cibacron Blue F3GA and lipase on ultra-fine cellulose fibrous membrane. J Membr Sci 348:21–27
111. Jarujamrus P, Tian J, Li X et al (2012) Mechanisms of red blood cells agglutination in antibody-treated paper. Analyst 137:2205–2210

112. Martinez AW, Phillips ST, Whitesides GM et al (2010) Diagnostics for the developing world: microfluidic paper-based analytical devices. Anal Chem 82:3–10
113. Dungchai W, Chailapakul O, Henry CS (2009) Electrochemical detection for paper-based microfluidics. Anal Chem 81:5821–5826
114. Coltro WKT, de Jesus DP, da Silva JAF et al (2010) Toner and paper-based fabrication techniques for microfluidic applications. Electrophoresis 31:2487–2498
115. Liu H, Crooks RM (2011) Three-dimensional paper microfluidic devices assembled using the principles of origami. J Am Chem Soc 133:17564–17566
116. Khan MS, Thouas G, Shen W et al (2010) Paper diagnostic for instantaneous blood typing. Anal Chem 82:4158–4164
117. Fu E, Lutz B, Kauffman P et al (2010) Controlled reagent transport in disposable 2D paper networks. Lab Chip 10:918–920
118. Hu L, Choi JW, Yang Y et al (2009) Highly conductive paper for energy-storage devices. Proc Natl Acad Sci U S A 106:21490–21494
119. Hu L, Pasta M, Mantia FL et al (2010) Stretchable, porous, and conductive energy textiles. Nano Lett 10:708–714
120. Hu L, Wu H, Mantia FL et al (2010) Thin, flexible secondary Li-ion paper batteries. ACS Nano 4:5843–5848
121. Pushparaj VL, Shaijumon MM, Kumar A et al (2007) Flexible energy storage devices based on nanocomposite paper. Proc Natl Acad Sci U S A 104:13574–13577
122. Gimenez AJ, Yáñez-Limón JM, Seminario JM (2011) Paper-based photoconductive infrared sensor. J Phys Chem C 115:18829–18834
123. Nie Z, Nijhuis CA, Gong J et al (2010) Electrochemical sensing in paper-based microfluidic devices. Lab Chip 10:477–483
124. Weng Z, Su Y, Wang DW et al (2011) Graphene-cellulose paper flexible supercapacitors. Adv Energy Mater 1:917–922
125. Anderson RE, Guan J, Ricard M et al (2010) Multifunctional single-walled carbon nanotube–cellulose composite paper. J Mater Chem 20:2400–2407
126. Zheng G, Hu L, Wu H et al (2010) Paper supercapacitors by a solvent-free drawing method. Energy Environ Sci 4:3368–3373
127. Nystrom G, Razaq A, Strømme M et al (2009) Ultrafast all-polymer paper-based batteries. Nano Lett 9:3635–3639
128. Callone E, Fletcher JM, Carturan G et al (2008) A low-cost method for producing high-performance nanocomposite thin-films made from silica and CNTs on cellulose substrates. J Mater Sci 43:4862–4869
129. Li J, Möhwald H, An Z et al (2005) Molecular assembly of biomimetic microcapsules. Soft Matt 1:259–264
130. An Z, Möhwald H, Li J (2006) pH Controlled permeability of lipid/protein biomimetic microcapsules. Biomacromol 7:580–585
131. He Q, Cui Y, Li J (2009) Molecular assembly and application of biomimetic microcapsules. Chem Soc Rev 38:2292–2303
132. Jia Y, Li J (2015) Molecular assembly of schiff base interactions: construction and application. Chem Rev 115:1597–1621
133. Li J, Jia Y, Dong W et al (2014) Transporting a tube in a tube. Nano Lett 14:6160–6164
134. Chen B, Jia Y, Zhao J et al (2013) Assembled hemoglobin and catalase nanotubes for the treatment of oxidative stress. J Phys Chem C 117:19751–19758
135. Zhao J, Fei J, Gao L et al (2013) Bioluminescent microcapsules: applications in activating a photosensitizer. Chem Eur J 19:4548–4555

Part IV
Molecular Assembly of Motor Proteins and Artificial Micro-/Nanomotors

Chapter 9
Directional Transportation of Assembled Molecular Linear Motors

Ning Zhang and Xiang-Dong Li

Abstract The transport and localization of organelles within a eukaryotic cell are dependent on a complex network of protein filaments called the cytoskeleton and the associated proteins, including actin filaments and myosins. Myosin is a molecular motor which has ATPase activity and is able to convert the energy from ATP into mechanical movement along actin filaments. To transport cargo efficiently and to prevent the futile hydrolysis of ATP, the motor activity of myosin must be tightly regulated. In the past two decades, considerable progress has been achieved in understanding new myosins and regulatory mechanism of myosin motors. This chapter will present a general view of myosin family and focus on the structure, function, and regulation of three myosins, i.e., smooth muscle myosin-2, myosin-5, and myosin-19.

Keywords Actin · ATP · Molecular motor · Myosin · Regulatory mechanism

9.1 Introduction

The eukaryotic cell has an elaborate internal membrane system, including a number of functionally distinct membrane-bounded compartments (organelles). Cells use the cytoskeleton and associated molecular motors to move, position, and segregate their organelles to ensure their homeostasis [1, 2]. Intracellular transport is an

N. Zhang · X.-D. Li (✉)
Group of Cell Motility and Muscle Contraction, State Key Laboratory of Integrated Management of Insect Pests and Rodents, Institute of Zoology, Chinese Academy of Sciences, Beijing 100101, China
e-mail: lixd@ioz.ac.cn

N. Zhang
e-mail: zhangning@ioz.ac.cn

X.-D. Li
University of Chinese Academy of Sciences, Beijing 100049, China

J. Li (ed.), *Supramolecular Chemistry of Biomimetic Systems*,
DOI 10.1007/978-981-10-6059-5_9

important cellular process, which is essential for various cellular physiological functions including endocytosis, extocytosis, cytokinesis, cell motility, muscle contraction, movement of mRNA, and organelles. Intracellular transport is dependent on microtubule-based, long-range movements and actin-based, short-range movements. The former are driven by kinesin and dynein, and the latter by myosin. Myosins are actin-based molecular motors which are able to efficiently convert the chemical free energy of adenosine triphosphate (ATP) hydrolysis to mechanical force and movement. This chapter will present a general view of myosin family and focus on the structure, function, and regulation of three myosins, i.e., smooth muscle myosin-2 (SmM), myosin-5 (Myo5), and myosin-19 (Myo19).

9.2 Structure, Classification, and Function of Myosin

Phylogenetic analysis of the motor domain divided the myosin superfamily members into 35 classes [3]. The human genome contains 38 myosin genes, belonging to 12 classes based on sequence homology of the motor domain and tail domain. Figure 9.1 shows an unrooted phylogenetic tree and predicted structure of all myosin encoded by human myosin genes [4].

In general, myosins are composed of three domains. At the N terminus is the motor domain containing both an ATP-binding site and the binding region for filamentous actin, linking to a repeated cycle of ATPase activity during its movement along actin filaments. Two surface loops named loop 1 and loop 2 may be important for both ATP and actin binding. Loop 1 (25–50 kDa) is near the ATP-binding site: substitutions within this region in myosin-2 have led to alteration in ADP release rates [5]. Loop 2 (50–20 kDa) has been implicated in myosin–actin interaction: substitutions within this domain in myosin-2 have resulted in altered myosin-actin affinity and actin-activated ATPase rates of the myosin [5]. Analysis of the vertebrate myosin-2 sequences suggested that loop 1 is relatively conserved, while loop 2 may have a class-specific function [5].

The motor domain is followed by a neck that consists of a variable number of IQ motifs with the consensus sequence IQXXXRGXXXR, which serve as the binding sites for light chains of calmodulin (CaM) or the CaM-like proteins. The neck domain complexed with light chains is also referred to as the lever arm, which is able to amplify small nucleotide-dependent changes that originate at the active site, allowing a large power stroke to occur following ATP hydrolysis.

In contrast to the conserved motor domain, the tail domains of myosin are highly variable. Nearly half of the known myosin classes, including myosin-2, -5, -6, -7, -8, -14, and -18, contain coiled-coil structures in the tail. Therefore, the myosins in those classes are predicted to form a homodimer via the coiled-coil structure. Indeed, biochemical analysis and electron microcopy observation show that some of those classes, including class-2, -5, -18, form homodimer. On the other hand,

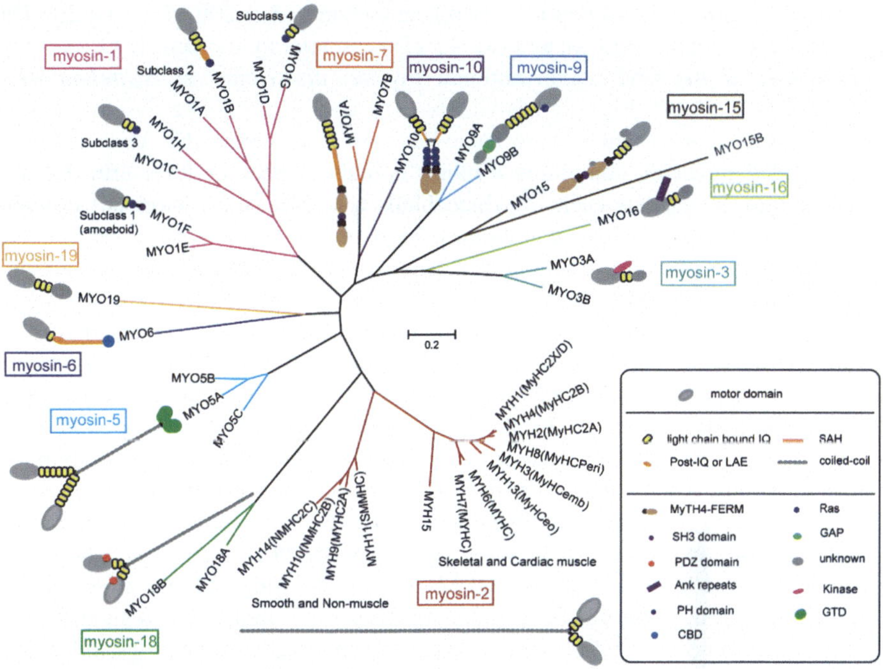

Fig. 9.1 The myosin family tree in humans. The 38 genes encoding myosins in humans are divided into 12 different classes. To generate the tree, the sequences of the motor domain for all of these myosins were aligned using ClustalW and the alignment was used to generate a tree using a maximum likelihood approach (MEGA5.0)

some classes (e.g., myosin-6 and -7) contain very short coiled-coil structure in the tail and cannot form stable dimer.

In addition to the coiled-coil structure, a number of class-specific structures are found in myosin tail regions. For example, SH3 domain which participates in protein–protein interaction is found in the myosin-1s. MyTH4 domain (myosin-tail-homology domain 4) and FERM domain are presented in the tails of myosin-4, -7, and -15. The tail domain of myosin-5 and -11 contain a specific globular tail domain (named dilute domain). Myosin tail domain not only plays a role in regulating the function of the motor domain, but also serves as the binding site for myosin to interact with the cargoes.

Myosin-2 s are a large family of filament-forming conventional myosin which were discovered over 150 years ago. Myosin-2 is also named as conventional myosins and others classified as unconventional myosins. Myosin-2 is the most abundant protein in muscles, including skeletal, cardiac, and smooth muscles. In muscles, myosin-2 together with actin makes up the major contractile proteins. The thick bipolar myosin filaments and thin actin filaments slide toward each other causing the muscle contraction. Importantly, myosin-2 also present in nonmuscle cells and involved in cell migration and control of cell shape [6]. Myosin-2 molecules are comprised of three pairs of peptides: two heavy chains of ~200 kDa, two 20 kDa regulatory light chains (RLCs), and two 17 kDa essential light chains (ELCs).

Myosin-1 is the first unconventional myosin discovered in *Acanthamoeba castellanii* in 1973 [7]. Myosin-1 is widely expressed in vertebrate tissues and has eight different isforms which were named as a to h in human [8]. Myosin-1 is a single-headed molecule containing a motor domain, a neck or lever arm domain (three or four IQ repeats each of which binds a CaM), and a short tail domain interacting with phospholipid membranes. Myosin-1 is involved in many cellular processes, such as the ultrastructure of intestinal microvilli, transcription regulation, cell adhesion, and so on.

Myosin-5 is one of the most ancient members of the myosin superfamily, distributing from lower eukaryotes, such as yeast, to vertebrates. Myosin-5 functions as a cargo-transporting motor, moving various cargoes along actin filaments to the specific destination in cells. In the *Drosophila melanogaster* compound eye, myosin-5 plays two distinct roles in response to light stimulation: transport of pigment granules to the rhabdomere base to decrease light exposure and transport of rhodopsin-bearing vesicles to the rhabdomere base to compensate for the rhodopsin loss during light exposure [9]. In vertebrate, there are three distinct subclasses of myosin-5, named Myo5a, Myo5b, and Myo5c among which Myo5a is so far the best characterized in terms of cellular function and molecular regulation. In mammal, myosin-5 was not only involved in transporting of pigment granules, but also in the movement of receptors in neurons. Mutations in Myo5a could lead to Griscellis syndrome and that in Myo5b could cause the retraction of microvillus in small intestine.

Myosin-6 plays a particularly important role within the cell, as it is the only myosin moving toward the minus end of actin filament. Myosin-6 exists as a monomer in isolation; however, cargo binding could induce the formation of homodimer to fulfill a number of specialized cell biological functions [10]. Myosin-6 participates in the formation of stereocilia in cells of the auditory system [11], membrane internalization [12], and delivery of membrane to the leading edge in migratory cells [13]. In humans, defects in myosin-6 also cause deafness and an inherited form of hypertrophic cardiomyopathy [14, 15]. Importantly, myosin-6 is dramatically up-regulated expressed in breast, lung, prostate, ovary, and gastresophagus carcinoma cells, therefore, it is regarded as an early marker of cancer development, aggressiveness, and cancer–cell invasion [13, 16–19].

9.3 Characterization of Myosin

Myosins are molecular machines that convert the chemical energy derived from ATP hydrolysis into mechanical work such as muscle contraction and cellular motility. Since the discovery of myosin more than 100 years ago, many techniques have been developed for myosin research. Here, we will introduce several commonly used techniques in characterization of myosin motor.

9.3.1 ATPase Activity

The most basic characterizations of myosin are the ATPase activity and the activation of ATPase activity by actin. Myosin motors are mechanoenzymes which can efficiently convert the energy from ATP hydrolysis to mechanical work, such as contraction of muscles, the transport of organelles, and the generation of cytoskeletal tension. Therefore, the motor domain of myosin could be considered as an ATPase, which works with actin filaments to generate force or movement.

Although myosins share a core of conserved residues in their motor domain, containing the ATP-binding site and the actin-binding site, their ATPase activities are quite variable, ranging from ~0.13 to ~390 s^{-1} [20]. Several myosins, such as Drosophila myosin-20, do not have detectable ATPase activity and cannot produce force or movement [21, 22].

In general, myosin's ATPase activity is correlated with its motor activity. Because it is relatively easier to measure the ATPase activity than to measure other myosin-specific activity, such as in vitro actin-gliding activity, the ATPase activity is generally used to compare the motor activity of different myosins. For the same myosin, its ATPase activity is varied under different conditions (i.e., Ca^{2+} concentration, phosphorylation, ionic concentration). It is convenient to use the ATPase activity as indicator to investigate the regulation of myosin. For example, Myo5a has higher ATPase activity in the presence of micromolar levels of Ca^{2+} than in its absence, indicating that the motor function of Myo5a is activated by Ca^{2+}.

9.3.2 Kinetic Analysis of Myosin ATPase Activity and Duty-Ratio

Myosins are involved in a plethora of physiological processes with different kinetic and structural adaptations. However, the rate and equilibrium constants that define the flux through the kinetic cycle (Fig. 9.2) vary among different myosins. Myosins

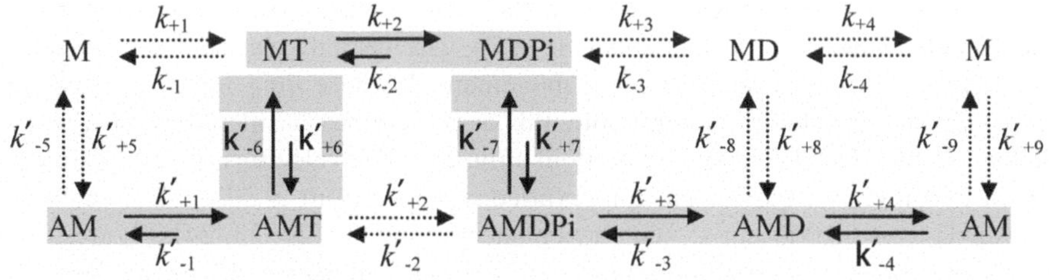

Fig. 9.2 Kinetic cycle of myosin ATPase. *M* myosin, *A* actin, *T* ATP, *D* ADP, *Pi* phosphoric acid. The length of the *arrow* represents the relative rate of the reaction. The main reaction of the kinetic cycle for myosin in the present of actin was shown in *shadow*

alone and myosins bound with ADP have a high affinity for actin, while that bound with ATP or ADP-Pi have a low affinity.

The duty-ratio is defined as the fraction of the time that a motor domain spends attached to an actin filament during an ATPase cycle. Myosins with high-duty ratios can maintain continuous attachment to an actin filament and can move along the filament on their own (thus processive), whereas myosins with low-duty ratios usually oligomerize into a large assembly in order to produce continuous motility (the individual molecules are nonprocessive). The duty-ratio can be calculated based on the kinetic rates of myosin ATPase.

For most myosin ATPases, the release rates for either phosphate (Pi) or ADP is the rate-limiting step, which limits the overall ATPase activity and is commonly associated with transitions between strong and weak actin-binding states. The rate-limiting step for the ATPase of most myosin-2 is Pi release (k'_{+3}). Thus, those myosins spend most time during the kinetic cycle in the weak actin-binding state, thus preventing the interference between the assemble myosin motors. In contrast, the rate-limiting step for Myo5a ATPase is ADP release (k'_{+4}), and therefore Myo5a spends most (~70%) of the time to bind tightly to actin. Furthermore, the two heads of Myo5a are strain-dependent, such that ADP release from the attached leading head is strongly suppressed as long as the trailing head is still attached to actin. Those properties enable individual Myo5a to continually move along actin filaments for several steps.

The rates of phosphate release and ADP release can be directly measured using stopped-flow apparatus. The rate-limiting step of myosin ATPase can be determined by comparing the phosphate and ADP release rates with the steady-state ATPase activity. Moreover, measurement of the steady-state ATPase activity in the presence of ADP can also indicate whether the rate-limiting step is ADP release or not. If the rate-limiting step is ADP release, the steady-state ATPase activity will be strongly inhibited by ADP.

9.3.3 In Vitro Actin-Gliding Assay

The motor activity of myosins can be assessed by an in vitro actin-gliding assay. In the assay, myosin molecules attached to the glass surface drive the movement of fluorescently labeled actin filaments (Fig. 9.3). The velocity can be determined by quantifying the rate of actin filament movement. Because drag force imposed on actin filament by solution is negligible, this assay can be considered as an in vitro analog of unloaded shortening velocity in a muscle. The velocities of actin filament driven by different myosins are quite varied, ranging from ~20 nm s^{-1} (such as myosin-1) to ~60,000 nm s^{-1} (such as plant myosin-11) [20, 23, 24]. Figure 9.3b shows the movement of fluorescently labeled actin filaments driven by Myo5a, which was fixed on nitrocellulose-coated glass.

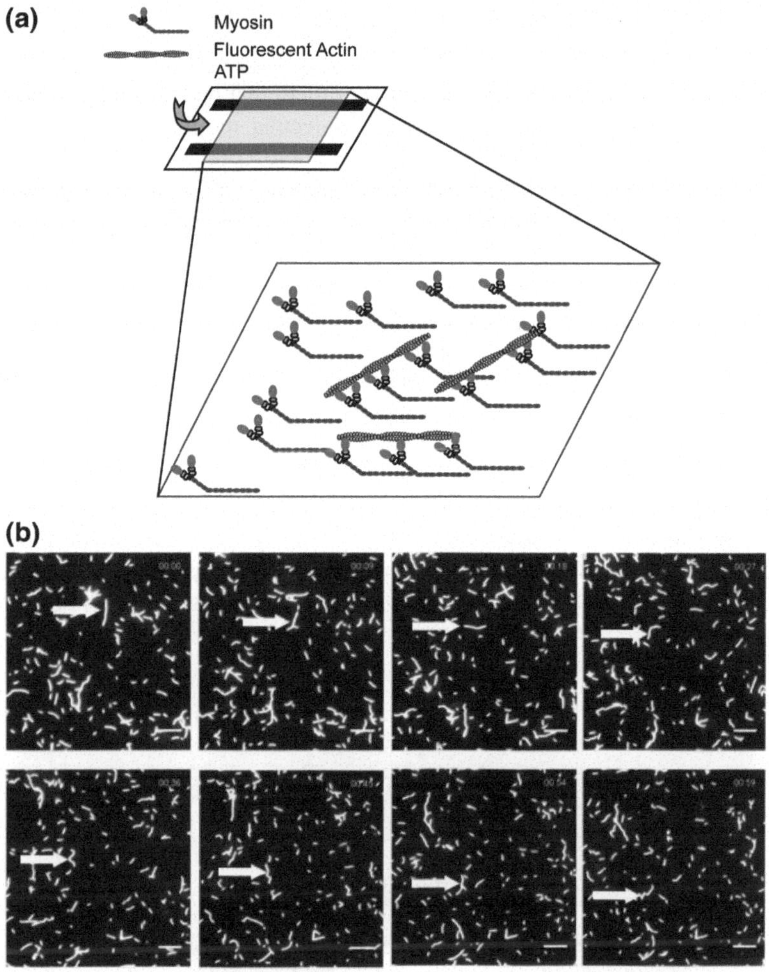

Fig. 9.3 The gliding of actin driven by Myo5a. **a** Diagram of actin-gliding assay. **b** The movement of actin filaments driven by Myo5a. The *arrows* show the change in the position of Rhodamin phalloidin labeled actin filament. Myo5a was fixed on nitrocellulose-coated glass. The experiment was conducted in 150 mM KCl. Scale bars, 5 μm

In the standard assay, myosin molecules are first attached onto a glass surface of a coverslip either directly or via a specific antibody which recognizes the tail domain of myosin, and then fluorescently labeled actin filaments are flowed into the coverslip. Actin filament movement is observed under an inverted fluorescence microscope. In addition, several modified in vitro actin-gliding assays have developed to determine the directionality and processivity of myosin (see below).

9.3.4 Single Molecular Motility Assay

The single molecular motility assay is an experiment that investigates the properties of individual myosin molecules. The single molecular motility assay is conducted under a total internal reflection fluorescence microscope (TIRFM). By using a TIRFM, a restricted region of the specimen immediately adjacent to the glass–water interface (less than 200 nm from the glass surface) is illuminated by the evanescent field, thus greatly reducing the background noise (Fig. 9.4a).

In a single molecular motility assay, actin filaments and myosin are labeled with different dyes. In a coverslip, the fluorescence-labeled actin filaments are first fixed on the surface of the glass via an actin-binding protein or an *N*-ethylmaleimide-modified myosin (*N*-ethylmaleimide modification destroys myosin motor function but does not affect the actin-binding property of myosin). Myosin molecules which have been labeled with a fluorescent moiety (such as Alexa-488) or a fluorescent protein (such as GFP) are then introduced into the coverslip. In the presence of ATP, processive myosin molecules will move processively along actin filaments. Figure 9.4b shows the movement of Cy3B labeled Myo5a on Alexa 488-phalloidin labeled actin filament. Many important parameters, such as step size, run length, and velocity, can be obtained by analyzing the behaviors of individual myosin molecules.

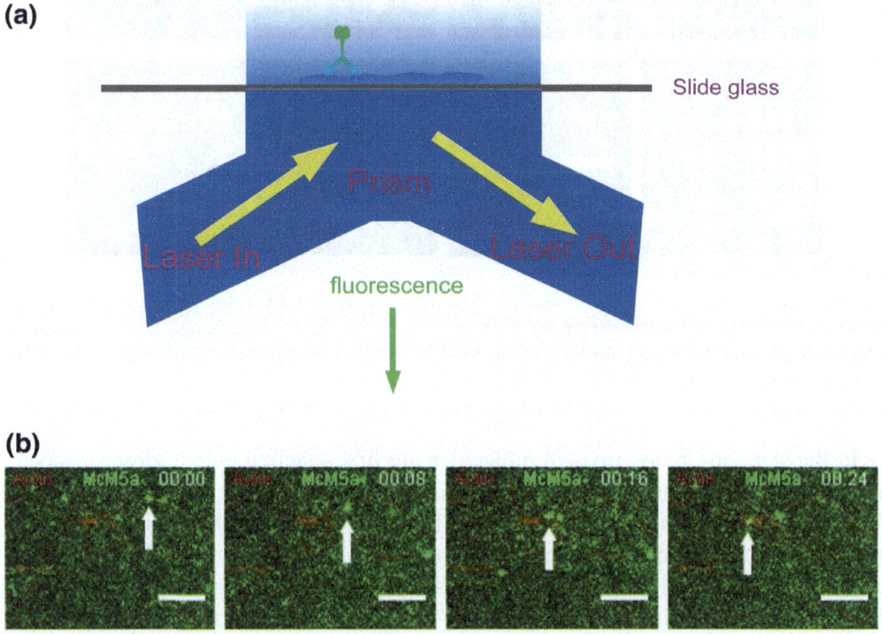

Fig. 9.4 The movement of Myo5a on actin filament. **a** Diagram of single molecular motility assay. **b** The movement of Myo5a along actin filaments. The *arrows* show the position of Cy3B labeled Myo5a along actin filament. Myo5a was shown in green, while actin filaments were shown in red. The experiment was conducted in 25 mM KCl. Scale bars, 5 μm

9.3.5 *Processivity*

Processivity is the ability of an individual myosin molecule to successively move along actin track for several steps without dissociating from the track. Thus, a processive myosin motor is more suitable for cargo transport than the unprocessive myosin. Nonprocessive myosin motors usually oligomerize into a large assembly in order to produce continuous motility (the individual molecules are nonprocessive). Nonprocessive myosins are commonly associated with low-duty ratio, which means the fraction time it spends in a state of high affinity for actin binding during an ATP hydrolysis cycle is short. A processive myosin usually is a high-duty ratio motor.

The best way to determine processivity of myosins is using single molecular assay to directly observe the successive movement of individual myosin molecule along actin filaments. Another way is analyzing the actin-gliding velocities in the presence of different densities of myosin molecules. The actin-gliding velocity is largely independent of the density of processive myosin molecules, but decreases with that of nonprocessive myosin molecules.

Vertebrate Myo5a is the first identified and the best characterized processive myosin. Single-molecule study shows that Myo5a walks in a hand-over-hand mechanism with a 36 nm step size before dissociating from actin [25]. Figure 9.5 shows the pathway of the processive movement of Myo5a along actin. Myo5a dimerizes to form a two-head motor through its coiled-coil sequences in its tail domain, allowing the molecule to walk along actin filaments by alternating the positions of the leading and trailing heads. Long run lengths are achieved by the two-gated heads: the two heads coordinated with each other by intermolecular strain to ensure that the leading head does not dissociate prematurely.

9.3.6 *Directionality*

All subunits in an actin filament point toward the same end of the filament. Therefore, actin filament has polarity: the end that possesses an actin subunit that has its ATP-binding site exposed is called the minus end, while the opposite end where the ATP-binding site is contacts the adjacent subunit is called the plus end. It is generally believed that most actin filaments have their plus-ends located near the plasma membrane.

All myosins so far characterized, except myosin-6, move toward the plus end of actin filament. It is likely that myosin-6 plays a unique role in cells. Structural analysis and mutagenesis show that a unique insert of $\sim$38 amino acids located between the converter and the lever arm is the sole determinant of directionality [26, 27].

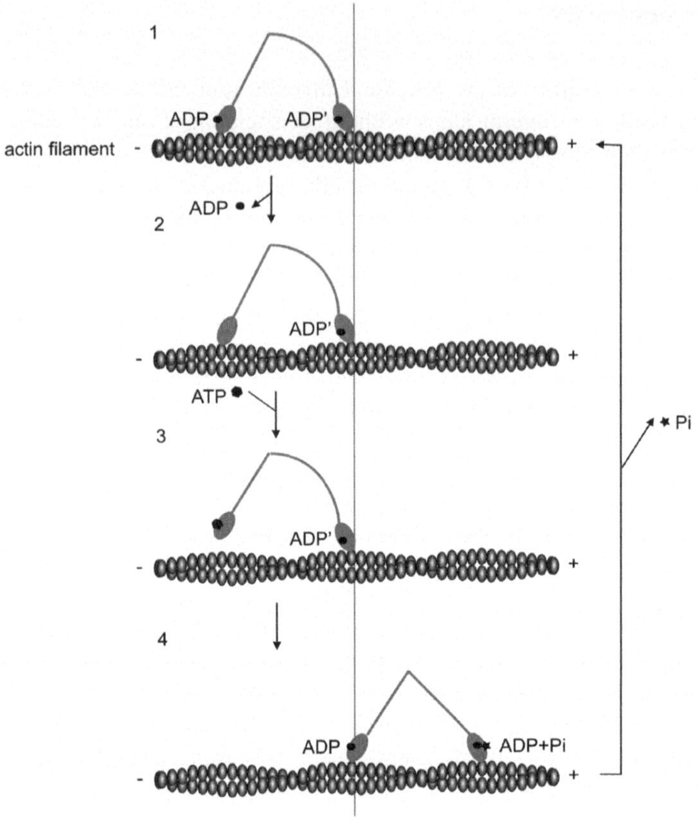

Fig. 9.5 Pathway of the processive movement of Myo5a along actin filament. *1* Myo5a probably dwells in a state with ADP bound to both heads. *2* The two heads exert intramolecular strain on each other, so that ADP is first released from the trailing head. *3* This head subsequently binds ATP and rapidly dissociates from actin. *4* The attached head undergoes a power stroke, positioning the new leading head (bound to ADP and Pi) to find a forward binding site via a thermally driven search. Upon binding to actin, the leading head rapidly releases Pi and establishes a strong binding conformation, which brings Myo5a to the same state as in 1 but translated forward by 36 nm

The directionality of myosin can be determined using a modified in vitro actin-gliding assay using the actin filaments that are polarity-labeled with different fluorescent probes [28]. Thus, preparation of the polarity-labeled actin filaments is crucial for successful determination of the directionality of myosin. The polarity-labeled actin filaments can be prepared by incubating the fluorescence-labeled actin filament fragments with G-actin. Those actin filament fragments function as the seeds in inducing the elongation of actin filaments at the plus-ends.

9.4 Smooth Muscle Myosin-2 (SmM)

Human genome contains 13 myosin-2 genes, i.e., seven skeletal muscle myosin-2s, two cardiac muscle myosin-2s, one smooth muscle myosin-2 (SmM), and three nonmuscle myosin-2s (Fig. 9.1). Similar to other myosin-2s, SmM is a hexameric molecule composed of two heavy chains, two ELCs and two RLCs (Fig. 9.6). Each molecule has two globular head domains that are the sites of enzymatic activity (i.e., force generation). Within each head the two myosin light chains bind to an extended α-helix of the heavy chain that functions as a "lever arm" amplifying small movements generated within the myosin head. Beyond the light chain domain, the myosin molecule dimerizes by virtue of an α-helical coiled-coil.

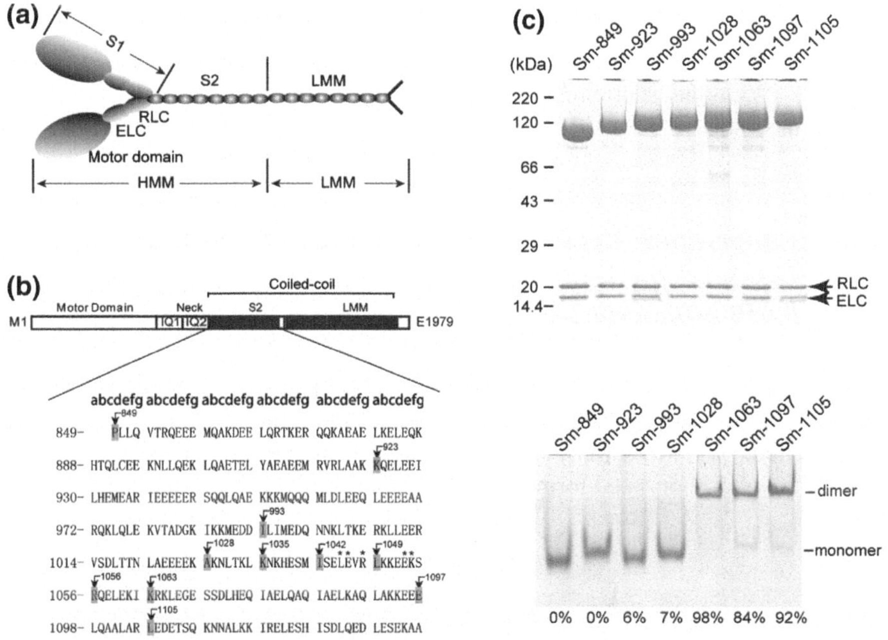

Fig. 9.6 SmM structure. **a** Diagram of SmM structure. **b** The S2 coiled-coil sequence. The amino acid residues with arrows in the sequence of the S2 coiled-coil indicate the points of SmM truncation. The positions of each amino acid residue in the coiled-coil heptad repeats are indicated by lower case letters a–g in the *top line*. For clarity, the diagram at the *top* was not drawn to scale. Reprinted with permission from Ref. [67]. Copyright 2013, American Chemical Society

9.4.1 The Key Coiled-Coil Region Stabilizes the Double-Headed Structures of SmM

SmM molecules form a double-headed structure via the C-terminal coiled-coil of the heavy chain, and that SmM constructs without coiled-coil, i.e., S1, are single-headed [29–31]. Previous work showed that the coiled-coil between residues 1025 and 1109 (heptads 25 and 37) are essential for the formation of a double-headed structure [30, 31]. In those studies, the distinction between a single-headed and a double-headed structure was based on a single criterion, i.e., the migration rate of SmM in native polyacrylamide gel electrophoresis (PAGE). However, the effect of electrophoresis on the stability of the double-headed structure of SmM is not known.

We investigated the dimerization of the truncated SmM using three complementary approaches, i.e., native PAGE, metal-shadowed electron microscopy (EM), and size-exclusion chromatography (Table 9.1). We found that the coiled-coil segment of 849–1056 enable SmM to form a stable double-headed structure (Fig. 9.7). With the shortening of the coiled-coil, the double-headed structure becomes less stable.

9.4.2 Cooperation Between the Two Heads of SmM Is Essential for Full Activation of the Motor Function by Phosphorylation

The motor activity of SmM is activated by phosphorylation of the RLC [32, 33]. In other words, the unphosphorylated form of SmM is inactive, i.e., it has low actin-activated ATPase activity and is incapable to move actin filaments in vitro, whereas the phosphorylated form is active in both respects. Thus, the regulation of

Table 9.1 Percentage of the double-headed structure of SmM constructs

SmM construct	Native PAGE[a]	Electron Microscopy[b]	Size exclusive chromatography[c]
Sm-849	0 (4)	0 (269)	0
Sm-923	0 (3)	8.5 (387)	14.7
Sm-993	4.9 ±4.3 (4)	46.5 (318)	24.0
Sm-1028	10.7 ±7.7(4)	64.1 (222)	68.3
Sm-1063	98.4 ±2.0 (4)	96.8 (292)	100

[a]Based on native PAGE

[b]Based on electron microscopy

[c]Based on size exclusive chromatography

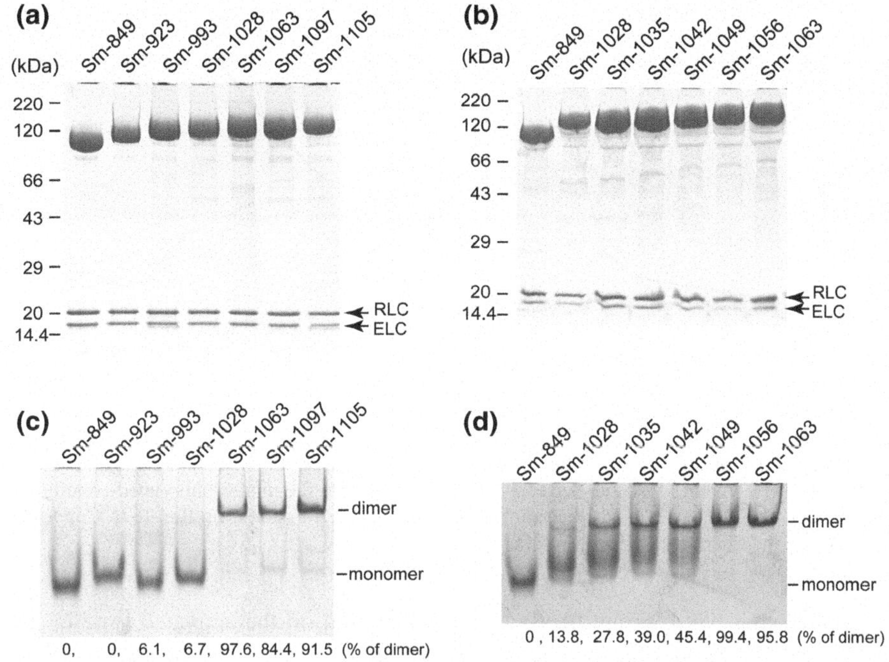

Fig. 9.7 Dimerization of the truncated SmM. **a** Purified SmM constructs were analyzed by SDS-PAGE and native PAGE. **a**, **b** SDS-PAGE shows the components of the purified SmM constructs. *Arrows* indicate the RLC and the ELC copurified with SmM heavy chain. **c**, **d** Native PAGE shows the dimerization (single-headed structure vs. double-headed structure) of the purified SmM constructs. The *number below the native gels* indicates the estimated percentage of double-headed structure using densitometry. Reprinted with permission from Ref. [67]. Copyright 2013, American Chemical Society

SmM represents two issues: the inhibited state of the unphosphorylated SmM and the activated state of the phosphorylated SmM.

It is widely accepted that the inhibited state of the unphosphorylated SmM requires the head-head interaction and the head-tail interaction. Early studies showed that, in unphosphorylated state, the S1 produced by limited protease treatment of tissue-isolated SmM has higher ATPase activity than the HMM [29, 34–36]. We investigated the regulation of recombinant SmM having truncated tail. In the unphosphorylated state, the ATPase activity of SmM increased as the coiled-coil length shortened (Fig. 9.8). When the coiled-coil was longer than 179 aa (i.e., for Sm-1028, -1063, -1097, and -1105), the ATPase activity reached the minimum in the unphosphorylated state. Together with the findings that Sm-1028 formed a labile double-headed structure and that Sm-1063, -1097, and -1105 formed stable double-headed structures (Fig. 9.7 and Table 9.1), these results indicate that the double-headed structure is essential for the inhibition state of the unphosphorylated SmM.

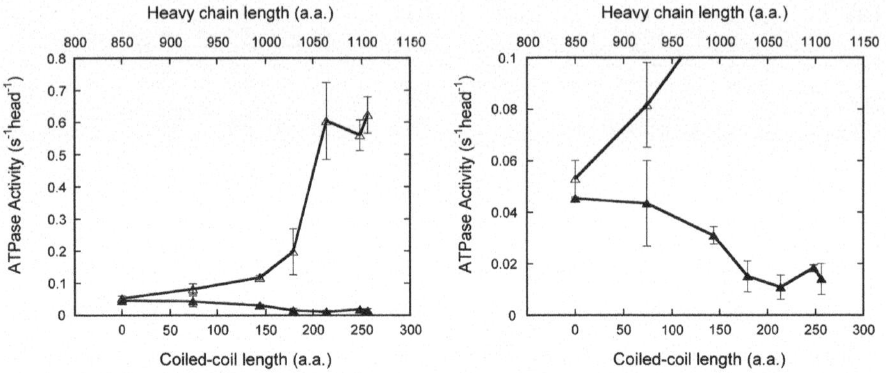

Fig. 9.8 Relationship between the actin-activated ATPase activity and coiled-coil length. The actin-activated ATPase activities of SmM constructs are plotted against the length of the coiled-coil, which are 0, 74, 144, 179, 214, 248, and 256 amino acids for Sm-849, -923, -993, 1028, -1063, -1097, and -1105, respectively. *Filled symbols* are the unphosphorylated conditions; *open symbols* the phosphorylated conditions. Reprinted with permission from Ref. [67]. Copyright 2013, American Chemical Society

On the other hand, it had been quite controversial on the activated state of the phosphorylated SmM. Sata et al. reported that recombinant S1 has the same ATPase activity, regardless of the state of phosphorylation, as the phosphorylated HMM [31]. However, Konishi et al. found that the ATPase activity of recombinant S1 is substantially lower than that of phosphorylated HMM [37]. We measured the ATPase activity of the phosphorylated SmM having truncated coiled-coil. Under phosphorylation conditions, the ATPase activity of SmM decreased as the coiled-coil length shortened (Fig. 9.8). The ATPase activities of stable HMMs (Sm-1063, Sm-1097, and Sm-1105) under phosphorylation conditions were significantly higher than unstable HMM (Sm-1028, -993, -923) and S1 (Sm-849).

Those results indicate that the stable double-headed structure is essential for the full-activated state of the phosphorylated SmM, suggesting the cooperation between the two heads of the phosphorylated SmM. This scenario is consistent with the results obtained from asymmetric SmM-HMM, which contains a wild-type head and a mutant head having no motor activity. Under phosphorylation conditions, asymmetric SmM-HMM has significantly lower ATPase activity and in vitro actin-gliding activity than the wild-type SmM-HMM [38].

Taken together, the stable double-headed structure is essential not only for the inhibited state of the unphosphorylated SmM, but also for the full-activated state of the phosphorylation SmM. Those results indicate that cooperation between the two heads of smooth muscle myosin is essential for full regulation of SmM by phosphorylation.

9.5 Myosin-5

Myosin-5, a motor protein that transports cellular cargoes toward the plus end of actin filaments, is distributed widely from yeast to mammals. Vertebrates contain three isoforms of myosin-5, named Myo5a, Myo5b, and Myo5c [39–41], which are expressed in different tissues and exhibit distinct cargo specificities. Myo5a heavy chain contains the N-terminal motor domain, followed by the neck region and the tail domain (Fig. 9.9a). The motor domain contains the ATP- and actin-binding site and is capable of converting energy from ATP hydrolysis into mechanical work. The neck region consists of six IQ motifs with the consensus sequence of

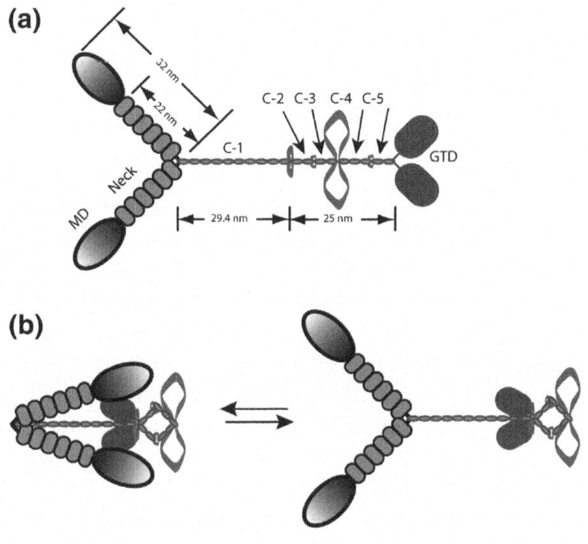

Fig. 9.9 Proposed model for the inhibited state of Myo5a. **a** Predicted structure of Myo5a based on its sequence. At the N terminus is the motor domain (MD) containing the ATP and actin-binding sites. The motor domain is followed by a neck that consists of six IQ motifs, which act as the binding sites for calmodulin or myosin light chains. The next 500 amino acids are predicted to form five coiled-coils (C-1 to C-5) separated by several flexible regions. The last 400 amino acids form a GTD. **b** Proposed structure of Myo5a in the inhibited state. At low Ca^{2+} conditions, GTD binds to the C-terminal end of the coil-1. The coil-2 to coil-5 region forms a compact structure that is visualized as a small globular domain. The binding of the motor domain to the GTD associated at the C-terminal end of the first coiled-coil stabilizes an isosceles triangle conformation, and prevents the conformational changes of the motor domain during the ATP turnover cycle; thus inhibiting the ATPase activity of motor domain. The activated state can be achieved by cargo binding and/or elevation of Ca^{2+}. Cargo binding to the tail domain may interfere with the interaction between GTD and head, thus disrupting the triangular shape. High Ca^{2+} stimulates the actin-activated ATPase activity and induces an extended conformation of Myo5a, presumably through the calmodulin bound to the IQ motifs. Reprinted with permission from Ref. [49]. Copyright 2006, American Society for Biochemistry and Molecular Biology

IQXXXRGXXXR, the binding sites for calmodulin (CaM) or CaM-like light chains. The neck region functions as a lever arm to amplify the small conformational change in the motor domain into large movement. The tail domain consists of two distinct regions. The proximal 500 amino acids of the tail form a series of coiled-coils separated by several flexible regions, and the distal 400 amino acids form a globular tail domain (GTD).

Vertebrate Myo5a is so far the best characterized motor protein among the myosin superfamily [42, 43]. One of the most important questions is how the motor function of Myo5a is regulated. If Myo5a were constantly active, then the high concentrations of Myo5a found in brain would consume substantial amounts of ATP. Therefore, it is desirable that the mechano-enzymatic activity of Myo5a is regulated in vivo.

9.5.1 Tail-Inhibition Model

The ATPase activities of tissue-isolated Myo5a [39, 44] and baculovirus-expressed Myo5a [45, 46] are well regulated, the ATPase activity of Myo5a being significantly increased by micromolar concentrations of Ca^{2+}. Sedimentation velocity analysis showed that the Myo5a undergoes a Ca^{2+}-induced conformational transition from 14 S to 11 S (Fig. 9.10) [44–46]. Electron microscopy revealed that at low ionic strength, Myo5a has an extended conformation in high Ca^{2+} whereas it forms a folded shape in the presence of EGTA, in which the tail domain is folded back toward the head [44]. The conformational transition is closely correlated with activation of the ATPase activity of Myo5a [45]. On the other hand, truncated Myo5a without the GTD is not regulated by Ca^{2+} [47] and does not undergo a large conformational transition like Myo5a [45, 46].

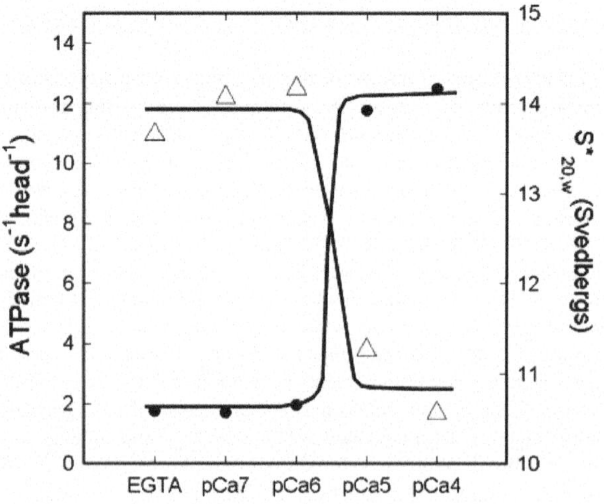

Fig. 9.10 Effects of Ca^{2+} on the ATPase activity and sedimentation coefficient of Myo5a. *Closed circle* represents the ATPase activity, while *open triangular* represents sedimentation coefficient. Reprinted with permission from Ref. [45]. Copyright 2004, Elsevier Inc

Based on above findings, the tail-inhibition model for the regulation of Myo5a was proposed (Fig. 9.9b): Myo5a in the inhibited state is in a folded conformation such that the tail interacts with the head and inhibits its motor activity; cargo binding, high Ca^{2+}, and/or phosphorylation may reduce the interaction between the head and the tail, thus activating the motor activity [44–46]. The tail-inhibition model is likely conserved among other types of Myo5a. Later studies show that vertebrate Myo5b and Myo5c and Drosophila Myo5 are also regulated by their tails in a Ca^{2+}-dependent manner [9, 48].

The GTD is the inhibitory domain of Myo5a, because Myo5a-HMM is inhibited and folded by exogenous GTD [49, 50]. The strong inhibition of motor activity by the GTD requires not only the two-headed structure, but also the presence of an intact first long coiled-coil segment of the tail [49]. The inhibition of Myo5a motor function by the GTD is very sensitive to ionic strength, suggesting that ionic interactions play a key role in the interaction between the head and the GTD. Consistent with this notion, we identified a conserved acidic residue in the motor domain (Asp136) and two conserved basic residues in the GTD (Lys1706 and Lys1779) as critical residues for this interaction (Fig. 9.11) [51]. Moreover, mutagenesis of the yeast Myo5 (Myo2p) shows that spatial and temporal regulation of Myo5a in vivo by a head-to-tail interaction is critical for the normal delivery functions of the motor [52].

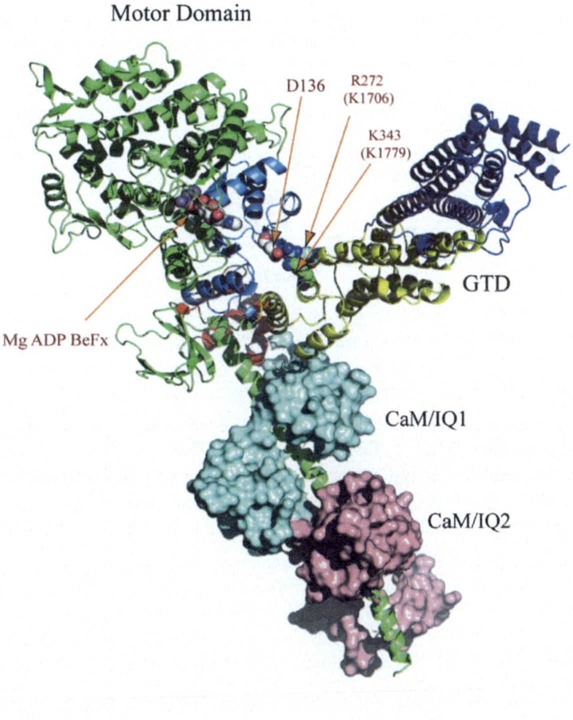

Fig. 9.11 Model for interaction between the GTD and the motor domain of Myo5a in the post-rigor conformation. The GTD-binding pocket consists of the N-terminal domain (*blue*), converter (*orange*), and the CaM (*cyan*) in IQ1. Subdomain-II (*yellow*) of the GTD interacts with the motor domain. The distances between the oxygen atom (O) of D136 in the motor domain and the nitrogen atom (N) of K272 and the nitrogen atom (N) of R343 in Myo2p-GTD (homolog residues of K1706 and K1779 in mouse myosin Va, respectively) are ≈4 Å. Reprinted with permission from Ref. [51]. Copyright 2008, National Academy of Sciences

9.5.2 Regulation of Myo5a by Cargo-Binding Protein

Myo5a is responsible for the transportation and localization of a number of vesicles, including melanosomes in melanocytes (for a review, see [1]). Because the tail of Myo5a not only functions as a cargo-binding site but also serves as a key regulatory component of Myo5a, Sellers and colleagues proposed that the binding of cargo to the tail might activate the motor activity of Myo5a [44]. Consistent with this prediction, we found that Mlph directly stimulates the actin-activated ATPase activity of Myo5a [53]. Recently, Trybus and colleagues demonstrated at the single-molecule level that Mlph significantly increases the number of processively moving Myo5a molecules [54]. Moreover, structural studies [55, 56] have shown that Mlph-GTBDP, a 26-residue peptide in the middle portion of Mlph (residues 176–201) binds to a clef in subdomain-1 (SD-1) of the GTD (Fig. 9.12). We found that Mlph-GTBDP stimulates the ATPase activity of Myo5a by inhibiting the interaction between the head and tail (Fig. 9.13).

Biochemical analysis and the crystal structures of the Myo5a-GTD/Mlph-GTBDP complex show that the GTD uses two distinct regions to interact with the

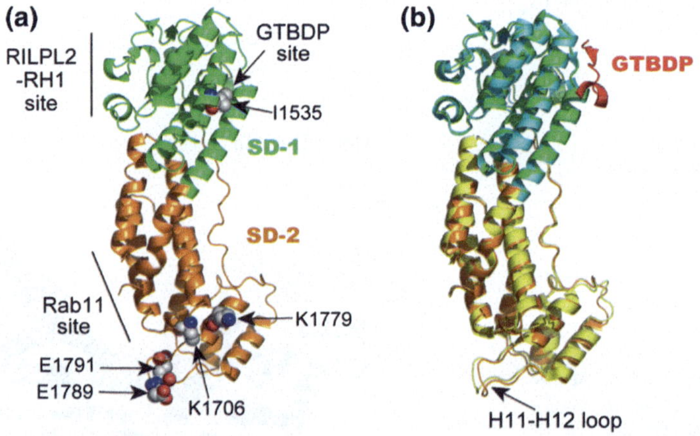

Fig. 9.12 Structural comparison of Myo5a-GTD and Myo5a-GTD/Mlph-GTBDP. **a** Ribbon representation of the human Myo5a-GTD structure (PDB ID: 4LX1) showing binding sites for Mlph-GTBDP, RILPL2-RH1, and Rab11. Te residues mutated in our study are shown as spheres. **b** Overlap of the crystal structures of the human Myo5a-GTD (PDB ID: 4LX1) and Myo5a-GTD/Mlph-GTBDP complex (PDB ID: 4LX2). Te overlap reveals a relatively large conformation change in the H11-H12 loop upon Mlph-GTBDP binding. Subdomain-1 (SD-1) and subdomain-2 (SD-2) are shown in *green* and *orange* for the apo-GTD and *cyan* and *yellow* for the GTD/Mlph-GTBDP complex. Reprinted with permission from Ref. [68]. Copyright 2015, the Nature Publishing Group

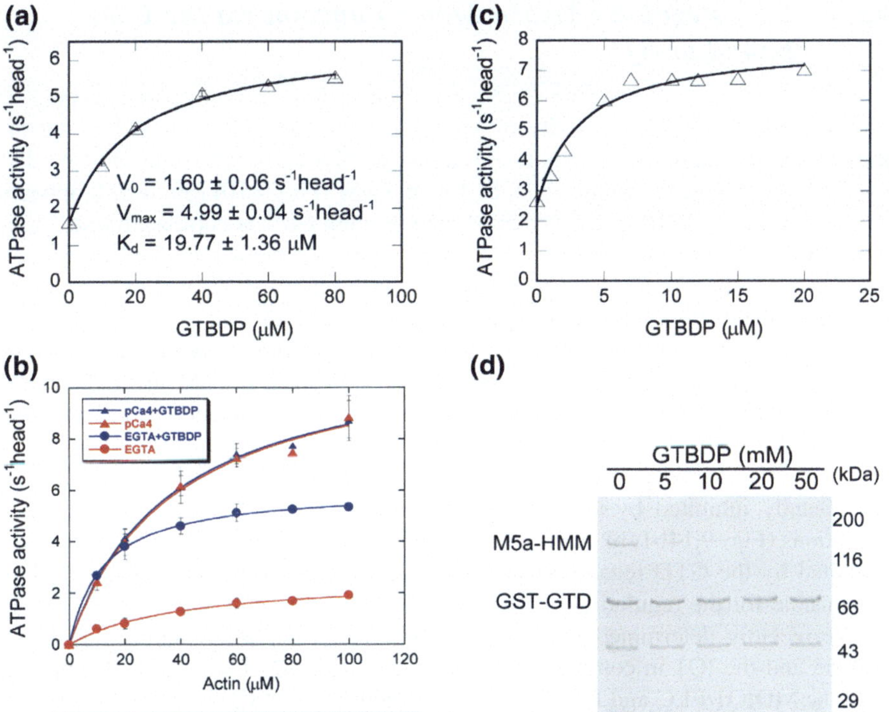

	V_{max} ($s^{-1}head^{-1}$)	K_{actin} (μM)
EGTA	2.75 ± 0.21	45.68 ± 15.89
EGTA + GTBDP	6.05 ± 0.41	11.52 ± 4.35
pCa4	12.60 ± 2.10	43.07 ± 12.10
pCa4 + GTBDP	12.25 ± 1.43	39.91 ± 8.98

Fig. 9.13 Mlph-GTBDP stimulates the ATPase activity of Myo5a by inhibiting the interaction between the head and GTD. **a** Mlph-GTBDP stimulates the ATPase activity of Myo5a. Stimulation of the ATPase activity of Myo5a by Mlph-GTBDP was fit to a hyperbolic equation. **b** Actin dependence of the ATPase activity of Myo5a in the absence or presence of Mlph-GTBDP. ATPase assays under EGTA or pCa4 conditions were conducted in the absence or presence of 60 μM Mlph-GTBDP. **c** Mlph-GTBDP attenuates the GST-GTD-mediated inhibition of Myo5a-HMM ATPase activity. ATPase assays were conducted in the presence of 40 μM actin and 0.5 mM GST-GTD. **d** Mlph-GTBDP inhibits the interaction between Myo5a-HMM and the GTD. GST pull-down of GST-GTD with Myo5a-HMM was performed in the presence of the indicated concentration of Mlph-GTBDP. The resulting samples were analyzed by SDS-PAGE and visualized by CBB staining. Reprinted with permission from Ref. [68]. Copyright 2015, the Nature Publishing Group

head of Myo5a and Mlph-GTBDP. Given the geometry of the GTD and the size of Mlph-GTBDP, it is unlikely that Mlph-GTBDP sterically blocks the interaction between the GTD and the head of Myo5a. Instead, it is likely that Mlph-GTBDP allosterically inhibits the interaction between the GTD and the head of Myo5a, thus stimulating the ATPase activity of Myo5a.

9.5.3 *Ca^{2+} Regulates Myo5a Motor Function via the CaM Bound to IQ1*

The fact that the ATPase activity of Myo5a is stimulated by micromolar concentrations of Ca^{2+} and CaM binds to IQ motifs of Myo5a heavy chain indicate that Ca^{2+} regulates Myo5a function via the bound CaM (Fig. 9.10) [39, 45]. It appears that activation of ATPase activity of Myo5a by high Ca^{2+} is correlated with Ca^{2+}-induced dissociation of CaM from a single specific IQ motif [57]. Several groups identified the specific IQ motif to be the second IQ motif (IQ2) and proposed that Ca^{2+}-dependent regulation of Myo5a is via the CaM in IQ2 [47, 58–60]. This hypothesis seems plausible since the regulation of scallop myosin-2 and smooth muscle myosin-2 is initiated from the regulatory light chain (a CaM-like light chain) bound to IQ2 [33, 42, 61]. However, there is no direct evidence to support this hypothesis.

The ATPase activity of various truncated Myo5a lacking various IQ motifs, was significantly inhibited by the GTD in the absence of Ca^{2+}, but not in the Ca^{2+} conditions (Fig. 9.14b1, b2, and b3). By contrast, the ATPase activity of MD is not inhibited by the GTD regardless of Ca^{2+} (Fig. 9.14b4). Thus, the CaM in IQ1 is responsible for the regulation of Myo5a by Ca^{2+}.

We recently determined the crystal structure of Myo5a containing the motor domain and the IQ1 in complex with Ca^{2+}-CaM (MD-IQ1/Ca^{2+}-CaM). Compared with the MDIQ1/ELC and IQ1-2/apo-CaM structures, MD-IQ1/Ca^{2+}-CaM displays a different conformation in IQ1/CaM with little difference in the motor domain (Fig. 9.15). Ca^{2+} binding induces the flip-flop of apo-CaM and Ca^{2+}-CaM relative to the IQ1 (Fig. 9.16). This structural comparison immediately suggests the dissociation of CaM from IQ1 during Ca^{2+} transition. In other words, the binding of Ca^{2+} to apo-CaM in the IQ1 first induces the dissociation of CaM from the IQ1, and then Ca^{2+}-CaM rebinds to the IQ1. Conversely, the dissociation of Ca^{2+} from Ca^{2+}-CaM in the IQ1 first induces the dissociation of CaM from the IQ1, and then apo-CaM rebinds to the IQ1. However, pull-down assay and single-molecule assay indicate that neither the binding of Ca^{2+} to IQ1/apo-CaM nor the dissociation of Ca^{2+} from IQ1/Ca^{2+}-CaM is accompanied by the dissociation of CaM from the IQ motif (Fig. 9.17).

It is likely that an in situ conformational change of CaM occurs during the Ca^{2+} transition. We proposed that during the Ca^{2+} transition the C-lobe continuously associates with the IQ1, and the N-lobe rotates around the IQ1 to bind to the new position in the IQ1, thus placing the interlobe linker on the opposite side of the IQ1.

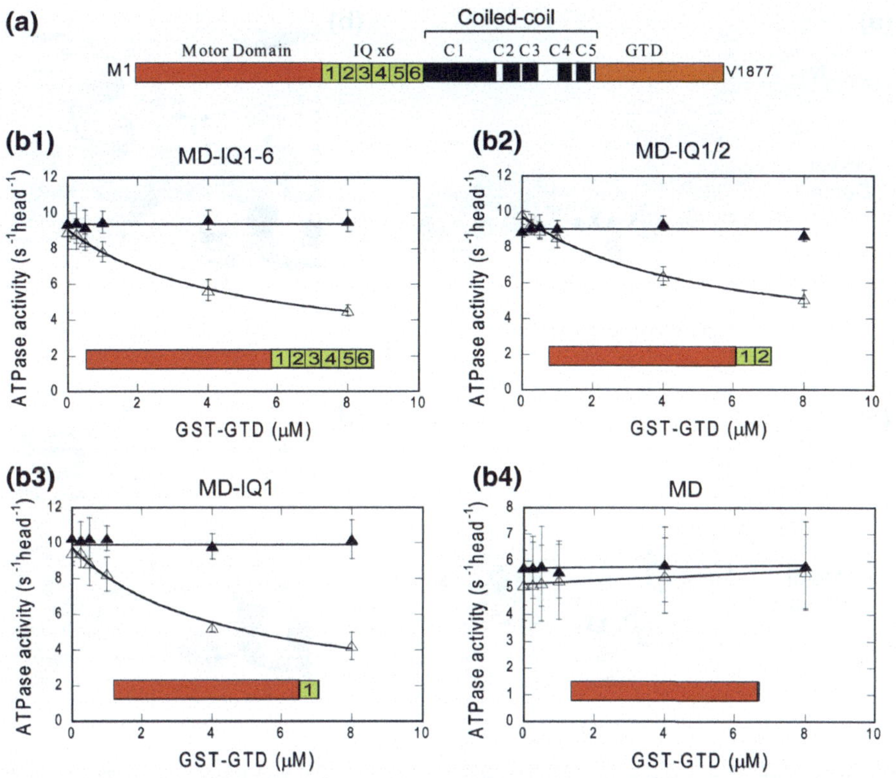

Fig. 9.14 Effects of IQ motif deletions on the inhibition of Myo5a ATPase activity by GTD. The actin-activated ATPase activities of the truncated Myo5a were measured in EGTA (*open triangles*) and pCa4 conditions (*closed triangles*). Reprinted with permission from Ref. [69]. Copyright 2012, the American Society for Biochemistry and Molecular Biology

9.6 Myosin-19

Myosin-19 (Myo19) is a vertebrate specific unconventional myosin and is one of the least studied myosins among the 25 unconventional myosins identified in human. Based upon the deduced amino acid sequence, it was predicted that Myo19 consists of a motor domain, a neck region containing three IQ motifs, and a short tail domain (Fig. 9.18a) [62]. Cell biology experiments revealed that Myo19 is strongly associated with mitochondria and plays a role in the transport of mitochondria along actin filament in human cells [63].

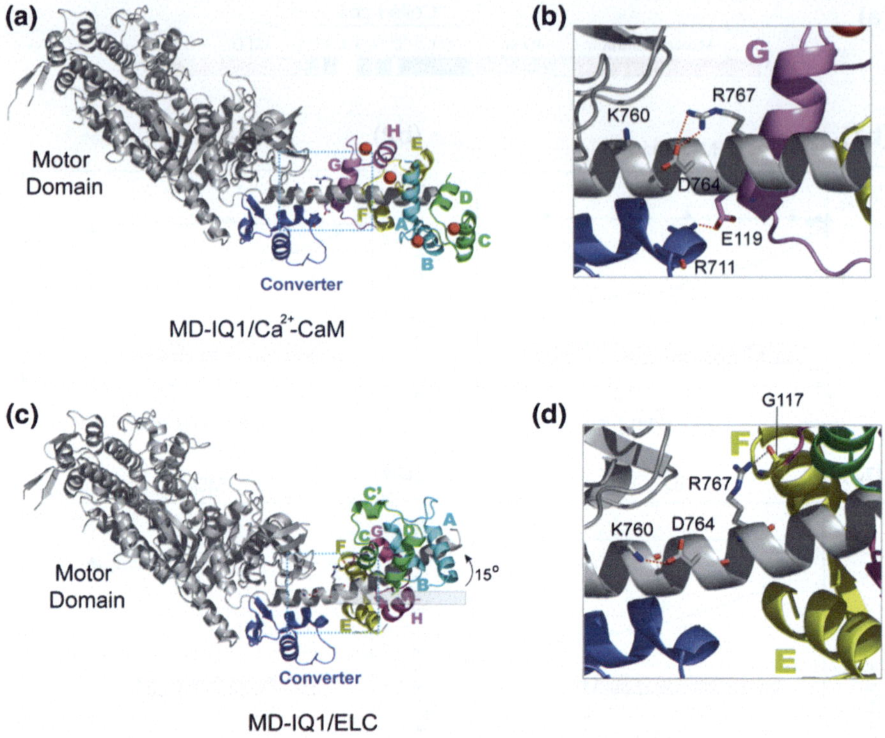

Fig. 9.15 Structural comparison of MD-IQ1/Ca^{2+}-CaM and MD-IQ1/ELC. **a** The overall structure of MD-IQ1/Ca^{2+}-CaM. **b** An enlarged view of the boxed area in A showing detailed interactions between the heavy chain and Ca^{2+}-CaM. **c** The overall structure of MD-IQ1/ELC (PDB ID code: 1OE9). The structure MD-IQ1/ELC has been superimposed on the motor domain of MD-IQ1/Ca^{2+}-CaM shown in **a**. The helices of ELC are colored as CaM in A, except the corresponding helix C of Ca^{2+}-CaM is spliced into helices C and C′. **d** An enlarged view of the *boxed area* in **c** showing detailed interactions between the heavy chain and the ELC [70]

9.6.1 The Light Chains of Myo19 are RLC9 and RLC12b

Since CaM is a common light chain bound to the IQ motif of unconventional myosin identified so far, we coexpressed Myo19 truncated constructs with CaM in sf9 cells and purified Myo19 with Anti-FLAG agarose. We found that the CaM copurified with Myo19-1IQ was barely detectable and the CaM copurified with Myo19-2IQ or Myo19-3IQ was far below stoichiometry. The low-than-expected amount of CaM associated with Myo19 was not due to insufficient CaM in sf9 cells, as similar amount of baculovirus encoding CaM was sufficient to saturate six IQ motifs of Myo5a. Thus, we expected that the light chain other than CaM binds to the IQ motifs of Myo19.

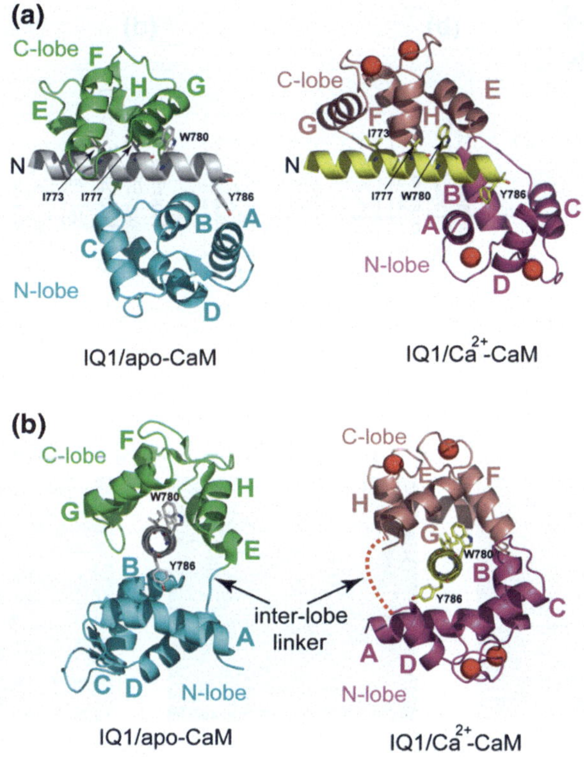

Fig. 9.16 The Ca^{2+} transition induces CaM to flip-flop around the IQ1. **a** Views of the structures of the Myo5a IQ1/apo-CaM (PDB ID code: 2IX7) and IQ1/Ca^{2+}-CaM with the IQ1 superimposed. The N-lobe and C-lobe of apo-CaM are colored *cyan* and *green*, respectively. The N-lobe and C-lobe of Ca^{2+}-CaM are colored *magenta* and *salmon*, respectively. The helices of CaM are designated A–H. The N terminus of the IQ1 is indicated. Four hydrophobic anchor residues (Ile773, Ile777, Trp780, and Tyr786) are shown as *sticks*. Compared with apo-CaM, the N-lobe of Ca^{2+}-CaM moves ~10 Å toward the C terminus of the IQ1. **b** Views of A with a 90° rotation around the vertical axis. The IQ1 is in a nearly vertical orientation with its N and C termini at the *bottom* and the *top*, respectively. The interlobe linker of Ca^{2+}-CaM, shown as a *dashed red line*, is located on the opposite side of the IQ1 in comparison with apo-CaM [70]

Therefore, we prepared the crude extract of myosin light chains from mouse kidney and mixed the purified Myo19-2IQ with the crude extract of myosin light chains from kidney, and re-purified Myo19-2IQ. The re-purified Myo19-2IQ contains two distinct ~20 kDa bands in addition to the ~17 kDa band. CaM gel shift assay shows that the ~17 KDa band displayed a motility shift in response to Ca^{2+} during SDS-PAGE, indicating that the ~17 KDa band was CaM. Mass spectrometry analysis indicates that the ~20 kDa bands are the RLCs of nonmuscle

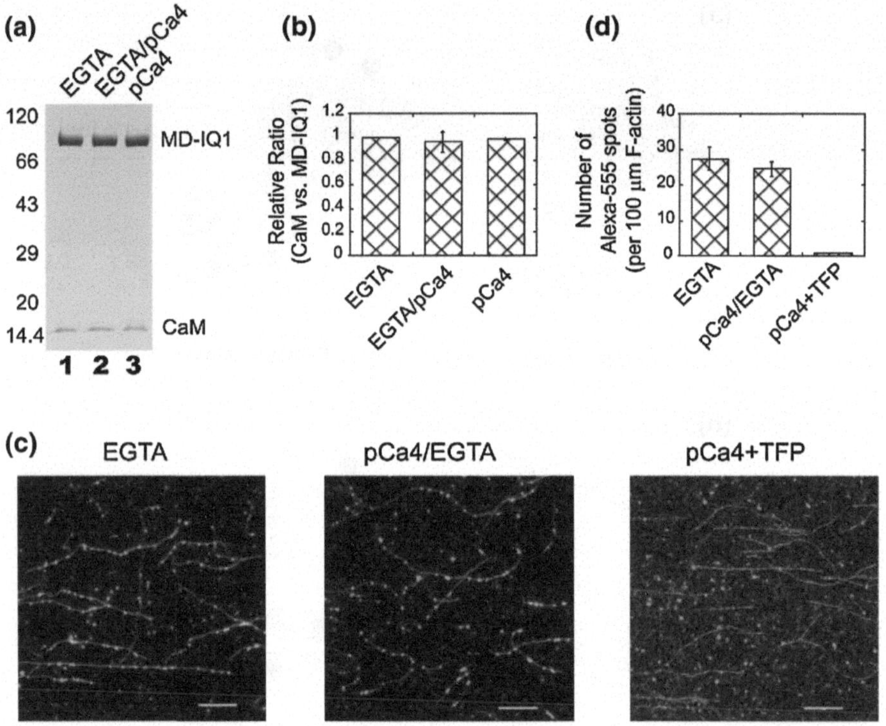

Fig. 9.17 CaM continuously associates with the Myo5a MD-IQ1 during the Ca^{2+} transition. **a** Purified MD-IQ1 bound to anti-Flag beads was washed twice alternately with Ca^{2+}-free (EGTA) solution and pCa4 solution, eluted by Flag peptide, and subjected to SDS/PAGE analysis (lane 2). Control samples were prepared similarly except they were washed only with Ca^{2+}-free solution (lane 1) or with pCa4 solution (lane 3). The protein bands in SDS/PAGE were visualized by Coomassie blue staining. **b** The relative ratio of CaM to the MD-IQ1. **c** Effects of Ca^{2+} treatment on the association of Alexa-555-labeled CaM with the MD-IQ1. The MD-IQ1 in complex with Alexa-555-labeled CaM was introduced into a chamber preabsorbed with Alexa-488-labeled F-actin and subjected to the following three treatments sequentially: (i) EGTA (rinsing with 50 µL EGTA buffer); (ii) pCa4/EGTA (rinsing with 50 µL pCa4 buffer followed immediately by 50 µL EGTA buffer); and (iii) pCa4+TFP (rinsing with 50 µL pCa4 buffer containing 100 µM TFP). The images (512 × 512 pixels each) of the Alexa-555-labeled MD-IQ1 (*green*) and Alexa-488-labeled F-actin (*red*) after each treatment were recorded. Scale bars, 5 µm. **d** The number of Alexa-555 fluorescent spots that colocalized with Alexa-488-labeled F-actin after each treatment. Values are means ± SD of six different images [70]

myosin-2. The upper ~20 kDa band is RLC9 and the lower one is RLC12a or RLC12b. Further studies show that RLC9 binds to the first and the third IQ motifs of Myo19 and RLC12b binds to the second IQ motif (Fig. 9.18b).

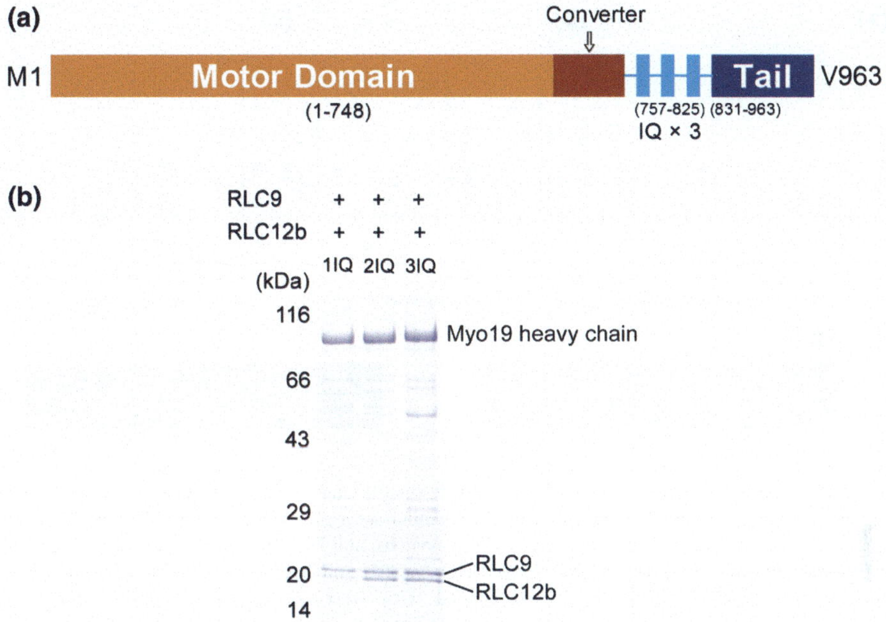

Fig. 9.18 Myo19 structure. **a** Predicted structure of Myo19. **b** SDS-PAGE of purified Myo19-1IQ, -2IQ, and -3IQ coexpressed with RLC9 and RLC12b. Reprinted with permission from Ref. [28]. Copyright 2014, the American Society for Biochemistry and Molecular Biology

9.6.2 *Myo19 is a Plus-End-Directed Molecular Motor*

Sequence alignment revealed that, comparing with Myo5a, Myo19 contains two unique inserts in a region of the motor domain known as the converter. It is known that the minus-end-directed movement of myosin-6 is dictated by a unique insert of ~38 amino acids located between the converter and the lever arm [27, 64]. Therefore, it is of interest to determine the directionality of Myo19 movement.

We found that most of the filaments (>95%) in the presence of Myo19-1IQ exhibited plus-end-directed movements, similar to that of Myo5a, a known plus-end-directed motor (Fig. 9.19). These results indicate that Myo19 is a plus-end-directed motor and the unique insert in the converter of Myo19 plays a role other than reverse direction of movement of the lever arm.

9.6.3 *Myo19 is a High-Duty Ratio Molecular Motor*

ADP release from actomyosin is the rate-limiting step for ATP hydrolysis cycle in high-duty ratio motors, such as Myo5a and myosin-6 [64, 65]. It has been shown

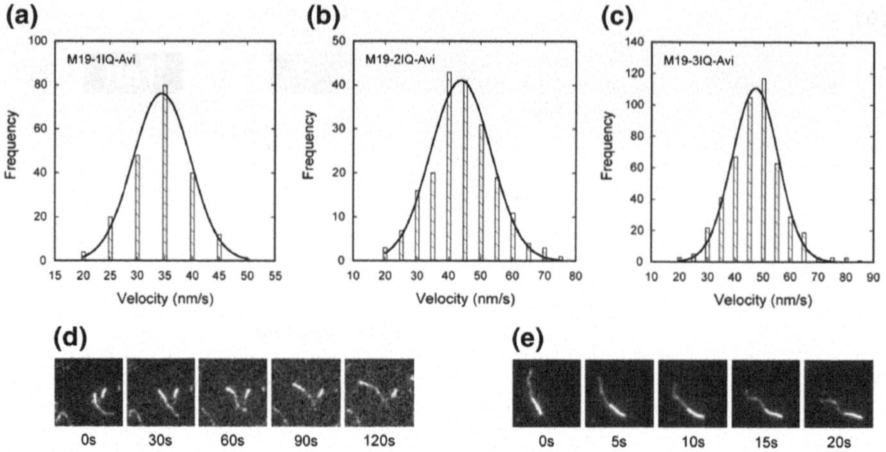

Fig. 9.19 Actin-gliding activity of Myo19 and its directionality. The *solid line* shows a fit to a single Gaussian curve. **a–c** Histogram of multimolecular actin-gliding velocity of Myo19 truncated constructs, i.e., Myo19-1IQ-Avi (**a**), Myo19-2IQ-Avi (**b**), and Myo19-3IQ-Avi (**c**). **d**, **e** Direction of the actin filament translocated by Myo19-1IQ-Avi (**d**) and Myo5a (**e**). Actin filaments were dual-fluorescence-labeled. Plus end was labeled with Alex-488 (*dim*) and minus end with Rhodamine (*bright*). Actin filaments were moved toward the minus end by Myo19-1IQ-Avi and Myo5a. Reprinted with permission from Ref. [28]. Copyright 2014, the American Society for Biochemistry and Molecular Biology

that the steady-state ATPase activity of Myo5a and myosin-6 are strongly inhibited by ADP, whereas that of low-duty ratio motor such as skeletal muscle myosin-2 is only slightly inhibited by ADP [64, 66]. Similar to that of Myo5a but different from that of skeletal muscle myosin-2, the ATPase activity of Myo19 was strongly inhibited by ADP (Fig. 9.20a). Moreover, ADP strongly inhibited the actin-gliding activities of Myo19 and Myo5a, but only slightly inhibited that of skeletal muscle myosin-2 (Fig. 9.20b).

These results suggest that, similar to that of Myo5a, ADP release is likely the rate-limiting step in the ATP hydrolysis cycle of acto-Myo19. Therefore, we measured ADP release rate by following the fluorescence decrease of mant-ADP from acto-Myo19 upon ATP binding in a stopped-flow instrument. The transient was best fit to single-exponential kinetics, and the rate of mant-ADP dissociation was $\sim 8\ s^{-1}$ (Fig. 9.20), which was comparable to the V_{max} of the steady-state ATPase activity, indicating that ADP release is a rate-limiting step in the ATP hydrolysis cycle of acto-Myo19 and Myo19 is a molecular motor with high-duty ratio.

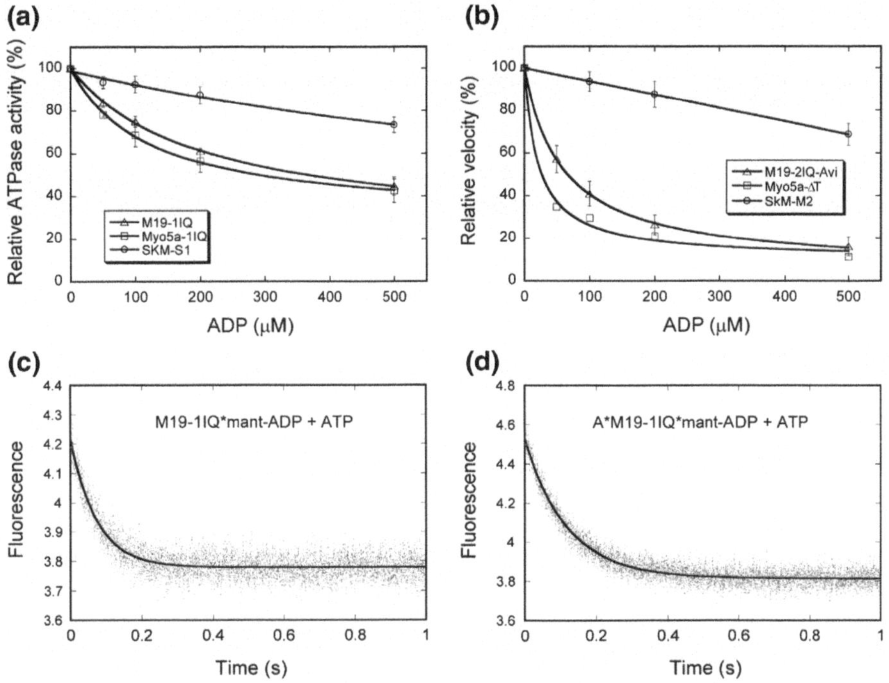

Fig. 9.20 ADP-off is the rate-limiting step for the motor activity of Myo19. **a** Effects of ADP on actin-activated ATPase activity of Myo19-1IQ, Myo5a-1IQ, and skeletal muscle myosin S1 (SkM-S1). The inhibition of ATPase activities by ADP was fitted with a hypophobic equation. **b** Effects of ADP on actin-gliding activity of Myo19-1IQ, Myo5a, and skeletal muscle myosin (SkM-M2). Actin-gliding activity was measured in the presence of 500 μM ATP and 0–500 μM ADP. **c**, **d** Dissociation of mant-ADP from Myo19-1IQ (**c**) and from acto-Myo19-1IQ (**d**). The *smooth line* is the fit to single-exponential kinetics, with Kobs of 13.65 s^{-1} (**c**) and 8.37 s^{-1} (**d**). Reprinted with permission from Ref. [28]. Copyright 2014, the American Society for Biochemistry and Molecular Biology

9.7 Concluding Remarks and Future Perspectives

Myosins are involved in a number of cellular pathways within a eukaryotic cell. In the past two decades, great progress has been made toward understanding the molecular mechanism of myosin motor. However, many prominent questions remain to be answered. First, the regulation mechanism of myosin motor function is far from established. Although in vitro experiments clearly show that Myo5a is regulated by Ca^{2+}, it is still lack in vivo evident supporting the role of Ca^{2+} in regulating myosin motor function. The structure of myosin in the inhibited state remains to be elucidated at atomic resolution, while the tail-inhibition model of myosin regulation is generally accepted. Second, it remains to be determined how multiple myosin motors transport vesicles coordinately. In most cases, a single vesicle contains ensembles of molecular motors. It is not clear how cargo is efficiently delivered. Third, it is not known how vesicles are transferred from microtubule-based motor (kinesin and dynein) to actin-based motor (myosin), and vise verse. The transport of intracellular vesicles is dependent on both microtubule-based motor and actin-based motor, but little is known about the underlined mechanism. Above questions could be addressed using single molecular techniques and cryo-EM. Those experimental information will not only provide a clearer understanding of the myosin mechanism but also facilitate the application of myosin motor in nanotechnology.

References

1. Hammer JA, Sellers JR (2012) Walking to work: roles for class V myosins as cargo transporters. Nat Rev Mol Cell Bio 13(1):13–26. doi:10.1038/nrm3248
2. Arden SD, Tumbarello DA, Butt T, Kendrick-Jones J, Buss F (2016) Loss of cargo binding in the human myosin VI deafness mutant (R1166X) increases actin filament binding. Biochem J. doi:10.1042/bcj20160571
3. Odronitz F, Kollmar M (2007) Drawing the tree of eukaryotic life based on the analysis of 2,269 manually annotated myosins from 328 species. Genome Biol 8(9):R196. doi:10.1186/gb-2007-8-9-r196
4. Li J, Lu Q, Zhang M (2016) Structural basis of cargo recognition by unconventional myosins in cellular trafficking. Traffic. doi:10.1111/tra.12383
5. Goodson HV, Warrick HM, Spudich JA (1999) Specialized conservation of surface loops of myosin: evidence that loops are involved in determining functional characteristics. J Mol Biol 287(1):173–185. doi:10.1006/jmbi.1999.2565
6. Pollard TD, Weihing RR (1974) Actin and myosin and cell movement. CRC Crit Rev Biochem 2(1):1–65
7. Pollard TD, Korn ED (1973) Acanthamoeba myosin. I. Isolation from Acanthamoeba castellanii of an enzyme similar to muscle myosin. J Biol Chem 248(13):4682–4690
8. McConnell RE, Tyska MJ (2010) Leveraging the membrane—cytoskeleton interface with myosin-1. Trends Cell Biol 20(7):418–426. doi:10.1016/j.tcb.2010.04.004
9. Ji HH, Zhang HM, Shen M, Yao LL, Li XD (2015) The motor function of Drosophila melanogaster myosin-5 is activated by calcium and cargo-binding protein dRab11. Biochem J 469:135–144. doi:10.1042/Bj20141330
10. Phichith D, Travaglia M, Yang Z, Liu X, Zong AB, Safer D, Sweeney HL (2009) Cargo binding induces dimerization of myosin VI. Proc Natl Acad Sci UAS 106(41):17320–17324. doi:10.1073/pnas.0909748106
11. Pylypenko O, Song L, Shima A, Yang ZH, Houdusse AM, Sweeney HL (2015) Myosin VI deafness mutation prevents the initiation of processive runs on actin. P Natl Acad Sci USA 112(11):E1201–E1209. doi:10.1073/pnas.1420989112
12. Spudich G, Chibalina MV, Au JSY, Arden SD, Buss F, Kendrick-Jones J (2007) Myosin VI targeting to clathrin-coated structures and dimerization is mediated by binding to disabled-2 and PtdIns(4,5)P-2. Nat Cell Biol 9 (2):176-U167. doi:10.1038/ncb1531
13. Buss F, Kendrick-Jones J (2011) Multifunctional myosin VI has a multitude of cargoes. P Natl Acad Sci USA 108(15):5927–5928. doi:10.1073/pnas.1103086108
14. Mohiddin SA, Ahmed ZM, Griffith AJ, Tripodi D, Friedman TB, Fananapazir L, Morell RJ (2004) Novel association of hypertrophic cardiomyopathy, sensorineural deafness, and a mutation in unconventional myosin VI (MYO6). J Med Genet 41(4):309–314

15. Hegan PS, Lanahan AA, Simons M, Mooseker MS (2015) Myosin VI and cardiomyopathy: left ventricular hypertrophy, fibrosis, and both cardiac and pulmonary vascular endothelial cell defects in the Snell's Waltzer Mouse. Cytoskeleton 72(8):373–387. doi:10.1002/cm.21236
16. Yoshida H, Cheng WJ, Hung J, Montell D, Geisbrecht E, Rosen D, Liu JS, Naora H (2004) Lessons from border cell migration in the Drosophila ovary: A role for myosin VI in dissemination of human ovarian cancer. P Natl Acad Sci USA 101(21):8144–8149. doi:10.1073/pnas.0400400101
17. Naora H, Montell DJ (2005) Ovarian cancer metastasis: integrating insights from disparate model organisms. Nat Rev Cancer 5(5):355–366. doi:10.1038/nrc1611
18. Dunn TA, Chen SL, Faith DA, Hicks JL, Platz EA, Chen YD, Ewing CM, Sauvageot J, Isaacs WB, De Marzo AM, Luo J (2006) A novel role of myosin VI in human prostate cancer. Am J Pathol 169(5):1843–1854. doi:10.2353/ajpath.2006.060316
19. Al Olama AA, Kote-Jarai Z, Berndt SI, Conti DV, Schumacher F, Han Y, Benlloch S, Hazelett DJ, Wang Z, Saunders E, Leongamornlert D (2014) A meta-analysis of 87,040 individuals identifies 23 new susceptibility loci for prostate cancer. 46(10):1103–1109. doi:10.1038/ng.3094
20. Heissler SM, Sellers JR (2016) Various themes of myosin regulation. J Mol Biol. doi:10.1016/j.jmb.2016.01.022
21. Cao Y, White HD, Li XD (2014) Drosophila myosin-XX functions as an actin-binding protein to facilitate the interaction between Zyx102 and actin. Biochemistry 53(2):350–360. doi:10.1021/bi401236c
22. Guzik-Lendrum S, Heissler SM, Billington N, Takagi Y, Yang Y, Knight PJ, Homsher E, Sellers JR (2013) Mammalian myosin-18A, a highly divergent myosin. J Biol Chem 288(13):9532–9548. doi:10.1074/jbc.M112.441238
23. Wang F, Kovacs M, Hu A, Limouze J, Harvey EV, Sellers JR (2003) Kinetic mechanism of non-muscle myosin IIB: functional adaptations for tension generation and maintenance. J Biol Chem 278(30):27439–27448. doi:10.1074/jbc.M302510200
24. Ito K, Ikebe M, Kashiyama T, Mogami T, Kon T, Yamamoto K (2007) Kinetic mechanism of the fastest motor protein. Chara myosin. J Biol Chem 282(27):19534–19545. doi:10.1074/jbc.M611802200
25. Yildiz A, Forkey JN, McKinney SA, Ha T, Goldman YE, Selvin PR (2003) Myosin V walks hand-over-hand: Single fluorophore imaging with 1.5-nm localization. Science 300(5628):2061–2065. doi:10.1126/science.1084398
26. Wells AL, Lin AW, Chen LQ, Safer D, Cain SM, Hasson T, Carragher BI, Milligan RA, Sweeney HL (1999) Myosin VI is an actin-based motor that moves backwards. Nature 401(6752):505–508
27. Menetrey J, Bahloul A, Wells AL, Yengo CM, Morris CA, Sweeney HL, Houdusse A (2005) The structure of the myosin VI motor reveals the mechanism of directionality reversal. Nature 435(7043):779–785. doi:10.1038/nature03592
28. Lu ZK, Ma XN, Zhang HM, Ji HH, Ding H, Zhang J, Luo D, Sun YJ, Li XD (2014) Mouse Myosin-19 is a plus-end-directed, high-duty ratio molecular motor. J Biol Chem 289(26):18535–18548. doi:10.1074/jbc.M114.569087
29. Ikebe M, Hartshorne DJ (1985) Proteolysis of smooth muscle myosin by Staphylococcus aureus protease: preparation of heavy meromyosin and subfragment 1 with intact 20 000-dalton light chains. Biochemistry 24(9):2380–2387
30. Trybus KM, Freyzon Y, Faust LZ, Sweeney HL (1997) Spare the rod, spoil the regulation: necessity for a myosin rod. Proc Natl Acad Sci UAS 94(1):48–52
31. Sata M, Matsuura M, Ikebe M (1996) Characterization of the motor and enzymatic properties of smooth muscle long S1 and short HMM: role of the two-headed structure on the activity and regulation of the myosin motor. Biochemistry 35(34):11113–11118. doi:10.1021/bi960435s

32. Kamm KE, Stull JT (1989) Regulation of smooth muscle contractile elements by second messengers. Annu Rev Physiol 51:299–313
33. Lowey S, Trybus KM (2010) Common structural motifs for the regulation of divergent class II myosins. J Biol Chem 285(22):16403–16407. doi:10.1074/jbc.R109.025551
34. Marston SB, Taylor EW (1980) Comparison of the myosin and actomyosin ATPase mechanisms of the four types of vertebrate muscles. J Mol Biol 139(4):573–600. 0022-2836 (80)90050-9 [pii]
35. Greene LE, Sellers JR, Eisenberg E, Adelstein RS (1983) Binding of gizzard smooth muscle myosin subfragment 1 to actin in the presence and absence of adenosine 5'-triphosphate. Biochemistry 22(3):530–535
36. Taylor EW (1977) Transient phase of adenosine triphosphate hydrolysis by myosin, heavy meromyosin, and subfragment 1. Biochemistry 16(4):732–739
37. Konishi K, Kojima S, Katoh T, Yazawa M, Kato K, Fujiwara K, Onishi H (2001) Two new modes of smooth muscle myosin regulation by the interaction between the two regulatory light chains, and by the S2 domain. J Biochem 129(3):365–372
38. Li XD, Ikebe M (2003) Two functional heads are required for full activation of smooth muscle myosin. J Biol Chem 278(32):29435–29441. doi:10.1074/jbc.M301784200
39. Cheney RE, O'Shea MK, Heuser JE, Coelho MV, Wolenski JS, Espraafico EM, Forecher P, Larson RE, Mooseker MS (1993) Brain myosin-V is a two-headed unconventional myosin with motor activity. Cell 75:13–23
40. Rodriguez OC, Cheney RE (2002) Human myosin-Vc is a novel class V myosin expressed in epithelial cells. J Cell Sci 115(5):991–1004
41. Zhao LP, Koslovsky JS, Reinhard J, Bahler M, Witt AE, Provance DW Jr, Mercer JA (1996) Cloning and characterization of myr 6, an unconventional myosin of the dilute/myosin-V family. Proc Natl Acad Sci UAS 93(20):10826–10831
42. Ikebe M (2008) Regulation of the function of mammalian myosin and its conformational change. Biochem Biophys Res Commun 369(1):157–164. doi:10.1016/j.bbrc.2008.01.057
43. Sellers JR, Thirumurugan K, Sakamoto T, Hammer JA, Knight PJ (2008) Calcium and cargoes as regulators of myosin 5a activity. Biochem Bioph Res Co 369(1):176–181. doi:10.1016/j.bbrc.2007.11.109
44. Wang F, Thirumurugan K, Stafford WF, Hammer JA 3rd, Knight PJ, Sellers JR (2004) Regulated conformation of myosin V. J Biol Chem 279(4):2333–2336. doi:10.1074/jbc.C300488200
45. Li XD, Mabuchi K, Ikebe R, Ikebe M (2004) Ca2+-induced activation of ATPase activity of myosin Va is accompanied with a large conformational change. Biochem Bioph Res Co 315 (3):538–545. doi:10.1016/j.bbrc.2004.01.084
46. Krementsov DN, Krementsova EB, Trybus KM (2004) Myosin V: regulation by calcium, calmodulin, and the tail domain. J Cell Biol 164(6):877–886. doi:10.1083/jcb.200310065
47. Homma K, Saito J, Ikebe R, Ikebe M (2000) Ca(2+)-dependent regulation of the motor activity of myosin V. J Biol Chem 275(44):34766–34771. doi:10.1074/jbc.M003132200
48. Yao LL, Shen M, Lu ZK, Ikebe M, Li XD (2016) Identification of the isoform-specific interactions between the tail and the head of class V myosin. J Biol Chem 291(15):8241–8250. doi:10.1074/jbc.M115.693762
49. Li XD, Jung HS, Mabuchi K, Craig R, Ikebe M (2006) The globular tail domain of myosin Va functions as an inhibitor of the myosin Va motor. J Biol Chem 281(31):21789–21798. doi:10.1074/jbc.M602957200
50. Thirumurugan K, Sakamoto T, Hammer JA 3rd, Sellers JR, Knight PJ (2006) The cargo-binding domain regulates structure and activity of myosin 5. Nature 442(7099):212–215. doi:10.1038/nature04865
51. Li XD, Jung HS, Wang QZ, Ikebe R, Craig R, Ikebe M (2008) The globular tail domain puts on the brake to stop the ATPase cycle of myosin Va. P Natl Acad Sci USA 105(4):1140–1145. doi:10.1073/pnas.0709741105
52. Donovan KW, Bretscher A (2015) Head-to-tail regulation is critical for the in vivo function of myosin V. J Cell Biol 209(3):359–365. doi:10.1083/jcb.201411010

53. Li XD, Ikebe R, Ikebe M (2005) Activation of myosin Va function by melanophilin, a specific docking partner of myosin Va. J Biol Chem 280(18):17815–17822. doi:10.1074/jbc.M413295200
54. Sckolnick M, Krementsova EB, Warshaw DM, Trybus KM (2013) More than just a cargo adapter, melanophilin prolongs and slows processive runs of myosin Va. J Biol Chem 288 (41):29313–29322. doi:10.1074/jbc.M113.476929
55. Wei Z, Liu X, Yu C, Zhang M (2013) Structural basis of cargo recognitions for class V myosins. Proc Natl Acad Sci UAS 110(28):11314–11319. doi:10.1073/pnas.1306768110
56. Pylypenko O, Attanda W, Gauquelin C, Lahmani M, Coulibaly D, Baron B, Hoos S, Titus MA, England P, Houdusse AM (2013) Structural basis of myosin V Rab GTPase-dependent cargo recognition. P Natl Acad Sci USA 110(51):20443–20448. doi:10.1073/pnas.1314329110
57. Nascimento AA, Cheney RE, Tauhata SB, Larson RE, Mooseker MS (1996) Enzymatic characterization and functional domain mapping of brain myosin-V. J Biol Chem 271 (29):17561–17569
58. Trybus KM, Krementsova E, Freyzon Y (1999) Kinetic characterization of a monomeric unconventional myosin V construct. J Biol Chem 274(39):27448–27456
59. Koide H, Kinoshita T, Tanaka Y, Tanaka S, Nagura N, Horste GMZ, Miyagi A, Ando T (2006) Identification of the single specific IQ motif of myosin V from which calmodulin dissociates in the presence of Ca2+. Biochemistry 45(38):11598–11604. doi:10.1021/bi0613877
60. Trybus KM, Gushchin MI, Lui H, Hazelwood L, Krementsova EB, Volkmann N, Hanein D (2007) Effect of calcium on calmodulin bound to the IQ motifs of myosin V. J Biol Chem 282 (32):23316–23325. doi:10.1074/jbc.M701636200
61. Burgess SA, Yu S, Walker ML, Hawkins RJ, Chalovich JM, Knight PJ (2007) Structures of smooth muscle myosin and heavy meromyosin in the folded, shutdown state. J Mol Biol 372 (5):1165–1178. doi:10.1016/j.jmb.2007.07.014
62. Berg JS, Powell BC, Cheney RE (2001) A millennial myosin census. Mol Biol Cell 12 (4):780–794
63. Quintero OA, DiVito MM, Adikes RC, Kortan MB, Case LB, Lier AJ, Panaretos NS, Slater SQ, Rengarajan M, Feliu M, Cheney RE (2009) Human Myo19 is a novel myosin that associates with mitochondria. Curr Biol 19(23):2008–2013. doi:10.1016/j.cub.2009.10.026
64. De La Cruz EM, Ostap EM, Sweeney HL (2001) Kinetic mechanism and regulation of myosin VI. J Biol Chem 276(34):32373–32381. doi:10.1074/jbc.M104136200
65. De La Cruz EM, Wells AL, Rosenfeld SS, Ostap EM, Sweeney HL (1999) The kinetic mechanism of myosin V. Proc Natl Acad Sci UAS 96(24):13726–13731
66. De La Cruz EM, Sweeney HL, Ostap EM (2000) ADP inhibition of myosin V ATPase activity. Biophys J 79(3):1524–1529. doi:10.1016/S0006-3495(00)76403-4
67. Ma RN, Mabuchi K, Li J, Lu Z, Wang CL, Li XD (2013) Cooperation between the two heads of smooth muscle myosin is essential for full activation of the motor function by phosphorylation. Biochemistry 52(36):6240–6248. doi:10.1021/bi400554s
68. Yao L-L, Cao Q-J, Zhang H-M, Zhang J, Cao Y, Li X-d (2015) Melanophilin stimulates Myosin-5a motor function by allosterically inhibiting the interaction between the head and tail of Myosin-5a. Sci Rep-UK
69. Lu ZK, Shen M, Cao Y, Zhang HM, Yao LL, Li XD (2012) Calmodulin Bound to the First IQ Motif Is Responsible for Calcium-dependent Regulation of Myosin 5a. J Bio Chem 287 (20):16530–16540. doi:10.1074/jbc.M112.343079
70. Shen M, Zhang N, Zheng S, Zhang WB, Zhang HM, Lu Z, Su QP, Sun Y, Ye K (2016) Calmodulin in complex with the first IQ motif of myosin-5a functions as an intact calcium sensor. 113(40):E5812–E5820

Chapter 10
Reconstitution of Motor Protein ATPase

Mingjun Xuan, Yi Jia and Junbai Li

Abstract Molecular motor proteins are amazing biological units that are responsible for the transformation of the chemical or biological components into mechanical works. These molecular machines express stronger energy conversion than man-made systems, which inspired scientists to pursue the target of improved performance of current synthetic devices. Thus, it is significant to explore interesting features of biomolecular motors, and design the novel intelligent platforms that mixed motor proteins with synthetic materials. Biomimetic molecular assembly enables the possibility for the in vitro reconstruction of biomolecular motors, further provides a variety of functionalized strategies. In this chapter, we gave a detailed introduction for one of the most familiar biomolecular motors, adenosine triphosphatase (ATPase), and deepen the understanding of their working mechanism and clarified how to conjugate ATPase with the artificially synthetic materials. In addition, some promising examples and significant comments were highlighted to display reconstructed performance of ATPase during this exploring voyage.

Keywords ATPase · Proton gradient · Reconstruction · Magnetic field · Light intervention · Electric field · ATP synthesis

10.1 Introduction

Biomolecular machines such as myosin, kinesin, dynein, and adenosine triphosphate synthase (ATPase) attract great research interests since they were initially discovered and observed [1–3]. ATPase is a class of biomolecular motor protein that is extensively present in mitochondria, chloroplast, prokaryotic algae, and photosynthetic bacterium [4, 5]. ATPase plays a range of fascinating roles in the

M. Xuan · Y. Jia · J. Li (✉)
Beijing National Laboratory for Molecular Sciences (BNLMS),
CAS Key Lab of Colloid, Interface and Chemical Thermodynamics,
Institute of Chemistry, Chinese Academy of Sciences, Beijing, China
e-mail: jbli@iccas.ac.cn

J. Li (ed.), *Supramolecular Chemistry of Biomimetic Systems*,
DOI 10.1007/978-981-10-6059-5_10

energy metabolism, and is involved in physiological activity, catalytically synthesize basic energy unit, adenosine triphosphate (ATP). Various types of energy can be further generated from the hydrolysis of ATP molecules, and power most cellular motility, intracellular transportation, and physiological metabolic activity in the living organisms. ATPase has more than 100 types according to their distributed location and function, and mainly include Ca^{2+}-ATPase, H^{+}/K^{+}-ATPase, H^{+}-ATPase Na^{+}/K^{+}-ATPase, Mg^{2+}-ATPase, Zn^{2+}-ATPase, and so on [6–8]. These types of ATPases display great contribution in the exchange and transportation of various ions in biological entities, importantly balance the ion homeostasis in the intracellular organelles.

Proton-pumping ATPase is an energy-linked enzyme and can be responsible for the regulation of proton concentration, and play important role in phosphorylation. As the transmembrane protein, ATPase is served as a smart channel in biological membrane that the proton can go through [9]. The proton stream can trigger the conformational change of ATPase to spin, achieving energy conversion from the chemical energy into mechanical work [10]. Many researches demonstrate that energy-linked ATPase enzyme can be magically separated from biological entities and be incorporated into the synthetic systems for the improvement of their biomimetic properties [11]. Integration of photoactive components into ATPase-based system could endow the biohybrid materials with the function of light harvesting property and further energy conversion, resulting in the fabrication of light-responsive devices [12]. The ATPase incorporated synthetic materials own several inherent properties like the biological unit. Compare with the traditional materials, these novel biohybrid systems display amazing high-efficiency because of the introduction of natural components.

Miraculous structures in nature show their great interesting points, and exhibit much smarter than man-made materials. Actually, bionics not only promotes people to study the nature, but also enable the natural components to participate in the construction of hybrid materials [13]. In the microworld, these structures have their own advanced functions and elegant properties. Over past twenty years, scientists have devoted many efforts to develop novel biomimetic systems [14]. Several hybrid platforms made of synthetic materials and bioactive units have been presented, and they have significantly achieved the breakthrough on the modification and functionalization of traditional materials. In this chapter, we mainly focus on how to reconstitute F-type ATPase-based structures and explore their biomimetic performance in vitro. Molecular assembly of bionics has been proved quite useful in the design and preparation of functional bioengineered materials. This approach has great potential for the assembly of structured materials, and develops several assembled mechanism based on electrostatic interaction, hydrogen bond, covalent bond, and metal coordination [15]. The natural biomacromolecules extracted from biological entities significantly inspire the design of biohybrid materials and optimally integrate the biological properties with the synthetic materials. The appearance of ATPase enzyme-based hybrid materials meets the current energy topics to design the novel energy-linked materials and devices, significantly open a new perspective to perform the energy capture and further conversion.

10.2 ATPase Family: Types, Structure, and Components

ATPase is a class of transmembrane proteins made of several structured subunits, and widely exists in the living organisms [16]. According to the distribution, these types of ATPase are given peculiar properties and specific functions, and are responsible for the various ions transport and exchange, biological signal transmission, and energy transduction [17]. In cellular physiological process, ATPase enables many necessary metabolites for cellular metabolism to import, simultaneously take charge of the export of toxins and wastes. Such a kind of protein-based biomolecular machines are of vital importance in the biological entities. ATPase is a big family and has its own category, and can be systematically divided into three types: P-type ATPase, V-type ATPase, and F-type ATPase (Table 10.1).

P-type ATPases are a large group of evolutionarily related pumps to transport ions that widespread across many biological entities from bacteria to human, mainly reside in bacteria, fungi, archaea, and eukaryotic plasma membrane and organelles [18]. At present, P-type ATPases including more than 20 different types of structures have been discovered [19]. All P-type ATPases have a central domain with α-helical structure and perform the transportation of many metal ions (Mg^{2+}, Ca^{2+}, Na^+, K^+, Zn^{2+}, Cd^{2+}, Cu^+, Hg^{2+}) [20–25]. During the process of cations across the entire membrane, P-type ATPase uses its own two conformational regiments, called E_1 and E_2, to maintain the electrochemical gradient [26]. The generated conformational change effectively regulates the ions stream, significantly prevents the formation of back-flow state when the membrane opens the channel for ions to go through. Actually, the ions transportation process is accompanied with the energy consumption due to the change of conformational regiments that are triggered by the ATP hydrolysis. The P-type ATPase acts a smart housekeeper in the regulation of ions transportation during transmembrane behavior.

As one of the membrane-associated enzymes, vacuolar-type H^+-ATPase (V-ATPase) plays an important role in a series of vesicles-mediated physiological activities [27, 28]. V-type ATP is a multimeric complex-like enzyme with remarkably diverse features and functions that resides in eukaryotic organisms and

Table 10.1 ATPase system

Main types	Category	Distribution	Driving force
P-type ATPase (E_1E_2-ATPase)	Ca^{2+}-ATPase H^+/K^+-ATPase Mg^{2+}-ATPase Cu^+-ATPase Zn^{2+}-ATPase Na^+/K^+-ATPase	Bacteria, fungi, archaea, eukaryotic plasma membrane and organelles	ATP hydrolysis
V-type ATPase (V_1V_0-ATPase)	H^+-ATPase	Eukaryotic vacuoles	ATP hydrolysis
F-type ATPase (F_1F_0-ATPase)	ATP synthase (H^+-ATPase)	Mitochondria, chloroplasts, bacterial plasma membranes	Proton gradient

their organelles [29]. In the eukaryotic cells, V-type ATPases are energy-consumed proton pumps and are responsible for the acidification of cellular interior and the proton across the plasma membrane. During the proton delivery, ATP hydrolysis miraculously triggers the rotation of V-type ATPase to form a proton gradient, effectively regulates the pH value in the intracellular compartment and vesicular cell organelles [30]. V-ATPases are mainly composed of two structured parts, V_1 domain and V_2 domain [31, 32]. The V_1 domain contains eight different subunits that are responsible to propel the rotation of V-type ATPase via ATP hydrolysis. The V_0 domain has six different subunits that can transport protons to pass the membrane. Moreover, plasma membrane-associated V-type ATPase can be employed to participate in the type identification for various cells [33]. The subunits of the V-ATPase reside in certain cells that have a tissue-specific manner to express, resulting in the given information for the detection of cell types. Several exampled cells such as macrophages, tumor cells, osteoclasts, and insect goblet cells can be exactly distinguished because of that they have specific functions and process in the renal acidification, pH homeostasis, and tumor invasion.

F-type ATPases are usually known as the F_0F_1 ATPase, and also called ATP synthase (ATPase for short in the following text) or proton-translocating ATPase [34, 35]. ATPases are extensively found in chloroplast thylakoid membrane, mitochondrial inner membrane, and bacterial plasma membrane [36]. F-type ATPase is an energy-linked enzyme coupled with oxidative phosphorylation that perform fantastic task to produce the energy molecule, ATP, for cellular consumption [37]. Compared with P-type ATPase and V-type ATPase, F-type ATPase transports proton through the generated proton gradient across the biological membrane. More importantly, there is no energy consumption during transport process. Upon the electrochemical gradient, the rotation of ATPase is immediately triggered as the form of mechanical work, meanwhile starts the proton flux delivery. In exceptional circumstances, ATPase also can reversely rotate when the ATP concentration outside of membrane is pretty high [38]. In some bacteria, sodium ions can be instead of proton as the "fuel" to drive the rotation of ATPase [39].

Similar to V-type ATPase, F_0F_1 ATPase is also a transmembrane protein composed of two separate domains, extrinsic portion F_1 and transmembrane portion F_0 (Fig. 10.1a) [40]. F_1 portion is a stator constructed by five subunits, α, β, γ, δ, and ε, which have many catalytic sites that are responsible for the ATP synthesis or ATP hydrolysis. F_0 portion is a hydrophobic membrane-bond part across the biological membrane that contains three subunits named a, b, and c, and regulate the proton flow to start the rotation. The two portions are connected together by the bridge of central stalk (γ and ε). During the ATP synthesis, this rotary protein motor effectively transfers adenosine diphosphate (ADP) and phosphate into ATP molecules, exhibits marvelous conversion efficiency compare to the current man-made machines. The proton flows generate an electrochemical gradient through F_0 portion that drives rotation of the F_0 portion and the connected stalk (Fig. 10.1b), inducing the conformational changes of F_1 portion that drives the catalytic synthesis of ATP [41]. Remarkably, this complicated enzyme can also perform ATP

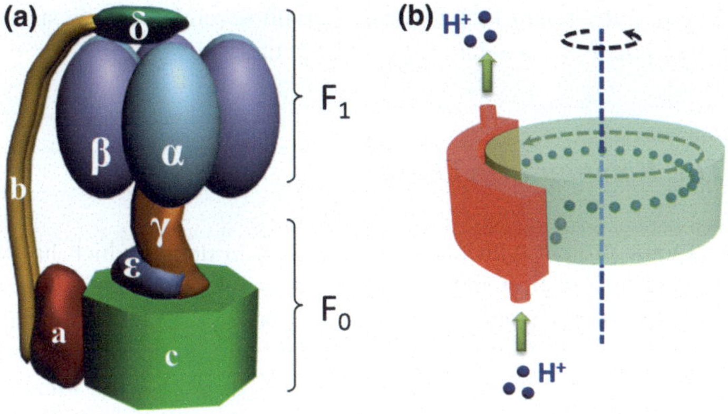

Fig. 10.1 The schematic structure of **a** F_0F_1-ATPase and **b** the rotary proton channel (F_0 portion)

hydrolysis to drive the reverse rotation that pumps proton against the electrochemical gradient.

There are a lot of ATPases enriched in the thylakoid membrane of chloroplast and inner membrane of mitochondria. As energy-linked protein, ATPase not only participate the oxidative phosphorylation, but also carry out the conversion between chemical energy and mechanical works. In the plant cells, ATPase mainly couple with photosynthesis proteins to achieve various tasks or assistant process, such as ATPase synthesis, water photolysis, proton delivery, and electron transfer [42]. Mitochondria are the energy factory that possess high density of ATPase, and conduct the continuous ATP synthesis for the energy supply in cells [43]. Some prokaryotic algal and photosynthetic bacteria also can use their own ATPase to acquire the energy and further conversion [44].

10.3 In Vitro Assembly of ATPase Enzyme Mediated Devices

The bionics has been proved helpful since it was used to manufacture nature-designed structures. The biomimetic concept is not only learning from nature, but also inspire scientists employ the natural components to modify and improve the current assembled systems [45–47]. Recently, many reports talk about how to fabricate the biohybrid materials and devices composed of the naturally biological units and artificially synthetic systems. These introduced biological units have eye-catching work efficiency and favorable biocompatibility, which can significantly enhance the performance of traditional biomimetic materials and devices. Inspired from this, ATPase enzymes possess great potential to participate in the fabrication of rotation-based nanomachines. The scientists significantly introduce

ATPase enzyme as the rotary motor to couple with several synthetic structures and achieve the biological components triggered rotation.

10.3.1 The Rotation of ATPase

ATPase is a motor protein with great investigation in detail, which has well definition in structure information and rotary mechanism. In 1997, Noji et al. [48] split F_1F_0-ATPase into two parts, and retained the F_1 portion composed of γ subunit and $\alpha\beta$-hexamer. Subsequently, the bare γ subunit is coupled with streptavidin-conjugated actin filament, and the opposite side of $\alpha\beta$-hexamer is immobilized on the substrate by His-Tag (Fig. 10.2a). In the presence of ATP molecules, F_1-ATPase starts to hydrolyze ATP due to no proton channel (F_0 portion) and no proton concentration gradient (Fig. 10.2b). Thus, the fluorescent actin filament can spin because of that generated energy of ATP hydrolysis drives the rotation of γ subunit.

View from the membrane side, the filament is rotated in an anticlockwise manner more than 100 revolutions (Fig. 10.2c). This is the first time to use F_1-ATPase motor for powering the rotation of employed structure, and also achieve the direct observation of single ATPase rotation. This approach opens a bright research aspect for biomolecular motor, and inspires the scientists devote many efforts to deeply investigate the rotary behavior of ATPase-based synthetic devices.

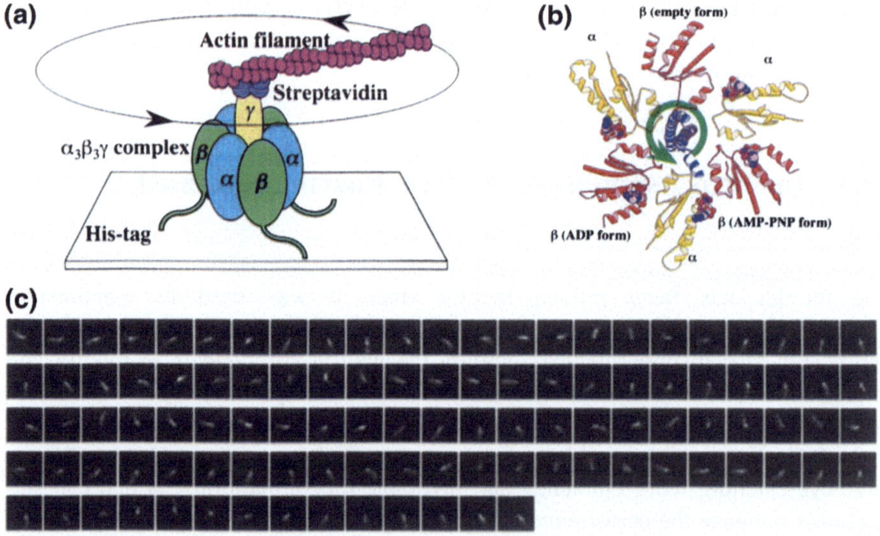

Fig. 10.2 **a** The schematic diagram of F_1-ATPase powered rotation of actin filament. **b** The crystal structure of F_1-ATPase. **c** The time-lapse images of the rotating actin filament. Reproduced with permission from Ref. [48]. Copyright 1997, Nature Publishing Group

To use the rotation torque generated by ATPase, many synthetic nanoengineered structures have been introduced in this rotary platform for the further investigation. Several biohybrid nanodevices based on F_1F_0-ATPase which are sequentially developed and fabricated, and significantly promote the progress of molecular machines. Emergent fabrication techniques enable the desired structure feasible to acquire, for example, Soong et al. [49] prepare F_1-ATPase-based rotator (Fig. 10.3a). This rotary nanodevice is immobilized on the Ni post. The Ni rod as the nanopropeller is assembled on the γ subunit side of ATPase via specific conjugation of biotin-streptavidin. Subsequently, ATP generated assay is immediately added to power the rotation of this nanodevice. To simplify the modified process of ATPase motor, the attached nanopropeller is directly conjugated on the bottom of F_0 portion without dismantling any accessories of ATPase motor (Fig. 10.3b) [50]. The $\alpha\beta$-hexamer side is still connected with the substrate by His-Tag. Accompany with ATP molecules, this F_1F_0 ATPase exhibits graceful rotary performance in powering inorganic propeller.

In the previous study, the rod-like structures play important role on the act of attached propeller. Meanwhile, some other types of fantastically employed structures have been used to couple F_1-ATPase, such as gold nanoparticle, dimeric particles and DNA double-strand linked gold nanorod [51–56]. These featured propellers equipped ATPases can be performed for further mechanical exploration and special tasks. As Fig. 10.3c shown, this rotary nanodevice performs the maximum work per 120° step, which is approximately equal to the thermodynamical maximum work that can be extracted from a single ATP hydrolysis under a broad

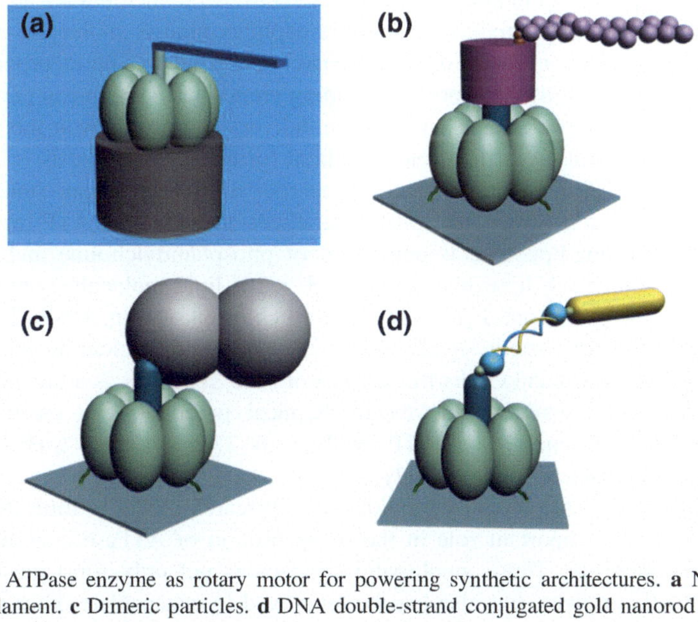

Fig. 10.3 ATPase enzyme as rotary motor for powering synthetic architectures. **a** Nickel rod. **b** Actin filament. **c** Dimeric particles. **d** DNA double-strand conjugated gold nanorod

range of conditions [55]. During the reversible rotation, the external torque and the chemical potential of ATP hydrolysis are both precisely controlled by the discrete 120° steps. Furthermore, the ATPase motor can also couple with double-strand DNA to develop the rotary biosensor. DNA double stranded is connected with the F_1-ATPase to form a biosensor for the detection of the specific bridged gold nanorods (Fig. 10.3d), which enable the sensitive discrimination depend on the rotation, resulting in the sensitive limitation of one zeptomole [56]. Thus, these rotary nanodevices based on the manipulation of propeller components and conjugated technology can help to develop novel features of rotary nanodevices and enhance their complexity and functions, as well as inspire many biological applications.

10.3.2 ATP Synthesis in ATPase Incorporated Liposomes

Since the investigation of protein motors, F_1F_0-ATPase is the most familiar motor because of specific feature and structure. Due to proton gradient-responsive rotation, the ATPase motors have been employed to perform miraculous rotation through conjugating with the biomacromolecules. Liposome is a classical vesicle that can be composed of several polymers and lipids [57], and have well biocompatibility for the hybridization with the ATPase. Usually, an enclosed space is formed by liposomes to generate the proton gradient for the activation of ATPase rotation. Like the biological membrane in cells, the naturally isolated space is formed by lipid membranes, and effectively provides the proper condition for the generation of proton gradient. To construct biomimetic membrane, an artificial vesicles made of lipid mixture, porphyrin-naphthoquinone molecular triad (1, C-P-Q), and lipophilic quinone (Qs) is prepared and used for the incorporation of ATPase (Fig. 10.4a) [58]. This proton-pumping photocycle system can perform the photon-induced electron transfer and proton delivery in liposomes in the presence of visible light, resulting in the proton gradient for the ATP synthesis.

All F-types ATPase have the specific properties in transmembrane function and proton delivery, generating the proton gradient to produce ATP molecules. A designed 120 nm liposome is composed of phosphatidylcholine and 5 mol% phosphatidic acid. Each liposome contain 1.3×10^5 lipid molecules, valinomycin ($\sim$800 molecules), and one F_1F_0-ATPase from chloroplast (Fig. 10.4b) [59]. This chemiosmotic model system significantly demonstrates the scientific relationship between H^+/ATP ratio and Gibbs free energy of ATP synthesis as a function of the changed pH and transmembrane electrochemical potential, and calculates the standard Gibbs free energies of ATP synthesis is 37 ± 2 kJ/mol (pH 8.45) and 36 ± 3 kJ/mol (pH 8.05), respectively.

Liposome-based sealed-cavity provides ideal research model to form the proton gradient, and play important role in the reconstitution of ATPase enzyme due to well biocompatibility. These amphipathic molecules not only form the hermetic structure for themselves, but also can be employed to encapsulate other materials

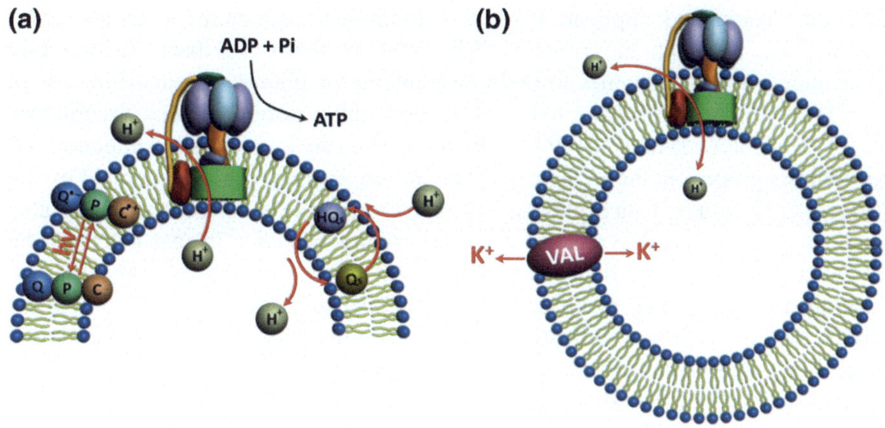

Fig. 10.4 ATPase enzymes incorporated vesicles for ATPase synthesis. **a** ATP production in C-P-Q molecules-based artificial photosynthetic membrane. **b** A chemiosmotic system for ATP synthesis

for the regulation of proton concentration. With the development of biotechnology, several bioactive molecules could be introduced to enhance the diversity of ATPase mediated liposomes, such as functional proteins, polysaccharide, polymers, and DNA molecule. Many controllable mechanisms could be used to adjust permeability of biomembrane, and control the transmembrane behavior of the protons.

10.3.3 ATPase-Linked Polymeric Structures

Due to rapid development of modern biotechnology and nanoscience, it is possible to design and fabricate biohybrid functional materials. Layer-by-layer (LbL) assembled capsules are promising candidates and draw much attention in constructing hybrid systems [60–64]. LbL assembled capsules with adjusted size, shape, wall thickness, and permeability were prepared by alternative assembly of multilayer materials on particle templates and subsequently dissolving core templates [65, 66]. The hollow structure endows the assembled capsules to be an ideal transporter for diverse cargos, such as DNA, proteins, and drug molecules [67]. Because of the above remarkable advantages, it is expected that the assembled capsules can be considered as functional container to reconstitute rotary motor-based biomimetic systems.

In 2007, Li et al. firstly report the LbL assembled polyelectrolyte microcapsules coated by ATPase incorporated lipid bilayers that is used for ATP synthesis [68]. Compared to the previously reported strategy, this lipid membrane modified microcapsules provide an approach to simulate the real cell membrane, and also make the design and application of new biomimetic structural materials to be possible. As a membrane protein, F_0F_1-ATP synthase has been successfully reconstituted in the liposomes to build biomimetic membrane systems (Fig. 10.5a) [68]. The F_0F_1-ATP synthase hybrid LbL assembled poly(allylamine hydrochloride) (PAH)/poly(acrylic acid) (PAA) microcapsules are employed to simulate ATP biosynthesis process in the living cells. The proton gradient can be generated by the change of pH values, which plays a key role to generate ATP molecules and also decide the rate of ATP synthesis. ATP generating behavior in real cells can be simulated and reproduced by combining ATPase with the proton gradient. Consequently, F_0F_1-ATP synthase hybrid microcapsules can retain the biological activity of F_0F_1-ATPase, indicating that many other bioactive membrane-bound proteins can also be reconstituted using this method.

Glucose is always used in F_0F_1-ATPase microcapsules to obtain continuous proton gradient for ATP generation by the enzymatic catalysis of glucose oxidase (GOD) [69, 70]. The oxidation of glucose can produce gluconic acid and generate a proton gradient. Thus, the proton gradient can be maintained for a longer time by adding glucose and sustaining drives the biological synthesis of ATP molecules from the microcapsules. The microcapsules with ATP generation that are assembled

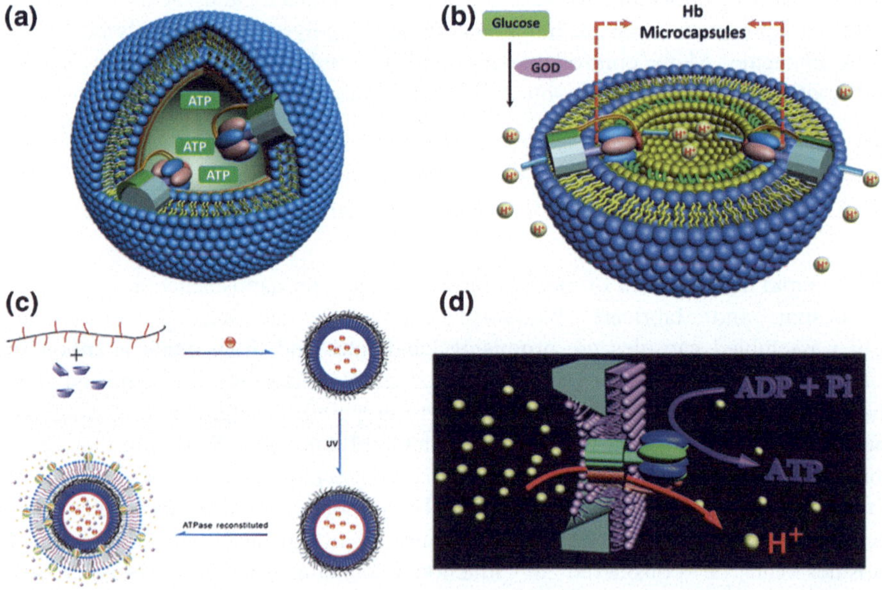

Fig. 10.5 The reconstitution of ATPase in the LbL assembled polymeric structures. **a** The F_0F_1 ATPase in lipid-coated PAH/PAA microcapsules for ATP synthesis. **b** The reconstitution of F_0F_1-ATPase in lipid-coated Hb microcapsules. **c** GOD loaded polymeric microcapsule as the bioreactor for ATP biosynthesis. Reprinted with the permission from Ref. [72]. Copyright 2013, American Chemical Society; **d** ATPase incorporated polymeric nanoporous films for pH-responsive ATP synthesis. Reprinted with the permission from Ref. [73]. Copyright 2011, Royal Society of Chemistry

by glutaraldehyde (GA) cross-linked hemoglobin (Hb) and F_0F_1-ATPase (Fig. 10.5b) [71]. Proton gradient between the interior and exterior of the F_0F_1-ATPase-microcapsule is generated by adding glucose and GOD solution, and it promotes the rotation of ATPase and induces the synthesis of ATP. To acquire the continuous proton gradient across the microcapsules, GOD as building block is directly used to fabricate F_0F_1-ATPase incorporated microcapsules. Once the glucose is injected, a proton gradient between the inside and outside of microcapsules is continuously generated by the catalytic hydrolysis of glucoses. This approach reveals that the amount of ATP production was continuously increased along with the continuous proton gradient. In other case, polymeric microcapsules can contribute their hollow structure to load GOD molecules that instead of being as the assembled materials (Fig. 10.5c) [72]. In addition, the two-dimensional polymeric film can also be incorporated with F_1F_0-ATPase to perform ATP synthesis (Fig. 10.5d) [73]. This smart film can effectively generate the proton gradient according to the manipulated pH value of two separated sides, resulting in the activation of ATP synthesis.

Hence, these rotating biological molecular motors, F_0F_1-ATPase, which can be assembled in the lipid membrane-coated microcapsules or two-dimensional film. Due to a proton gradient generation, the process of ATP production can be successfully performed. It is possible to use F_0F_1-ATPase-based systems for the storage of ATP and provide biological energy on demand.

10.4 External Field Modulated ATP Synthesis

ATP molecule is the basic energy unit in the biological entities, which enables physiological activity work on the regular running mode. The ATPases are a type of smart system in the biological entities and know how and when to start ATP synthesis or ATP hydrolysis on their own benefit. However, a critical issue gets great concern is how to control the in vitro performance of ATPase on demand. Rapid development of nanotechnology provides it possible to achieve the fabrication and manipulation in micro/nanoscale, and highly meet the requirement of ATPase modification and functionalization. Unlike in vivo, in vitro assembly of ATPase is not smart enough, resulting in that many driven mechanisms are used to regulate work performance of ATP synthesis. Currently, several external fields such as the magnetic field, light, and electric field, that promise an untouching approach to trigger on/off the ATP synthesis. This is really favorable to preserve the bioactivity of ATPase proteins. More importantly, it is a simple and valid method to achieve precisely remote manipulation.

10.4.1 Magnetic Field-Driven ATP Synthesis

Magnetic field is a most common phenomenon generated by the permanent magnet or electromagnetic effect. This inspires the fabrication and development of several magnetic materials composed of elements Fe, Co, and Ni, further develops the magnetism-responsive telecontrolled manner. Thus, scientists use magnetic field to propel the rotation of ATPase and produce ATP molecules [74–78]. During whole process, there is no proton gradient across the ATPase motor, and directly achieve the transfer from mechanical rotation into chemical energy.

Similar to the up-mentioned ATPases-powered rotator and propeller, ATPases are upside-down immobilized again on the substrate by His-Tag conjugation. Subsequently, many structured magnetic materials such as particle, dimeric particle, and rod, were conjugated on the side of γ subunit via biotin-streptavidin-biotin approach. The magnetic particle modified F_1-ATPase can be rotated by using six steerable electromagnets, achieving a modulated ATP synthesis (Fig. 10.6a) [79]. In physiological conditions, ATP synthesis is an energy-consumed reaction that requires 80–100 pN nm energy. Using external magnetic field trigger ATPase for ATP synthesis that needs only about 30 pN nm of free energy to produce one molecule ATP because of no external energy consumption in local medium. Compared with the magnetic field manipulated by magnet, magnetic tweezers have better performance in flexible operation and high precision. The single-molecule manipulation and micromanufacture are integrated together to study the mechanochemical transformation of this system. The magnetic bead modified F_1-ATPase is immobilized in femtoliter-sized hermetic chamber to synthesize ATP by using the rotating magnetic field (Fig. 10.6b) [80]. A clockwise rotating magnetic tweezers can trigger the ATP synthesis as usual. Interestingly, the "switch off" magnetic tweezers can immediately end the rotation, and induce ATPase rotate in anticlockwise to perform ATP hydrolysis due to high concentration of ATP molecules. It demonstrates that the "switching on/off" ATP synthesis/hydrolysis of ATPase motors can be flexibly regulated by the manipulation of magnetic tweezers.

Magnetic field manipulated rotation also provides a strong tool to study biological behaviors or processes of ATPase during ATP hydrolysis. Inorganic phosphate (Pi) release, ADP release, and ATP binding are the important status and independent processes, respectively. The single-molecule imaging technology can directly observe the conformational change of ATPase to elucidate the above-mentioned status. Scientists use the fluorescence-marked ATP molecule to research what happen in ATP hydrolysis process (Fig. 10.6c) [81]. To develop and improve this rotary nanosystem, many efforts have been devoted to fabricate series of magnetism-responsive propellers for the rotatory motors. A designated propeller is assembled by plastic bead and Au capped Ni rod which is connected on the side of γ subunit (Fig. 10.6d) [82]. Compared with other conjugated structures, this specific propeller can significantly enhance the resolution of rotary ATPase motor, and provide a strategy for the research of torque behavior during ATP synthesis.

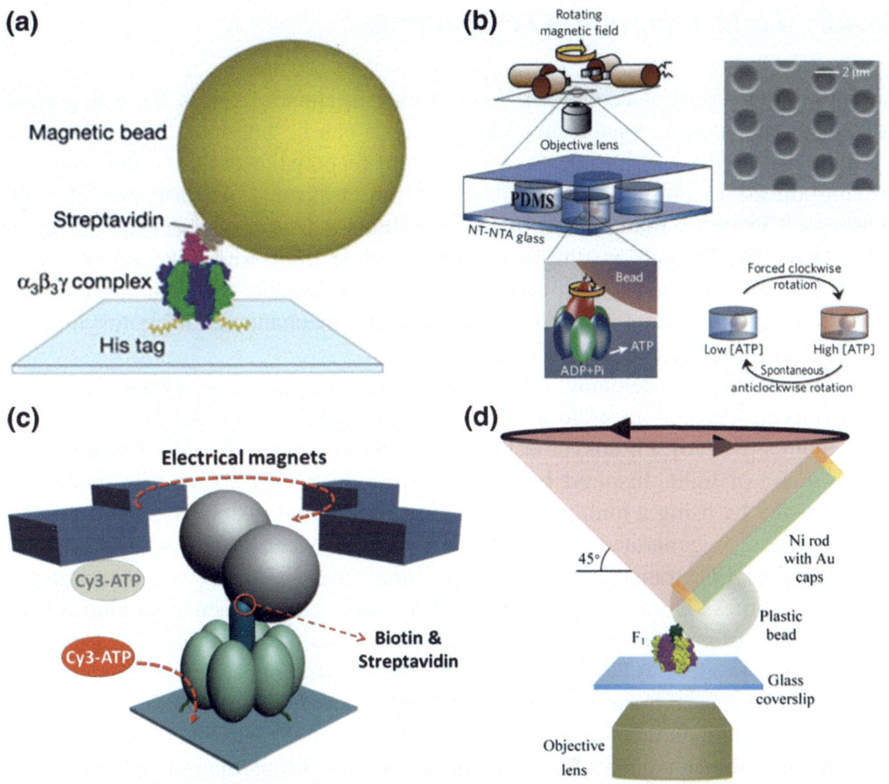

Fig. 10.6 The magnetic field triggered "switch on/off" ATP synthesis. **a** Magnetically driven rotation of F_1-ATPase for ATP synthesis. Reprinted with the permission from Ref. [79] Copyright 2004, Nature Publishing Group; **b** Rotary magnetic tweezers manipulated ATP production. Reprinted with the permission from Ref. [80]. Copyright 2005, Nature Publishing Group; **c** Rotary magnetic field mediated approach for the behavior research of ATP hydrolysis. **d** The torque composed of plastic bead-connected Ni rod for the magnetically manipulated rotation. Reprinted with the permission from Ref. [82]. Copyright 2010, American Chemical Society

The magnetic field propelled rotation of ATPase enzyme significantly achieves the regulation of ATP synthesis. More importantly, rotary process of ATPase enzyme can be successfully operated in clockwise or anticlockwise without proton gradient, resulting in a controlled transform between ATP synthesis and ATP hydrolysis. With the development of biotechnology and nanotechnology, high precise manipulation in single molecule gives the design sparks of diverse nanostructures for conjugation of ATPase rotary motor, and enables the energy conversion from mechanical rotation into chemical power.

10.4.2 Light Triggered ATP Synthesis

Sunlight plays an irreplaceable role in the living organism [83]. In the living plant cells or photosynthetic bacterium, ATP synthesis can be intelligently regulated by sunlight. Inspired by chloroplasts, several light-responsive proteins and enzymes are introduced to couple with ATPase for the controlled performance of ATP synthesis. Currently, many novel materials exhibit light-responsive behavior upon given wavelength, such as ultraviolet light, visible light, and near-infrared light [84–86]. Reconstituting smart platforms in vitro to simulate the natural biological process is very important to deeply understand the mechanisms of photosynthesis, and develop biomimetic materials.

To achieve the "on-demand" ATP synthesis and rotation, many efforts have been devoted to design the "open/close" feature of ATP synthesis for ATPase assembled architectures [87–90]. Photosynthesis in green plants can regulate ATP synthesis in the presence of light. Inspired by this, a light-induced ATP synthesis process is reconstructed by using a multiprotein inlaid polymersome system (Fig. 10.7a) [91]. ATP molecule is generated by coupled reactions between light-responsive bacteriorhodopsin (BR), transmembrane proton pump, and F_0F_1-ATPase, reconstituted in polymersomes. These inserted BR protein provide a light sensitive channel for the regulation of proton concentration in polymerliposome, resulting in the rotation of ATPase to produce ATP. This artificially hybrid proteopolymersome have great potential application in several fields ranging from the investigation of cellular physiology to the synthesis and assembly of bioinspired materials.

To acquire the enhanced ATPase synthesis in vitro, Montemagno and coworkers present an in vitro artificial photosynthesis platform that conjugates the necessary enzymes of the Calvin cycle with a nanosize photophosphorylation system engineered into a foam architecture by using the Túngara frog surfactant protein Ranaspumin-2 (Fig. 10.7b) [92]. This unique protein allows lipid vesicles and the coupled enzyme to concentrate in the microscale interlaced foam nets, which transforming the photon-derived chemical energy into carbon fixation and sugar production. Light-induced ATP synthetic process usually occurs in the thylakoid membrane of green plants. A light sensitive protein, photosystem II (PSII), is employed to construct F_0F_1-ATPase inserted proteoliposome-coated PSII-based microspheres by using molecular assembly approach (Fig. 10.7c) [12]. Upon light illumination, PSII can split water into protons, oxygen, and electrons and can generate a proton gradient for ATPase to produce ATP. This biomimetic platform gives a strategy to simulate the photophosphorylation process, and may facilitate the development of ATP-driven devices by remote light control. In addition, ATPase-based rotary system can be used in the biomedical aspect. This assembled rotary platform can be introduced to enhance thrombolysis along with the urokinase against the thrombus [93]. The δ-subunit-free F_0F_1-ATPase motor is obtained by reconstructing an original chromatophore, which is extracted from Rhodospirillum rubrum. The removal of δ-subunit-free aims to enhance the rotation of F_0F_1-ATPase motor. Upon the light illumination, the photosynthesis center triggers the generation of proton that results in the rotation of ATPase

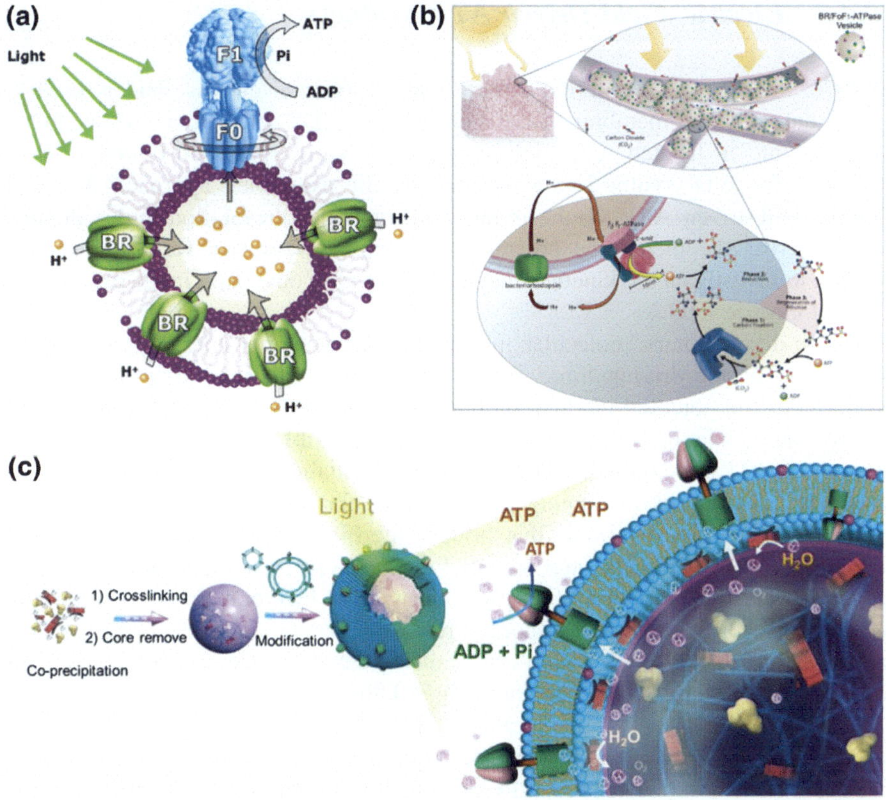

Fig. 10.7 Light triggered the rotation of ATPase enzyme for ATP synthesis. **a** Bacteriothodopsin (BR) mediated ATP synthesis in the presence of light. Reprinted with the permission from Ref. [91]. Copyright 2005, American Chemical Society; **b** Ranaspumin-2 protein-based light-to-bioenergy converted platform. Reprinted with the permission from Ref. [92]. Copyright 2010 American Chemical Society; **c** Lipids-photosystem II-$CaCO_3$ particle for the light-responsive ATPase synthesis. Reprinted with the permission from Ref. [12]. Copyright 2016, American Chemical Society

motor protein. After the targeting modification of anti-fibrinogen antibody, this ATPase-based nanodevice can act as a drill to mechanically destroy the thrombosis. Along with the urokinase, the introduced ATPase significantly accelerates the dissolution of thrombus in the presence of light.

Like the manipulation of magnetic field, light-induced ATP synthesis is a biological friendly approach that has the flexible controllability and high precision in manipulation. This strategy uses remote control to trigger on/off the rotation of ATPase enzyme without needing close contact. Significantly, it could remain the bioactivity of ATPase enzyme and enhance their work performance in vitro. The light triggered rotation of ATPase opens a door to regulate molecular machine for ATP synthesis, and devotes great contribution to develop the light-responsive materials and nanodevices.

10.4.3 Electric Field-Driven ATP Synthesis

In 1976, the first demonstration reports that an external electric field is used to impact the thylakoid membranes in the presence of ADP and ^{32}Pi and results in the formation of ^{32}P labeled ATP molecules [94]. Usually, the reconstituted ATPase structures for ATP synthesis are reported by the assembled approach. External electric field-driven ATP synthesis has both high time resolution and a flexible control across the membrane potential [95, 96]. Thus, it is a valid strategy for ATPase to investigate the kinetic behavior and energy conversion in the presence of external electric field.

Proton-driven rotary molecular motors can use potential energy generated by proton flux as the driving force. The transmembrane behavior of protons could produce electric potential gradient on both sides of the membrane. Proton delivery through the membrane is dependent on the F_0 portion of F_1F_0-ATPase, inducing the rotation of ATPase motor. Usually, F_0 portion contains a-subunit, b-subunit, and c-subunit. The ring-like c-subunit is responsible for the channel of proton transportation across the membrane that bridges the high potential and low potential (Fig. 10.8a) [97]. The generated potential gradient can generate a motive force to start the rotary transportation of proton. During cross membrane process of proton, a-subunit of ATPase plays an important role to generate an electric field. Like the particle in the electric field, proton can be manipulated by electric field in c-subunit to move from an electrode to another (Fig. 10.8b).

In the living organisms, membrane-bond ATPase always couples with several external fields to achieve the mechanical rotation and ATP synthesis or hydrolysis. Electric field is a better approach to simulate special environment and check the

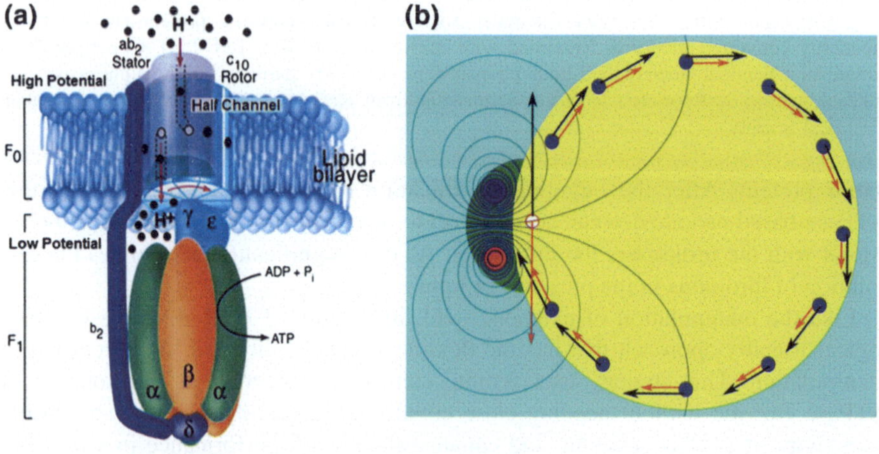

Fig. 10.8 The mechanism of electric field-driven ATP synthesis. **a** Lateral view of F_1F_0-ATPase. **b** The distribution of electric field in a-subunit and c-subunit. Reprinted with the permission from Ref. [97]. Copyright 2013, Miller et al.

in vitro performance of ATPase-based complex. A rotary electric field is introduced to trigger the rotation of ATPase conjugated dimeric polystyrene beads with the diameter of 460 nm (Fig. 10.9) [98]. This specially designed propeller is a dielectric and biologically attached to the side of γ-subunit. The F_1 portion is fixed on the Ni-NTA-coated coverslip substrate and accompanied with a rotary electric field. The rotating electric field is composed of four electrodes that own a frequency of 10 MHz is generated by using sinusoidal voltages. Due to the dielectric feature, the dimeric polystyrene beads can be rotated by the electrical rotary field. Then, the peak-to-peak voltage of the applied sinusoidal voltage is increased (Fig. 10.9 upper left) that results in dimeric polystyrene beads attached F_1-ATPase rotate faster. Moreover, the rotary direction can be controlled by the reversing rotation of external electric field. The above strategy is the first electric field-driven ATPase-based rotary device.

At present, electric field-driven ATPase rotation is still in challenge so that many requisite efforts have been attributed to develop the optimized devices and deeply study the rotary mechanism. It is important to effectively use the generated potential energy for the further energy conversion, and reveal the detail of rotary molecular machine. To enhance the stability in the presence of electric field, many realistic models and auxiliary theories should be further developed. Also, diverse of attached structures can be designed and matched for ATPase to perform electric field-driven rotation. Electric field for the manipulation of ATPase rotation has smart features and great potential to achieve advanced transform from the electric energy into chemical energy or mechanical work.

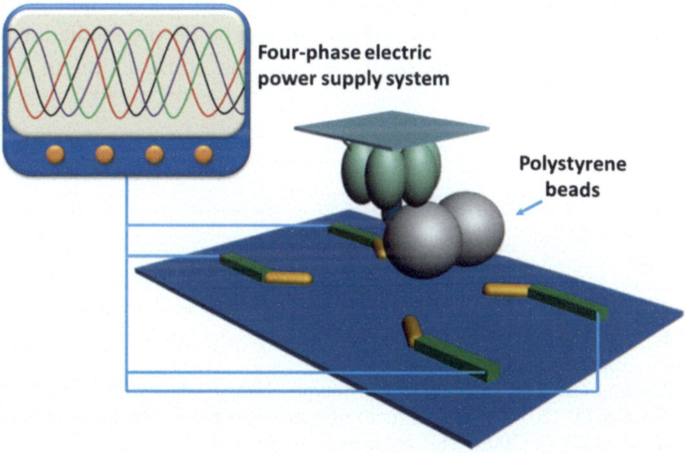

Fig. 10.9 The rotary electric field triggered rotation of F_1-ATPase motor conjugated polystyrene beads

10.5 Conclusions and Perspectives

Biomimetic materials-based assembly has attracted great concerns because of great biocompatibility and programmable feature on benefit. ATPase motor is a regulatory protein distributed in biological membrane that is responsible for the proton delivery across membrane. The molecular assembly of ATPase reconstitution employs bioactive enzyme and synthetic material to achieve construction of biohybrid materials. Based on the rotation of ATPase, many rotary nanodevices are miraculously designed to perform energy transformation between chemical energy and mechanical rotation, further develop external field manipulated approaches to regulate the rotation. In this chapter, detail introduction and promising examples are mainly discussed, and it strengthens the understanding of the biomolecular rotary motor.

In vitro reconstitution of ATPase inspires research sparks of bioactive matters and ultimately develops diverse assembled materials with fascinating functions. Scientists have successfully introduced biological macromolecules and synthetic structures as the attached propellers for ATPase. Consider the regulation of rotation, several strategies such as the magnetic field, light, and electric field that have been used to switch on/off the rotation of ATPase. Due to the own properties of propellers, manipulation of external fields possesses giant priority in controlling the rotation direction of ATPase with clockwise and anticlockwise, and then perform the conversion between ATP synthesis and ATP hydrolysis. Thus, this is significant to enhance the operated precision and enables the design of nanosize smart structures to make it possible to carry out various biological tasks. However, ATPase-based rotary systems suffer from technological barriers such as biological stability and bioactivity so that long-time work in vitro is expectant performance. These present limitations as breakthrough points can be deeply studied in the future research, and provide basic supports to achieve the construction of biomimetic devices in diverse applications.

References

1. van den Heuvel MGL, Dekker C (2007) Motor proteins at work for nanotechnology. Science 317:333–336
2. Hess H, Bachand GD, Vogel V (2004) Powering nanodevices with biomolecular motors. Chem Eur J 10:2110–2116
3. Hess H, Bachand GD (2005) Biomolecular motors. Nano Today 8:22–29
4. Junge W, Müller DJ (2011) Seeing a molecular motor at work. Science 333:704–705
5. Schliwa M, Woehlke G (2003) Molecular motors. Nature 422:759–765
6. Rensing C, Mitra B, Rosen BP (1997) The *zntA* gene of *Escherichia coli* encodes a Zn (II)-translocating P-type ATPase. Proc Natl Acad Sci USA 94:14326–14331
7. MacLennan DH, Brandl CJ, Korczak B, Green NM (1985) Amino-acid sequence of a $Ca^{2+} + Mg^{2+}$-dependent ATPase from rabbit muscle sarcoplasmic reticulum, deduced from its complementary DNA sequence. Nature 316:696–700

8. Murata T, Yamato I, Kakinuma Y, Leslie AGW, Walker JE (2005) Structure of the rotor of the V-type Na^+-ATPase from *Enterococcus hirae*. Science 308:654–659
9. Stock D, Leslie AGW, Walker JE (1999) Molecular architecture of the rotary motor in ATP synthase. Science 286:1700–1705
10. Wang H, Oster G (1998) Energy transduction in the F_1 motor of ATP synthase. Nature 396:279–282
11. Yasuda R, Noji H, Yoshida M Jr, Kinosita K, Itoh H (2001) Resolution of distinct rotational substeps by submillisecond kinetic analysis of F_1-ATPase. Nature 410:898–904
12. Feng X, Jia Y, Cai P, Fei J, Li J (2016) Coassembly of photosystem II and ATPase as artificial chloroplast for light-driven ATP synthesis. ACS Nano 10:556–561
13. Pfeifer R, Lungarella M, Iida F (2007) Self-organization, embodiment, and biologically inspired robotics. Science 318:1088–1093
14. Dreyfus R, Baudry J, Roper ML, Fermigier M, Stone HA, Bibette J (2005) Microscopic artificial swimmers. Nature 437:862–865
15. Jia Y, Li J (2015) Molecular assembly of schiff base interactions: construction and application. Chem Rev 115:1597–1621
16. Junge W, Sielaff H, Engelbrecht S (2009) Torque generation and elastic power transmission in the rotary F_0F_1-ATPase. Nature 459:364–370
17. Møller JV, Juul B, le Maire M (1996) Structural organization, ion transport, and energy transduction of P-type ATPases. Biochim Biophys Acta 1286:1–51
18. Morsomme P, Chami M, Marco S, Nader J, Ketchum KA, Goffeau A, Rigaud A (2002) Characterization of a hyperthermophilic P-type ATPase from *Methanococcus jannaschii* expressed in yeast. J Biol Chem 277:29608–29616
19. Chan H, Babayan V, Blyumin E, Gandhi C, Hak K, Harake D, Kumar K, Lee P, Li T, Liu H, Lo T, Meyer CJ, Stanford S, Zamora KS (2010) The P-types ATPase superfamily. J Mol Microbiol Biotechnol 19:5–104
20. von Ballmoos C, Wiedenmann A, Dimroth P (2009) Essentials for ATP synthesis by F_1F_0 ATP synthases. Annu Rev Biochem 78:649–672
21. Veshaguri S, Christensen SM, Kemmer GC, Ghale G, Møller MP, Lohr C, Christensen AL, Justesen BH, Jørgensen IL, Schiller J, Hatzakis NS, Grabe M, Pomorski TG, Stamou D (2016) Direct observation of proton pumping by a eukaryotic P-type ATPase. Science 351:1469–1473
22. Ramirez A, Heimbach A, Gründemann J, Stiller B, Hampshire D, Cid LP, Goebel I, Mubaidin AF, Wriekat A, Roeper J, Al-Din A, Hillmer AM, Karsak M, Liss B, Woods CG, Behrens MI, Kubisch C (2006) Hereditary parkinsonism with dementia is caused by mutations in *ATP13A2*, encoding a lysosomal type 5 P-type ATPase. Nat Genet 38:1184–1191
23. Patrick ML, Aimanova K, Sanders HR, Gill SS (2006) P-type Na^+/K^+-ATPase and V-type H^+-ATPase expression patterns in the osmoregulatory organs of larval and adult mosquito *Aedes aegypti*. J Exp Biol 209:4638–4651
24. Rensing C, Fan B, Sharma R, Mitra B, Rosen BP (2000) Cop A: an *Escherichia coli* Cu (I)-translocating P-type ATPase. Proc Natl Acad Sci USA 97:652–656
25. Axelsen KB, Palmgren MG (1998) Evolution of substrate specificities in the P-type ATPase superfamily. J Mol Evol 46:84–101
26. Bublitz M, Morth JP, Nissen P (2012) P-type ATPase at a glance. J Cell Sci 124:2515–2519
27. Weng X, Huss M, Wieczorek H, Beyenbach KW (2003) The V-type H^+-ATPase in malpighian tubules of *Aedes aegypti*: localization and activity. J Exp Biol 206:2211–2219
28. Beyenbach KW, Wieczorek H (2006) The V-type H^+ ATPase: molecular structure and function, physiological roles and regulation. J Exp Biol 209:577–589
29. Graham LA, Flannery AR, Stevens TH (2003) Structure and assembly of the yeast V-ATPase. J Bioenerg Biomembr 35:301–312
30. Recchi C, Chavrier P (2006) V-ATPase: a potential pH sensor. Nat Cell Biol 8:107–109

31. Sagermann M, Stevens TH, Matthews BW (2001) Crystal structure of the regulatory subunit H of the V-type ATPase of *Saccharomyces cerevisiae*. Proc Natl Acad Sci USA 98:7134–7139
32. Hinton A, Bond S, Forgac M (2009) V-ATPase functions in normal and disease processes. Pflugers Arch-Eur J Physiol 457:589–598
33. Marshansky V, Futai M (2008) The V-type H^+-ATPase in vesicular trafficking: targeting, regulation and function. Curr Opin Cell Biol 20:415–426
34. Lipfert J, van Oene MM, Lee M, Pedaci F, Dekker NH (2015) Torque spectroscopy for the study of rotary motion in biological systems. Chem Rev 115:1449–1474
35. Buchachenko AL, Kuznetsov DA, Breslavskaya NN (2012) Chemistry of enzymatic ATP synthesis: an insight through the isotope window. Chem Rev 112:2042–2058
36. Okuno D, Iino R, Noji H (2011) Rotation and structure of F_0F_1-ATP synthase. J Biochem 149:655–664
37. Abrahams JP, Leslie AGW, Lutter R, Walker JE (1994) Structure at 2.8 Å resolution of F_1-ATPase from bovine heart mitochondria. Nature 370:621–628
38. Jr Kinosita K, Adachi K, Itoh H (2004) Rotation of F_1-ATPase: how an ATP-driven molecular machine may work. Annu Rev Biophys Biomol Struct 33:245–268
39. Mesbah NM, Wiegel J (2011) The Na^+-translocating F_1F_0-ATPase from the halophilic, alkalithermophile *Natranaerobius thermophilus*. Biochim Biophys Acta 1807:1133–1142
40. Schalley CA, Beizai K, Vögtle F (2001) On the way to rotaxane-based molecular motors: studies in molecular mobility and topological chirality. Acc Chem Res 34:465–476
41. Uchihashi T, Iino R, Ando T, Noji H (2011) High-speed atomic force microscopy reveals rotary catalysis of rotorless F_1-ATPase. Science 333:755–758
42. Pain D, Blobel G (1987) Protein import into chloroplasts requires a chloroplast ATPase. Proc Natl Acad Sci USA 84:3288–3292
43. Senior AE (1973) The structure of mitochondrial ATPase. Biochim Biophys Acta 301:249–277
44. Sowa Y, Rowe AD, Leake MC, Yakushi T, Homma M, Ishijima A, Berry RM (2005) Direct observation of steps in rotation of the bacterial flagellar motor. Nature 437:916–919
45. Shu Y, Yue J, Ou-Yang Z (2010) F_0F_1-ATPase, rotary motor and biosensor. Nanoscale 2:1284–1293
46. Omote H, Sambonmatsu N, Saito K, Sambongi Y, Iwamoto-Kihara A, Yanagida T, Wada Y, Futai M (1999) The γ-subunit rotation and torque generation in F_1-ATPase from wide-type or uncoupled mutant *Escherichia coli*. Proc Natl Acad Sci USA 96:7780–7784
47. Yasuda R, Noji H Jr, Kinosita K, Yoshida M (1998) F_1-ATPase is a highly efficient molecular motor that rotates with discrete 120° steps. Cell 93:1117–1124
48. Noji H, Yasuda R, Yoshida M Jr, Kinosita K (1997) Direct observation of the rotation of F_1-ATPase. Nature 386:299–302
49. Soong RK, Bachand GD, Neves HP, Olkhovets AG, Craighead HG, Montemagno CD (2000) Powering an inorganic nanodevice with a biomolecular motor. Science 290:1555–1558
50. Sambongi Y, Iko Y, Tanabe M, Omote H, Iwamoto-Kihara A, Ueda I, Yanagida T, Wada Y, Futai M (1999) Mechanical rotation of the c subunit oligomer in ATP synthase (F_0F_1): direct observation. Science 26:1722–1724
51. Suzuki T, Tanaka K, Wakabayashi C, Saita E, Yoshida M (2014) Chemomechanical coupling of human mitochondrial F_1-ATPase motor. Nat Chem Biol 10:930–936
52. Hayashi S, Ueno H, Shaikh AR, Umemura M, Kamiya M, Ito Y, Ikeguchi M, Komoriya Y, Iino R, Noji H (2012) Molecular mechanism of ATP hydrolysis in F1-ATPase revealed by molecular simulations and single-molecule observations. J Am Chem Soc 134:8447–8454
53. Bachand GD, Soong PK, Neves HP, Olkhovets A, Craighead HG, Montemagno CD (2001) Precision attachment of individual F_1-ATPase biomolecular motors on nanofabricate substrates. Nano Lett 1:42–44
54. Furuike S, Nakano M, Adachi K, Noji H Jr, Kinosita K, Yokoyama K (2011) Resolving stepping rotation in *thermus thermophilus* H^+-ATPase/synthase with an essentially drag-free probe. Nat Commun 2:233

55. Toyabe S, Watanabe-Nakayama T, Okamoto T, Kudo S, Muneyuki E (2011) Thermodynamic efficiency and mechanochemical coupling of F_1-ATPase. Proc Natl Acad Sci USA 108:17951–17956
56. York J, Spetzler D, Xiong F, Frasch WD (2008) Single-molecule detection of DNA via sequence-specific links between F_1-ATPase motors and gold nanorod sensors. Lab Chip 8:415–419
57. Safinya CR, Ewert KK (2012) Liposomes derived from molecular vases. Nature 489:372–374
58. Steinberg-Yfrach G, Rigaud JL, Durantini EN, Moore AL, Gust D, Moore TA (1998) Light-driven production of ATP catalysed by F_0F_1-ATP synthase in an artificial photosynthetic membrane. Nature 392:479–482
59. Turina P, Samoray D, Gräber P (2003) H^+/ATP ratio of proton transport-coupled ATP synthesis and hydrolysis catalysed by CF_0F_1-liposomes. EMBO J 22:418–426
60. De Koker S, Hoogenboom R, De Geest BG (2012) Polymeric multilayer capsules for drug delivery. Chem Soc Rev 41:2867–2884
61. Donath E, Sukhorukov GB, Caruso F, Davis SA, Möhwald H (1998) Novel hollow polymer shells by colloid-templated assembly of polyelectrolytes. Angew Chem Int Ed 37:2202–2205
62. De Cock LJ, De Koker S, De Geest BG, Grooten J, Vervaet C, Remon JP, Sukhorukov GB, Antipina MN (2010) Polymeric multilayer capsules in drug delivery. Angew Chem Int Ed 49:6954–6973
63. Peyratout CS, Dähne L (2004) Tailor-made polyelectrolyte microcapsules: from multilayers to smart containers. Angew Chem Int Ed 43:3762–3783
64. Tong W, Song X, Gao C (2012) Layer-by-layer assembly of microcapsules and their biomedical applications. Chem Soc Rev 41:6103–6124
65. Shao J, Xuan M, Dai L, Si T, Li J, He Q (2015) Near-infrared activated nanocalorifiers in microcapsules: vapor bubble generation for in vivo enhanced cancer therapy. Angew Chem Int Ed 54:12782–12787
66. Shao J, Xuan M, Si T, Dai L, He Q (2015) Biointerfacing polymeric microcapsules for in vivo near-infrared light-triggered drug release. Nanoscale 7:19092–19098
67. Such GK, Johnston APR, Caruso F (2011) Engineered hydrogen-bonded polymer multilayers: from assembly to biomedical applications. Chem Soc Rev 40:19–29
68. Duan L, He Q, Wang K, Yan X, Cui Y, Möhwald H, Li J (2007) Adenosine triphosphate biosynthesis catalyzed by F_0F_1 ATP synthase assembled in polymer microcapsules. Angew Chem Int Ed 46:6996–7000
69. He Q, Duan L, Qi W, Wang K, Cui Y, Yan X, Li J (2008) Microcapsules containing a biomolecular motor for ATP biosynthesis. Adv Mater 20:2933–2937
70. Duan L, Qi W, Yan X, He Q, Cui Y, Wang K, Li D, Li J (2009) Proton gradients produced by glucose oxidase microcapsules containing motor F_0F_1-ATPase for continuous ATP biosynthesis. J Phys Chem B 113:395–399
71. Qi W, Duan L, Wang K, Yan X, Cui Y, He Q, Li J (2008) Motor protein CF_0F_1 reconstituted in lipid-coated hemoglobin microcapsules for ATP synthesis. Adv Mater 20:601–605
72. Li J, Wang Y, Ha W, Liu Y, Ding L, Li B, Zhang S (2013) Cyclodextrin-based microcapsules as bioreactors for ATP biosynthesis. Biomacromol 14:2984–2988
73. Dong H, Nie R, Hou X, Wang P, Yue J, Jiang L (2011) Assembly of F_0F_1-ATPase into solid state nanoporous membrane. Chem Commun 47:3102–3104
74. Hirono-Hara Y, Ishizuka K Jr, Kinosita K, Yoshida M, Noji H (2005) Activation of pausing F_1 motor by external force. Proc Natl Acad Sci USA 102:4288–4293
75. Saita E, Suzuki T Jr, Kinosita K, Yoshida M (2015) Simple mechanism whereby the F_1-ATPase motor rotates with near-perfect chemomechanical energy conversion. Proc Natl Acad Sci USA 112:9626–9631
76. Adachi K, Oiwa K, Yoshida M, Nishizaka T Jr, Kinosita K (2012) Controlled rotation of the F_1-ATPase reveals differential and continuous binding changes for ATP synthesis. Nat Commun 3:1022
77. Gaspard P, Gerritsma E (2007) The stochastic chemomechanics of the F_1-ATPase molecular motor. J Theor Biol 247:672–686

78. Craighead H (2006) Future lab-on-a-chip technologies for interrogating individual molecules. Nature 442:387–393
79. Itoh H, Takahashi A, Adachi K, Noji H, Yasuda R, Yoshida M Jr, Kinosita K (2004) Mechanically driven ATP synthesis by F_1-ATPase. Nature 427:465–468
80. Rondelez Y, Tresset G, Nakashima T, Kato-Yamada Y, Fujita H, Takeuchi S, Noji H (2005) Highly coupled ATP synthesis by F_1-ATPase single molecules. Nature 433:773–777
81. Adachi K, Oiwa K, Nishizaka T, Furuike S, Noji H, Itoh H, Yoshida M Jr, Kinosita K (2007) Coupling of rotation and catalysis in F_1-ATPase revealed by single-molecule imaging and manipulation. Cell 130:309–321
82. Palanisami A, Okamoto T (2010) Torque-induced slip of the rotary motor F_1-ATPase. Nano Lett 10:4146–4149
83. Bang M, Chisholm P (2009) Living sunlight: how plants bring the earth to life. Science 326:1483
84. Danielson E, Golden JH, McFarland EW, Reaves CM, Weinberg WH, Wu X (1997) A combinatorial approach to the discovery and optimization of luminescent materials. Nature 389:944–948
85. Grinberg I, West DV, Torres M, Gou G, Stein DM, Wu L, Chen G, Gallo EM, Akbashev AR, Davies PK, Spanier JE, Rappe AM (2013) Perovskite oxides for visible-light-absorbing ferroelectric and photovoltaic materials. Nature 503:509–512
86. Xuan M, Wu Z, Shao J, Dai L, Si T, He Q (2016) Near infrared light-powered Janus mesoporous silica nanoparticle motors. J Am Chem Soc 138:6492–6497
87. Inoue K, Ono H, Abe-Yoshizumi R, Yoshizawa S, Ito H, Kogure K, Kandori H (2013) A light-driven sodium ion pump in marine bacteria. Nat Commun 4:1678
88. Hoffmann A, Hildebrandt V, Heberle J, Büldt G (1994) Photoactive mitochondria: *in vivo* transfer of a light-driven proton pump into the inner mitochondrial membrane of *Schizosaccharomyces pombe*. Proc Natl Acad Sci USA 91:9367–9371
89. Luo T, Soong R, Lan E, Dunn B, Montemagno C (2005) Photo-induced proton gradients and ATP biosynthesis produced by vesicles encapsulated in a silica matrix. Nature Mater 4:220–224
90. Diez M, Zimmermann B, Börsch M, König M, Schweinberger E, Steigmiller S, Reuter R, Felekyan S, Kudryavtsev V, Seidel CAM, Gräber P (2004) Proton-powered subunit rotation in single membrane-bound F_0F_1-ATP synthase. Nat Struct Mol Biol 11:135–141
91. Choi H, Montemagno CD (2005) Artificial organelle: ATP synthesis from cellular mimetic polymersomes. Nano Lett 5:2538–2542
92. Wendell D, Todd J, Montemagno C (2010) Artificial photosynthesis in ranaspumin-2 based foam. Nano Lett 10:3231–3236
93. Duan X, Liu L, Jiang W, Yue J (2015) Visible thrombolysis acceleration of a nanomachine powered by light-driving F_0F_1-ATPase motor. Nanoscale Res Lett 10:1–10
94. Vinkler C, Korenstein R (1982) Characterization of external electric field-driven ATP synthesis in chloroplasts. Proc Natl Acad Sci USA 79:3183–3187
95. Bauermeister H, Schlodder E, Gräber P (1988) Electric field-driven ATP synthesis catalyzed by the membrane-bound ATP-synthase from chloroplasts. Ber Bunsenges Phys Chem 92:1036–1039
96. Kaim G, Dimroth P (1998) Voltage-generated torque drives the motor of the ATP synthase. EMBO J 17:5887–5895
97. Jr Miller JH, Rajapakshe KI, Infante HL, Claycomb JR (2013) Electric field driven torque in ATP synthase. PLoS ONE 8:e74978
98. Watanabe-Nakayama T, Toyabe S, Kudo S, Sugiyama S, Yoshida M, Muneyuki E (2008) Effect of external torque on the ATP-driven rotation of F_1-ATPase. Biochem Bioph Res Co 366:951–957

Chapter 11
Controlled Molecular Assembly Toward Self-propelled Micro-/Nanomotors

Xiankun Lin, Zhiguang Wu and Qiang He

Abstract Micro-/nanomotors (MNMs) are able to propel themselves in fluids through converting different energies from environment into kinetic energy. Recently, layer-by-layer (LbL) assembly, a versatile assembly approach, has been employed to access MNMs with advantages such as regulated motion, stimuli-response properties, and multifunctionality. In this chapter, we review the recent progress on controlled fabrication, motion control, and biomedical applications of MNMs based on controlled molecular assembly. Through integrating diverse functional building blocks such as nanoparticles, enzymes, and metal shells, MNMs with various structures (e.g., hollow capsules and nanotubes) have been prepared, and the control over the on/off state of the MNM motion has been realized. In addition, we also discuss a special type of MNMs which is derived from the combination of as-assembled biological aggregates and artificial nanostructures. These MNMs can be driven by bubble recoil, irradiation by near-infrared light, and ultrasonic fields. We have also demonstrated the potential applications of these assembled MNMs in biomedical fields such as targeted drug delivery, photothermal therapy, and detoxification.

Keywords Micro-/nanomotor · Layer-by-layer assembly · Self-propulsion · Gold nanoshell · Near-infrared light · Cell membrane · Drug delivery · Photothermal therapy · Detoxification

---

X. Lin · Z. Wu · Q. He (✉)
Key Laboratory of Microsystems and Microstructures Manufacturing, Ministry of Education, Micro/Nanotechnology Research Center, Harbin Institute of Technology, Harbin 150080, China
e-mail: qianghe@hit.edu.cn

X. Lin
e-mail: xiankunlin@hit.edu.cn

Z. Wu
e-mail: zhiguangwu@hit.edu.cn

J. Li (ed.), *Supramolecular Chemistry of Biomimetic Systems*,
DOI 10.1007/978-981-10-6059-5_11

11.1 Introduction

Nature has created exact, elegant architectures such as biological motors to overcome the Brownian motion and finish the controlled autonomous movement for achieving complex and important tasks with high efficiency [1, 2]. One of the typical examples is the F_0F_1-synthase, a rotary motor, which is in charge of the energy storage through catalytically synthesizing the universal "fuel" molecule adenosine triphosphate (ATP) from adenosine diphosphate and inorganic phosphate by using the proton or sodium gradient across the membranes. Other examples are kinesin and dynein motors which transport cargos from one end of the microtubules to the other. The mechanical energy of these motors derives from the chemical energy provided by the ATP hydrolysis. Inspired by autonomous movement in the biological systems [3], diverse synthetic molecular machines such as molecular elevators and rotary motors have been fabricated sophisticatedly with the controlled motion at the molecular level, which has been awarded the Nobel prize in 2016 [4, 5]. On the other hand, to achieve the tasks such as long-distance transport and high-efficiency separation of nanoscale objects, faster synthetic motors with the motion displacements at the nano-, micro- and macroscopic levels are highly desired [6, 7]. As a response to such expectation, synthetic MNMs with different architectures and propulsion mechanisms have been developed as one of the most exciting yet challenging areas in the field of nanotechnology. During the past decade, enormous progresses have been achieved toward the fabrication and applications of MNMs which are tiny devices capable of performing self-propelled motion in fluids through harnessing different types of energies into mechanical movement [8, 9]. MNMs show numerous potential applications in diverse fields, such as directed drug transportation [10–14], isolation of biological targets [15], diagnostics [16], and environmental remediation [17–19], and may afford the revolutionary solutions for such fields due to their self-propulsion ability.

In 2002, the first chemically driven motor was proposed conceptually by Prof. Whitesides et al. [20], despite the size of the motor was on the centimeter scale. The motor, a hemicylindrical polydimethylsiloxane (PDMS) plate with a platinum-coated porous glass, is self-propelled by the release of the oxygen bubbles that are produced by the platinum-catalyzed decomposition of hydrogen peroxide (H_2O_2). Unlike macroscopic objects such as a ship or a swimming person, the MNMs is at low Reynolds number regimes in which the viscous drag and the Brownian diffusion are the determinant effects [9]. In 2004 and 2005, Prof. A. Sen and Prof. T.E. Mallouk in Penn State University [21], and Prof. G.A. Ozin in Toronto University [22] achieved independently the breakthrough by preparing first the catalytic MNMs at the microscale, which indicates that the energy derived from chemical reactions could be employed to propel the micro- and nanoscale objects in fluids. During the following decade, a vast variety of MNMs have been explored, and the field of MNMs has been growing into a hot topic, as shown by the International Symposium on Micro- and Nanomachines which is held every 2 years since 2012.

The architectures of MNMs mainly include nanowires/nanorods, nanotubes, and Janus spheres which usually possess the asymmetric structures for producing a net driven force. As an important type of MNMs, the catalytic MNMs are powered by in situ chemical reactions in which catalyst and fuel are usually necessary. The fuels could be the fluid in which the MNMs swim or a part of the MNMs, and their properties and concentrations have a considerable influence on the motion of catalytic MNMs. The MNMs could also be driven by external fields such as magnetic, electric, and acoustic fields, as well as the light irradiation. These physical triggers allow for the remote and rapid control over the motor motions. Some motion mechanisms, such as bubble propulsion, interfacial tension gradients, self-electrophoresis, self-diffusiophoresis, self-thermophoresis, osmotic propulsion, ultrasound propulsion, and magnetically driven propulsion, have been proposed recently [23].

Diverse approaches have been employed for the controlled fabrication and multifunctionality of MNMs, as reviewed by Pumera [24]. Among these progresses, the combination of self-assembly with the self-propelled MNMs offers a powerful, promising strategy [25–28]. In 2012, the pioneering works from van Hest's group [29] and He's group [30] initialized molecular assembly-derived approaches for the biomimetic production of MNMs. Controlled molecular assembly, regarding as a typical bottom-up nanofabrication strategy, has contributed significant advances to various fields, ranging from novel nature-mimicking nanostructures to stimuli-responsive materials [31]. It aims at molecular manufacturing into ordered nanostructures with defined morphologies from various building blocks by using noncovalent interactions such as electrostatic interactions, hydrogen bonding, and hydrophobic interactions. The use of controlled assembly techniques can not only provide the efficient control over the sizes, geometries, and components of the resulting MNMs, but also conveniently enhance the functionalities of MNMs by integrating different building blocks such as polymers, nanoparticles, proteins, inorganic or organic functional molecules, and artificial or natural vesicles. Besides the feasibility on multifunctionality, controlled molecular assembly could be beneficial for realizing the massive production, regulated motion, stimuli-response properties, and biocompatibility of MNMs [25]. Self-assembled MNMs with defined structures can be prepared and engineered in a versatile and simple manner, and can be used for the applications such as encapsulation, transportation, separation, and sensing with high efficiency in diverse fields, especially in the biomedicine fields.

11.2 Construction of Layer-by-Layer Assembled MNMs

Well-defined polyelectrolyte multilayers (PEMs) can be readily obtained by depositing alternately the positively and negatively charged polyelectrolytes onto the solid substrates for certain recycles [32]. Depending on the structures of the employed substrates, different assembled micro-/nanostructures could be produced. Hollow capsules can be prepared by using sacrificial colloidal particles [33], while the porous templates will result in LbL-assembled nanotubes [34]. In recent years, the spectrum of the building blocks, morphologies, and the types of the noncovalent interactions for LbL assemblies have been extended tremendously [35]. The multiple advantages of LbL approaches make them as promising tools for the controlled fabrication of self-propelled objects. The structural feature of LbL assemblies, such as the wall thicknesses, the overall sizes, and the morphologies, can be modulated by controlling the assembly process and the geometric feature of the templates. Moreover, diverse components including polymers, nanoparticles, proteins, lipids, and inorganic or organic functional molecules could be conveniently integrated into the LbL-assembled structures. To endow the LbL-assembled capsules, nanotubes or other structures with the ability of self-propulsion, introducing the functional components such as catalytic nanoparticles and making the assemblies be asymmetric are the two key steps. In principle, the LbL assemblies should have an asymmetric structure intrinsically, or should be equipped asymmetrically with the catalysts, which may require the assistance of some “top-down” methods.

In 2012, Prof. He and his coworkers described the first MNM based on LbL-assembled microcapsules (Fig. 11.1) [30], which realizes the combination of the self-propelled motor and the smart cargo in one microscale object. The PEM capsules with the diameter of ca. 8 μm are prepared by the LbL assembly of polystyrene sulfonate (PSS)/polyallylamine hydrochloride (PAH) on the sacrificial SiO_2 templates. In the following step, dendritic Pt nanoparticles (NPs) that have a large specific surface area are coated onto the prepared capsules by the microcontact printing technique (μCP). The dendritic Pt NPs display the high catalytic reactivity and should be beneficial for achieving the high-speed motion of the motors. In detail, a PDMS stamp loaded with a Pt NP ink is placed on top of the self-assembled monolayer of the PEM-covered SiO_2 spheres. After that, the monolayer is disassembled and SiO_2 templates are dissolved to form the hollow asymmetric capsules modified with dendritic Pt NPs, that is, the Janus capsule MNMs. The Pt NPs on the Janus capsule motors are able to catalytically decompose hydrogen peroxide into water and oxygen. The oxygen bubbles accumulate and eventually desorb from the surface of the catalytic NPs, which in turn pushes the

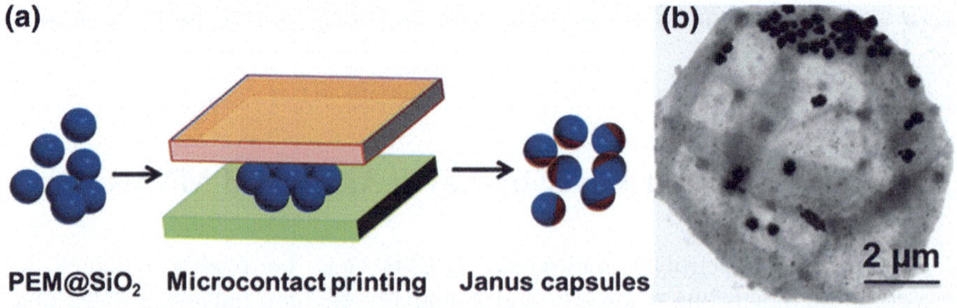

Fig. 11.1 **a** Fabrication process and **b** transmission electron microscopic (TEM) image of dendritic Pt NP-modified Janus capsule MNMs *via* LbL assembly and μCP. Reprinted with permission from Ref. [30]. Copyright 2012 American Chemical Society

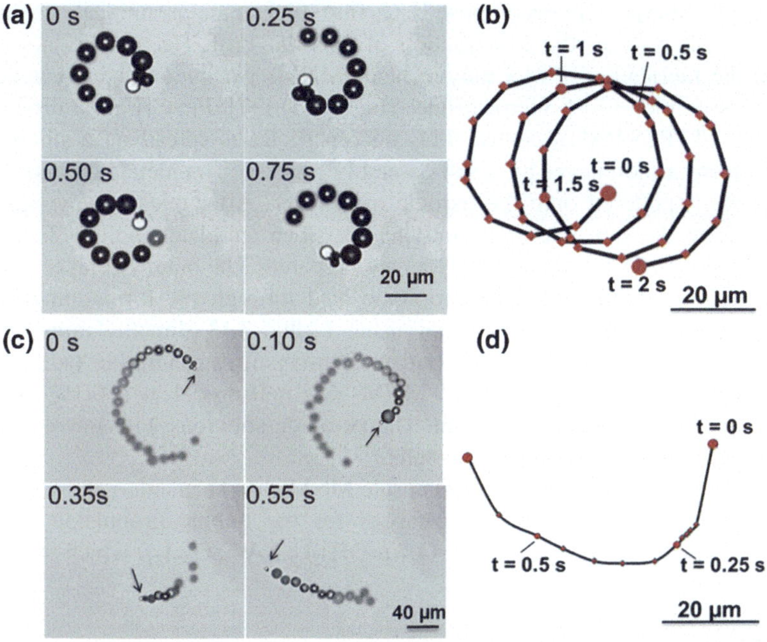

Fig. 11.2 **a**, **c** Time-lapse images and **b**, **d** the corresponding trajectories of high-speed motion of the Janus capsule motors in 15% H_2O_2 recorded by an optical microscope, showing the circular motion (**a**, **b**) and the spiral motion (**c**, **d**). Reprinted with permission from Ref. [30]. Copyright 2012 American Chemical Society

Janus motors to move forward. The capsule MNMs in a 15% H_2O_2 solution exhibit two kinds of typical trajectories: circular and spiral motion with a speed of ca. 140 and 110 μm/s, respectively (Fig. 11.2). The Janus capsule MNMs are self-propelled with an ultrafast speed of above 1 mm/s (equivalent to 125 body lengths/s) in a 30% H_2O_2 solution, representing a strong driving force of 75 pN which is calculated according to the Stokes' law.

Besides combining with the μCP technique, the metal catalysts could also be coated onto the top half of the LbL-assembled microcapsules by sputter coating under high vacuum [36]. To deposit firmly the catalytic metal layers onto the PEMs, an immediate Ti or Ni layer is required to serve as an adhesive layer. The combination of sputter coating with LbL assembly shows multiple advantages. The multilayered structures, which are consisting of different metal layers with the defined thicknesses, could be constructed through designing the process and conditions of sputter coating. Moreover, the sputter coating method is more convenient than the μCP technique, due to the fact that the metal atoms are used directly and the synthesis of metal NPs is not necessary in the sputter coating.

To realize the applications of MNMs in biological fields such as controllable drug delivery and release, preparing the MNMs from biocompatible and biodegradable materials is highly desirable. Catalase is a biocompatible enzyme that is capable to catalytically decompose H_2O_2 into oxygen and is an ideal substitution for the Pt catalysts in the construction of MNMs. Prof. He and his coworkers realized the fabrication of biocatalytic Janus MNMs *via* immobilizing catalase on the LbL-assembled PEM microcapsules (Fig. 11.3) [37]. First of all, a drop of the dispersion of $(PSS/PAH)_5$-coated SiO_2 microparticles is spread on a silicon substrate to form a monolayer by a self-assembly process. The metals Ni and Au are subsequently sputtered onto the particle monolayer. After releasing the spherical particles from the substrate and removing the silica templates, hollow Janus capsules covered partly by the metal layers are prepared. The outer Au layer could be further modified with 3-mercaptopropionic acid through the formation of Au–S bonding, and coupled with catalase by using 1-ethyl-3-[3-(dimethylamino)propyl] carbodiimide hydrochloride (EDC) and *N*-hydroxysulfosuccinimide (sulfo-NHS). These catalase-immobilized capsule MNMs can self-propel at a 0.1% peroxide solution at physiological temperature. The peroxide solutions with low concentrations show the lower toxicity to the cells.

The family of the LbL-assembled capsule MNMs has been enlarged by using the defined inorganic clusters as the catalysts for the bubble propulsion. A Ru^{IV} polyoxometalate $Na_{10}[Ru_4(H_2O)_4(\mu\text{-}O)_4(\mu\text{-}OH)_2(\gamma\text{-}SiW_{10}O_{36})_2]$, which shows the

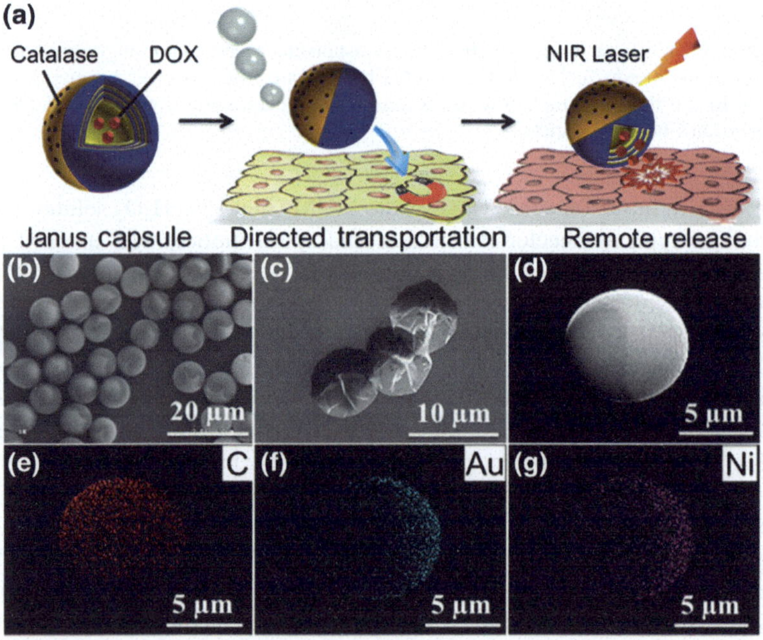

Fig. 11.3 **a** Scheme illustrating the structure of the biohybrid Janus capsule motors and its applications in the directed transportation and remote release of the drug DOX under the NIR irradiation. **b**, **c** SEM images of the motors **b** before and **c** after removal of the SiO_2 microparticles. **d**–**g** SEM image and the corresponding EDX mapping of a motor with the spherical template reserved. Reprinted with permission from Ref. [37]. Copyright 2014 American Chemical Society

catalytic ability to decompose H_2O_2 into water and oxygen, has been introduced into the PEM capsules for the construction of MNMs [38]. The tetrasubstituted Ru^{IV} polyoxometalate is a charged cluster with well-defined structures, and can be sandwiched easily into the multilayer wall of the LbL-assembled microcapsules *via* the electrostatic interactions. A calcium carbonate microparticle doping with a dextran–rhodamine is employed as the sacrificial template for the preparation of the hollow capsules, which leads to the encapsulation of dextran–rhodamine in the cavity of the capsules. The dextran–rhodamine behaves as a fluorescent tag to favor the observation of the motor motion. The resulting MNMs show linear or circular trajectories with a maximum speed of ca. 25 μm/s.

Although exploring new driven mechanisms with biocompatible fuels such as water has been proved to be one way toward the practical applications of MNMs in biomedical fields, fuel-free propulsion with the energy sources provided by external physical triggers such as ultrasound, light, electrical fields, and magnetic fields is highly expected [24]. Light is one of the powerful physical triggers with impressive advantages such as rapid and remote response. Particularly, the near-infrared light (NIR) is of special interest in the biomedical applications due to its substantial penetration and minimal absorption in tissues. The fuel-free propulsion of the LbL-assembled capsule MNMs has been activated by using NIR as the energy source [39, 40]. Through employing the LbL assembly assisted by the μCP technology, the light-driven capsule MNMs consisting of biodegradable, natural polysaccharides chitosan (CHI) and sodium alginate (ALG) have been prepared by Prof. He's group. Another multilayer prepared by the LbL assembly of CHI, ALG, and gold nanorods (GNRs) is used as the ink for the μCP step. The multilayered patch with GNRs is anchored firmly onto the $(CHI/ALG)_5$ capsule through electrostatic attraction. The Janus structure has been investigated by using TEM and two-photon confocal laser scanning microscopy. The UV–vis spectrum indicates that the GNRs possess the surface plasmon resonance absorption at 850 nm, suggesting the resulting capsule MNMs could show a photothermal effect to response the NIR irradiation. The NIR illumination with a laser power of 9.6 J/cm^2 results in the autonomous movement of the capsule MNMs with a speed of 23.27 μm/s, due to the temperature gradients produced by the photothermal effect of GNRs.

The LbL-assembled microcapsules are a versatile platform for the construction of MNMs driven by the catalytic reactions or the light irradiation, through integrating different functional components such as Pt NPs, Pt caps, catalase, polyoxometalates, and GNRs. Besides the MNMs based on the spherical microcapsules, the MNMs with different morphologies are also highly desirable, due to their diversified movement behaviors and advantages in various applications. The LbL technologies allow for the fabrication of assembled structures with different features expediently. Recently, the MNMs based on LbL-assembled nanotubes, shells [42], and two-dimensional plates [43, 44] have been fabricated successfully. As an important type of MNMs, tubular MNMs propelled by the bubble recoil (so-called "nanorockets"), show interesting movement behaviors and propelled mechanisms [45]. The PEM nanotubes derived from nanoporous template-assisted LbL assembly have been transformed into nanorockets by Prof. He and his coworkers recently (Fig. 11.4) [41]. In the preparation process of the polymer-based multilayer nanorockets, positively charged CHI and negatively charged ALG are alternatively absorbed into track-etched porous polycarbonate (PC) membranes in the first step, and poly(diallyldimethylammonium chloride)-stabilized Pt NPs are subsequently assembled into the asymmetric pore channels of the PC template. The asymmetric channels could lead to the asymmetric structures of the LbL-assembled nanotubes. The nanorocket has two openings with the different sizes, which can force the produced oxygen bubbles to move toward and eventually release from the large opening rather than the small one. The nanorockets exhibit four typical trajectories, including straight, circular, curved, and self-rotating motions. The nanorockets have reached a speed of ca. 74 μm/s (10 body lengths/s).

In the following study, the same group fabricated a biodegradable LbL-assembled nanorocket consisting of bovine serum albumin (BSA) and poly-L-lysine (PLL) (Fig. 11.5) [46]. This study shows that the cavities of the assembled nanotubes could be employed to accommodate various functional components for the

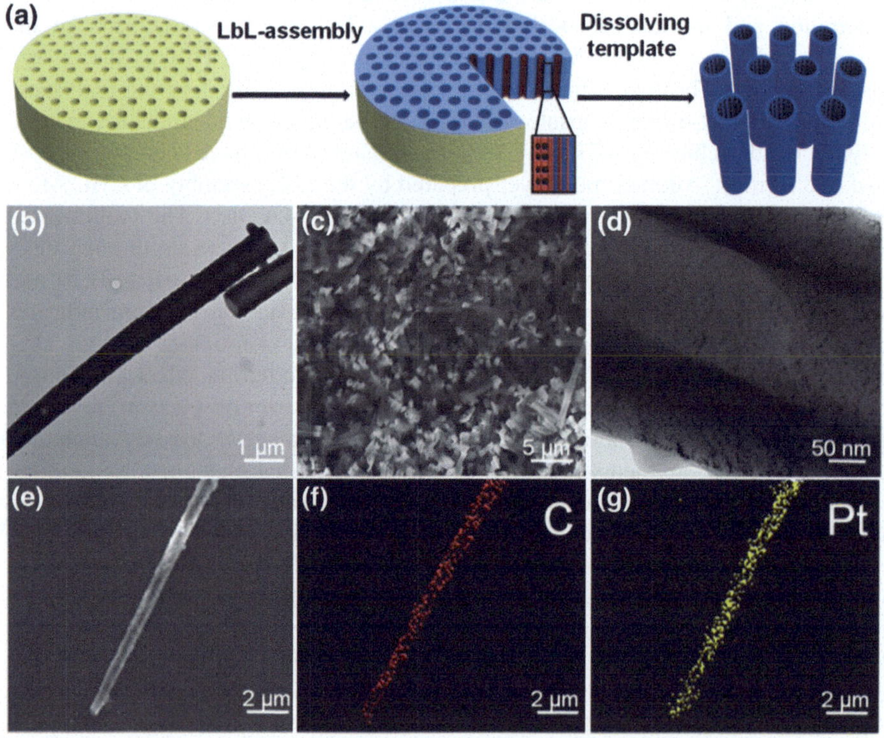

Fig. 11.4 **a** Scheme illustrating the fabrication of self-propelled PEM nanorockets *via* the template-assisted LbL assembly. **b**, **d** TEM and **c** SEM images of the nanorockets. **e**–**g** EDX mapping of a nanorocket. Reproduced from Ref. [41] by permission of John Wiley & Sons Ltd.

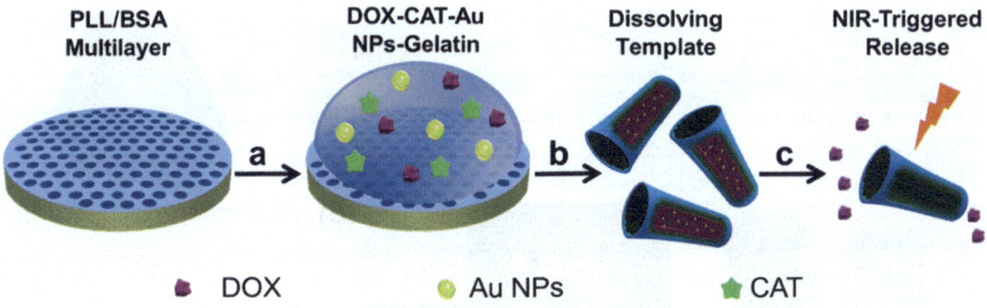

Fig. 11.5 Scheme illustrating the template-assisted LbL fabrication of $(PLL/BSA)_{10}$ nanorockets which is filled with the gelatin hydrogel containing DOX, catalase (CAT), and Au NPs, and the triggered release of drugs under the NIR irradiation. Reprinted with permission from Ref. [46]. Copyright 2015 American Chemical Society

multifunctionalization and encapsulation applications of the nanorockets. Several components can be incorporated into the nanorockets in one step through filling the cavities with the aqueous mixed solution containing gelatin, catalase, Au NPs, and doxorubicin (DOX) at 40 °C, and then cooling the filled nanorockets to 4 °C. This cooling process induces the phase transfer of the thermal-sensitive gelatin from a solution into a hydrogel, resulting in the encapsulation of those components in the nanorockets. The nanorockets can reach a velocity of ca. 4 μm/s in a 0.5% H_2O_2 solution at 37 °C. The framework materials of the nanorockets are biodegradable proteins and polypeptides. The biodegradability of rockets has been evaluated by using α-chymotrypsin, a digestive enzyme in human pancreatic juice. After incubation with α-chymotrypsin overnight, the nanorockets could be degraded into tiny thin pieces.

Through introducing the gold nanoshell into the PEM nanotubes, He's group has achieved the transformation of the nanotubes into the highly efficient NIR-powered rockets (Fig. 11.6) [47]. The NIR-propelled rockets consist of polymer multilayer framework with the gold nanoshell inside and possess an asymmetric cylindrical geometry with two openings. The diameters of the openings are about 5 and 5.5 μm, respectively. These tubular MNMs are able to move at a maximum speed of ca. 158 μm/s with an almost straight trajectory under the external NIR irradiation. The rocket propulsion is ascribed to a self-thermophoresis mechanism in which a local temperature gradient, associated with plasma resonance adsorption of gold nanoshell and the asymmetric structure of the rockets, leads to the more powerful movement of the rockets comparable to chemically propelled rockets. Theoretical simulation confirms that the produced thermal gradient around the asymmetric rocket results in a net force along the long axis of the rocket, and the force drives the rocket to move toward the small opening. More importantly, the rockets show the efficient propulsion in cell media under the NIR illumination, which overcomes the disadvantages of bubble-propelled MNMs on the speed decrease owing to the fouling of the catalysts by proteins in cell medium and should be beneficial to the biomedicine applications of tubular MNMs.

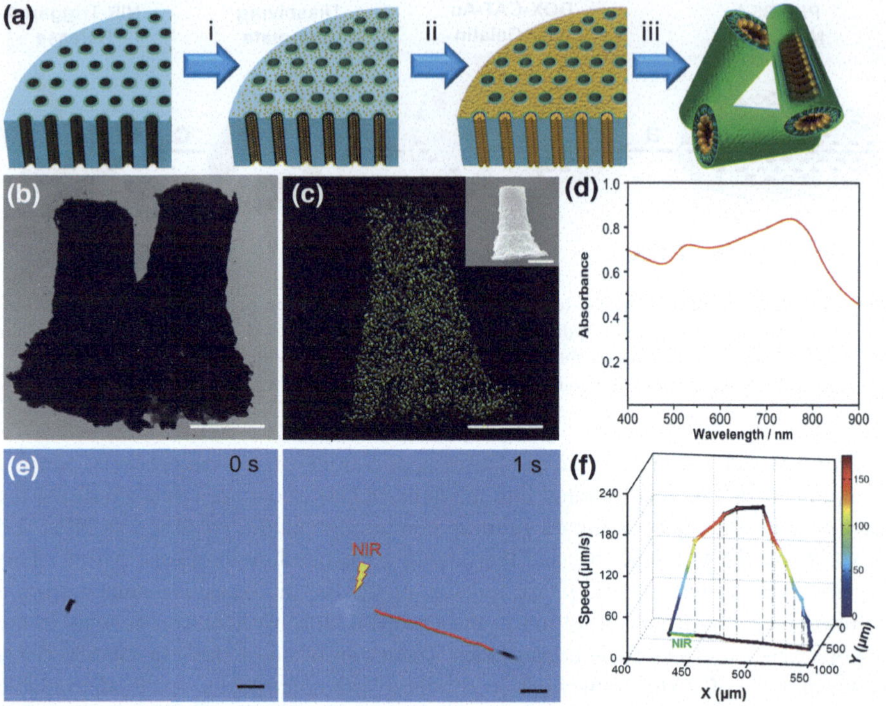

Fig. 11.6 **a** Fabrication of NIR-driven PEM rockets: (i) LbL assembly and deposition of Au NPs into the template channels; (ii) Transition of Au NPs into Au nanoshells by a surface seeding growth method; (iii) Releasing the rockets by dissolving the templates. **b** TEM image, **c** EDX mapping with the corresponding SEM image (inset), and **d** UV–vis spectrum of the rockets. Scale bar: 2 μm. **e** Time-lapse images and the corresponding speed distribution showing the movement of a rocket under NIR irradiation. Scale bar: 20 μm. Reproduced from Ref. [47] by permission of John Wiley & Sons Ltd.

11.3 Construction of MNMs Derived from Natural Aggregates

Cell membranes are special, functional aggregates created by nature. The incorporation of cell membranes and synthetic carriers is a promising biointerfacing strategy which improves the design of the carriers for diverse biomedical applications [48–51]. Combining as-assembled biological aggregates with artificial nanoscale components also provides a hopeful route to fabricating the MNMs with outstanding biocompatibility. Besides the assembly at molecular level, the controllable assembly also facilitates the integration of cell membranes onto the surface of MNMs. Diverse cell membrane-based MNMs have been fabricated [52–55], and lipid bilayer membranes have been coated onto mesoporous silica nanomotors with a diameter of ca. 75 nm for targeted drug delivery [56].

One of the fabrication strategies of the cell-based MNMs is to equip or remold the natural cells with artificial building blocks. Recent studies have demonstrated that red blood cells (RBCs) can be transformed into functional MNMs through loading citrate-stabilized Fe_3O_4 NPs inside by a hypotonic hemolysis process at 4 °C and an isotonic treatment at 37 °C [52]. The aggregates of magnetic nanoparticles could locate asymmetrically within the RBC MNMs, and this asymmetric distribution enables the magnetic alignment and the guidance of MNMs under acoustic propulsion. The RBC-based motors possess remarkable biocompatibility verified by the macrophage uptake study and the fact that they can travel for extended periods in undiluted blood without biofouling. Other components such as drugs and quantum dots can also be encapsulated into the RBC MNMs to achieve the multifunction integration [53].

Another strategy to access biocompatible, cell-based MNMs is to cover the prepared MNMs with the derivatives from natural cells. Mallouk's group has demonstrated the ultrasound-triggered motion of gold nanowires based on a self-acoustophoresis mechanism [57]. Recently, RBC membrane-derived vesicles have been fused onto ultrasound-powered gold nanowires to form a cell membrane-camouflaged nanomotors which are also known as motor sponges [54]. The RBC membrane-derived vesicles with the diameter of 50–100 nm are prepared from fresh RBCs, while the gold nanowires are synthesized by template-assisted electrochemical deposition, and subsequently modified with citric acid to endow themselves with negative charges (Fig. 11.7). The gold nanowire possesses a concave-shaped end which is thought to be critical to the acoustic propulsion. The concave end induces a pressure gradient, and the gradient drives the motor movement upon an ultrasound field. For the fusion process, the negatively charged gold nanowires were incubated with RBC membrane vesicles under the ultrasound treatment. The electrostatic repulsion between the negative charged gold nanowire motors and RBC vesicles results in the "right-side-out" orientation of the membranes. The RBC membrane-coated nanomotors even display an effective propulsion with a speed of 14 μm/s in the undiluted blood for a prolonged period, indicating that the protein-induced biofouling effect has been obviated. Due to the combined effect of toxin absorption and enhanced mass transport, the motor sponges can achieve the rapid neutralization of membrane-damaging toxins. Both the cell membrane camouflage and the acoustic propulsion make the MNMs be more bio-friendly for the biomedical applications.

RBC-based motors driven by the chemical reactions have also been explored. An RBC membrane-modified Janus magnesium particle has been fabricated to serve as a water-powered cell-mimicking motor [55]. The components, including ALG, Fe_3O_4 NPs, Au NPs, and RBC membrane vesicles, have been asymmetrically coated onto the exposed surface of magnesium microparticles which are partially embedded into a Parafilm. The resulting MNMs can self-propel efficiently in both water and various biological media. These RBC-based water-powered MNMs display the impressive detoxification capability for both protein toxins and nerve agent simulants.

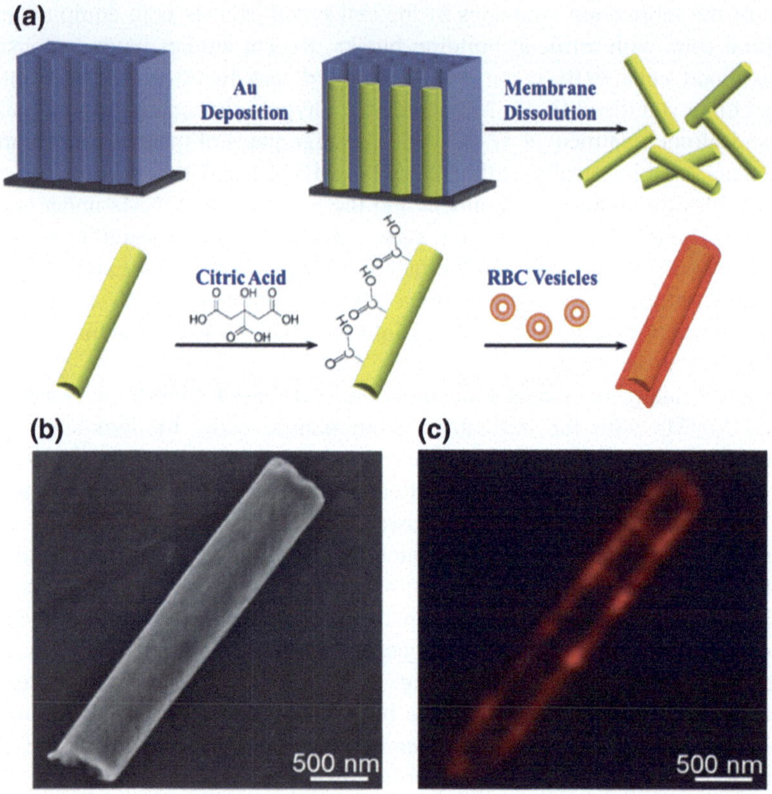

Fig. 11.7 Schematic illustration of the preparation process (**a**), SEM image (**b**), and fluorescent image (**c**) of motor sponges that are synthesized through fusing RBC membranes onto the gold nanowire motors. RBC membranes were stained with Rhodamine B. Reproduced from Ref. [54] by permission of John Wiley & Sons Ltd.

11.4 On-Demand Motion of Self-assembled MNMs

The regulation of the movement of self-assembled MNMs is essential to fulfill the requirement of various applications and complex tasks such as efficiently transporting cargos or separating targeted objects. Advanced control over the motion speed, direction, and on/off states is significantly crucial for guiding these controlled assembled motors to swim in a predetermined manner.

The speed of catalytic MNMs can be modulated through controlling the fuel concentrations, the medium temperatures, and the catalyst activities. The bubble-propelled MNMs can accelerate with the increase of the fuel concentration under certain conditions. Prof. He and his coworkers found that the velocity of LbL-assembled nanorockets increases over 14 times with an increase of H_2O_2 concentration from 1 to 15%, which is accompanied with the increase of the release

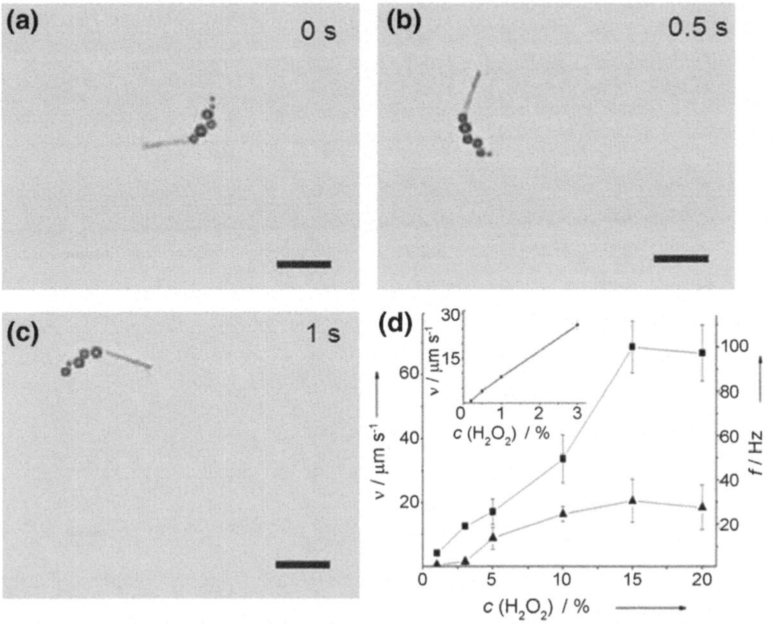

Fig. 11.8 **a–c** Time-lapse images showing the movement of a nanorocket in 15% H_2O_2 at 22 °C. Scale bar: 10 μm. **d** The dependence of the nanorocket speed (*filled triangle*) and the bubble frequency (*filled square*) on H_2O_2 concentration at 22 °C. Inset: The dependence of the nanorocket speed on H_2O_2 concentration at 37 °C. Reproduced from Ref. [41] by permission of John Wiley & Sons Ltd.

frequency of oxygen bubbles from 2 to 30 Hz (Fig. 11.8) [41]. The increase of the reactant concentration accelerates the reaction rate and the production of oxygen, which exerts the increased driven force on the MNMs and results in the faster motor movement. Moreover, as shown in Fig. 11.8d, raising the solution temperature is also an efficient way to accelerate the nanorockets, which should be ascribed to the increase of the rate of mass transport and catalytically chemical reactions.

The motor movement can be navigated through introducing magnetic components such as Fe_3O_4 NPs and Ni layers into MNMs. Prof. He and his coworkers have integrated negatively charged Fe_3O_4 NPs into the multilayer walls of the capsule MNMs *via* the electrostatic interactions before assembling the Pt NPs by μCP [30]. These Fe_3O_4 NP-modified capsule MNMs can swim toward the direction of an external magnetic field. Fe_3O_4 NPs have also been introduced into the polymer nanorockets in the LbL-assembled process for navigation [41]. The as-assembled nanorockets can move toward the HeLa cells in a targeted manner under the remote guide by a magnetic field (Fig. 11.9). In addition, the nanorockets are able to adhere to and even pierce the cancer cells. The magnetic navigation can also be achieved by coating the MNMs with the Ni layer [37].

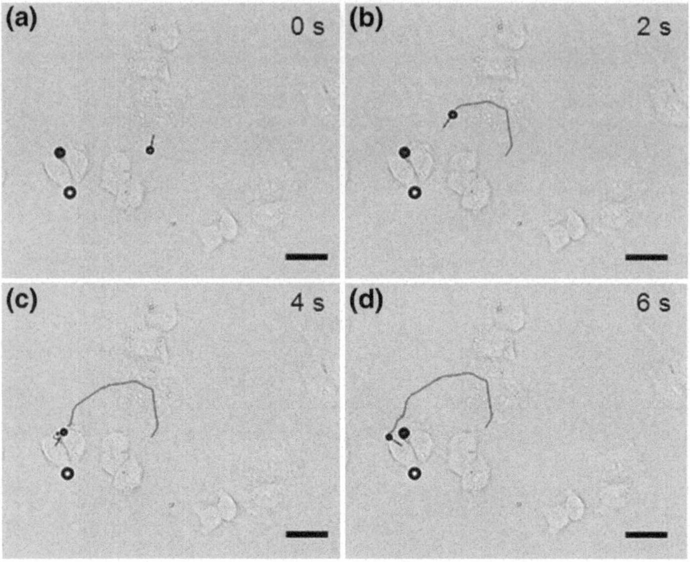

Fig. 11.9 Time-lapse images showing that a nanorocket modified with Fe_3O_4 NPs was magnetically guided toward a HeLa cell. Scale bar: 20 μm. Reproduced from Ref. [41] by permission of John Wiley & Sons Ltd.

Controlling the launch and stop of the MNM motions effectively and remotely could improve the applications of MNMs with on-demand loading, transport, and release. Prof. He and his coworkers realized the "on/off" motion control of PEM nanorockets triggered by the NIR laser (Fig. 11.10) [58]. The Au nanoshell is employed as the functional component due to its plasmon resonance absorption in the NIR region and capability to dissipate the absorbed photons into thermal energy by the so-called photothermal effect. First of all, catalytic Pt NPs are assembled on the inner surface of the $(PSS/PAH)_{20}$ tubes, and then citrate-stabilized Au NPs with negative charges are absorbed onto the outer surface of the nanotubes due to the electrostatic attractions. Finally, the Au NPs are transformed into Au nanoshells by the addition of NH_2OH and $HAuCl_4$. The motion behavior of the rockets is observed and recorded by using an optical microscope. Due to the low fuel decomposition rate and the poor reactant diffusion, the nanorockets remain immobile in the 0.1% (v/v) H_2O_2 solution. When projecting a focused NIR laser at 780 nm onto the selected motor, the movement onset is triggered with a maximum speed of ca. 62 μm/s. The theoretical modeling reveals the temperature increase derived from the photothermal effect under NIR irradiation. The raised temperature in the vicinity of the nanorockets accelerates the catalytic decomposition reaction and the mass transport, as well as the release of the generated bubbles from surface.

In the same way, the motion of bubble-propelled capsule MNMs can be modulated by employing the photothermal effect of Pt layers [36]. For the on-demand control over the motion states, the Cr and Pt layers are integrated onto the Janus

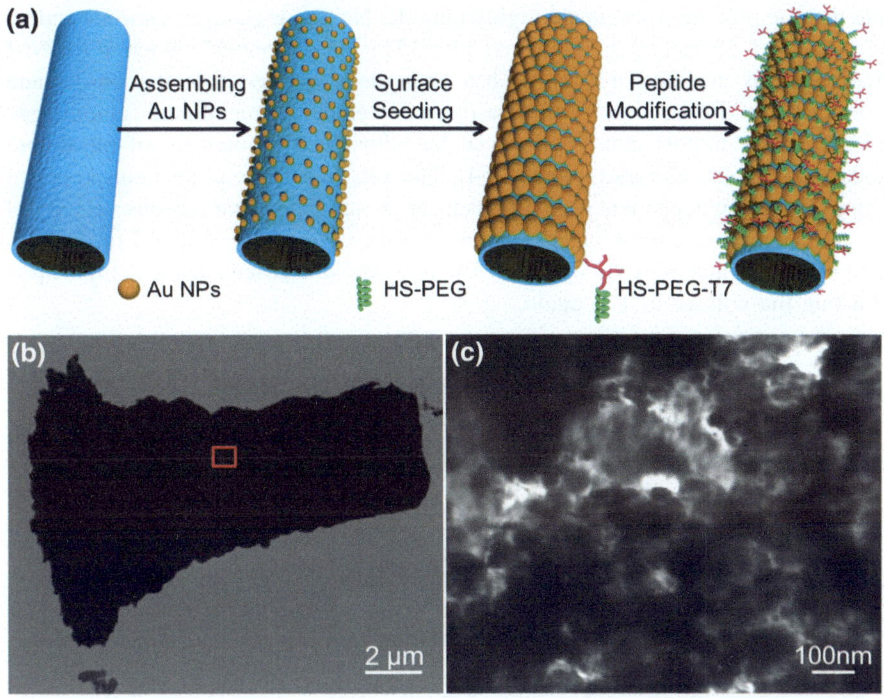

Fig. 11.10 **a** Fabrication process, **b** TEM image, and **c** enlarged TEM image of the Pt NP-modified nanorockets covered with a thin Au nanoshell and a mixed monolayer. Reprinted with permission from Ref. [58] Copyright 2014 American Chemical Society

$(PSS/PAH)_5$ capsules in sequence by sputter coating. The Cr layer serves as an adhesive layer to realize the firm bond between the polyelectrolyte walls and the catalytic Pt layer. By irradiating using a focused NIR laser with the wavelength of 808 nm at room temperature, the motion onset of a capsule MNM can be activated within 0.3 s at 0.1% H_2O_2, showing a maximum speed of ca. 220 μm/s and the travel duration of 17 s. The introduction of the functional building blocks with the photothermal effect has offered a convenient way to control the motion states of catalytic MNMs in a remote manner.

Unlike the catalytic propulsion, when the MNMs are powered by the external physical fields, the energy suppliers could also serve as the motion rulers and provide a rapid and reversible control over the motion states of MNMs. He's group has demonstrated the "on/off" motion control of gold nanoshell-functionalized rockets through the "on/off" switch of the NIR illumination just recently (Fig. 11.11) [47]. The movement of the light-propelled rockets was decelerated and stopped in the several seconds when the NIR irradiation is switched off, and resumed with the ultrafast speeds upon switching the NIR laser on. It is worth to note that the polymer rockets are highly durable. No major change in speeds and no destruction of the rockets are observed after 30 cycles of "on/off" switch of the NIR irradiation. In the similar way, it has been demonstrated that the on-demand acceleration and deceleration of ultrasound-powered cell membrane-camouflaged gold nanowire motors can be achieved by adjusting the intensity of ultrasound fields in a simple and rapid manner [54]. The velocity of the ultrasound-propelled motors is usually dependent on the transducer power. The instantaneous velocity of the cell membrane-camouflaged nanomotors increases from 4 to 13 μm/s upon increasing the transducer voltage from 1 to 3 V, and decreased back to 4 μm/s upon adjusting the voltage to 1 V again.

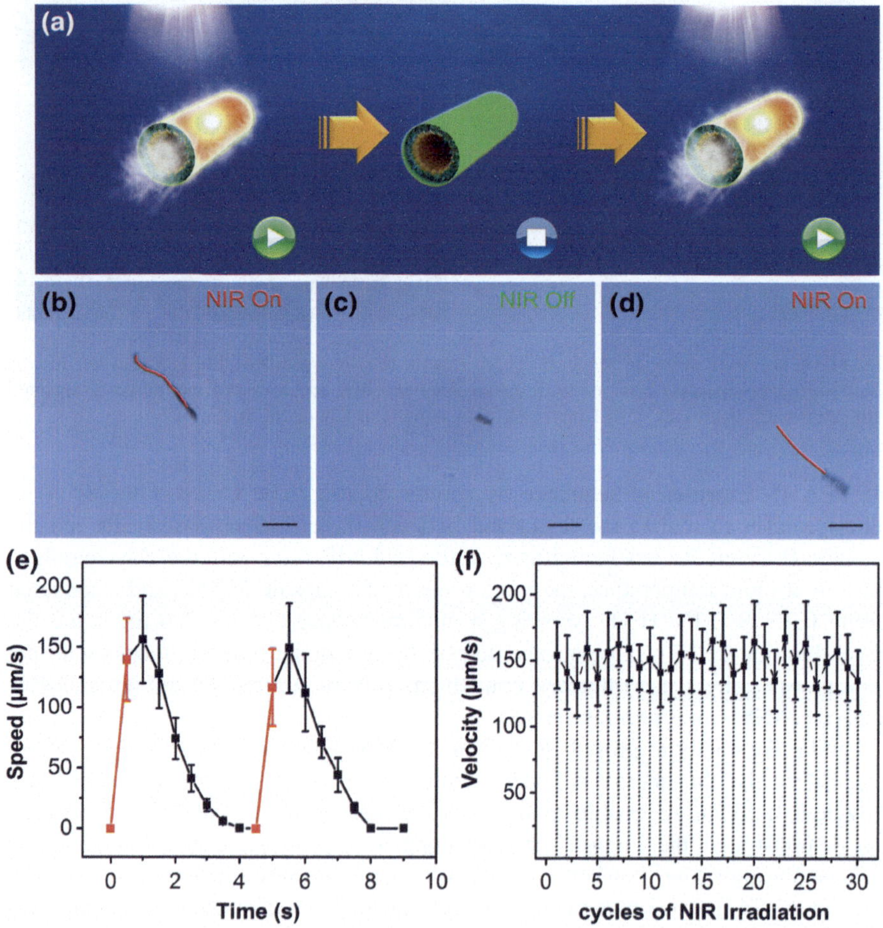

Fig. 11.11 NIR-modulated "on/off" motion of a rocket with an Au nanoshell inside: **a** schematic illustration; **b–d** time-lapse images (Scale bar: 20 μm); E) speed change in two "on/off" cycles; **f** Velocity of the rockets under the NIR irradiation which is switched on/off for 30 cycles. Reproduced from Ref. [47] by permission of John Wiley & Sons Ltd.

11.5 Biomedical Applications of Self-assembled MNMs

Comparing with common functional micro-/nanoparticles, MNMs overcome the Brownian motion and possess the advanced feature of the autonomous movement, which could allow the applications involving mixing, detection, transport, and separation to be achieved with higher efficiencies. One of the very important advantages of controllable assembly methods is that functional components can be easily incorporated into the assemblies without significantly changing their structures and properties, which makes the controlled assembled MNMs be excellent candidates as smart cargos for drug delivery in the biomedicine fields. Moreover, when the components with the navigation functions are introduced, self-propelled MNMs are able to achieve advanced tasks in a targeted manner. The studies on the applications of self-assembled MNMs in drug delivery, photothermal therapy, and detoxification have been done in the recent years.

The hollow structures and the multilayer walls of the LbL-assembled capsules and nanotubes not only are able to serve as versatile frameworks for the fabrication of multifunctional MNMs, but also provide the capability for drug loading and transportation. Through combining the features such as self-propulsion, targeted recognition, navigated movement, and drug encapsulation, the LbL-assembled MNMs can work as the smart drug carriers. The biohybrid Janus $(PSS/PAH)_5$ capsule MNMs can be fabricated through partially coating the capsules with Cr, Ni, and Au layers and immobilizing catalase onto the surface of the Au coating [37]. The resulting capsule MNMs maintain the response ability of the permeability of the microcapsule walls very well. In detail, the capsule walls of the MNMs can still be switched from a closed state to a permeable one through adding ethanol, which has been demonstrated by the encapsulation and release of fluorescein isothiocyanate-dextrans with different molecular weights, investigated by the confocal laser scanning microscopy. Due to the permeability response, the DOX could also be loaded efficiently inside the capsule MNMs. The MNMs can move toward the targeted HeLa cells guided by an external field, and then the encapsulated DOX can be released around the cancer cells through destroying the capsule structures by using the NIR irradiation in a remotely controlled manner (Fig. 11.12). The NIR illumination-induced drug release is attributed to the heat-induced rupture of the hollow capsules derived from the photothermal effect of the Au layer. As shown by this contribution, LbL assembly has provided a simple, powerful way to achieve the multifunctionality of MNMs and to access self-propelled carriers for smart drug delivery and release.

The LbL-assembled nanorockets with Pt NPs on the inside surface have also been used as drug carriers through assembling Fe_3O_4 NPs and DOX in the multilayer walls [41]. As mentioned above, the guided nanorockets can move toward and then bound to the targeted HeLa cells. Some nanorockets can even partially penetrate into the cancer cells. The integrated Fe_3O_4 NPs is sensitive to the ultrasound field. Upon exposed to the ultrasonic treatment for several seconds, the nanorockets begin to collapse, and the encapsulated DOX is released at the same

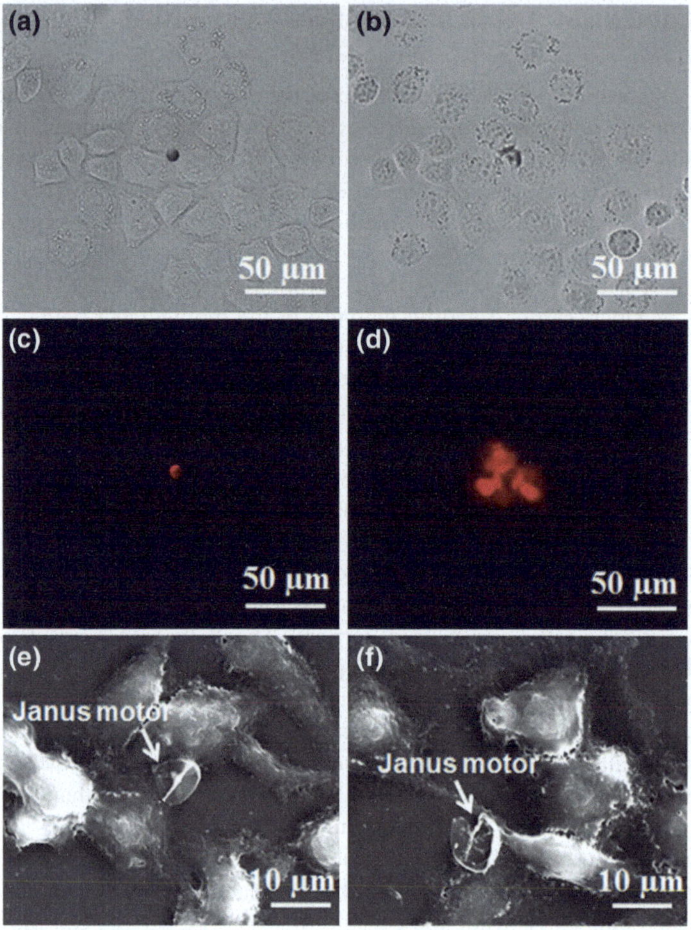

Fig. 11.12 Differential interference contrast (DIC) (**a**, **b**), fluorescence images (**c**, **d**), and SEM images (**e**, **f**) of Janus capsule MNMs attaching to HeLa cells before (**a**, **c**, **e**) and after (**b**, **d**, **f**) the NIR illumination, showing the release of DOX with the red fluorescence. Reprinted with permission from Ref. [37]. Copyright 2014 American Chemical Society

time, which leads to the modest apoptosis of HeLa cells as shown by the fluorescent images. The hydrogel-filled nanorockets can also perform drug delivery and release [46]. As previously stated, the mixed hydrogel containing catalase, Au NPs, DOX, and gelatin has been filled into the LbL-assembled nanotubes to construct the biodegradable rockets. When exposing the nanorockets to the NIR illumination, the dramatic temperature increase derived from the photothermal effect can lead to the gel−sol transformation of the mixed hydrogel and eventually the release of the accommodated DOX.

LbL-assembled MNMs with Au nanoshells could also serve as self-propelled, targeted agents for the photothermal therapy [47, 58]. Due to the Au–S chemistry and the photothermal effect, the NIR-responsive Au nanoshells show the multiple functions including immobilizing the cell-targeted molecules that have a thiol group, controlling the motion "on/off" state of bubble-propelled nanorockets, driving the nanorockets directly, and serving as a photothermal agent. The Au nanoshells on the outer surface of the nanorockets powered by the bubble release have been modified by the mixed self-assembled monolayer of thiol-modified poly (ethylene glycol) (HS-PEG) and thiol-modified peptide (HS-CH_2COO-$(CH_2CH_2O)_7$-HRPYIAH, HS-PEG-T7) (Fig. 11.10) [58]. Due to the specific recognition of HS-PEG-T7 with some tumor cells and the antifouling effect of HS-PEG, the introduction of the peptide and HS-PEG into the nanorockets promotes the binding with HeLa cells, and evades the nonspecific recognition with normal cells. When irradiating the nanorocket that is bound to HeLa cells with a NIR laser at the power of 13 mW/μm^2, the dramatic increase of the temperature induced by the photothermal effect leads to the apoptosis of the cancer cells (Fig. 11.13). Analogously, the NIR-driven tubular MNMs with the Au nanoshells inside can also behave as self-propelled agents for the photothermal therapy [47].

Cell membrane-camouflaged MNMs have shown the capability for effectively absorbing and neutralizing toxins recently [54, 55]. The high-coverage RBC membranes on the MNMs not only share the biological functions of natural RBCs against the protein fouling to endow the MNMs with efficient propulsion in

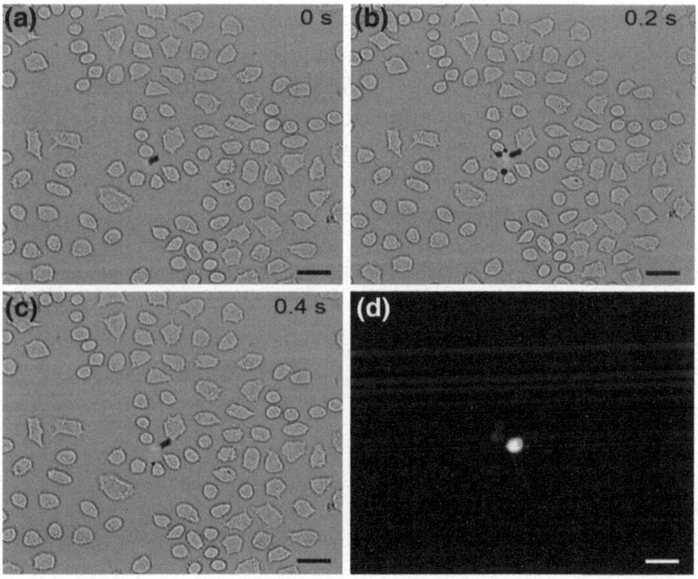

Fig. 11.13 a–c Time-lapse images of the motion of Au shell-covered nanorockets toward HeLa cells promoted by a NIR laser in 0.1% H_2O_2. **d** Fluorescence image indicating the apoptosis of HeLa cells after irradiating the adhesive rocket with a NIR laser at the power of 13 mW/μm^2. Scale bar: 20 μm. Reprinted with permission from Ref. [58] Copyright 2014 American Chemical Society

biological media but also are able to serve as decoys to absorb and neutralize various toxins. By camouflaging themselves to be like natural RBCs, the ultrasound-propelled nanomotors are able to absorb melittin, a cell membrane-damaging toxin [54]. The ultrasound propulsion promotes the interaction of the MNMs with melittin. RBC membrane-coated water-powered Janus MNMs also show the applications in detoxification to remove the α-toxin or the methyl paraoxon from albumin-rich solution [55]. The bubble propulsion derived from the water-based chemical reactions enhance the detoxification efficiency. The detoxification efficiency can be evaluated by using the relative hemolysis which decreases from 78 to 9% when the self-propelled MNMs with a concentration of 0.8 mg/mL are added.

11.6 Summary

In this chapter, we reviewed the recent progress on the controlled fabrication, the regulated motion, and biomedical applications of MNMs based on the controlled molecular assembly (especially layer-by-layer assembly) assisted by several "top-down" technologies. Due to the advantages of self-assembly, diverse functional components, such as Pt NPs, enzymes, biomacromolecules, magnetic metal NPs (or layers), and Au nanoshells, could behave as building blocks and be introduced into the assemblies. Such integration achieves the transformation from the self-assembled structures into the MNMs with the advanced features including self-population, biocompatibility, biodegradability, controlled motion, and loading ability. Some proof-of-concept studies have demonstrated the potential applications of the self-assembled MNMs in biomedical fields, such as targeted delivery and remotely controlled release of drugs as well as the photothermal therapy of cancer cells. Controlled molecular assembly has provided a great opportunity for the biomimic design of MNMs toward the future applications in biomedical fields.

Acknowledgements This work was supported by the National Nature Science Foundation of China (21674029, 21573053, and 21603047).

References

1. Li JB, He Q, Yan XH (2011) Molecular assembly of biomimetic systems. Wiley-VCH, Weinheim
2. Spudich JA (2011) Molecular motors, beauty in complexity. Science 331(6021):1143–1144. doi:10.1126/science.1203978
3. van den Heuvel MGL, Dekker C (2007) Motor proteins at work for nanotechnology. Science 317(5836):333–336. doi:10.1126/science.1139570
4. Badjić JD, Balzani V, Credi A, Silvi S, Stoddart JF (2004) A molecular elevator. Science 303 (5665):1845–1849. doi:10.1126/science.1094791

5. Fletcher SP, Dumur F, Pollard MM, Feringa BL (2005) A reversible, unidirectional molecular rotary motor driven by chemical energy. Science 310(5745):80–82. doi:10.1126/science.1117090
6. Ozin GA, Manners I, Fournier-Bidoz S, Arsenault A (2005) Dream nanomachines. Adv Mater 17(24):3011–3018. doi:10.1002/adma.200501767
7. Sengupta S, Ibele ME, Sen A (2012) Fantastic voyage: designing self-powered nanorobots. Angew Chem Int Ed 51(34):8434–8445. doi:10.1002/anie.201202044
8. Wang J (2013) Nanomachines: fundamentals and applications. Wiley-VCH, Weinheim
9. Sánchez S, Soler L, Katuri J (2015) Chemically powered micro- and nanomotors. Angew Chem Int Ed 54(5):1414–1444. doi:10.1002/anie.201406096
10. Wang J, Gao W (2012) Nano/Microscale motors: biomedical opportunities and challenges. ACS Nano 6(7):5745–5751. doi:10.1021/nn3028997
11. Patra D, Sengupta S, Duan W, Zhang H, Pavlick R, Sen A (2013) Intelligent, self-powered, drug delivery systems. Nanoscale 5(4):1273–1283. doi:10.1039/C2NR32600K
12. Guix M, Mayorga-Martinez CC, Merkoçi A (2014) Nano/micromotors in (bio)chemical science applications. Chem Rev 114(12):6285–6322. doi:10.1021/cr400273r
13. Abdelmohsen LKEA, Peng F, Tu YF, Wilson DA (2014) Micro- and nano-motors for biomedical applications. J Mater Chem B 2(17):2395–2408. doi:10.1039/c3tb21451f
14. Gao W, Wang J (2014) Synthetic micro/nanomotors in drug delivery. Nanoscale 6 (18):10486–10494. doi:10.1039/c4nr03124e
15. Balasubramanian S, Kagan D, Hu C-MJ, Campuzano S, Lobo-Castañon MJ, Lim N, Kang DY, Zimmerman M, Zhang LF, Wang J (2011) Micromachine-enabled capture and isolation of cancer cells in complex media. Angew Chem Int Ed 50(18):4161–4164. doi:10.1002/anie.201100115
16. Chalupniak A, Morales-Narvaez E, Merkoçi A (2015) Micro and nanomotors in diagnostics. Adv Drug Deliv Rev 95:104–116. doi:10.1016/j.addr.2015.09.004
17. Soler L, Magdanz V, Fomin VM, Sánchez S, Schmidt OG (2013) Self-propelled micromotors for cleaning polluted water. ACS Nano 7(11):9611–9620. doi:10.1021/nn405075d
18. Gao W, Wang J (2014) The environmental impact of micro/nanomachines. A review. ACS Nano 8(4):3170–3180. doi:10.1021/nn500077a
19. Moo JGS, Pumera M (2015) Chemical energy powered nano/micro/macromotors and the environment. Chem Eur J 21(1):58–72. doi:10.1002/chem.201405011
20. Ismagilov RF, Schwartz A, Bowden N, Whitesides GM (2002) Autonomous movement and self-assembly. Angew Chem Int Ed 41(4):652–654. doi:10.1002/1521-3773(20020215)41:4<652:AID-ANIE652>3.0.CO;2-U
21. Paxton WF, Kistler KC, Olmeda CC, Sen A, St. Angelo SK, Cao Y, Mallouk TE, Lammert PE, Crespi VH (2004) Catalytic nanomotors: autonomous movement of striped nanorods. J Am Chem Soc 126(41):13424–13431. doi:10.1021/ja047697z
22. Fournier-Bidoz S, Arsenault AC, Manners I, Ozin GA (2005) Synthetic self-propelled nanorotors. Chem Commun 4:441–443. doi:10.1039/B414896G
23. Wang W, Duan WT, Ahmed S, Mallouk TE, Sen A (2013) Small power: autonomous nano- and micromotors propelled by self-generated gradients. Nano Today 8(5):531–554. doi:10.1016/j.nantod.2013.08.009
24. Wang H, Pumera M (2015) Fabrication of micro/nanoscale motors. Chem Rev 115(16):8704–8735. doi:10.1021/acs.chemrev.5600047
25. Lin XK, Wu ZG, Wu YJ, Xuan MJ, He Q (2016) Self-propelled micro-/nanomotors based on controlled assembled architectures. Adv Mater 28(6):1060–1072. doi:10.1002/adma.201502583
26. Wu ZG, Lin XK, Si TY, He Q (2016) Recent progress on bioinspired self-propelled micro/nanomotors via controlled molecular self-assembly. Small 12(23):3080–3093. doi:10.1002/smll.201503969
27. Tu YF, Peng F, Adawy A, Men YJ, Abdelmohsen LKEA, Wilson DA (2016) Mimicking the cell: bio-Inspired functions of supramolecular assemblies. Chem Rev 116(4):2023–2078. doi:10.1021/acs.chemrev.5b00344

28. Dong B, Zhou T, Zhang H, Li CY (2013) Directed self-assembly of nanoparticles for nanomotors. ACS Nano 7(6):5192–5198. doi:10.1021/nn400925q
29. Wilson DA, Nolte RJM, van Hest JCM (2012) Autonomous movement of platinum-loaded stomatocytes. Nat Chem 4(4):268–274. doi:10.1038/nchem.1281
30. Wu YJ, Wu ZG, Lin XK, He Q, Li JB (2012) Autonomous movement of controllable assembled janus capsule motors. ACS Nano 6(12):10910–10916. doi:10.1021/nn304335x
31. Whitesides GM, Grzybowski B (2002) Self-assembly at all scales. Science 295(5564): 2418–2421. doi:10.1126/science.1070821
32. Decher G (1997) Fuzzy nanoassemblies: toward layered polymeric multicomposites. Science 277(5330):1232–1237. doi:10.1126/science.277.5330.1232
33. Donath E, Sukhorukov GB, Caruso F, Davis SA, Möhwald H (1998) Novel hollow polymer shells by colloid-templated assembly of polyelectrolytes. Angew Chem Int Ed 37(16):2201–2205. doi:10.1002/(SICI)1521-3773(19980904)37:16<2201:AID-ANIE2201>3.0.CO;2-E
34. Ai SF, Lu G, He Q, Li JB (2003) Highly flexible polyelectrolyte nanotubes. J Am Chem Soc 125(37):11140–11141. doi:10.1021/ja0356378
35. Richardson JJ, Björnmalm M, Caruso F (2015) Technology-driven layer-by-layer assembly of nanofilms. Science 348(6233):aaa2491. doi:10.1126/science.aaa2491
36. Wu YJ, Si TY, Lin XK, He Q (2015) Near infrared-modulated propulsion of catalytic Janus polymer multilayer capsule motors. Chem Commun 51(3):511–514. doi:10.1039/c4cc07182d
37. Wu YJ, Lin XK, Wu ZG, Möhwald H, He Q (2014) Self-propelled polymer multilayer janus capsules for effective drug delivery and light-triggered release. ACS Appl Mater Interfaces 6 (13):10476–10481. doi:10.1021/am502458h
38. del Mercato LL, Carraro M, Zizzari A, Bianco M, Miglietta R, Arima V, Viola I, Nobile C, Soraru A, Vilona D, Gigli G, Bonchio M, Rinaldi R (2014) Catalytic self-propulsion of supramolecular capsules powered by polyoxometalate cargos. Chem Eur J 20(35): 10910–10914. doi:10.1002/chem.201403171
39. Xuan MJ, Shao JX, Lin XK, Dai LR, He Q (2015) Light-activated janus self-assembled capsule micromotors. Colloids Surf A 482:92–97. doi:10.1016/j.colsurfa.2015.04.032
40. Wu YJ, Si TY, Shao JX, Wu ZG, He Q (2016) Near-infrared light-driven Janus capsule motors: fabrication, propulsion, and simulation. Nano Res 9(12):3747–3756. doi:10.1007/s12274-016-1245-0
41. Wu ZG, Wu YJ, He WP, Lin XK, Sun JM, He Q (2013) Self-propelled polymer-based multilayer nanorockets for transportation and drug release. Angew Chem Int Ed 52(27): 7000–7003. doi:10.1002/anie.201301643
42. Lin ZH, Wu ZG, Lin XK, He Q (2016) Catalytic polymer multilayer shell motors for separation of organics. Chem Eur J 22(5):1587–1591. doi:10.1002/chem.201503892
43. Gai MY, Frueh J, Hu N, Si TY, Sukhorukov GB, He Q (2016) Self-propelled two dimensional polymer multilayer plate micromotors. Phys Chem Chem Phys 18(5):3397–3401. doi:10.1039/c5cp07697h
44. Gai MY, Frueh J, Si TY, Hu N, Sukhorukov GB, He Q (2016) The collision phenomena of Janus polymer micro-plate motors propelled by oscillating micro-bubbles. Colloids Surf A 510:113–121. doi:10.1016/j.colsurfa.2016.04.042
45. Li JX, Rozen I, Wang J (2016) Rocket science at the nanoscale. ACS Nano 10(6):5619–5634. doi:10.1021/acsnano.6b02518
46. Wu ZG, Lin XK, Zou X, Sun JM, He Q (2015) Biodegradable protein-based rockets for drug transportation and light-triggered release. ACS Appl Mater Interfaces 7(1):250–255. doi:10.1021/am507680u
47. Wu ZG, Si TY, Gao W, Lin XK, Wang J, He Q (2016) Superfast near-infrared light-driven polymer multilayer rockets. Small 12(5):577–582. doi:10.1002/smll.201502605
48. Xuan MJ, Shao JX, Dai LR, He Q, Li JB (2015) Macrophage cell membrane camouflaged mesoporous silica nanocapsules for in vivo cancer therapy. Adv Healthcare Mater 4 (11):1645–1652. doi:10.1002/adhm.201500129

49. Gao CY, Lin ZH, Jurado-Sánchez B, Lin XK, Wu ZG, He Q (2016) Stem cell membrane-coated nanogels for highly efficient in vivo tumor targeted drug delivery. Small 12(30):4056–4062. doi:10.1002/smll.201600624
50. Gao CY, Wu ZG, Lin ZH, Lin XK, He Q (2016) Polymeric capsule-cushioned leukocyte cell membrane vesicles as a biomimetic delivery platform. Nanoscale 8(6):3548–3554. doi:10.1039/c5nr08407e
51. He WP, Frueh J, Wu ZW, He Q (2016) Leucocyte membrane-coated janus microcapsules for enhanced photothermal cancer treatment. Langmuir 32(15):3637–3644. doi:10.1021/acs.langmuir.5b04762
52. Wu ZG, Li TL, Li JX, Gao W, Xu TL, Christianson C, Gao WW, Galarnyk M, He Q, Zhang LF, Wang J (2014) Turning erythrocytes into functional micromotors. ACS Nano 8 (12):12041–12048. doi:10.1021/nn506200x
53. Wu ZG, de Avila BEF, Martin A, Christianson C, Gao WW, Thamphiwatana SK, Escarpa A, He Q, Zhang LF, Wang J (2015) RBC micromotors carrying multiple cargos towards potential theranostic applications. Nanoscale 7(32):13680–13686. doi:10.1039/c5nr03730a
54. Wu ZG, Li TL, Gao W, Xu TL, Jurado-Sánchez B, Li JX, Gao WW, He Q, Zhang LF, Wang J (2015) Cell-membrane-coated synthetic nanomotors for effective biodetoxification. Adv Funct Mater 25(25):3881–3887
55. Wu ZG, Li JX, de Ávila BEF, Li TL, Gao WW, He Q, Zhang LF, Wang J (2015) Water-powered cell-mimicking janus micromotor. Adv Funct Mater 25(48):7497–7501. doi:10.1002/adfm.201503441
56. Xuan MJ, Shao JX, Lin XK, Dai LR, He Q (2014) Self-propelled janus mesoporous silica nanomotors with sub-100 nm diameters for drug encapsulation and delivery. ChemPhysChem 15(11):2255–2260. doi:10.1002/cphc.201402111
57. Wang W, Castro LA, Hoyos M, Mallouk TE (2012) Autonomous motion of metallic microrods propelled by ultrasound. ACS Nano 6(7):6122–6132. doi:10.1021/nn301312z
58. Wu ZG, Lin XK, Wu YJ, Si TY, Sun JM, He Q (2014) Near-infrared light-triggered "on/off" motion of polymer multilayer rockets. ACS Nano 8(6):6097–6105. doi:10.1021/nn501407r

Part V
Hierarchical Dendrimer, Polyoxometalates Complexes and Inorganic-organic Hybrid Systems

Chapter 12
Functional Dendrimer-Based Vectors for Gene Delivery Applications

Lingdan Kong and Xiangyang Shi

Abstract Poly (amidoamine) (PAMAM) dendrimers are a class of highly branched, monodispersed, synthetic macromolecules with abundant terminal functional groups, and have significant advantages over other cationic polymers as gene delivery vectors due to their well-defined structure, the possibility of facile surface modification, and capacity of carrying large gene segments. The surface amine groups of dendrimers can be conjugated with functional molecules (e.g., hydrophobic moieties, β-cyclodextrin, polyethylene glycol, etc.), and targeting ligands (e.g., folic acid, arginine-glycine-aspartic peptide), while the unique interior of dendrimers affords their uses to form dendrimer-entrapped gold nanoparticles. These modifications render the dendrimer-based vectors with an ability for targeted and enhanced gene delivery, including pDNA and siRNA delivery. In this chapter, we review some recent advances made in multifunctional poly(amidoamine) dendrimer-based nanoparticles for gene delivery applications.

Keywords PAMAM dendrimers · Surface modification · Gold nanoparticles · Gene delivery · Gene silencing

12.1 Introduction

Gene therapy has been considered as a promising approach for cancer therapy due to the fact that exogenous therapeutic genes are able to be delivered to target cells to correct or compensate the genetic defect and abnormality [1–3]. The common

L. Kong · X. Shi (✉)
State Key Laboratory for Modification of Chemical Fibers and Polymer Materials,
College of Chemistry, Chemical Engineering and Biotechnology,
Donghua University, Shanghai 201620, People's Republic of China
e-mail: xshi@dhu.edu.cn

X. Shi
CQM-Centro de Química da Madeira, Universidade da Madeira,
Campus da Penteada, 9000-390 Funchal, Portugal

J. Li (ed.), *Supramolecular Chemistry of Biomimetic Systems*,
DOI 10.1007/978-981-10-6059-5_12

method of gene therapy including plasmid DNA (pDNA) therapy and small interfering RNA (siRNA) therapy have been intensively studied in treating serious diseases such as cancer and genetic disorders [4, 5]. The foreign gene segments or siRNA delivered into cells could express normal proteins and repair cell defects [4, 6], or destroy the function of cognate mRNA and knockdown the expression of disease-causing proteins, respectively [7, 8]. Although the pDNA and siRNA have been used in gene therapy, the easy enzymatic degradation and limited cell membrane permeability of naked pDNA and siRNA have seriously restricted their in vivo therapeutic applications [9–11]. The success of gene therapy is largely dependent on the development of an ideal delivery system that can selectively and efficiently deliver genetic materials to target cells without causing any associated pathogenic effects [12, 13]. Therefore, the main issue of gene therapy is not the cellular expression of an exogenous gene itself, but the development of safe and efficient gene delivery systems [14].

It is well known that the gene delivery systems include both viral and nonviral vector systems, and viral vectors usually exhibit high transfection efficiency [15]. However, the safety problems raised by the toxicity, oncogenicity, and immunogenicity of the viral vectors greatly hamper their routine use in both basic research laboratories and clinical settings [16–19]. Hence, nonviral delivery systems have continuously received considerable attention because they can be structurally varied, are relatively safe, and have an ability to carry large and diverse genetic materials into cells [14]. Cationic polymers are the commonly used nonviral gene vectors that have been extensively investigated due to their synthetic controllability and multivalent-functionalized surface amine groups, as well as their ability to compact nucleic acid [14, 20–22]. Recent advances in pDNA and siRNA delivery systems including cationic poly-L-lysine, polyethylenimine, diethylaminoethyl-dextran, and chitosan have been proven to be able to transfect genes to different cell lines [21, 23, 24].

Poly(amidoamine) (PAMAM) dendrimers are a class of high branched synthetic macromolecules with narrow molecular weight (Mw) distribution and easily functionalized amine-terminated surface [25–27]. These structural features make PAMAM dendrimer a promising candidate as pDNA or siRNA vector, especially the abundant surface amine groups render PAMAM dendrimers with a strong capacity in compressing pDNA or siRNA to form polyplexes [28–31]. However, some obstacles such as transient gene expression, low transfection efficiency, and high cytotoxicity limit their practical applications [23]. Much effort has been devoted to enhance their gene delivery efficiency and specificity while simultaneously decreasing their cytotoxicity. It has been reported that partial PEGylation [32–35], acetylation [36], alkylation [37, 38], and peptide-conjugation [39, 40] of PAMAM dendrimers can greatly enhance their gene delivery efficiency or specificity, and reduce their cytotoxicity. Therefore, through appropriate surface functionalization of dendrimers, a highly efficient and less toxic nonviral gene delivery vector may be developed for various biomedical applications [41–43]. The aim of this chapter is to give an overview of the recent advances related to the design of functional dendrimer-based vector systems for gene delivery applications.

12.2 Alkylated PAMAM Dendrimers for Gene Delivery

In order to improve the cellular uptake of materials, hydrophobic modification on the surface of dendrimers appears to be crucial, which helps cross the lipid bilayer on the cell surface, thus realizing the safe and efficient delivery of gene. In our previous report, we synthesized hydrophobically modified PAMAM dendrimers by partially reacting the periphery dendrimer amines with 1, 2-epoxyhexane or 1, 2-epoxydodecane [44]. The formed hydrophobically modified generation 5 (G5) dendrimers (denoted as $G5.NH_2–C_6$ or $G5.NH_2–C_{12}$) were used to complex two different plasmid DNAs (pDNAs) encoding luciferase (Luc) and enhanced green fluorescent protein (EGFP), respectively, for gene transfection studies.

Cytotoxicity assay data reveal that at a relatively high concentration (above 2000 nM), the cytotoxicity of the vectors follows the order of $G5.NH_2–C_{12}$ > $G5.NH_2–C_6$ > $G5.NH_2$. This is likely due to the fact that at a relatively high concentration, the hydrophobic long alkyl chain of dendrimers is able to strongly interact with the hydrophobic cell membrane, thereby enhancing the hole formation on the cell membranes [45].

Subsequently, both Luc and EGFP gene expression assays were used to investigate the gene transfection efficiency of G5 dendrimers in HeLa cells (a human cervical carcinoma cell line). The results showed that in all cases, the Luc gene transfection efficiency decreased with the N/P ratio. And all dendrimeric vectors possessed much higher gene transfection efficiency at the studied N/P ratios than the control cells without transfection treatment and cells transfected with naked pDNA. Under the N/P ratio of 2.5 and 5, the Luc gene transfection efficiency followed the order of $G5.NH_2–C_{12}$ > $G5.NH_2–C_6$ > $G5.NH_2$. In particular, at an N/P ratio of 1, the Luc expression value of $G5.NH_2$-C_{12} was around four times higher than that of $G5.NH_2$, also much higher than that of the commercial lipofectamine 2000 vector. This suggests that the partial hydrophobic modification of G5 dendrimers enables an enhanced interaction between the vector and cell membrane, thereby significantly enhancing the gene delivery efficiency, in agreement with the literature [37]. However, when compared to the work reported by Santos et al. [37], where the higher gene delivery efficiency was shown by the vectors containing the shortest hydrophobic chains (12 versus 14 or 16 carbon alkyl chains), in this study we show that the gene delivery efficiency of the $G5.NH_2–C_{12}$ dendrimers is much higher than that of the $G5.NH_2–C_6$ dendrimers. The trend was apparently reversed, although the used numbers of carbon alkyl chains were different.

The gene transfection efficiency of the hydrophobically modified G5 dendrimers was further qualitatively assessed by confocal microscopic imaging of the EGFP gene expression in HeLa cells. At the N/P ratio of 1, all vectors have better gene transfection efficiency than at other N/P ratios. These results were consistent with the Luc assay data. Taken together, both quantitative Luc assay and qualitative EGFP expression confirmed the potential to use both $G5.NH_2–C_6$ and $G5.NH_2–C_{12}$ dendrimers as vectors for gene delivery applications. Our results suggest that partial modification of G5 dendrimers with hydrophobic alkyl chains might be an alternative approach to develop dendrimer-based nonviral vectors for enhanced gene delivery applications.

12.3 Au DNEPs for Gene Delivery

Gold nanoparticles (AuNPs) have been identified as a suitable platform for drug/gene delivery due to their unique physicochemical properties, such as size- and shape-dependent optical properties, high surface area to volume ratio, and rich surface chemistry allowing for facile modification with different functionalities [46–52]. Functional AuNPs delivery vehicles, with a positively charged surface, can be used to compress negatively charged genetic materials such as plasmid DNA (pDNA) [42, 53–56] and siRNA [57–61]. It has been reported that lysine dendron-functionalized AuNPs are 28-fold superior to polylysine in reporter gene expression [62]. The super gene transfection performance of the lysine dendron-functionalized AuNPs is likely due to the biomimetic design of the particles that has a size more or less similar to the nucleosome core proteins ($\sim$6 nm) having a large proportion of basic residues (lysine and arginine) that form electrostatic bonding with the phosphate backbone of DNA [63]. In our previous work [42], we designed the use of dendrimer-entrapped Au NPs (Au DENPs) for enhanced gene delivery. The advantages of the Au NP entrapment within the dendrimer interior stem from two aspects: (1) the entrapped AuNPs are able to neutralize some of the dendrimer terminal amines due to the amine stabilization of AuNPs, decreasing the dendrimer cytotoxicity; and (2) the existence of AuNPs helps to reserve the three-dimensional (3-D) spherical shape of dendrimers, thereby significantly improving the DNA or siRNA compaction ability of the dendrimers.

To prove our above hypothesis, amine-terminated G5 PAMAM dendrimers ($G5.NH_2$) were used as templates to synthesize AuNPs with different Au atom/dendrimer molar ratios (25:1, 50:1, 75:1, and 100:1, respectively) [42] (Scheme 12.1). For simplicity, the $G5.NH_2$ dendrimers were denoted as S0, while the Au DENPs with the Au atom/dendrimer molar ratio at 25:1, 50:1, 75:1, and 100:1 were denoted as S25, S50, S75, and S100, respectively, in the naming system.

The gene transfection efficiency of the Au DENP vectors was investigated using firefly Luc gene expression (Fig. 12.1). The gene transfection efficiency is dependent on both the composition of the used Au DENP vectors prepared with different Au atom/dendrimer molar ratios and the selected N/P ratios. The most efficient vector is S25 that has its peak value at an N/P ratio of 2.5:1 for all three different cell lines. For COS-7 cells, the Luc expression reaches 4.27×10^6 RLU/mg protein, while the Luc expression is only 3.42×10^4 RLU/mg protein using S0 as a vector. It appears that at the N/P ratio of 2.5:1, the Luc gene transfection efficiency of the S25 vector is almost 125 times higher than that of the S0 vector. For the other cell lines, at the same N/P ratio, the relative Luc expressed using S25 as a vector is

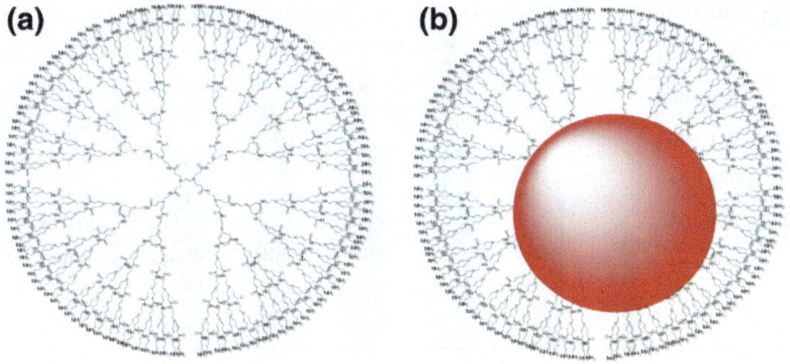

Scheme 12.1 Schematic illustration of the structures of G5.NH_2 dendrimers (**a**) and Au DENPs (**b**) prepared using G5.NH_2 dendrimers as templates. Reprinted from Ref. [42], Copyright 2012, with permission from Elsevier

153 times (for 293T cells) and 358 times (for HeLa cells) higher than that using S0 as a vector, respectively.

It is interesting to note that the gene transfection efficiency of Au DENPs is not increased with the increase of the Au atom/dendrimer molar ratio, or with the size of the AuNPs entrapped within the dendrimer templates. S25 seems to be the best vector in terms of the Luc transfection efficiency. At the N/P ratio of 2.5:1, both S25 and S50 are able to significantly enhance the gene delivery when compared with S0 without AuNPs ($p < 0.01$ or $p < 0.001$). For the Au DENPs prepared with the increased Au atom/dendrimer molar ratios, the formed AuNP cores have a larger size, which requires more dendrimer terminal amines to stabilize the AuNPs, consequently resulting in a lower binding affinity to pDNA and gene transfection efficiency than those of S25.

Surface modification of dendrimers is reported to cause significant changes in their biological properties, including cytotoxicity and the cell membrane permeation ability [64, 65]. Therefore, it is vital to confirm whether the changes of the physical shape and surface primary amine density of Au DENPs could change the transcellular pathways and localization of the vector by intracellular trafficking and colocalization assay (Fig. 12.2). In the fluorescent microscopic images of cells transfected with all Au DENPs/pDNA complexes, lysosomes were labeled by Lysotracker green (green), Cy3-labeled pDNA (red) were used to localize the polyplexes, and DAPI was used to stain the nuclei (blue). We can see that in the merged images that the polyplexes are well colocalized within the lysosomes and they are all around the nuclei after 2 h of transfection, which means that Au DENPs with different compositions follows the same way as G5.NH_2 dendrimers in the intracellular trafficking pathway and fate.

Results clearly show that Au DENPs with an appropriate composition (Au atom/dendrimer molar ratio = 25:1) enable enhanced gene delivery, with a gene transfection efficiency more than 100 times higher than the G5.NH_2 dendrimers without AuNPs entrapped. It is believed that the entrapment of AuNPs within

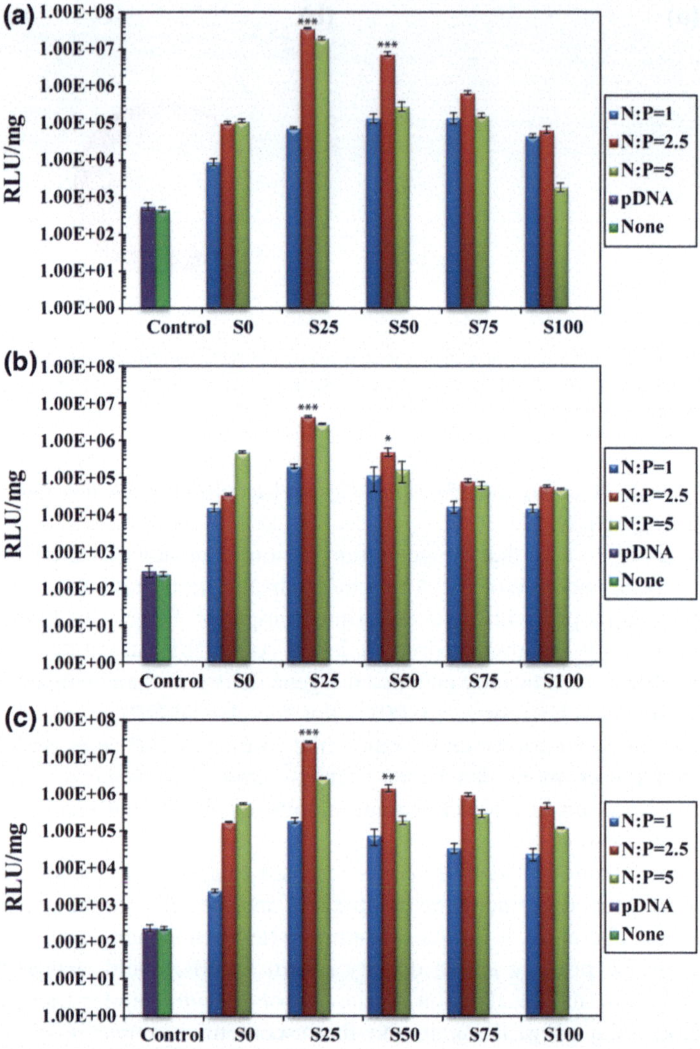

Fig. 12.1 Luciferase gene transfection efficiency of Au DENPs/DNA polyplexes determined in HeLa (**a**), COS-7 (**b**), and 293T (**c**) cells at the N/P ratios of 1:1, 2.5:1, and 5:1, respectively. Transfection was performed at a dose of 1 μg/well of DNA (mean ± SD, $n = 3$). Cells without treatment (None) and cells treated with vector-free pDNA (pDNA) were used as controls. Statistical differences between Au DENPs (S25, S50, S75, and S100, respectively) versus $G5.NH_2$ dendrimers (S0) at an N/P ratio of 2.5:1 was compared and indicated with (*) for $p < 0.05$, (**) for $p < 0.01$, and (***) for $p < 0.001$, respectively. Reprinted from Ref. [42], Copyright 2012, with permission from Elsevier

dendrimer templates helps preserve 3-D spherical shape of dendrimers, enabling high compaction of DNA to form smaller particles, and consequently resulting in enhanced gene delivery. With a lower cytotoxicity of Au DENPs than that of the

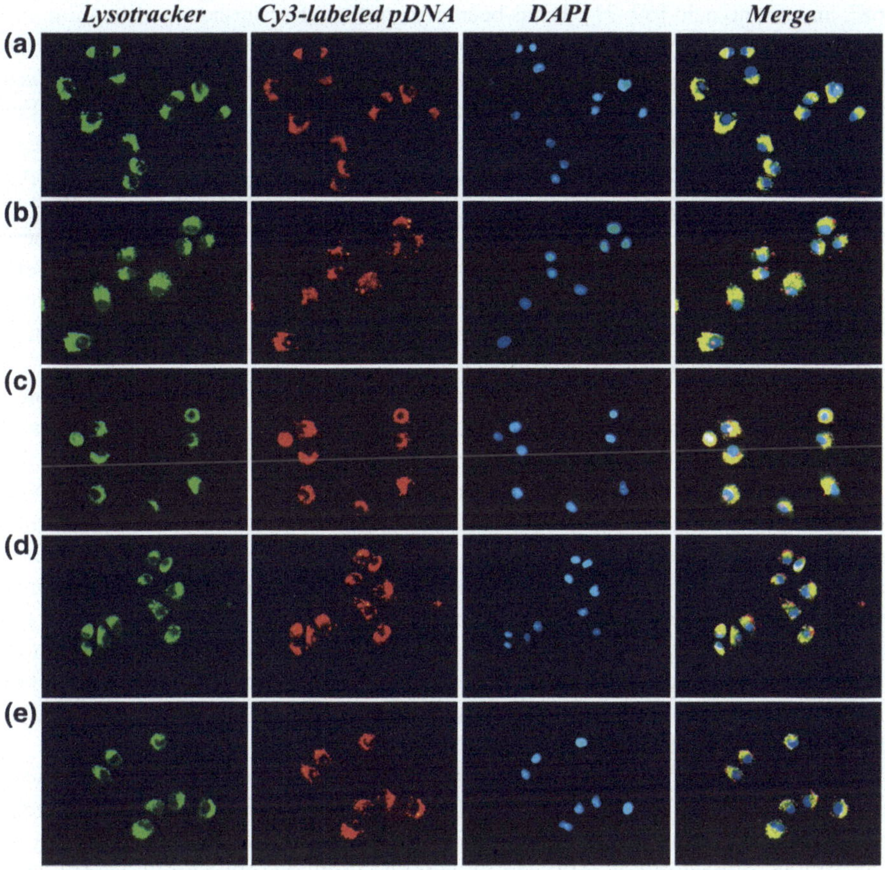

Fig. 12.2 Fluorescence microscopic images (400×) of intracellular trafficking and localization of the Cy3-labeled pDNA with carriers of S0 (**a**), S25 (**b**), S50 (**c**), S75 (**d**), and S100 (**e**) in COS-7 cells recorded after 2 h of gene transfection (*green* Lysotracker green used to label lysosomes; *red* Cy3-labeled pDNA; *blue* DAPI stained cell nuclei). Reprinted from Ref. [42], Copyright 2012, with permission from Elsevier

G5 amine dendrimers and enhanced gene delivery capability, Au DENPs may be used as a new family of gene delivery vectors for various gene therapy applications.

12.3.1 Acetylated Au DNEPs for Gene Delivery

In order to improve the gene delivery efficiency and lower the toxicity, much effort has been devoted to the modification of PAMAM dendrimers. The surface amine groups of dendrimers can be conjugated with hydrophilic polymers [64, 66, 67], or modified with small molecules, such as acetic anhydride [7], cell specific ligands

[68], and amino acid [69–71]. It has been proven that after surface modification, the cytotoxicity of PAMAM dendrimers can be dramatically reduced due to the decreased density of positive charges [34, 35, 37, 38, 72]. Moreover, surface modification may impact the gene transfection efficiency of dendrimers. For example, Waite et al. showed that a modest acetylation (approximately 20%) of PAMAM dendrimers was able to maintain the siRNA delivery efficiency comparable to the unmodified dendrimers, and increased degree of acetylation resulted in the reduced siRNA delivery efficiency [31]. In addition, PAMAM dendrimers modified with L-arginine have been proven to be able to improve the gene delivery efficiency [69, 70]. These studies imply that via an appropriate surface modification, PAMAM dendrimers are able to be afforded with enhanced gene delivery efficiency and simultaneously have reduced toxicity.

Our previous work has shown that Au DENPs prepared using amine-terminated G5 PAMAM dendrimers as templates display higher gene transfection efficiency than G5 dendrimers without the entrapment of AuNPs [42]. To eliminate the possible cytotoxicity resulting from the amine groups on the surface of Au DENPs for safe gene delivery applications, here we synthesized a series of partially acetylated Au DENPs and systematically evaluated the effect of the partial acetylation on their performance in gene transfection. Although acetylation of dendrimer terminal amine groups has been proven to be an effective way to decrease the cytotoxicity of dendrimers, a high degree of acetylation modification may result in reduced gene delivery efficiency due to the decreased density of amine groups left on the dendrimer surface [31]. Therefore, the molar ratios of acetic anhydride to dendrimer were controlled in a range of 5:1–30:1.

The cytotoxicity of partially acetylated and non-acetylated Au DNEPs was evaluated by MTT viability assay of HeLa cells treated with the vectors at different concentrations. It could be seen that all vectors did not have apparent cytotoxicity at low vector concentrations (500 nM or below). With the increase of vector concentration, the cell viability gradually decreased; however, the partially acetylated Au DENPs displayed much less cytotoxicity than non-acetylated ones due to the decreased surface positive charge, in agreement with the literature [7]. We also note that at the high vector concentrations (1000–3000 nM), the cell viability increases with the acetylation degree of the Au DENPs vectors, and $\{(Au^0)_{25}\text{-G5.NH}_2\text{-Ac}_{30}\}$ with the highest acetylation degree exhibits the lowest cytotoxicity at all the tested concentrations. Compared with amine-terminated $\{(Au^0)_{25}\text{-G5.NH}_2\}$, $\{(Au^0)_{25}\text{-G5.NH}_2\text{-Ac}_{30}\}$ caused a significant increase in cell viability at the concentration of 2000 nM ($p < 0.01$). At the highest vector concentration of 3000 nM, partially acetylated Au DENPs (both $\{(Au^0)_{25}\text{-G5.NH}_2\text{-Ac}_{20}\}$ and $\{(Au^0)_{25}\text{-G5.NH}_2\text{-Ac}_{30}\}$) show significantly improved cell viability when compared with $\{(Au^0)_{25}\text{-G5.NH}_2\}$ ($p < 0.01$ and $p < 0.001$, respectively). Therefore, the partially acetylated Au DENPs show advantages over non-acetylated ones in gene delivery applications due to their relatively low cytotoxicity.

The gene transfection efficiency of the acetylated and non-acetylated Au DENP vectors was investigated by luciferase activity assay after gene expression in HeLa cells. The gene transfection efficiency is largely dependent on the selected N/P ratio, and it seems that the acetylation degree of Au DENPs did not significantly impact the gene transfection efficiency under the same N/P ratios. All vectors displayed highest Luc activity at the N/P ratio of 2.5:1, therefore this N/P ratio was selected as an optimal one for further study. We believe that the acetylated Au DENPs have advantages in gene delivery due to their less cytotoxicity. Our study suggests that under certain degree of partial acetylation, the reduced amine density on the surface of Au DENPs may slightly compromise the DNA compaction ability, but does not significantly impact the gene delivery efficiency under the optimized N/P ratio (2.5:1). The gene transfection efficiency of acetylated Au DENP vectors was further qualitatively assessed by microscopic observation of the EGFP gene expression. All Au DENPs vectors enabled high EGFP expression, corroborating the Luc activity assay results. With the proven less cytotoxicity of the partially acetylated Au DENPs than that of non-acetylated Au DENPs by cell viability assay, the developed partially acetylated Au DENPs may serve as promising vectors for safe gene delivery applications with non-compromised gene transfection efficiency.

12.3.2 PEGylated Au DNEPs for Gene Delivery

Polyethylene glycol (PEG) is a class of macromolecules with good biocompatibility, non-immunogenicity, and high antifouling property [72–74]. Previous studies have shown that surface PEGylation could enhance the gene delivery efficiency and reduce the cytotoxicity of PAMAM dendrimers [27, 66, 75]. For example, Qi et al. reported that generation 5 PAMAM ($G5.NH_2$) or generation 6 PAMAM ($G6.NH_2$) dendrimers partially conjugated with PEG monomethyl ether (*m*PEG) could dramatically facilitate intramuscular gene delivery in neonatal mice [35]. In another study, Tang et al. [27] showed that intramuscular delivery of GFP-siRNA using PEG-modified $G5.NH_2$ dendrimers could significantly suppress the expression of EGFP gene in both transient adenovirus infected C57BL/6 mice and EGFP transgenic mice. It is believed that with the PEGylation of dendrimer periphery, both the water solubility of the polymer/DNA polyplexes and the intracellular release of DNA molecules can be enhanced, thereby resulting in enhanced gene delivery efficiency [3, 76]. In addition to the *m*PEG conjugation with terminal periphery amine groups of dendrimers, a rational design towards the interior space of PAMAM dendrimers is also feasible in reducing the cytotoxicity and improving the gene transfection efficiency of dendrimers [28, 42, 43]. In either case of dendrimer surface PEGylation or interior Au NP entrapment, the positive charge of the dendrimer surface amines is able to be compromised, thereby having improved cytocompatibility.

Inspired by the fact that both of the exterior modified *m*PEG and the interior entrapped AuNPs are able to decrease the cytotoxicity and increase the gene delivery efficiency of PAMAM dendrimers, we herein attempted to develop a unique PAMAM-based nonviral gene delivery vector by simultaneously modifying *m*PEG onto the dendrimer surface and entrapping AuNPs within the dendrimer

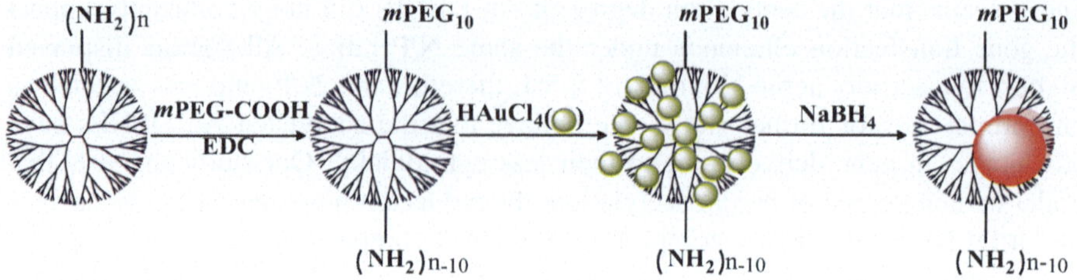

Scheme 12.2 Schematic illustration of the preparation of partially PEGylated Au DENPs. Reproduced from Ref. [77] by permission of The Royal Society of Chemistry

interior (Scheme 12.2) [77]. The pDNA/siRNA transfection efficiency was optimized by altering the molecular weight of *m*PEG (Mw = 2 K or 5 K) and the gold salt/dendrimer molar ratio (25:1 or 50:1) of the Au DENPs. The partially PEGylated Au DENPs were employed to deliver Luc reporter gene, EGFP gene, or B-cell lymphoma-2 (Bcl-2) siRNA to HeLa cells in vitro. The major point of this work is to utilize Au DENPs with different PEGylation chain lengths and different Au contents to explore their efficacy in pDNA and siRNA delivery.

Similar to the previous characterization methods, the viability of cells treated with the G5.NH_2 dendrimers was lower than 60% at a concentration at 3000 nM, while cells treated with the partially PEGylated Au DENPs showed a viability higher than 80% at the same concentration. DNA transfection efficiency of all vectors was investigated by Luc activity assay. Compared with HeLa cells treated with the G5.NH_2 dendrimer/pDNA polyplexes or PEGylated G5.NH_2 dendrimer/pDNA polyplexes, the Luc expression of HeLa cells treated with the partially PEGylated Au DENPs/pDNA polyplexes was remarkably increased (Fig. 12.3a). It could be seen that partially PEGylated Au DENPs/pDNA polyplexes showed the highest Luc expression at the N/P ratio of 5:1, and {$(Au^0)_{50}$-G5.NH_2-*m*PEG2K} vector enabled the highest Luc expression within cells at various N/P ratios. At the N/P ratio of 5:1, the average light unit of {$(Au^0)_{50}$-G5.NH_2-*m*PEG2K}-transfected HeLa cells was 7.39×10^7 RLU/mg protein, which was 292 times higher than that of cells transfected with the G5.NH_2 dendrimers (2.53×10^5 RLU/mg protein). Therefore, the N/P ratio of 5:1 was used for further study. Undoubtedly, {$(Au^0)_{50}$-G5.NH_2-*m*PEG2K} possesses a significantly higher Luc gene transfection efficiency than Au DENPs without PEGylation modification, implying that both the exterior surface *m*PEG modification and the interior Au NP entrapment significantly contribute to the enhanced gene delivery efficiency. The DNA transfection efficiency of all vectors at the optimal N/P ratio of 5:1 was further qualitatively validated by confocal microscopic observation of the EGFP gene expression (Fig. 12.3b). Clearly, there was no obvious EGFP expression in cells treated with PBS, naked pDNA or G5.NH_2/pDNA complexes. In sharp contrast, HeLa cells incubated with all PEGylated Au DENPs/pDNA polyplexes showed a remarkably increased EGFP expression. The confocal microscopic observation further indicates that {$(Au^0)_{50}$-G5.NH_2-*m*PEG2K} vector has excellent

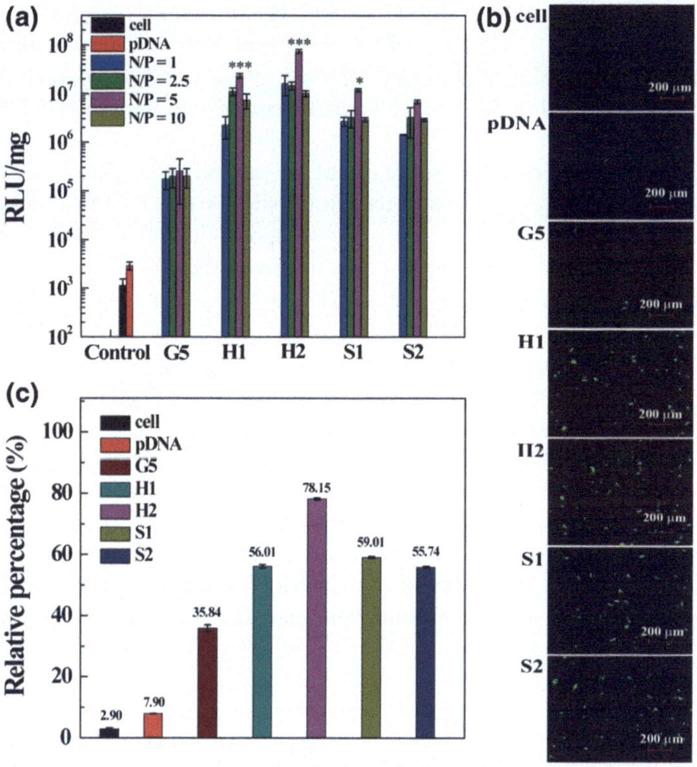

Fig. 12.3 Gene transfection and intracellular uptake efficiency of different vector/pDNA polyplexes in HeLa cells. **a** Luc gene transfection efficiency at N/P ratios of 1:1, 2.5:1, 5:1 and 10:1, respectively (mean ± SD, $n = 3$). Statistical differences between PEGylated Au DENPs/pDNA polyplexes (H1, H2, S1, and S2, respectively) versus G5.NH_2/pDNA polyplexes at an N/P ratio of 5:1 was compared. **b** Confocal microscopic images of HeLa cells treated with different vector/EGFP-pDNA polyplexes at the N/P ratio of 5:1. **c** Flow cytometry measurement of HeLa cells incubated with different vector/Cy3-labeled pDNA polyplexes at the N/P ratio of 5:1 for 2 h. Transfection was performed at a dose of 1 μg/well of DNA (mean ± SD, $n = 3$). Cells without treatment (cell) and cells treated with vector-free pDNA (pDNA) were used as controls. The denotations are as follows: H1 for {$(Au^0)_{25}$-G5.NH_2-*m*PEG2K}, H2 for {$(Au^0)_{50}$-G5.NH_2-*m*PEG2K} S1 for {$(Au^0)_{25}$-G5.NH_2-*m*PEG5K}, and S2 for {$(Au^0)_{50}$-G5.NH_2-*m*PEG5K}, respectively. Reproduced from Ref. [77] by permission of The Royal Society of Chemistry

gene transfection efficiency, in accordance with the quantitative Luc activity assay results.

Likewise, cellular uptake capability of the polyplexes was necessary for enhanced gene delivery and expression. Cy3-labeled pDNA was selected as a probe to compare the cellular uptake of different PEGylated Au DENPs/DNA polyplexes using flow cytometry (Fig. 12.3c). It can be found that about 78.15% HeLa cells display Cy3-derived red fluorescence signal when {$(Au^0)_{50}$-G5.NH_2-*m*PEG2K} was used as the vector, while the populations of the red fluorescent cells are 56.01,

59.01, and 55.74%, when $\{(Au^0)_{25}$-G5.NH_2-*m*PEG2K$\}$, $\{(Au^0)_{25}$-G5.NH_2-*m*PEG5K$\}$, and $\{(Au^0)_{50}$-G5.NH_2-*m*PEG5K$\}$ were used as the vectors, respectively. As expected, HeLa cells treated with all of the synthesized vectors show a higher population of red fluorescent cells than those treated with the G5.NH_2 vector (35.84%).

Under the optimized conditions, the use of the partially PEGylated Au DENPs as vectors for siRNA delivery was explored. Bcl-2 siRNA was able to knockdown the expression of the Bcl-2 protein, a class of anti-apoptotic defense protein related to the multiple drug resistance in cancer cells. In this study, Bcl-2 siRNA was chosen as a model siRNA to evaluate the siRNA delivery capacity of the partially PEGylated Au DENPs. The siRNA transfection efficiency was finally comparatively studied by western blot analysis of Bcl-2 protein expression (Fig. 12.4). The Bcl-2 protein expression level in HeLa cells treated with G5.NH_2/siRNA polyplex was set at 100%. Clearly, similar to Luc gene delivery, HeLa cells treated with $\{(Au^0)_{50}$-G5.NH_2-*m*PEG2K$\}$/Bcl-2 siRNA polyplex displayed the lowest Bcl-2 protein expression (15%). The Bcl-2 protein expression levels are 56, 40, and 26% when $\{(Au^0)_{25}$-G5.NH_2-*m*PEG2K$\}$, $\{(Au^0)_{25}$-G5.NH_2-*m*PEG5K$\}$, and $\{(Au^0)_{50}$-G5.NH_2-*m*PEG5K$\}$ were used as the vectors, respectively.

These results prove that both of the interior entrapped AuNPs and exterior conjugated PEG chains can enhance the gene transfection efficiency of the G5.NH_2

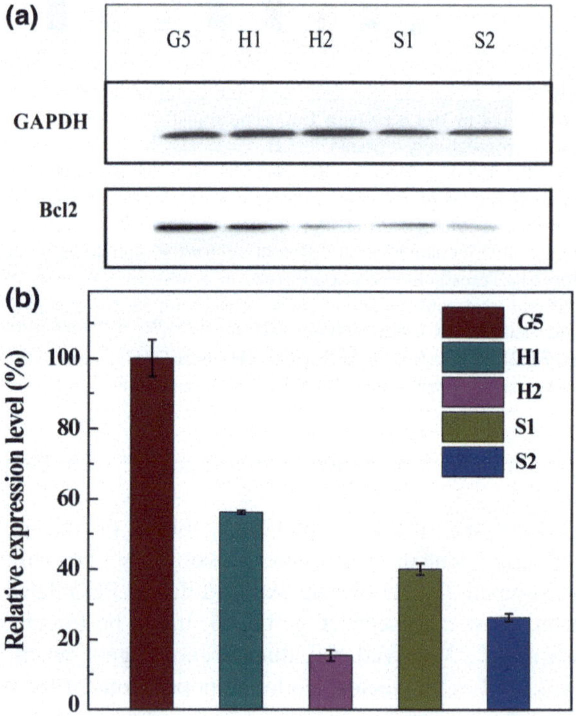

Fig. 12.4 Transfection efficiency of different vector/siRNA polyplexes in HeLa cells. **a** Protein expression levels of Bcl-2 detected by western blot assay. **b** The quantification of relative expression level (%) of Bcl-2 relative to GAPDH (mean ± SD, $n = 3$). The expression of Bcl-2 in G5.NH_2/Bcl-2 siRNA-treated HeLa was set to be 100%. Reproduced from Ref. [77] by permission of The Royal Society of Chemistry

dendrimers. It is interesting to note that the proton sponge effect originated from the tertiary amine groups of PAMAM dendrimers or Au DENPs plays a key role in the endosomal escape of the vector/gene complexes [42, 78]. Although the dendrimers were subjected to surface partial PEGylation modification and interior Au NP entrapment, the proton sponge effect is believed not to be compromised and still makes the vector/pDNA or vector/siRNA polyplexes escape from the endosomes. Overall, $\{(Au^0)_{50}$-G5.NH$_2$-*m*PEG2K$\}$ is demonstrated to be the best vector for enhanced DNA or siRNA delivery among all the PEGylated Au DENPs.

12.3.3 *β-CD-Modified Au DENPs for Gene Delivery*

It is well known that cyclodextrins (CD) possess low toxicity [79, 80], non-immunogenicity and excellent hydrophobic cavity [81–83]. CD has been used as an enhancer in both viral and nonviral oligonucleotide delivery due to its binding with nucleic acids, facilitating increased gene stability against nuclease [84–86]. Meanwhile, positively charged CD has been explored as a new vector for gene delivery [87, 88]. Furthermore, various polycations covalently linked with CD have improved biocompatibility, water solubility, and gene transfection efficiency [89]. The enhanced gene delivery efficiency using CD-conjugated polymers could be due to the possible mechanisms that CD-polymer conjugates might increase the release of pDNA or the pDNA complex from endosomes following cellular uptake, although it is still unclear to what degree the membrane disruptive ability of CD-polymer conjugates contributes to the enhancing effect on gene transfer activity [85]. For instance, Gonzalez et al. [90] synthesized β-cyclodextrin (β-CD)-containing polymers for enhanced gene delivery applications, and showed that β-CD was able to be used as a biocompatibility- and solubility-enhancing moiety. It is reasonable to hypothesize that β-CD-modified G5 PAMAM dendrimers with AuNPs loaded within their interior may be developed as a highly efficient gene delivery vector.

In Qiu et al.'s study [91], an approach to using β-CD-modified dendrimer-entrapped AuNPs (Au DENPs) for enhanced gene delivery applications was presented (Scheme 12.3). G5.NH$_2$-β-CD was first synthesized by modification of G5.NH$_2$ dendrimer with N,N'-carbonyldiimidazole-activated β-CD according to the literature [92]. The gene transfection efficiency of the Au DENPs-β-CD vector was evaluated by transfecting both Luc reporter gene and EGFP gene to 293T cells in vitro.

According to the characterization of ^{1}H NMR, the number of β-CD molecules conjugated on each G5 dendrimer was calculated to be 8.4, which decreased the cytotoxicity of G5 dendrimers to some extent. The cytotoxicity of the materials follows the order of Au DENPs-β-CD < Au DENPs < G5.NH$_2$. Subsequent gene transfection experiment proved that at the N/P ratios of 5:1 and 10:1, the Luc expression using Au DENPs-β-CD vector reached 6.2×10^6 and 5.9×10^6 RLU/mg, respectively, significantly higher than that using G5.NH$_2$ and Au DENPs

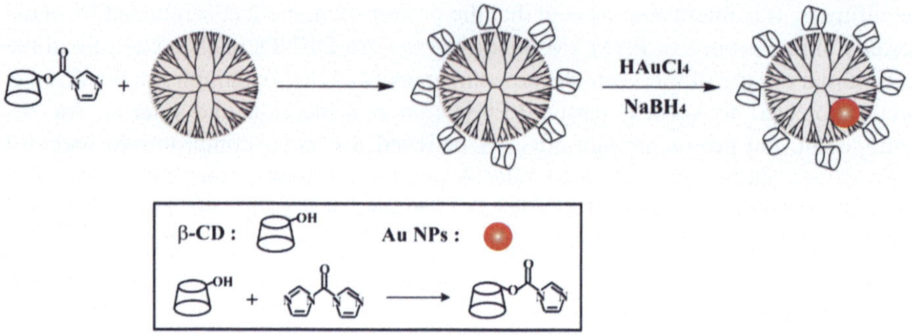

Scheme 12.3 Schematic representation of the synthesis of Au DENPs-β-CD. Reproduced from Ref. [91] by permission of The Royal Society of Chemistry

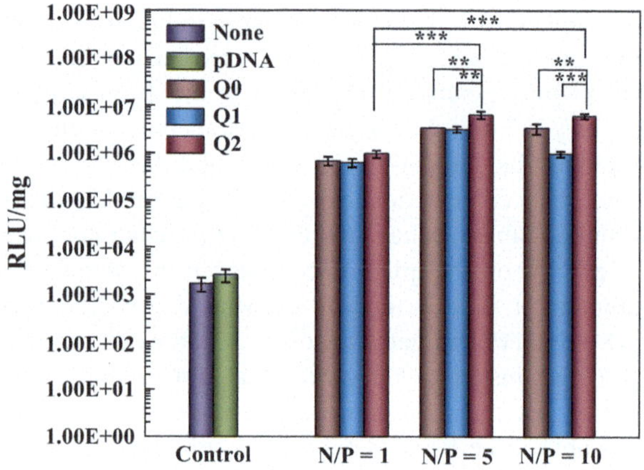

Fig. 12.5 Luc gene transfection efficiency in 293T cells transfected using Q0 (G5.NH_2), Q1 (Au DENPs), and Q2 (Au DENPs-β-CD) vectors at the N/P ratios of 1:1, 5:1, 10:1, respectively. Cells without treatment were used as controls. Data are shown with mean ± SD (n = 3). Reproduced from Ref. [91] by permission of The Royal Society of Chemistry

vectors ($p < 0.01$) (Fig. 12.5). The result of EGFP expression was in accordance with the above data, showing the enhanced gene transfection efficiency of the Au DENPs-β-CD. Overall, our results imply that the surface conjugation of β-CD and the interior entrapment of AuNPs render the G5 dendrimers with less cytotoxicity than Au DENPs without β-CD conjugation, and enable more efficient cellular gene delivery than Au DENPs without β-CD conjugation.

12.4 Targeted Au DENPs for Gene Delivery

In order to realize targeted gene delivery, it is necessary to modify the dendrimer surface with a targeting ligand that can target to cells specifically via a receptor-mediated manner. Various targeting ligands such as folic acid (FA) [56, 93], peptides [28, 94], antibodies [95, 96], lactobionic acid (LA) [97, 98], and hyaluronic acid (HA) [99, 100], can be modified onto the surface of dendrimers for gene delivery. In this part, we focus on some recent developments of the Au DENP-based vectors for targeted gene delivery applications.

12.4.1 FA-Targeted Au DENPs for Gene Delivery

As one of the most studied cancer-targeting ligands, FA has been known to target FA receptors (FAR) that are overexpressed in several human carcinomas including breast, ovary, endometrium, kidney, lung, head, and neck, brain, and myeloid cancers [101–103]. In our study, we reported the use of FA-modified Au DENPs (Au DENPs-FA) for targeted gene delivery applications [43]. First, amine-terminated G5 dendrimers ($G5.NH_2$) were covalently modified with FA. The formed FA modified G5 dendrimers ($G5.NH_2$-FA) were then used as templates to synthesize Au DENPs-FA. For comparison, Au DENPs without FA were used as a control.

Interestingly, Au DENPs-FA seemed to be less toxic than Au DENPs especially at high concentrations. This was presumably due to the fact that the surface modification of FA moieties onto the dendrimer surface might be able to alleviate the strong electrostatic interaction between the particles and the cells, although both vectors were measured to have a similar number of primary amines on the particle surfaces.

We used both firefly Luc and EGFP gene expression assays to investigate the targeted gene transfection efficiency of Au DENPs-FA vector in FAR-overexpressing HeLa cells. Figure 12.6a shows the gene transfection efficiency of both Au DENPs and Au DENPs-FA vectors as a function of the N/P ratio. It can be seen that both vectors have much higher transfection efficiency than that of the control cells without transfection treatment and cells transfected with naked pDNA, however, the transfection efficiency of both vectors is lower than that of the lipofectamine 2000 vector. We also show that the modification of FA onto Au DENPs affords Au DENPs-FA with much higher gene delivery efficiency compared to Au DENPs vector without FA at the N/P ratios of 1, 2.5, and 5 ($p < 0.001$). The gene transfection efficiency of both vectors is largely dependent on the N/P ratio of the polyplexes. At the N/P ratios of 2.5 and 5, the Luc expression value for Au DENPs-FA vector reached 2.19×10^6 and 1.99×10^6, respectively, significantly higher than that of Au DENPs without FA and the N/P ratio of 2.5 gave rise to the highest Luc transfection efficiency for the FA-targeted Au DENPs. These results suggest that the FA-modification onto the Au DENPs is able to render the vector with significantly high gene transfection efficiency to FAR-overexpressing HeLa cells, although both Au DENPs with or without FA display no appreciable difference in the binding affinity to pDNA. It should be noted that the entrapment of Au core NPs within dendrimers is essential to have improved gene delivery efficiency. Using $G5.NH_2$-FA dendrimers without the core AuNPs as a vector, the gene delivery efficiency is much less than that using the Au DENPs-FA vector.

This further suggests that the entrapment of Au core NPs are able to render the 3-D spherical shape of dendrimers, significantly improving the gene delivery efficiency, in agreement with our previous results [104].

The targeted gene transfection efficiency of Au DENPs-FA vector was also confirmed by EGFP expression assay. In this case, the gene delivery efficiency was determined by flow cytometry and reported as mean fluorescence intensity, which represents the amount of the EGFP expression. As shown in Fig. 12.6b, at the N/P ratios of 2.5 and 5, the Au DENPs-FA vector have significantly higher gene delivery efficiency than Au DENPs vector without FA-modification, corroborating with the Luc transfection assay data. Our results clearly suggest that the developed Au DENPs-FA vector was able to specifically deliver genes to FAR-overexpressing cancer cells via a receptor-mediated targeting pathway.

12.4.2 RGD-Targeted Au DENPs for Gene Delivery

The cell surface integrin receptors which are overexpressed in many types of cells including human mesenchymal stem cells (hMSCs), glioblastoma cells, ovarian cancer cells, and breast cancer cells [105–107] have been shown to have a high affinity to bind arginine-glycine-aspartic (Arg-Gly-Asp, RGD) peptide [35]. Therefore, RGD peptide has been identified as a promising targeting ligand for different biomedical applications [108]. Dendrimers modified with RGD peptide have been demonstrated to have binding specificity to integrin-overexpressing cancer cells for drug delivery [109, 110] and specific gene delivery [40] applications.

In one of our studies, we developed the use of RGD-modified Au DENPs for highly efficient and specific gene delivery to stem cells, where G5 dendrimers

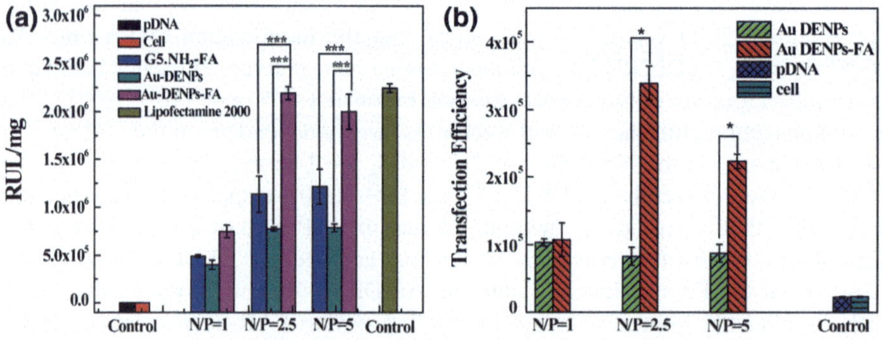

Fig. 12.6 a Luciferase gene transfection efficiency of Au DENPs/pDNA and Au DENPs-FA/pDNA polyplexes determined in HeLa cells at the N/P ratios of 1:1, 2.5:1, and 5:1, respectively. **b** EGFP gene transfection efficiency of Au DENPs/pDNA and Au DENPs-FA/pDNA polyplexes determined in HeLa cells at the N/P ratios of 1:1, 2.5:1, and 5:1, respectively. Reproduced from Ref. [43] by permission of The Royal Society of Chemistry

modified with RGD via a poly(ethylene glycol) (PEG) spacer or with PEG monomethyl ether were used as templates to entrap AuNPs [28]. The RGD-modified PEGylated dendrimers and the respective well characterized Au DENPs were used as vectors to transfect hMSCs with pDNA carrying both the EGFP and the luciferase (pEGFPLuc) reporter genes, as well as pDNA encoding the human bone morphogenetic protein-2 (hBMP-2) gene.

In spite of the modification of PEG and RGD moieties onto the surface of dendrimers and internal entrapment of AuNPs, the hydrodynamic diameter (approximately 150–200 nm) and surface potential (around 20 mV) of the particles are appropriate for gene delivery. The prepared RGD-modified PEGylated Au DENPs own better biocompatibility than other groups in the given concentration range. This may be attributed to the fact that the modification of G5.NH_2 dendrimers via surface PEGylation or interior AuNPs entrapment is beneficial to improve the cytocompatibility of the dendrimer-based vectors. In vitro gene transfection efficiency of pEGFPLuc reporter genes demonstrated by quantitative Luc activity assay and qualitative evaluation by fluorescence microscopy revealed that PEG-RGD-modified Au DENPs could transfer pEGFPLuc DNA to hMSCs successfully and possessed the highest delivery efficiency at an N/P ratio of 2.5.

Based on the transfection performance of pEGFPLuc, we selected an N/P ratio of 2.5:1 to evaluate the possibility of using the developed vector systems to transfect hMSCs with a pDNA carrying the hBMP-2 reporter gene. From Fig. 12.7a, it was clear that the non-transfected cells did not have appreciable hBMP-2 expression. In contrast, the hBMP-2 protein was expressed in cells transfected with the different dendrimer-based vectors. The hBMP-2 gene transfection efficiency followed the order of $\{(Au^0)_{25}$-G5.NH_2-$(PEG\text{-}RGD)_{10}$-$m PEG_{10}\}$ DENPs (K4) > $\{(Au^0)_{25}$-G5.NH_2-$m PEG_{20}\}$ DENPs (K2) > G5.NH_2-$(PEG\text{-}RGD)_{10}$-$m PEG_{10}$ dendrimers (K3) > G5.NH_2-$m PEG_{20}$ dendrimers (K1) > G5.NH_2 dendrimers, similar to the above pEGFPLuc gene transfection results. The highest hBMP-2 gene transfection efficiency using the K4 vector should be due to the fact that the PEG-RGD modification of dendrimers and the entrapment of AuNPs within the dendrimers render the vector with RGD-mediated targeting effect and well-maintained 3-D conformation of dendrimers, respectively.

To confirm the hBMP-2 gene transfection-induced stem cell osteogenic differentiation, the activity of ALP, which was a membrane-bound enzyme secreted early in bone formation and had been identified to be an important early marker of osteogenesis [2, 35, 111], was analyzed. Figure 12.7b showed that the ALP activity increased with the cell culture time for hMSCs transfected with a given vector/pDNA complex. Additionally, the ALP activity of hMSCs on day 21 was significantly higher than that on days 7 and 14. At all time points, the ALP activity for the transfected hMSCs was obviously higher than that for the non-transfected cells. K4 displayed the highest ALP activity when compared with the other vectors due to the surface PEG-RGD modification and the entrapped AuNPs.

Osteocalcin was synthesized and secreted by osteoblasts, and had been identified as an important marker of late-stage osteogenic differentiation [112]. Both non-transfected and transfected cells displayed a low production rate of osteocalcin

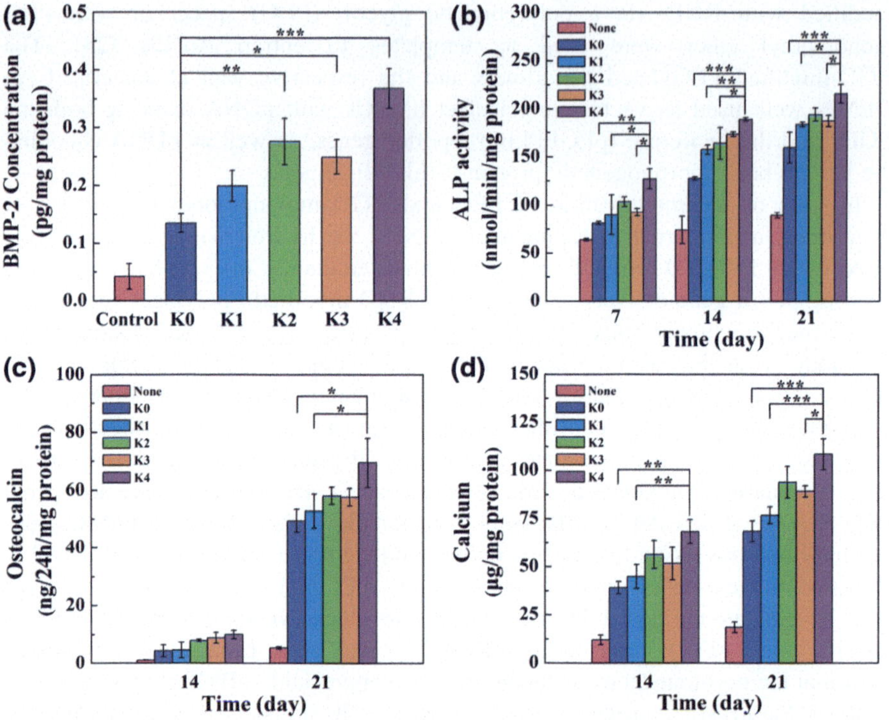

Fig. 12.7 **a** hBMP-2 expression in hMSCs 3 days post-transfection via different vectors; **b** Time course of ALP activity of hMSCs transfected with different vector/pDNA polyplexes; **c** Osteocalcin content secreted by hMSCs transfected with different vector/pDNA polyplexes at different culture times; **d** Calcium deposition on the extracellular matrix of the hMSCs transfected with different vector/pDNA polyplexes after being cultured for 14 and 21 days. All data were presented as mean ± SD (n = 3). Non-transfected cells were used as control. Reprinted with the permission from Ref. [28]. Copyright 2015 American Chemical Society

on day 14 (Fig. 12.7c). On day 21, a distinct increase in the level of osteocalcin secretion was detected for all transfected hMSCs, suggesting that hMSCs after transfection with the hBMP-2 gene using different dendrimer-based vectors are able to differentiate into the osteoblast lineage. Also, relative to G5.NH_2 dendrimers, other four vector-transfected hMSCs were all observed to have visibly higher levels of osteocalcin secretion on day 21. Similarly, the K4 enabled the transfected hMSCs to have the highest osteocalcin secretion when compared to all other vectors, corroborating the ALP activity data.

Calcium deposition was another indicator that could be used to further characterize the degree of hMSCs osteogenic differentiation (Fig. 12.7d). The results clearly demonstrated higher quantities of calcium production in the transfected hMSCs cultures than in the non-transfected cultures. With time, the deposition of calcium progressively increased from day 14 to day 21, in agreement with work previously reported in the literature [113]. On both day 14 and day 21, hMSCs transfected by the K4 vector displayed the highest calcium content when compared to the other vectors. These results further verified that all dendrimer-based vectors can deliver exogenous genes and successfully induce the hMSCs osteogenic differentiation, with K4 being the most effective vector. Our study suggests that the gene delivery efficiency is largely dependent on the composition and surface modification of the vector. The surface modification with PEG-RGD and the entrapment of AuNPs render the dendrimer platform not only with targeting specificity to recognize integrin-expressing hMSCs, but also with the well-maintained 3-D conformation to have improved DNA compaction ability, thus affording the dendrimer-based vector with high gene transfection efficiency and specificity.

12.5 Conclusion and Outlooks

This chapter gives a brief literature overview of the recent advances in the use of dendrimer-based nanopaticles for gene delivery. To realize enhanced and specific gene delivery applications, dendrimers can be surface modified via alkylation, acetylation or PEGylation, can be surface modified with targeting ligands (e.g., FA, RGD peptide, and can be entrapped with AuNPs.

Although much effort has been devoted to the development of dendrimer-based NPs for gene delivery, this area of research still remains largely unexplored and needs to be further expanded. In view of the abundant amine groups on the dendrimer surface, antineoplastic drugs (e.g., doxorubicin or paclitaxel) may be conjugated, achieving combinational gene therapy and chemotherapy, significantly enhancing the efficacy of cancer therapy. In addition, due to the excellent antifouling property of zwitterionic materials, it is believed that further modification of zwitterionic materials onto the surface of dendrimers may be an effective approach to improve or optimize the transfection efficiency of the materials. Moreover, Au DENPs possess CT imaging effect when the molar ratio of Au/dendrimer achieves the certain proportion, hence, the combination of CT imaging and gene therapy is worthy of development. Besides, various photothermal agents may be modified with Au DENPs, thereby realizing the combinational photothermal therapy and gene therapy. Overall, All these challenges will stimulate scientists to develop multifunctional dendrimer-based NPs for gene delivery and other biological applications.

Acknowledgments This research is financially supported by the Science and Technology Commission of Shanghai Municipality (17540712000), National Science Foundation of China (81761148028, 21773026), and the Fundamental Research Funds for the Central Universities. X. Shi also acknowledges the support by FCT-Fundação para a Ciência e a Tecnologia (project PEst-OE/QUI/UI0674/2013, CQM, Portuguese Government funds) and by ARDITI-Agência Regional para o Desenvolvimento da Investigação Tecnologia e Inovação through the project M1420-01-0145-FEDER-000005-Centro de Química da Madeira-CQM+ (Madeira 14-20)

References

1. Cavazzana-Calvo M, Hacein-Bey S, Basile CD, Gross F, Yvon E, Nusbaum P, Selz F, Hue C, Certain S, Casanova JL, Bousso P, Le Deist F, Fischer A (2000) Gene therapy of human severe combined immunodeficiency (SCID)-X1 disease. Science 288(5466): 669–672. doi:10.1126/science.288.5466.669
2. Jeong JH, Kim SW, Park TG (2003) Novel intracellular delivery system of antisense oligonucleotide by self-assembled hybrid micelles composed of DNA/PEG conjugate and cationic fusogenic peptide. Bioconjugate Chem 14(2):473–479
3. Yang ZR, Wang HF, Zhao J, Peng YY, Wang J, Guinn BA, Huang LQ (2007) Recent developments in the use of adenoviruses and immunotoxins in cancer gene therapy. Cancer Gene Ther 14(7):599–615. doi:10.1038/sj.cgt.7701054
4. Barhoumi A, Huschka R, Bardhan R, Knight MW, Halas NJ (2009) Light-induced release of DNA from plasmon-resonant nanoparticles: Towards light-controlled gene therapy. Chem Phys Lett 482(4):171–179
5. Biswas S, Torchilin VP (2013) Dendrimers for siRNA delivery. Pharmaceuticals 6(2): 161–183
6. Scholz C, Wagner E (2012) Therapeutic plasmid DNA versus siRNA delivery: common and different tasks for synthetic carriers. J Controll Release 161(2):554–565
7. Kolhatkar RB, Kitchens KM, Swaan PW, Ghandehari H (2007) Surface acetylation of polyamidoamine (PAMAM) dendrimers decreases cytotoxicity while maintaining membrane permeability. Bioconjugate Chem 18(6):2054–2060
8. Whitehead KA, Langer R, Anderson DG (2009) Knocking down barriers: advances in siRNA delivery. Nat Rev Drug Discovery 8(2):129–138
9. Nguyen K, Dang PN, Alsberg E (2013) Functionalized, biodegradable hydrogels for control over sustained and localized siRNA delivery to incorporated and surrounding cells. Acta Biomater 9(1):4487–4495
10. Paroo Z, Corey DR (2004) Challenges for RNAi in vivo. Trends Biotechnol 22(8):390–394
11. Tang J, Chen JY, Liu JY, Luo M, Wang YJ, Wei XW, Gao X, Wang BL, Liu YB, Yi T (2012) Calcium phosphate embedded PLGA nanoparticles: a promising gene delivery vector with high gene loading and transfection efficiency. Int J Pharm 431(1):210–221
12. Li S, Huang L (2000) Nonviral gene therapy: promises and challenges. Gene Ther 7:31–34
13. Verma IM, Weitzman MD (2005) Gene therapy: twenty-first century medicine. Annu Rev Biochem 74:711–738
14. Guillot-Nieckowski M, Eisler S, Diederich F (2007) Dendritic vectors for gene transfection. New J Chem 31:1111–1127
15. Anderson WF (1998) Human gene therapy. Nature 392:25–30
16. Kay MA, Glorioso JC, Naldini L (2001) Viral vectors for gene therapy: the art of turning infectious agents into vehicles of therapeutics. Nat Med 7:33–40
17. Luo D, Saltzman WM (2000) Synthetic dna delivery systems. Nat Biotechnol 18:33–37
18. Mancheño-Corvo P, Martin-Duque P (2006) Viral gene therapy. Clin Transl Oncol 8: 858–867
19. Thomas CE, Ehrhardt A, Kay MA (2003) Progress and problems with the use of viral vectors for gene therapy. Nat Rev Genet 4:346–358
20. Paleos CM, Tziveleka LA, Sideratou Z, Tsiourvas D (2009) Gene delivery using functional dendritic polymers. Expert Opin Drug Delivery 6:27–38
21. Shcharbin DG, Klajnert B, Bryszewska M (2009) Dendrimers in gene transfection. Biochemistry (Moscow) 74:1070–1079
22. Svenson S, Tomalia DA (2005) Dendrimers in biomedical applications–reflections on the field. Adv Drug Delivery Rev 57:2106–2129
23. Dufès C, Uchegbu IF, Schätzlein AG (2005) Dendrimers in gene delivery. Adv Drug Delivery Rev 57(15):2177–2202

24. Mintzer MA, Simanek EE (2009) Nonviral vectors for gene delivery. Chem Rev 109: 259–302
25. Kesharwani P, Iyer AK (2015) Recent advances in dendrimer-based nanovectors for tumor-targeted drug and gene delivery. Drug Discov Today 20(5):536–547
26. Menjoge AR, Kannan RM, Tomalia DA (2010) Dendrimer-based drug and imaging conjugates: design considerations for nanomedical applications. Drug Discov Today 15(5): 171–185
27. Tang Y, Li YB, Wang B, Lin RY, Van Dongen M, Zurcher DM, Gu XY, Banaszak Holl MM, Liu G, Qi R (2012) Efficient in vitro siRNA delivery and intramuscular gene silencing using PEG-modified PAMAM dendrimers. Mol Pharm 9(6):1812–1821
28. Kong LD, Alves CS, Hou WX, Qiu JQ, Möhwald H, Tomás H, Shi XY (2015) RGD peptide-modified dendrimer-entrapped gold nanoparticles enable highly efficient and specific gene delivery to stem cells. ACS Appl Mater Interfaces 7(8):4833–4843
29. Liu C, Liu XX, Rocchi P, Qu FQ, Iovanna JL, Peng L (2014) Arginine-terminated generation 4 PAMAM dendrimer as an effective nanovector for functional siRNA delivery in vitro and in vivo. Bioconjugate Chem 25(3):521–532
30. Merdan T, Kopeček J, Kissel T (2002) Prospects for cationic polymers in gene and oligonucleotide therapy against cancer. Adv Drug Deliv Rev 54(5):715–758
31. Waite CL, Sparks SM, Uhrich KE, Roth CM (2009) Acetylation of PAMAM dendrimers for cellular delivery of siRNA. BMC Biotechnol 9(1):38
32. Fant K, Esbjorner EK, Jenkins A, Grossel M, Lincoln P, Norden B (2010) Effects of pegylation and acetylation of pamam dendrimers on dna binding, cytotoxicity and in vitro transfection efficiency. Mol Pharm 7:1734–1746
33. He H, Li Y, Jia XR, Du J, Ying X, Lu WL, Lou JN, Wei Y (2011) PEGylated poly (amidoamine) dendrimer-based dual-targeting carrier for treating brain tumors. Biomaterials 32:478–487
34. Mishra S, Webster P, Davis ME (2004) PEGylation significantly affects cellular uptake and intracellular trafficking of non-viral gene delivery particles. Eur J Cell Biol 83:97–111
35. Qi R, Gao Y, Tang Y, He RR, Liu TL, He Y, Sun S, Li BY, Li YB, Liu G (2009) PEG-conjugated PAMAM dendrimers mediate efficient intramuscular gene expression. AAPS J 11(3):395–405
36. Gabrielson NP, Pack DW (2006) Acetylation of polyethylenimine enhances gene delivery via weakened polymer/DNA interactions. Biomacromol 7:2427–2435
37. Santos JL, Oliveira H, Pandita D, Rodrigues J, Pêgo AP, Granja PL, Tomas H (2010) Functionalization of poly(amidoamine) dendrimers with hydrophobic chains for improved gene delivery in mesenchymal stem cells. J Controlled Release 144:55–64. doi:10.1016/j.jconrel.2010.01.034
38. Shakhbazau A, Isayenka I, Kartel N, Goncharova N, Seviaryn I, Kosmacheva S, Potapnev M, Shcharbin DG, Bryszewska M (2010) Transfection efficiencies of PAMAM dendrimers correlate inversely with their hydrophobicity. Int J Pharm 383(1–2):228–235
39. Han L, Zhang AL, Wang HJ, Pu PY, Jiang XG, Kang CS, Chang J (2010) Tat-BMPs-PAMAM conjugates enhance therapeutic effect of small interference RNA on u251 glioma cells in vitro and in vivo. Hum Gene Ther 21:417–426. doi:10.1089/hum.2009.087
40. Santos JL, Pandita D, Rodrigues J, Pêgo AP, Granja PL, Balian G, Tomás H (2010) Receptor-mediated gene delivery using PAMAM dendrimers conjugated with peptides recognized by mesenchymal stem cells. Mol Pharm 7(3):763–774
41. Hou WX, Wen SH, Guo R, Wang SG, Shi XY (2015) Partially acetylated dendrimer-entrapped gold nanoparticles with reduced cytotoxicity for gene delivery applications. J Nanosci Nanotechnol 15(6):4094–4105
42. Shan YB, Luo T, Peng C, Sheng RL, Cao AM, Cao XY, Shen MW, Guo R, Tomas H, Shi XY (2012) Gene delivery using dendrimer-entrapped gold nanoparticles as nonviral vectors. Biomaterials 33(10):3025–3035. doi:10.1016/j.biomaterials.2011.12.045

43. Xiao TY, Hou WX, Cao XY, Wen SH, Shen MW, Shi XY (2013) Dendrimer-entrapped gold nanoparticles modified with folic acid for targeted gene delivery applications. Biomater Sci 1(11):1172–1180. doi:10.1039/c3bm60138b
44. Xiao TY, Cao XY, Hou WX, Peng C, Qiu JR, Shi XY (2015) Poly(amidoamine) dendrimers modified with 1,2-epoxyhexane or 1,2-epoxydodecane for enhanced gene delivery applications. J Nanosci Nanotechnol 15(12):10134–10140. doi:10.1166/jnn.2015.11693
45. Hong SP, Bielinska AU, Mecke A, Keszler B, Beals JL, Shi XY, Balogh L, Orr BG, Baker JR, Holl MMB (2004) Interaction of poly(amidoamine) dendrimers with supported lipid bilayers and cells: Hole formation and the relation to transport. Bioconjugate Chem 15 (4):774–782. doi:10.1021/bc049962b
46. Daniel MC, Astruc D (2004) Gold nanoparticles: Assembly, supramolecular chemistry, quantum-size-related properties, and applications toward biology, catalysis, and nanotechnology. Chem Rev 104:293–346
47. Duncan B, Kim CK, Rotello VM (2010) Gold nanoparticle platforms as drug and biomacromolecule delivery systems. J Controlled Release 148:122–127
48. Li D, Cui Y, Wang K, He Q, Yan X, Li J (2007) Thermosensitive nanostructures comprising gold nanoparticles grafted with block copolymers. Adv Funct Mater 17:3134–3140. doi:10.1002/adfm.200700427
49. Li DX, He Q, Cui Y, Li JB (2007) Fabrication of pH-responsive nanocomposites of gold nanoparticles/poly(4-vinylpyridine). Chem Mater 19:412–417. doi:10.1021/cm062290+
50. Li DX, He Q, Cui Y, Wang KW, Zhang XM, Li JB (2007) Thermosensitive copolymer networks modify gold nanoparticles for nanocomposite entrapment. Chem - Eur J 13:2224–2229. doi:10.1002/chem.200600839
51. Li D, He Q, Li J (2009) Smart core/shell nanocomposites: intelligent polymers modified gold nanoparticles. Adv Colloid Interface Sci 149:28–38
52. Li DX, He Q, Yang Y, Möhwald H, Li JB (2008) Two-stage pH response of poly (4-vinylpyridine) grafted gold nanoparticles. Macromolecules 41:7254–7256. doi:10.1021/ma800894c
53. Elbakry A, Wurster EC, Zaky A, Liebl R, Schindler E, Bauer-Kreisel P, Blunk T, Rachel R, Goepferich A, Breunig M (2012) Layer-by-Layer coated gold nanoparticles: size-dependent delivery of DNA into cells. Small 8(24):3847–3856. doi:10.1002/smll.201201112
54. Figueroa ER, Lin AY, Yan JX, Luo L, Foster AE, Drezek RA (2014) Optimization of PAMAM-gold nanoparticle conjugation for gene therapy. Biomaterials 35(5):1725–1734. doi:10.1016/j.biomaterials.2013.11.026
55. Kawano T, Yamagata M, Takahashi H, Niidome Y, Yamada S, Katayama Y, Niidome T (2006) Stabilizing of plasmid DNA in vivo by PEG-modified cationic gold nanoparticles and the gene expression assisted with electrical pulses. J Controlled Release 111(3):382–389. doi:10.1016/j.jconrel.2005.12.022
56. Xiao TY, Cao XY, Shi XY (2013) Dendrimer-entrapped gold nanoparticles modified with folic acid for targeted gene delivery applications. J Controll Release 172(1):e114–e115. doi:10.1016/j.jconrel.2013.08.275
57. Heinemann D, Schomaker M, Kalies S, Schieck M, Carlson R, Escobar HM, Ripken T, Meyer H, Heisterkamp A (2013) Gold nanoparticle mediated laser transfection for efficient siRNA mediated gene knock down. Plos One 8(3). doi:10.1371/journal.pone.0058604
58. Jiwaji M, Sandison ME, Reboud J, Stevenson R, Daly R, Barkess G, Faulds K, Kolch W, Graham D, Girolami MA, Cooper JM, Pitt AR (2014) Quantification of functionalised gold nanoparticle-targeted knockdown of gene expression in HeLa cells. PLoS ONE 9(6):e99458. doi:10.1371/journal.pone.0099458
59. Kirkland-York S, Zhang YL, Smith AE, York AW, Huang FQ, McCormick CL (2010) Tailored design of Au nanoparticle-siRNA carriers utilizing reversible addition—fragmentation chain transfer polymers. Biomacromol 11(4):1052–1059. doi:10.1021/bm100020x
60. Kong WH, Bae KH, Jo SD, Kim JS, Park TG (2012) Cationic lipid-coated gold ganoparticles as efficient and non-cytotoxic intracellular siRNA delivery vehicles. Pharm Res 29(2):362–374. doi:10.1007/s11095-011-0554-y

61. Mitra M, Kandalam M, Rangasamy J, Shankar B, Maheswari UK, Swaminathan S, Krishnakumar S (2013) Novel epithelial cell adhesion molecule antibody conjugated polyethyleneimine-capped gold nanoparticles for enhanced and targeted small interfering RNA delivery to retinoblastoma cells. Mol Vis 19:1029–1038
62. Ghosh PS, Kim CK, Han G, Forbes NS, Rotello VM (2008) Efficient gene delivery vectors by tuning the surface charge density of amino acid-functionalized gold nanoparticles. ACS Nano 2:2213–2218
63. Voet D, Voet JG (1995) Biochemistry, 2nd edn. Wiley, New York
64. Jevprasesphant R, Penny J, Jalal R, Attwood D, McKeown NB, D'emanuele A (2003) The influence of surface modification on the cytotoxicity of PAMAM dendrimers. Int J Pharm 252 (1):263–266
65. M. KK, B. KR, W. SP, D. EN, H. G (2006) Transport of poly (amidoamine) dendrimers across caco-2 cell monolayers: influence of size, charge and fluorescent labeling. Pharm Res 23(12):2818–2826
66. Luo D, Haverstick K, Belcheva N, Han E, Saltzman WM (2002) Poly (ethylene glycol)-conjugated PAMAM dendrimer for biocompatible, high-efficiency DNA delivery. Macromolecules 35(9):3456–3462
67. Wang X, Wu X, Fan W, Ding BY, Wang XY, Zhang W, Ding XY, Gao J, Zhu QG, Liu JY, Cai Z, Gao S (2012) Surface modification with pluronic P123 enhances transfection efficiency of PAMAM dendrimer. Macromol Res 20(2):162–167. doi:10.1007/s13233-012-0031-4
68. Shukla R, Thomas TP, Peters J, Kotlyar A, Myc A, Baker JR Jr (2005) Tumor angiogenic vasculature targeting with PAMAM dendrimer–RGD conjugates. Chem Commun 46:5739–5741
69. Choi JS, Nam K, Park J, Kim JB, Lee JK (2004) Enhanced transfection efficiency of PAMAM dendrimer by surface modification with L-arginine. J Controlld Release 99(3): 445–456
70. Kim JB, Choi JS, Nam K, Lee M, Park JS, Lee JK (2006) Enhanced transfection of primary cortical cultures using arginine-grafted PAMAM dendrimer, PAMAM-Arg. J Controll Release 114(1):110–117
71. Yu GS, Bae YM, Kim JY, Han J, Ko KS, Choi JS (2012) Amino acid-modified bioreducible poly(amidoamine) dendrimers: synthesis, characterization and in vitro evaluation. Macromol Res 20(11):1156–1162. doi:10.1007/s13233-012-0164-5
72. Wang W, Xiong W, Wan J, Sun X, Xu H, Yang X (2009) The decrease of PAMAM dendrimer-induced cytotoxicity by PEGylation via attenuation of oxidative stress. Nanotechnology 20:105103
73. Lee M, Kim SW (2005) Polyethylene glycol-conjugated copolymers for plasmid DNA delivery. Pharm Res 22(1):1–10
74. Peng C, Zheng LF, Chen Q, Shen MW, Guo R, Wang H, Cao XY, Zhang GX, Shi XY (2012) PEGylated dendrimer-entrapped gold nanoparticles for in vivo blood pool and tumor imaging by computed tomography. Biomaterials 33(4):1107–1119. doi:10.1016/j.biomaterials.2011.10.052
75. Xua L, Shena W, Wang B, Wang X, Liua G, Taoc Y, Qia R (2016) Efficient siRNA delivery using PEG-conjugated PAMAM dendrimers targeting vascular endothelial growth factor in a $CoCl_2$-induced neovascularization model in retinal endothelial cells. Curr Drug Delivery 13(10):590–599
76. Kim T, Seo HJ, Choi JS, Jang HS, Baek J, Kim K, Park JS (2004) PAMAM-PEG-PAMAM: novel triblock copolymer as a biocompatible and efficient gene delivery carrier. Biomacromol 5:2487–2492
77. Hou WX, Wei P, Kong LD, Guo R, Wang SG, Shi XY (2016) Partially PEGylated dendrimer-entrapped gold nanoparticles: a promising nanoplatform for highly efficient DNA and siRNA delivery. J Mater Chem B 4(17):2933–2943. doi:10.1039/c6tb00710d

78. Patil ML, Zhang M, Taratula O, Garbuzenko OB, He HX, Minko T (2009) Internally cationic polyamidoamine PAMAM-OH dendrimers for siRNA delivery: effect of the degree of quaternization and cancer targeting. Biomacromol 10(2):258–266
79. Davis ME, Brewster ME (2004) Cyclodextrin-based pharmaceutics: Past, present and future. Nat Rev Drug Discovery 3(12):1023–1035. doi:10.1038/nrd1576
80. Irie T, Uekama K (1997) Pharmaceutical applications of cyclodextrins.3. Toxicological issues and safety evaluation. J Pharm Sci 86(2):147–162. doi:10.1021/js960213f
81. Popielarski SR, Mishra S, Davis ME (2003) Structural effects of carbohydrate-containing polycations on gene delivery. 3. Cyclodextrin type and functionalization. Bioconjugate Chem 14(3):672–678. doi:10.1021/bc034010b
82. Szejtli J (1998) Introduction and general overview of cyclodextrin chemistry. Chem Rev 98 (5):1743–1753. doi:10.1021/cr970022c
83. Yao H, Ng SS, Tucker WO, Tsang YKT, Man K, Wang XM, Chow BKC, Kung HF, Tang GP, Lin MC (2009) The gene transfection efficiency of a folate-PEI600-cyclodextrin nanopolymer. Biomaterials 30(29):5793–5803. doi:10.1016/j.biomaterials.2009.06.051
84. Croyle MA, Roessler BJ, Hsu CP, Sun R, Amidon GL (1998) Beta cyclodextrins enhance adenoviral-mediated gene delivery to the intestine. Pharm Res 15(9):1348–1355. doi:10.1023/a:1011985101580
85. Kihara F, Arima H, Tsutsumi T, Hirayama F, Uekama K (2003) In vitro and in vivo gene transfer by an optimized alpha-cyclodextrin conjugate with polyamidoamine dendrimer. Bioconjugate Chem 14(2):342–350. doi:10.1021/bc025613a
86. Ortiz Mellet C, Garcia Fernandez JM, Benito JM (2011) Cyclodextrin-based gene delivery systems. Chem Soc Rev 40(3):1586–1608. doi:10.1039/c0cs00019a
87. Cryan SA, Holohan A, Donohue R, Darcy R, O'Driscoll CM (2004) Cell transfection with polycationic cyclodextrin vectors. Eur J Pharm Sci 21(5):625–633. doi:10.1016/j.ejps.2004.01.001
88. Mourtzis N, Paravatou M, Mavridis IM, Roberts ML, Yannakopoulou K (2008) Synthesis, characterization, and remarkable biological properties of cyclodextrins bearing guanidinoalkylamino and aminoalkylamino groups on their primary side. Chem - Eur J 14(14): 4188–4200. doi:10.1002/chem.200701650
89. Forrest ML, Gabrielson N, Pack DW (2005) Cyclodextrin-polyethylenimine conjugates for targeted in vitro gene delivery. Biotechnol Bioeng 89(4):416–423. doi:10.1002/bit.20356
90. Gonzalez H, Hwang SJ, Davis ME (1999) New class of polymers for the delivery of macromolecular therapeutics. Bioconjugate Chem 10(6):1068–1074. doi:10.1021/bc990072j
91. Qiu JR, Kong LD, Cao XY, Li AJ, Tan HR, Shi XY (2016) Dendrimer-entrapped gold nanoparticles modified with beta-cyclodextrin for enhanced gene delivery applications. RSC Adv 6(31):25633–25640. doi:10.1039/c6ra03839e
92. Wang H, Shao NM, Qiao SN, Cheng YY (2012) Host-guest chemistry of dendrimer-cyclodextrin conjugates: selective encapsulations of guests within dendrimer or cyclodextrin cavities revealed by NOE NMR techniques. J Phys Chem B 116(36):11217–11224. doi:10.1021/jp3062916
93. Sunoqrot S, Bugno J, Lantvit D, Burdette JE, Hong S (2014) Prolonged blood circulation and enhanced tumor accumulation of folate-targeted dendrimer-polymer hybrid nanoparticles. J Controll Release 191:115–122. doi:10.1016/j.jconrel.2014.05.006
94. Baigude H, Katsuraya K, Okuyama K, Uryu T (2004) Synthesis of structurally-controlled AIDS vaccine model with glyco-peptide dendrimer scaffolds. Macromol Chem Phy 205 (5):684–691. doi:10.1002/macp.200300097
95. Kobayashi H, Sato N, Saga T, Nakamoto Y, Ishimori T, Toyama S, Togashi K, Konishi J, Brechbiel MW (2000) Monoclonal antibody-dendrimer conjugates enable radiolabeling of antibody with markedly high specific activity with minimal loss of immunoreactivity. Eur J Nucl Med 27(9):1334–1339. doi:10.1007/s002590000293
96. Yoon HC, Lee D, Kim HS (2002) Reversible affinity interactions of antibody molecules at functionalized dendrimer monolayer: affinity-sensing surface with reusability. Anal Chim Acta 456(2):209–218. doi:10.1016/s0003-2670(02)00032-6

97. Fu FF, Wu YL, Zhu JY, Wen SH, Shen MW, Shi XY (2014) Multifunctional lactobionic acid-modified dendrimers for targeted drug delivery to liver cancer cells: investigating the role played by PEG spacer. ACS Appl Mater Interfaces 6(18):16416–16425. doi:10.1021/am504849x
98. Liu H, Wang H, Xu YH, Guo R, Wen SH, Huang YP, Liu WN, Shen MW, Zhao JL, Zhang GX, Shi XY (2014) Lactobionic acid-modified dendrimer-entrapped gold nanoparticles for targeted computed tomography imaging of human hepatocellular carcinoma. ACS Appl Mater Interfaces 6(9):6944–6953. doi:10.1021/am500761x
99. Bi XD, Liang AY, Tan Y, Maturavongsadit P, Higginbothem A, Gado T, Gramling A, Bahn HN, Wang Q (2016) Thiol-ene crosslinking polyamidoamine dendrimer-hyaluronic acid hydrogel system for biomedical applications. J Biomat Sci - Polymer E 27(8):743–757. doi:10.1080/09205063.2016.1159473
100. Zhan JZ, Wang L, Liu S, Chen JJ, Ren L, Wang YJ (2015) Antimicrobial hyaluronic acid/poly(amidoamine) dendrimer multi layer on poly(3-hydroxybutyrate-co-4-hydroxybutyrate) prepared by a layer-by-layer self-assembly method. ACS Appl Mater Interfaces 7(25):13876–13881. doi:10.1021/acsami.5b02262
101. Campbell IG, Jones TA, Foulkes WD, Trowsdale J (1991) Folate-binding protein is a marker for ovarian cancer. Cancer Res 51(19):5329–5338
102. Ross JF, Chaudhuri PK, Ratnam M (1994) Differential regulation of folate receptor isoforms in normal and malignant tissues in vivo and in established cell lines-Physiologic and clinical implications. Cancer 73(9):2432–2443. doi:10.1002/1097-0142(19940501)73:9<2432:aid-cncr2820730929>3.0.co;2-s
103. Weitman SD, Lark RH, Coney LR, Fort DW, Frasca V, Zurawski VR, Kamen BA (1992) Distribution of the folate receptor GP38 in normal and malignant cell lines and tissues. Cancer Res 52(12):3396–3401
104. Shan Y, Luo T, Peng C, Sheng R, Cao A, Cao X, Shen M, Guo R, Tomás H, Shi X (2012) Gene delivery using dendrimer-entrapped gold nanoparticles as nonviral vectors. Biomaterials 33:3025–3035
105. Cai WB, Chen XY (2006) Anti-angiogenic cancer therapy based on integrin alphavbeta3 antagonism. Adv Anticancer Agents Med Chem 6(5):407–428
106. Dechantsreiter MA, Planker E, Mathä B, Lohof E, Hölzemann G, Jonczyk A, Goodman SL, Kessler H (1999) N-methylated cyclic RGD peptides as highly active and selective $\alpha_v\beta_3$ integrin antagonists. J Med Chem 42(16):3033–3040
107. Gronthos S, Simmons PJ, Graves SE, G. Robey P (2001) Integrin-mediated interactions between human bone marrow stromal precursor cells and the extracellular matrix. Bone 28(2):174–181
108. Qiao Z, Shi XY (2014) Dendrimer-based molecular imaging contrast agents. Prog Polym Sci. doi:10.1016/j.progpolymsci.2014.1008.1002
109. He X, Alves C, Oliveira N, Rodrigues J, Zhu J, Bányai I, Tomás H, Shi X (2015) RGD peptide-modified multifunctional dendrimer platform for drug encapsulation and targeted inhibition of cancer cells. Colloids Surf B 125:82–89
110. Shukla R, Hill E, Shi XY, Kim JB, Muniz MC, Sun K, Baker JR (2008) Tumor microvasculature targeting with dendrimer-entrapped gold nanoparticles. Soft Matter 4(11): 2160–2163
111. Garg A, Tisdale AW, Haidari E, Kokkoli E (2009) Targeting colon cancer cells using PEGylated liposomes modified with a fibronectin-mimetic peptide. Int J Pharm 366(1): 201–210
112. Reseland JE, Syversen U, Bakke I, Qvigstad G, Eide LG, Hjertner Ø, Gordeladze JO, Drevon CA (2001) Leptin is expressed in and secreted from primary cultures of human osteoblasts and promotes bone mineralization. J Bone Miner Res 16(8):1426–1433
113. Wang SG, Castro R, An X, Song CL, Luo Y, Shen MW, Tomás H, Zhu MF, Shi XY (2012) Electrospun laponite-doped poly (lactic-co-glycolic acid) nanofibers for osteogenic differentiation of human mesenchymal stem cells. J Mater Chem 22(44):23357–23367

Chapter 13
Polyoxometalates and Their Complexes Toward Biological Application

Lixin Wu and Jing Liang

Abstract Polyoxometalates (POMs) are a type of inorganic polyanionic clusters bearing well-defined topologic architecture consisted of transition oxo-metalates. Due to their negatively charged features, various dimensions, acidity, and so forth, POMs also show specific functions in biological system. To understand the activity at molecular level, we start the discussion from the basic binding modes of POMs with biomolecules to the expression of the binding diversity on the crystallography, inhibition and hydrolysis of biomolecules. Moreover, the selective inhibition of POMs for biomolecules displays the potential roles in antitumor, antiviral, and antimicrobial activities. In the chapter, recent achievements concerning the applications of POMs on biological-related systems are summarized. The discussion involves the interaction of POMs with amino acids, peptides, and proteins, the co-crystallization of proteins with the help of POMs, the inhibitory effect of POMs on enzymes and some diseases, the mimetic enzyme functions of POMs for hydrolysis of peptides and proteins, the antiviral, antibacterial, and antitumoral activity of POMs, and their bio-imaging features.

Keywords Polyoxometalates · Proteins · Interaction · Crystallization · Hydrolysis · Inhibition · Antiviral, antibacterial, and antitumor activity · Bio-imaging

Polyoxometalates (abbreviated as POMs) in chemistry are a class of nano-sized inorganic polyanionic clusters, which consist of two or more transition metal ions such as tungsten, molybdenum, and vanadium, linked together through shared oxygen anions to constitute a closed three-dimensional framework. In these metal-oxide framework structures, those metal cations are generally in their highest oxidation state. The POM architectures can be incorporated by many metal and non-metal elements, inducing interesting diversity in compositions and configurations. Besides the chemical composition and cluster topology, the applications of

L. Wu (✉) · J. Liang
State Key Laboratory of Supramolecular Structure and Materials, College of Chemistry, Jilin University, Changchun, People's Republic of China
e-mail: wulx@jlu.edu.cn

J. Li (ed.), *Supramolecular Chemistry of Biomimetic Systems*,
DOI 10.1007/978-981-10-6059-5_13

POMs also receive extensive attention. Because of the mixed valence, POMs with intramolecular electron transfer, ring current, electron storage, and release properties can be used as electronic and protonic conductors [1–3]. Due to the strong acidity of proton counterions and/or the redox of coordination metals, POMs were used as catalysts for the synthesis and hydrolysis of esters, reduction of metal ions, oxidative polymerization of unsaturated compounds, and so forth in the past decades [4–8]. In addition, POMs also possess fantastic prospective applications in the fields of optics [9], magnetics [10], energy storage, and material science [11, 12]. In comparison to the mentioned potential applications in multidisciplinary fields, the significant utilizations of POMs appear in the biological activity as well. Due to the negatively charged features, various dimensions, strong acidity, and so forth, POMs are believed suitable for achieving specific functions in the living system. Based on the recent investigations, in the chapter we present a brief introduction about POMs' structures and fundamental properties, and then summarize the applications of POMs and their organic hybrid complexes in biological-related systems. The discussions involve the interaction of POMs with proteins, the crystallography of proteins under the support of POMs, the inhibiting effect of POMs, the mimetic enzyme effect of POMs, the antitumor, antibacterial and antiviral activity of POMs, and bio-imaging. To understand the activity of POMs at molecular level, we first elaborate the binding modes of POMs with amino acids, and then the proteins in solutions. Following the understanding of POMs' specific interactions under various conditions, the binding modes can be found in favorable for expression of the crystallography, inhibition, and hydrolysis of proteins. On the other hand, enzymes representing a type of proteins can be found inhibited by POMs, which are partially derived from the unique interactions. Further, the selective inhibition of POMs for enzyme functions demonstrates the potential role in antitumor, antiviral, and antimicrobial activities. In many cases, to enhance the interaction, selectivity, and specificity, POMs are modified with organic components and thus some POM complexes are introduced. Through the description in the chapter, the biological activity of POMs is outlined.

13.1 The Interaction of POMs with Amino Acids, Peptides, and Proteins

13.1.1 The Introduction of POMs

POMs that are known for almost two centuries are a kind of discrete anionic metal-oxygen clusters that possess a series of important features such as the diversity of dimensions and shapes, redox properties, high negative charges, and nucleophilicity. At present, the commonly used metal atoms in POMs are molybdenum, tungsten, vanadium, niobium, and tantalum. Nonmetals and non-transition metals like boron, aluminum and gallium, silicon, germanium, phosphorous, arsenic, and so forth, appear in

POMs as heteroatoms. Lacunary POMs perform the ligand to combine many other metal ions including rare earth metals to form substituted clusters displaying more additive properties. The well-known types of POMs are Keggin ($XM_{12}O_{40}^{n-}$), Silverton ($XM_{12}O_{42}$), Well-Dawson ($X_2M_{18}O_{62}^{n-}$), Waugh ($XM_9O_{32}^{n-}$), Lindqvist ($M_6O_{19}^{n-}$), Anderson-Evans ($XM_6O_{24}^{n-}$), and Finke [$M_4(XW_9)_2^{n-}$ or $M_4(X_2W_{15})_2^{n-}$] structures, where "X" denotes the heteroatom, "M" represents the coordination metal ion, and "n" means the number of charges in the chemical formula [13–16]. The periphery of POM structures spreads terminal and bridging oxygen atoms, providing the binding sites for organic groups and metal ions to enrich structural composition because in the state that one or two coordination atoms are taken out, the vacant positions exhibit reaction activity from surrounding oxygen atoms. These characteristics contribute to a variety of stereochemical structures. Some derivative POM architectures are summarized in Fig. 13.1. According to the ratio between metal ions and oxygen atoms, oxidation state of coordination atom and heteroatom, as well as topologic morphologies, the negative charges of POMs can be tuned in a pretty large range. For example, Lindqvist-type $[Mo_6O_{19}]^{2-}$ has two charges, while a number of charges on Anderson-type cluster $[MMo_6O_{18}(OH)_6]^{n-}$ and Keggin-type cluster $[MW_{12}O_{40}]^{n-}$ can be modulated via the change of central heteroatoms gradually. The similar changes can be found in another type of clusters. For more charged clusters, $[EuW_{10}O_{36}]^{9-}$, $[EuP_5W_{30}O_{110}]^{12-}$, and even $[(Mo_{132}O_{372}(H_2O)_{72}(CH_3COO)_{30})]^{42-}$ are also important members of POM family. Thus, the topology, active oxygen atoms, and negative charges in POM system provide

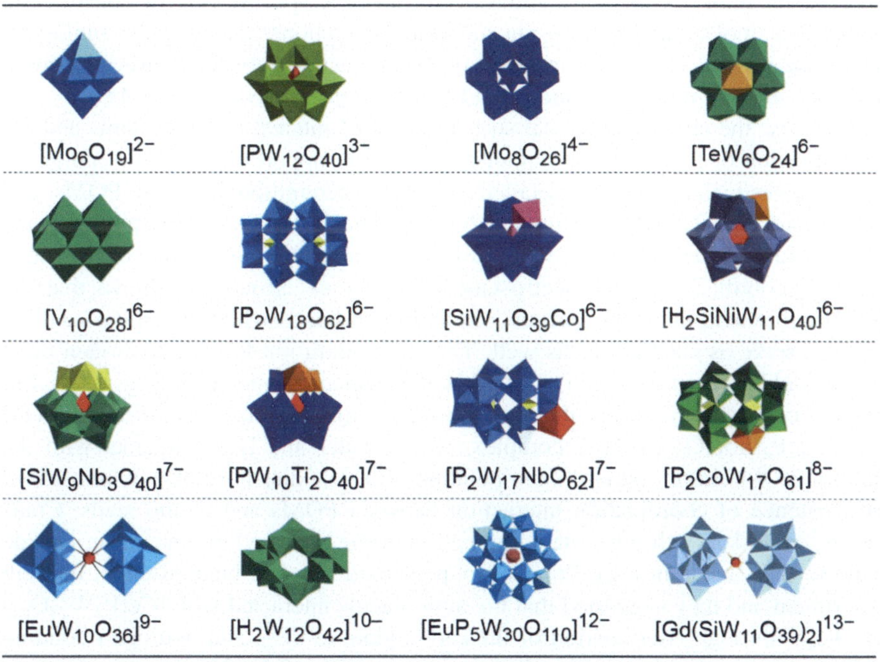

Fig. 13.1 Several representative architectures of biological POM clusters

various binding modes with metal ions and organic groups closing to biological nanostructure surfaces, through hydrogen bonding, coordination, electrostatic interaction, etc. Hence, on one hand, the anionic character of POMs naturally allows the association with cationic domains in proteins such as the areas with rich amine groups [17, 18]. On the other hand, the inorganic moieties can link organic molecules via covalent bonds, where the organic ligands substitute the oxo groups of POMs by directly bounding to the metallic center. The nucleophilic feature of the oxygen atoms located on the outside surface of POMs enhances the covalent interaction with electrophilic groups in organic units. Therefore, these characters of POMs give rise to rich binding possibility with biomolecules such as amino acids, peptides, and proteins.

13.1.2 The Interaction of POMs with Amino Acids

Amino acid is a general term for a class of organic molecules containing amino group at Alfa's position and carboxylic acid groups, and specifies on those natural molecules found in human body. They are basic units for proteins and other biological macromolecules, and the nutrition for animals. The proteins are constituted of twenty amino acids for the genetic code. The structure of amino acids has common characters, each containing at least one amino and one carboxyl group and they are connected to the same carbon atom. The substituted R group determines the type of amino acids. The amino acids play the specific role in constructing hierarchical molecular structure via multiple interactions, realizing the biological activity of protein. Thus, the exposition of the interaction between POMs and amino acids becomes significant to understand the biological activity of POMs.

Recently, the chiral and helical structures are of interest in bio-systems and the crystallization of framework compounds under the support of chiral molecules are widely investigated, but the detailed selective recognition between POMs and amino acids is relatively infrequent [20, 21]. The interaction between POM and amino acid is demonstrated to derive from the synergistic interactions of hydrogen bonding, coordination, and electrostatic force. Crans' group has shown that the POM–amino acid interaction occurs via hydrogen bonds while the amino acids or peptides serve as counterions as well. The ammonium ion forms a hydrogen bond with a double bridging oxygen atom of the decavanadate anion [19] (Fig. 13.2). Liu and his coworkers reported the first structural information of amino acid–POM $(Lys)_2H_6[P_2Mo_{18}O_{62}]\cdot 16H_2O$ complex, where water and lysine interact with the Dawson POM cluster via hydrogen bonding. Their studies have also demonstrated the existence of coordination interaction between POMs and amino acids. Crans and his coworkers exhibited the coordination of the hydroxyl groups on serine side chain to the vanadium [22]. Pope's group verified the coordination effect through experiment and they suggested that the amino acids interacted with $[Ce(P_2W_{17}O_{61})(H_2O)_x]^{7-}$ through coordination of the carboxylate to Ce^{3+} and hydrogen bonding of the ammonium cation to the adjacent oxygen (Fig. 13.3) [23]. In comparison to those unique results, Wu and his coworkers investigated the interactions of all 20

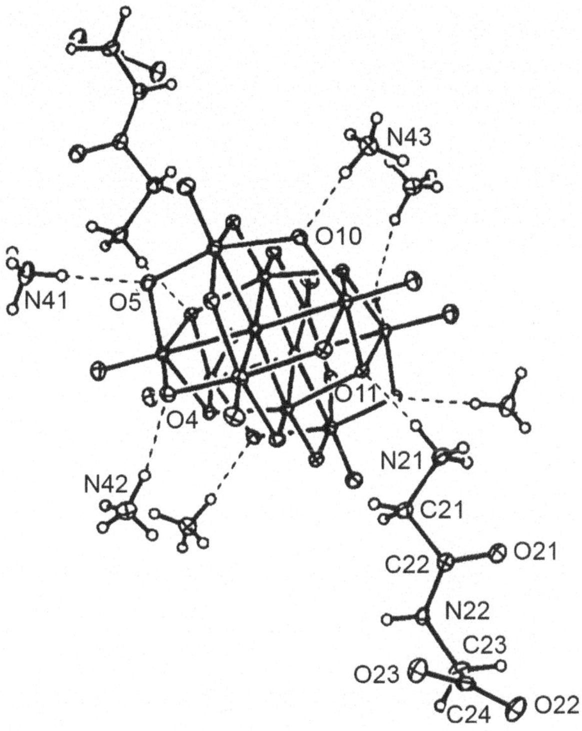

Fig. 13.2 Drawing of a portion of the structure of $(NH_4)_6$(Gly-Gly)-$V_{10}O_{28}\cdot 4H_2O$. Reprinted with the permission from Ref. [19]. Copyright 1994, American Chemical Society

amino acids with common POMs systematically. By using a luminescent POM, $K_{13}[Eu(SiW_{10}MoO_{39})_2]$, they observed that basic amino acids, lysine, arginine, and histidine strongly enhance the luminescent intensity, while acidic and neutral amino acids do not show apparent action to the POM's emission [24]. It is known that the electrostatic interaction from organic cations such as those cationic surfactants to polyanionic POMs yields a relative hydrophobic environment on the POM's surface, which restricts the luminescent quenching of POMs. According to the isoelectric points, basic amino acids possess positive charges at pH *ca.* 6 buffer solution. Under this condition, the positively charged residues from the basic amino acids are dominant and combine with POMs carrying thirteen negative charges. In contrary to this, neutral and acidic amino acids do not show any positive charges, so that no obvious interaction in aqueous solution could be observed distinctly. The NMR spectra demonstrated that the binding site was located at the residual amino groups of lysine whose isoelectric point is 9.7. Further, the controlled experiments exhibited that more negative surface charges on the POMs were beneficial for recognition of amino acids, indicating that the combination of POMs and amino acids was mainly modulated by electrostatic interaction (Fig. 13.3). In addition to this understanding, it could be predicted that the acidic and neutral

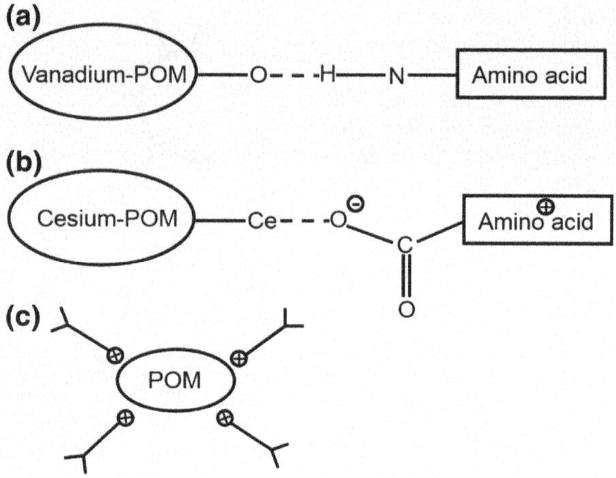

Fig. 13.3 The schematic types for the most frequent interactions occurred in POM and amino acid, where **a** denotes hydrogen bonds, **b** coordination interaction, and **c** electrostatic interaction

amino acids interact with POMs under the pH lower than each of their isoelectric points though the corresponding acidic environment normally does not appear in living biological system.

13.1.3 The Interaction of POMs with Peptides

Peptides are biological oligomers of amino acid monomers, which are prepared through the reaction of carboxyl group in one amino acid with the amine group in another. Like amino acids, the interactions of POMs with peptides also involve hydrogen bonding, electrostatic interaction, and coordination effect. As the early research, Crans et al. reported two model complexes for vanadium–peptides oxovanadium triethanolaminate and oxovanadium tri-2-propanolaminate, to reveal their interactions. The vanadium atoms in the complex displayed distorted trigonal bipyramidal coordination with the nitrogen atom and the oxo group in the axial positions, because the vanadium was displaced out of the plane of the triethanolaminate oxygens in the direction of the doubly bonded oxygen [25]. Li's group studied the self-assembly of cationic diphenylalanine peptide FF with a Keggin-type POM, $H_3PW_{12}O_{40}$. Because the strong acidity of the cluster used, its combination with the short peptide was similar to the neutralization but the formed complex was under predominantly driven by electrostatic interaction due to the protonation of amino group on the peptide. The formed hybrid spherical aggregations were found not only sensitive to pH and temperature, but also having adaptive encapsulation features for a variety of guest materials [26]. Wu and his coworkers recently reported the self-assembly of arginine/lysine-rich peptide from human papillomavirus (HPV) capsid protein and $Na_9[EuW_{10}O_{36}]\cdot 32H_2O$. In contrast to amino acids, the interaction of the peptide with the POM leads to the self-assembled nanospheres and the luminescence of europium-POM increases greatly. On the other hand, different from small amino acid molecules, the peptide provides multiple binding sites involving hydrogen bonding and electrostatic interaction, as confirmed by nuclear magnetic resonance (NMR) and isothermal titration calorimetry (ITC) spectra [27]. Further investigation demonstrated that other POMs, especially those with higher charges, have similar features during mixing with peptides, as seen from much stronger luminescent intensity enhancement of europium-substituted POM cluster $K_{13}[Eu(SiW_{10}MoO_{39})_2]\cdot 28H_2O$. Accompanying a higher luminescence enhancement of the POM, a two-step assembly was observed as well. At beginning, increasing the peptide concentration leads to the formation of strip-like aggregates and then bigger spherical assemblies (step I). Further increasing the ratio of peptide in the mixture solution resulted in the stronger association of the two components in the assemblies (step II) (Fig. 13.4). Due to the composition of the peptide used, the process of step I refers to the hydrophobic interaction from the C-termini, while step II involves the electrostatic interaction and hydrogen bonding. They suggested that the highly charged POMs might lead to synergistic interaction forces when binding with peptides or proteins, which have not been taken into consideration in the previous investigations [28].

13.1.4 The Interaction of POMs with Proteins

As proteins are biomacromolecules that consist of one or more polypeptides, their structures are normally divided into four levels in order to account for functions. The primary structure is the collating sequence of amino acids on the peptide chains, which is also the most basic structure of proteins. The secondary structure refers to the local spatial arrangement of atoms on the main peptide chains, which

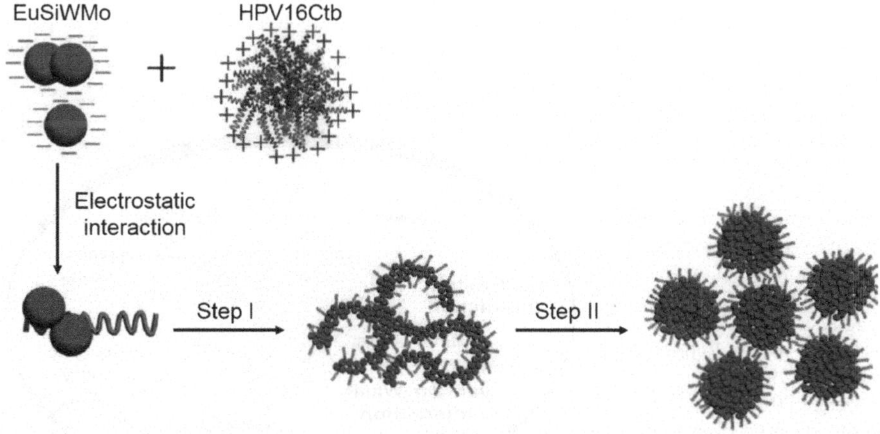

Fig. 13.4 The schematic drawings of process between POMs and peptides at different molar ratios. Reproduced from Ref. [28] by permission of John Wiley & Sons Ltd.

does not involve the conformation of side chains. A certain regular three-dimensional structure which is folded by secondary structure is called the tertiary structure. As a typical characteristic, in some protein structures, there exist hydrophobic sites and/or hydrophilic domains in caves and pockets and they become active regions. When the proteins consist of two or more independent peptide chains, the spatial structures from the chains linking through secondary bonds are called quaternary structure. In addition to the interactions deriving from the primary and secondary structures, the hydrophobic/hydrophilic interaction and the spatial matching of proteins with POMs are also investigated recently. In many cases, proteins can be regarded as the nano-sized self-assemblies and their interactions with POMs subject to several factors besides the local supramolecular force. Though the POMs can be simplified to be a rigid spherical structure, there is still crucial influence on the proteins' activity when they close to each other. These properties of POMs make them special candidates and attract much attention in exploiting new inorganic inhibitors and drugs. Following the understanding of amino acids and peptides, the fundamental interactions between proteins and POMs are divided into electrostatic interaction, hydrogen bonds, covalent bonds, and van der Waals interactions (Fig. 13.5) [29]. Actually, because of lacking crystal structure identification for the binary system, the precise spatial location of the two components is in general not very distinct, and described mostly to be an outline.

Due to the intrinsically negative charged features of POMs, electrostatic interaction often involves in the combination with proteins. Human serum albumin (HSA) is usually employed as a model for the evaluation of biological activity of POMs. As HSA possesses only one tryptophan, the combination from POMs was observed through monitoring the fluorescence change of tryptophan. The increase of quenching constant for tryptophan was regarded as the sign of HSA interacting with POMs while the binding location could be estimated [30–32]. As demonstrated in early studies, the negative charged POMs are believed to anchor the positive cavity surface of HSA. This speculation is supported by the fact that increasing the POMs' charges or decreasing pH value leads to the stronger combination of POMs with proteins [17, 33, 34]. Zhang and his coworkers studied the

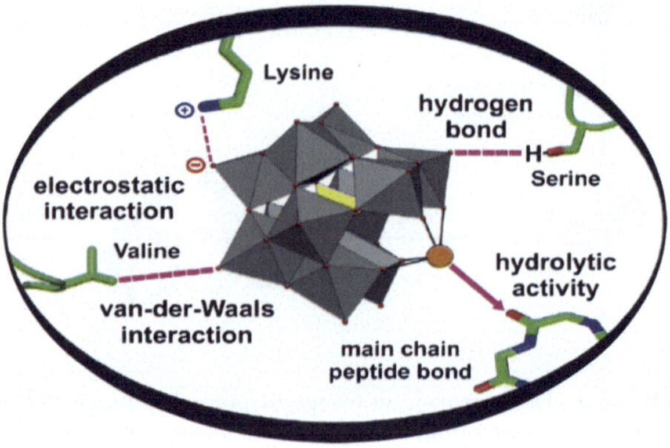

Fig. 13.5 Graphic showing the most frequent POM–protein interactions. A proteolytic active POM $[Ce(PW_{11}O_{39})_2]^{10-}$ consisting of a strong Lewis acid metal ion, Ce(IV), connected to a Keggin structure is used as illustrative model. Reprinted from Ref. [29], Copyright 2015, with permission from Elsevier

molecular interaction between HSA and two POMs, $[H_2W_{12}O_{40}]^{6-}$ (H_2W_{12}) and $[NaP_5W_{30}O_{110}]^{14-}$ (P_5W_{30}), via ITC, fluorescent and circular dichroism (CD) spectra. Their results supported the strong binding of POMs to the protein and it was demonstrated that the polyanions' size played an important role as well for the combination [35]. It should be noted that there was no specific binding site around POMs, the electrostatic interaction to proteins was not selective. Therefore, the intensity for the interaction was mainly related to the POMs' size and charges, and for the same POM, one can observe the interaction extending to other proteins, such as bovine serum albumin (BSA), histone H1, and so forth [18, 33]. This interaction was not only limited to solution, but can also be transferred onto interface, where the POMs such as $[EuW_{10}O_{36}]^{9-}$ and $[NaP_5W_{30}O_{110}]^{14-}$ patterned polymer film can be applied for the selective adsorption of proteins (BSA and hemoglobin) through modulation of pH value and POM charges [11]. Thus, as shown in Fig. 13.5, the basic lysine residues with positive charges in most situation afford positive protein surface, and could be always the binding site to the POMs while only the shape of architecture and the charge localization-induced unsymmetrical distribution make difference from proteins.

Hydrogen bonds are also a type of electrostatic interaction, thus they are contributed to the binding with POMs. However, the studies on this binding mode are pretty limited. Müller and his coworkers utilized a molybdenum/tungsten storage protein from Azotobacter vinelandii to investigate the driving force. It was found that $[W_3O_{10}H_xN_3]^{(6-x)-}$ indirectly connected with the protein through precisely confirmed hydrogen bonds via single X-ray structure analysis [36]. In the POMs–protein complex, hydrogen bonds are found surrounding different POMs but part of them is ascribed to solvent molecules. This type of solvent-mediated interaction brings an unexpected advantage, that is, the negatively charged POMs can bind with negatively charged proteins through the bridging hydrogen bonds [37]. Thus, the POMs can interact with the proteins that provide proton donors through hydrogen bonds like serine residues on the proteins (Fig. 13.5).

Covalent binding was employed to serve as interaction mode. The molybdenum storage protein has been reported playing a pouch to accommodate approximately 100 biologically relevant molybdenum or tungsten atoms [38]. This protein was observed to bind with the POM clusters covalently, in which some octamolybdates covalently bound to the nitrogen of histidine and oxygen of glutamic acid, forming local complex structure of $[Mo_8O_{26}O(Glu)N(His)H_n]^{n-5}$. Another example using covalent binding is that the storage protein binds with tungsten cluster to form a complex with the formula $[W_3O_{10}H_xN_3]^{(6-x)-}$. This small metal-oxide unit is consisted of three edge-sharing octahedra that are related by the crystallographic threefold axis; the N_3 in the formula represents three nitrogen atoms of the histidine [36]. For $[Mo_8O_{28}]^{8-}$ cluster, it was found to bind covalently with the hydroxyl oxygen of serine and the nitrogen from histidine in Nucleoside triphosphate diphosphohydrolase 1 protein [39]. These results indicate that the covalent interaction from POMs is feasible, which provide specific recognition with biomolecules.

Van der Waals force is also possible for POMs–protein interactions, though it is rarely applied. The limited study pointed that van der Waals interaction as the main binding force to drive the complexation of molybdenum storage protein with molybdenum clusters [40]. Among of which, three types of POMs, Mo_6, Mo_7, and Mo_{13}, are seated in the protein-formed pouch, and attach to nonpolar hydrophobic regions which have rich glycine, valine, and proline and a few of serine residues (Fig. 13.5).

13.2 The Crystallography of Proteins by POMs

13.2.1 POMs as Crystallization Additives

In order to understand various biological and pharmacological functions of biomacromolecules, it is important to carry out clear chemical and spatial structure of proteins through the crystallographic study. For the latter case, the most important work is to realize the protein crystallization, which is not an easy process because it is still empiric with luck and we have to face many complicated procedures and environmental factors. Fortunately, the introduction of additives such as some small molecules often has an unexpected effect on the achievement of protein crystallization. Among the trials, POMs are found as a promising candidate to raise the crystallization ability of proteins.

When POMs act as additives for proteins' crystallization, suitable combination with proteins and solubility in buffer solution are the requirements in crystallization process. POMs possess high solubility in water and polar organic solvents with unchanged framework structure, and the solubility can be modulated through the change of counterions [41]. Due to the positive charged portion of proteins, the electrostatic interaction from POMs provides favorable condition to ensure the complexation for co-crystallization. On the other hand, the POMs as nano-sized clusters have the dimension comparable to proteins or the domain of proteins, and thus some POMs exhibit shape functions. In special cases, POMs are anchored to the cavities of proteins and catalytic centers of enzyme, or trap the conformation of enzymes. Of course, the POM complexes with hydrophobic covering are used for the same purpose in aqueous solution at room temperature. For protein crystallization, the pH values are needed to control in the range of pH 2–10; stable POMs in the solution should be considered though most of them are stable under the conditions. Therefore, controlling the stability of POMs in the wide pH range is essential. Several techniques have been applied to determine the species of POMs in the experimental process, such as UV/vis, NMR, and ESI-MS [42, 43]. According to the above discussion, when choosing POMs as additives for protein crystallization, maintaining the integrity of POMs is required.

13.2.2 The Usage of POMs in Protein Crystallography

Since famous work by Yonath, the usage of POMs in protein crystallization attracted lots of attention [44]. Rompel and his coworkers succeeded in obtaining the structure of latent precursor form of mushroom tryosinase polyphenol oxidase 4 (PPO4), one of the six tyrosinase isoforms from Agaricus bisporus, from a single crystal [45]. They found that the protein crystallization can occur in the presence of a POM cluster, $Na_6[TeW_6O_{24}]\cdot 22H_2O$ as co-crystallization agent with low-salt concentration. In contrast, with $MgCl_2$ as additives, only wispy sea urchin-like microcrystals were obtained. Although several attempts have been utilized, such as alternation of pH, precipitation agent, and temperature both qualitatively and quantitatively, single crystals were not achieved. The substitution of POM cluster as additives, flat rod-shaped single crystals appeared (Fig. 13.6). The structural analysis proved that the Anderson cluster bound to positive regions of proteins through electrostatic interaction. Moreover, in the crystal packing cases, the unique topology and high negative charged feature make the cluster act as 'glue' layer connecting two electrostatically repulsive protein layers [46].

The utilization of POMs on protein crystallization focuses on several aspects: ability of protein crystallization by stabilizing conformation, rigidification of flexible regions of proteins, and enhancement of protein packing stability during crystallization process. For a typical example, those POMs with transition metal appendages such as Mo, V, and W were used as substrates or inhibitors for the assistance of enzyme crystallization [47, 48]. Deisenhofer et al. have reported that the addition of $Na_3PW_{12}O_{40}$ as a large anomalous scatterer improves the diffractive quality of lipoprotein receptor largely in single crystals, which were favorable for

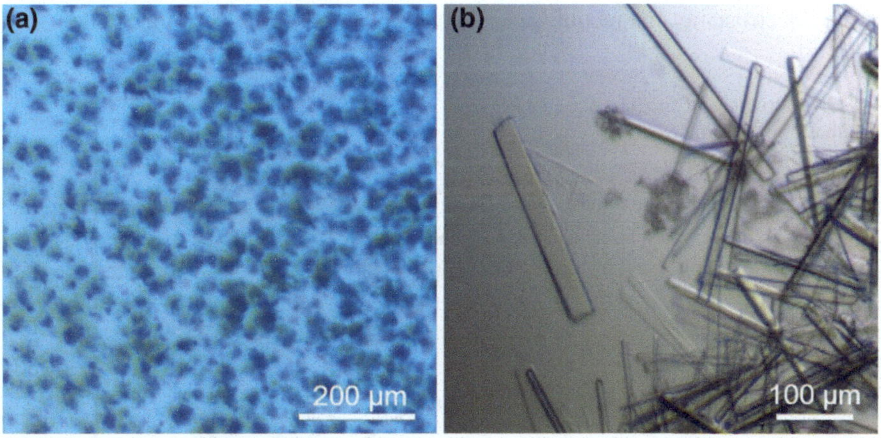

Fig. 13.6 Crystal images of PPO4 mushroom tyrosinase: **a** wispy microcrystals obtained using $MgCl_2$ as a crystallization additive, and **b** flat rod-shaped crystals obtained using the POM as a crystallization additive. Reproduced from Ref. [45] by permission of John Wiley & Sons Ltd.

the determination of crystal structure [49]. In another example, the vanadate (V) complex in a trigonal bipyramidal geometry makes the compound similar to a transition state of phosphoryl transfer reactions, which is a powerful technique in macromolecular crystallography [50]. The vanadates are known for high versatility in bond order and bond angles. The unique feature allows them not only to bind to the active site of phosphoryl transferring enzymes but also to adopt a form similar to the species in the transition state. Thus, POMs with vanadate are applied to bind the enzymes by closing to catalytic center while stabilizing the conformation at the site and driving the crystallization of substrate [49, 51–55].

The ribosomal subunit performs the decoding of genetic information during translation encounters serious problems due to their large size, complicated structure, inherent flexibility, and high conformational variability. Interestingly, Dawson POM $K_6[P_2W_{18}O_{62}]$ (P_2W_{18}) displays the ability to stabilize and rigidify the flexible subunit, by increasing the homogeneity of the activated form within the protein crystals, yielding a dramatic increase of the structural resolution [56]. In the structure of 30S ribosome, there are head, platform, and base domains. The head area is pretty flexible, which makes 30S ribosome difficult for crystallization (Fig. 13.7). The addition of Dawson cluster improved the rigidity greatly and in the formed protein crystal there are one single 30S particle and seven P_2W_{18} clusters in one asymmetric unit. The result demonstrated that the cluster provided a beneficial conformational controlling to the protein, where it frozes the whole vicinity of particle regions, which affected the head motion of protein and traps the structure in one conformation. Thus, the Dawson POM exhibited two functions: providing anomalous phasing power and increasing the structure resolution through stabilization of a selected functional conformation [57].

Some POMs were also used to enhance the stability and packing order of protein crystals by displaying the ability to link different monomers or regions during mediating the packing protocol of crystals [46, 51, 58, 59]. The linkage is attributed to the electrostatic interaction and hydrogen bonds between the POMs and

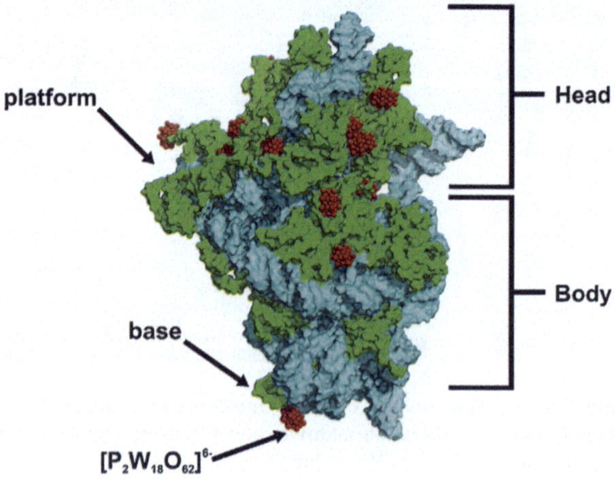

Fig. 13.7 Structure of the 30S ribosome in surface representation. Reprinted from Ref. [29], Copyright 2015, with permission from Elsevier

monomers. The combination of two protein units through the specific interactions with negatively charged POMs induces new crystal contact, which supports the lattice formation and final crystal growth. It has been reported that only in the presence of $Na_6[TeW_6O_{24}]$, the PPO4 protein crystals can grow up enough for the structural measurement. In the obtained crystal structure, POM locates at the place at which the positively charged regions of the two proteins locate [46]. This POM shows a broad ability in the crystallization process with other proteins such as hen egg-white lysozyme (HEWL), as investigated by the same group [58]. The HEWL co-crystallizes with $Na_6[TeW_6O_{24}]$ under the state closing to the region that the liquid–liquid phase separation occurs. The X-ray structure analysis exhibits that the Anderson cluster acts as the constitution of the crystal by locating between two or more symmetry-related protein chains. Thus, the POM as the charged counterpart neutralizes the electrostatic repulsion at protein–protein surfaces, inducing the formation of a stable crystal lattice. These investigations demonstrate that POMs have great potential as crystallization additives to promote both the formation of protein crystals and elucidation of their structures.

As it is normally hard for one protein to directly interact with another due to the less driving force between them during crystallization, the introduction of POMs plays the role of linker to combine protein layer like a glue to optimize the crystallization process. In this case, several crucial factors, such as negative charge number, charge distribution, dimension, symmetry, and topology, affect the final role of POMs in the process [49, 55, 60]. With the increase of charge numbers of POMs, the binding force with protein regions increases accordingly. Due to the binding site selection, the symmetry of POMs can affect the array of proteins to some extent. Additionally, the dimension of POMs is always one of the important factors during the co-crystallization with protein macromolecules. When a POM with proper size is selected, it becomes easier to link two incompatible proteins in a certain distance during the tight packing. As a result, the crystallization process is facilitated because the long-range force between the two units decreases while the short-range attraction increases. Therefore, the topologic matching between POM and protein promotes the synergistic assembly between different protein units, resulting in a denser crystal packing mode.

13.2.3 The Advantages of POMs on Protein Crystallization

Comparing with other common crystallization additives such as small molecules or ions, POMs display unique advantages for biomolecule crystallization. The POM family presents highly versatile negative charges, size, and shape, as well as rich interactions during connecting two or more protein monomers, many features of which small crystallization additives do not possess.

13.3 The Inhibiting Effect of POMs

Enzymes represent a type of biological polymers that have the function of catalysis. In the catalytic reactions participated by enzymes, the reactants known as the substrates are transformed to other molecules under ambient conditions. Almost all the cells' activities require the participation of enzymes while they do not consume themselves in the chemistry process. In contrast to those non-biological catalysts, natural enzymes are highly selective; only catalyze specific reactions for certain products. The activity of enzymes is usually influenced by the factors such as temperature, chemical environment, substrate concentration, electromagnetic wave, and the added molecules, of which some are activators increasing the activity but the others are inhibitors restraining the reaction.

13.3.1 Stabilization of POMs in Physiological Solution

POMs that are found biologically active have the ability on inhibition of enzymes' function. However, some POMs are both kinetically and thermodynamically unstable in aqueous solution and tend to convert into different clusters. In order to guarantee the initial application on biological system, several methods are adopted to improve POMs' stability in physiological condition. The metal ion substitution in a given cluster shows a convenient route to get different Keggin-type clusters with a broad tolerance from the required environment. With this approach, $[PW_{10}Ti_2O_{40}]^{7-}$ was prepared to have an increased hydrolytic stability even at the pH reaching *ca.* 7.6, compared with the cobalt- and titanium-substituted compound $[TiW_{11}CoO_{40}]^{7-}$ and the typical one $[PW_{12}O_{40}]^{3-}$, which only maintain their frameworks under an acidic condition [61–63]. Other strategies such as surface modifications to the POMs were adopted to get a higher hydrolytic stability and a lower toxicity (Fig. 13.8). As an example, the covalent linkage of organic ligands to POMs such as Anderson structures largely improves the stability of POMs at physiological condition, and the grafted POMs involve a condensation reaction between a functional group such as alcohols, carboxylates and amines, and metal-oxygen structure such as polyoxo-tungstate, -vanadate, and -molybdate. Amino acids, peptides, and carbohydrates have been grafted onto the POMs such as $[P_2W_{17}O_{61}\{SnCH_2CH_2COOH\}]^{7-}$ to realize the bio-functions [64–67]. Surface covering is also an efficient procedure to realize the organic modification of POMs. The encapsulation to the POMs with organic compounds such as dendrons and surfactants represents a general route to increase the stability of clusters, which provides not only hybrid building blocks for diverse self-assembly but also an artificial model with catalytic center locating at a hydrophobic environment similar to enzyme for catalytic organic reactions [68–70]. The nanocarriers formed by liposome- or starch-encapsulated POMs with high stability and low toxicity were prepared for enhanced performance on antitumor activity [71–74]. Besides small molecules, the polycationic biopolymer chitosan was

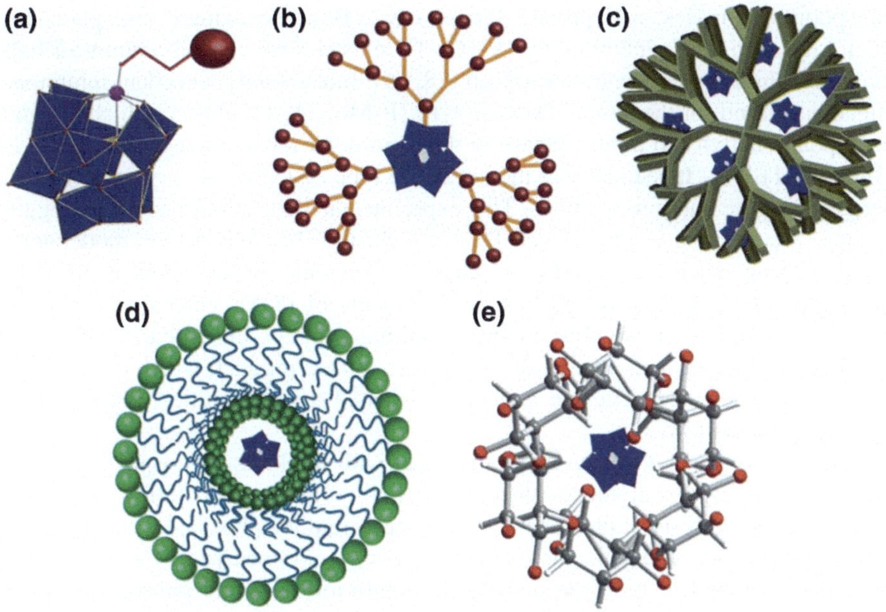

Fig. 13.8 Strategies for the functionalization of POMs; **a** covalent attachment of organic ligands (polymers), **b** encapsulation using dendrons, **c** encapsulation in dendrimers, **d** encapsulation in liposomes and **e** encapsulation in biopolymers (starch, chitosan). Reproduced from Ref. [77] by permission of John Wiley & Sons Ltd.

employed to encapsulate POMs such as $[Eu(\beta_2\text{-}SiW_{11}O_{39})_2]^{13-}$ and $[Cs \subset Eu_6As_6W_{63}O_{218}(H_2O)_{14}(OH)_4]^{25-}$ to perform POM's therapeutic property [75, 76].

13.3.2 *POMs as the Inhibitors of Enzymes*

POMs have been reported to have the functions of enzyme inhibitors on several enzyme families, but the systematic structure–activity relationship on POMs' inhibiting effects was believed to be rarely understood [77]. The main challenges focused on (1) most studies involved in a limited number of POMs with large structural variations; (2) the alternation of POMs' stability and the fragment formation interfered with the evaluation of activity; and (3) the mechanism of inhibition was rarely discussed. Here, we just summarize the limited types of enzymes inhibited by POMs.

Kinase is a kind of enzyme transferring phosphate group from high-energy donor molecules such as ATP to a specific target molecule, and the process is called phosphorylation. The bioactivity of different POMs by a noncompetitive fashion has been investigated for several kinases, for instance, hexokinase, phosphofructokinases, and protein kinase CK2. Both hexokinase and phosphofructokinase could be inhibited by $[V_{10}O_{28}]^{6-}$ cluster with a small amount of half maximal inhibitory concentration (IC_{50}) [78, 79]. Interestingly, excellent inhibitory effect was found for the simple Dawson-type $[P_2Mo_{18}O_{62}]^{6-}$ cluster, which had the most potent activity to inhibit kinase in the nanomolar range among tens of POMs (Table 13.1) [80]. It seemed that the structural type of POMs is very important in determining the inhibitory effect. The experimental data confirmed that Keggin clusters were inactive, the larger Dawson clusters became active moderately, whereas those much larger and more charged Preyssler clusters showed the most active inhibition. It was trusted that the fragments of POMs entered into several inhibitory moieties of proteins. In the inhibiting process, the active form of the POM is fragmented with a particular protein-stabilized product, which occurs only in the protein environment and cannot be isolated as free entities, playing their inhibitory effects [80].

The functions of phosphatases are opposite with the kinase. Phosphatases have the ability to carry out the dephosphorylation of the corresponding substrates through the hydrolysis of phosphate monoester to remove phosphate groups in the substrate molecules and generate the phosphate ions and free hydroxyls. The POMs were demonstrated to have the activity for inhibition on many phosphatases, for instance, acid phosphatase, alkaline phosphatase, protein tyrosine phosphatase, and phosphoglycerate mutase B, as shown in Table 13.1 [81–83]. Among the POMs, the most widely used one is $[V_{10}O_{28}]^{6-}$, and some of its complexes with organic groups are also active. At present, a mixed inhibition for alkaline phosphatase and competitive inhibition for acid phosphatase and protein tyrosine phosphatase were proposed though the mechanism of inhibitory effects was not dealt in detail.

Sulfotransferases are a type of transferase enzymes which catalyze the transfer of sulfo groups from donor molecules to acceptors such as alcohol or amine. Sialyltransferases are the enzymes that transfer sialic acid to nascent oligosaccharide. The two types of enzymes are located on the cell surfaces. The inhibitions of sulfotransferases and sialyltransferases have been realized by a set of POMs with a low IC_{50} value in nanomolar range [84, 85]. Several tungstate-type POMs, such as $[(GeTi_3W_9O_{37})_2O_3]^{14-}$, $[PTi_2W_{10}O_{40}]^{7-}$, and $[H_2SiNiW_{11}O_{40}]^{6-}$, were observed to have the behavior to inhibit the activity of Gal α 2, 3-sialyltransferase I (ST3Gal-I) at nanomolar levels. For noncompetitive inhibition on both donor and acceptor the IC_{50} of the best inhibitors was 0.2 nM. The isopolyvanadates display inhibition for both ST3Gal-I and Gal 3-O-sulfotransferase-2 (Gal3ST-2) at nanomolar levels (Table 13.1). The inhibitory effect of tungstate-type POMs on ST3Gal-I involves with reversible and electrostatic behaviors.

Nucleotidases are the type of hydrolytic enzymes that catalyze the hydrolysis of nucleotide into nucleoside and phosphate. A number of POMs have been reported to inhibit the activity of *ecto*-nucleoside triphosphate diphosphohydrolases (E-NTPDases types 1–3) with micromolar potency (Table 13.1) [61, 86]. The most potent POM cluster found in the present stage is $[TiW_{11}CoO_{40}]^{8-}$, which presents IC_{50} value of NTPDase1 with 0.14 μM, NTPDase2 with 0.91 μM, and NTPDase3 with 0.56 μM. The commercial POM $[H_2W_{12}O_{40}]^{6-}$ exhibits even a higher potency

Table 13.1 Selected enzymes inhibited by POMs

Enzyme	POMs	IC_{50}, K_i or K_d [μM]
Kinases		
Hexokinase	$[V_{10}O_{28}]^{6-}$	62
Phosphofructokinase	$[V_{10}O_{28}]^{6-}$	0.045
Protein kinase CK2	$[P_2Mo_{18}O_{62}]^{6-}$, $[P_5W_{30}O_{110}]^{15-}$	0.0014, 0.001
Phosphatases		
Acid phosphatase	$[V_{10}O_{28}]^{6-}$	Low on μM level
Alkaline phosphatase	$[V_{10}O_{28}]^{6-}$	*ca.* 5
Protein tyrosine phosphatase 1B	$[V_{10}O_{28}]^{6-}$	Low on μM level
Phosphoglycerate mutase B	$[V_{10}O_{28}]^{6-}$	*ca.* 10
Sulfotransferases		
Gal 3-*O*-sulfotransferase-2	$[(GeTi_3W_9O_{37})_2O_3]^{14-}$	0.02
	$[SiVW_{11}O_{40}]^{5-}$	0.05
	$[H_2V_{18}O_{44}(N_3)]^{14-}$	0.003
Sialyltransferases		
Gal α2,3-sialyltransferase I	$[(GeTi_3W_9O_{37})_2O_3]^{14-}$	0.003
	$[SiVW_{11}O_{40}]^{5-}$	0.002
	$[H_2V_{18}O_{44}(N_3)]^{14-}$	0.007
E-nucleotidasescto		
NTPDase1	$[H_2W_{12}O_{40}]^{6-}$, $[PW_{12}O_{40}]^{3-}$ $[Ti_2W_{10}PO_{40}]^{7-}$, $[TiW_{11}CoO_{40}]^{8-}$ $[Co_4(H_2O)_2(PW_9O_{34})_2]^{10-}$	0.14−3.5
NTPDase2	$[H_2W_{12}O_{40}]^{6-}$, $[PW_{12}O_{40}]^{3-}$ $[Ti_2W_{10}PO_{40}]^{7-}$, $[TiW_{11}CoO_{40}]^{8-}$ $[Co_4(H_2O)_2(PW_9O_{34})_2]^{10-}$ $[NaSb_9W_{21}O_{86}]^{18-}$	0.9−30
NTPDase3	$[H_2W_{12}O_{40}]^{6-}$, $[PW_{12}O_{40}]^{3-}$ $[Ti_2W_{10}PO_{40}]^{7-}$, $[TiW_{11}CoO_{40}]^{8-}$ $[Co_4(H_2O)_2(PW_9O_{34})_2]^{10-}$ $[NaSb_9W_{21}O_{86}]^{18-}$	0.6−9
Histone deacetylases		
Histone deacetylase	$[(nBu)Sn(OH)_3GeW_9O_{34}]^{4-}$	
Choline esterases		
Acetylcholine esterase	$[H_2W_{12}O_{42}]^{10-}$	0.3
	$[TeW_6O_{24}]^{6-}$	
Butyrylcholine esterase	$[(O_3PCH_2PO_3)_4W_{12}O_{36}]^{16-}$	0.18
Nucleases		
Ribonuclease	$[V_{10}O_{28}]^{6-}$	1.4
Polymerases		
Proteases		
HIV-1 protease	$[P_2W_{17}(NbO_2)O_{61}]^{7-}$	1.2−2.0
	$[NaP_5W_{30}O_{110}]^{14-}$	5.5

for NTPDase1 and 3. Interestingly, $[NaSb_9W_{21}O_{86}]^{18-}$ shows selectivity for NTPDases2 and 3 corresponding to NTPDase1.

POMs can also perform an inhibitor for histone deacetylases. The POM $[(n\text{-Bu})Sn(OH)_3GeW_9O_{34}]^{4-}$ demonstrated the activity on some proteins such as histone deacetylase with a low IC_{50} concentration [87]. It was estimated that the fragments derived from the degradation of the POM cluster might contribute to the inhibitory activity.

In terms of efficiency and selectivity, POMs have been found to be effective cholinesterase inhibitors. Iqbal and his workers elaborated that some isopolytungstates act as potent inhibitors of acetyl and butyrylcholinesterases. Among the polytungstates, $[H_2W_{12}O_{42}]^{10-}$ and $[TeW_6O_{24}]^{6-}$ were thought to have the most potent activity on acetylcholinesterase with the IC_{50} value about 0.3 μM. For $[(O_3PCH_2PO_3)_4W_{12}O_{36}]^{16-}$, it performs an effective and selective inhibitor for butyrylcholinesterase with IC_{50} value about 0.18 μM [88]. Proteases can be inhibited by a set of tungstate-type POMs through a noncompetitive process at low micromolar concentrations [89, 90]. Nucleases such as ribonuclease and polymerases can be inhibited by $[V_{10}O_{28}]^{6-}$ cluster with IC_{50} at micromolar level [91]. Thus, in general, POMs show the bio-functional potentials as powerful agents to inhibit various kinds of enzymes.

13.3.3 The Restriction of POMs for Self-assembly of Proteins

Besides acting as enzyme inhibitors, a significant inhibiting application of POMs is the prevention of aggregation of amyloid β-peptides (Aβ), which is associated with Alzheimer's disease (AD). The most common form of AD is dementia, which is characterized by a loss of brain function affecting memory, cognition, and behavior [92]. Due to the complexity of AD, its molecular mechanism of pathogenesis is not fully understood, but the known researches have proposed that the aggregation of Aβ into amyloid fibrils is crucial [93]. Qu and his coworkers have made important achievements on the prevention of aggregation of Aβ by using POMs as inhibitor. Through fluorescent measurement and transmission electron microscope (TEM) observation, they show that $K_7[PTi_2W_{10}O_{40}]$, $K_8[SiW_{11}O_{39}]$, $K_8[P_2CoW_{17}O_{61}]$, and $H_3[PMo_{12}O_{40}]$ present activity of Aβ inhibitor by using Aβ1–40 as a model. Among the used species, the highest inhibition is observed for Dawson-type $K_8[P_2CoW_{17}O_{61}]$ (Table 13.2). Through comparing with a number of Dawson, Keggin, and Anderson POMs bearing different chemical compositions and charges, the inhibitory efficiency is confirmed in relevant to the architecture types. With the increase of dimension and charge amount of POMs, inhibitory efficiency of Aβ reinforces greatly, indicating that electrostatic and size effects play an important role in the restriction of unwanted protein assembly due to the interaction of POM with Aβ. The fluorescent measurement by utilizing 4,4′-bis

Table 13.2 Inhibitory efficiency of POMs with respect to Ap fibrillization

	Structure type	Charges	$K_a^{[a]}(M^{-1})$	$IC_{50}^{[b]}(\mu M)$
$Na_5[IMo_6O_{24}]$	Anderson	5	–	No inhibition
$K_7[PTi_2W_{10}O_{40}]$	Keggin	7	4.65×10^5	39.04
a-Na_9H-	Keggin	10	4.55×10^5	19.85
$[SiW_9O_{34}]$				
$K_8[\beta\text{-}SiW_{11}O_{39}]$	Keggin	8	4.76×10^5	39.02
$K_8[P_2CoW_{17}O_{61}]$	Dawson	8	2.08×10^6	16.68
$H_3[PMo_{12}O_{40}]$	Keggin	3	$<1 \times 10^5$	No inhibition

Reproduced from Ref. [94] by permission of John Wiley & Sons Ltd.

(1-anilinonaphthalene 8-sulfonate) as competitive agent suggests that the cationic cluster histidine$_{13}$–histidine–glutamine–lysine$_{16}$ (HHQK) of Aβ could be the binding site for the POM inhibitors. The possible mechanism involves the following two facts: (1) the binding decreases the concentration of the free monomer and shifts the equilibrium away from fibrillization; (2) the interaction of the negative surface of POM and Aβ oligomer leads to unfavorable situation for nucleation and growth of fibril, which prevents the direct touch of monomers. This mechanism explains the depletion of sub- and near-critical oligomers and partial blockage of the kinetic pathway until the surface is saturated and new nuclei generates [94]. With this understanding, hybrid colloidal spheres were prepared through the self-assembly of Aβ15-20 peptide and POM $K_8[P_2CoW_{17}O_{61}]$. Compared with the isolated POM, the formed nanoparticles show enhanced target inhibition effect on amyloid aggregation in mice cerebrospinal fluid. Thus, the complex spheres with large size cover the hydrophobic part of the N-terminal, and then drive Aβ15-20 to bind with Aβ1-40, enhancing the antiaggregation activities [95]. They also extend the Dawson POMs to their derivatives. Although the electrostatic interaction is robust, the binding affinity with Aβ still needs to improve due to the competitive combination of haem with Aβ monomer, which may play another major role for AD. Thus, a series of transition metal-substituted Dawson-type POM derivatives with various well-known histidine-chelating metals such as Cu, Fe, Ni, Co, and Mn were selected to bind with chelating site on Aβ surface. The inhibitory effects were observed to follow the order POM-Ni > POM-Co > POM, indicating that besides the electrostatic attraction, the histidine-chelating also contributes clusters' inhibition [96].

Besides the misfolded Aβ peptide, the level of the reactive oxygen species (ROS) induced by Aβ and the movement of metal ions were considered [97–99]. An efficient nanozyme that exhibits protease activities was used for suppressing the aggregation of Aβ, superoxide dismutase (SOD)-like functionality for decreasing the toxicity of ROS, and metal ion chelation capabilities. The constitute of the nanozyme is an Au nanoparticle coating with above-mentioned Dawson-type POM and octapeptide (N-Ac-Cys-His-Sar-His-Sar-His-Sar-His-peptide, Sar = sarcosine,

His = histidine, Cys = cystine), abbreviated as AuNPs@POM-8pep. Here, Au nanoparticles help in penetration of the blood–brain barrier (BBB), while the peptide exhibits the SOD-like activity and metal ions' clearance, and the POM affords the inhibitor of Aβ (Fig. 13.9). Unexpectedly, Aβ monomers and aggregates were both hydrolyzed by AuNPs@POM-8pep and hydrolyzing positions are similar to those associated with native enzyme degradation. Recently, the intermediates induced by Cu in the fibril formation process are supposed as the main cause of cytotoxicity leading to neuronal death rather than the large fibril aggregates [100, 101]. Further, Cu was also considered to produce ROS through the formation of an Aβ/Cu complex. In the presence of AuNPs@POM-8pep, the Cu-induced Aβ oligomer without the formation of amorphous aggregates was observed due to the hydrolysis effect of multiple nanospheres. ROS has the ability to cause oxidative stress and triggers the damage to cellular components, such as DNA, lipids, and proteins. The AuNPs@POM-8pep nano-enzymes perform the activity like SOD to eliminate ROS [102–104].

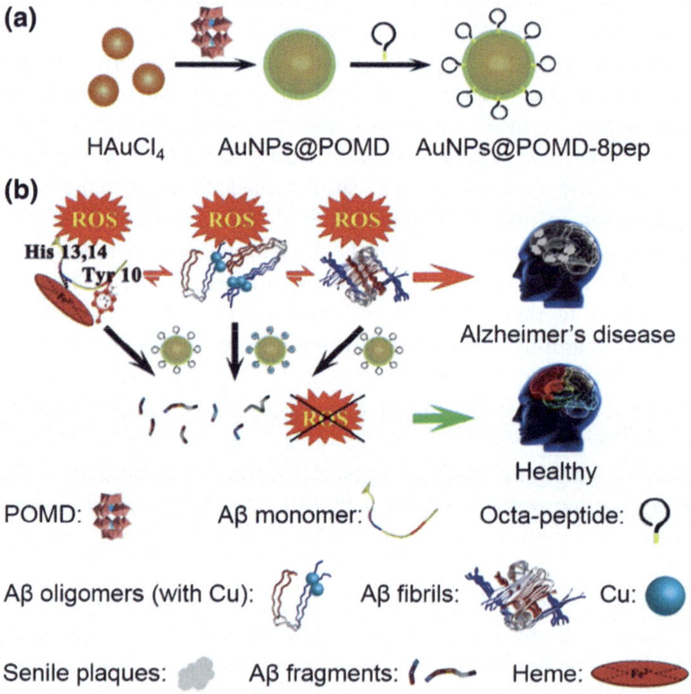

Fig. 13.9 Aβ pathways influenced by AuNPs@POM-8pep: **a** Synthetic route of the nanozyme; **b** AuNPs@POM-8pep acted as a multifunctional nanozyme to modulate multiple facets of Alzheimer's disease. Reprinted from Ref. [102], with kind permission from Springer Science + Business Media

13.4 The POMs and POM Complexes for Enzyme Mimicking

The regulation of protein hydrolysis which can be widely used in structure and protein function analysis, for example, the determination of the primary amino acid sequence, active site analysis, mapping of the metal and ligand binding sites and protein folding studies, is an important process in biotechnology and proteomics [105–108]. Some important agents, which are applied on protein hydrolysis, play a wide role for protein footprinting, proteomics, semi-synthesis of proteins, production of bioengineered, fusion proteins, and the development of therapeutic drugs [109–115]. Currently, proteolytic enzymes and chemical reagents are widely applied to protein cleavage in proteomics. However, most of the available proteases are not so satisfied because of the intrinsic disadvantages, for example, the proteolytic enzymes often cleave the protein at many sites, which leads to the formation of short fragments to be difficultly analyzed. And, they often make denaturation of the proteins resulting in the loss of structural information. Thus, in order to meet the demand for biotechnology, new chemical reagents with obvious selectivity and efficiency toward protein hydrolysis are highly desired. In this context, the metal-substituted POMs that can selectively bind to certain residues or regions of the protein are developed to promote hydrolysis near the binding site as one of the promising candidates.

13.4.1 Catalytic Hydrolysis of POMs for Peptides

Like the traditional protease, the hydrolysis of peptides and proteins via POMs also involves the immobilization of catalysts on the cleft sites of peptides to attack and cleave the peptide bonds to obtain the hydrolyzing products. Parac-Vogt's group reported peptide hydrolysis catalyzed by POMs initially [116]. Among a series of selected POMs, the Zr(IV)-substituted Dawson-type $K_{15}H[Zr(\alpha_2\text{-}P_2W_{17}O_{61})_2]\cdot 25H_2O$ acting as a catalyst for the hydrolysis of the peptide bond in glycylglycine (GG) exhibited excellent activity. GG was totally hydrolyzed to glycine (G) in the presence of equimolar amounts of $K_{15}H[Zr(\alpha_2\text{-}P_2W_{17}O_{61})_2]\cdot 25H_2O$ under pD 5.0 at 60 °C. However, no hydrolysis was detected in the presence of the lacunary α_2-$[P_2W_{17}O_{61}]^{10-}$ POM lacking Zr coordination, indicating the Zr ion as the active center. In addition to the hydrolysis of amide bond, the diketopiperazine also appeared in the whole process because the Zr-POM accelerated both reactions. The $K_{15}H[Zr(\alpha_2\text{-}P_2W_{17}O_{61})_2]$ cluster also displayed hydrolytic activity for a series of a Gly-X sequence dipeptides, where X donated alanine, valine, leucine, phenylalanine, lysine, histidine, cysteine, and so forth. Moreover, triglycine, tetraglycine, and pentaglycine were also examined to be hydrolyzed through adding this POM, yielding the final product of glycine. A brief hydrolysis mechanism was proposed (Fig. 13.10), in which the coordination amide carbonyl of GG to the Zr ion in POM

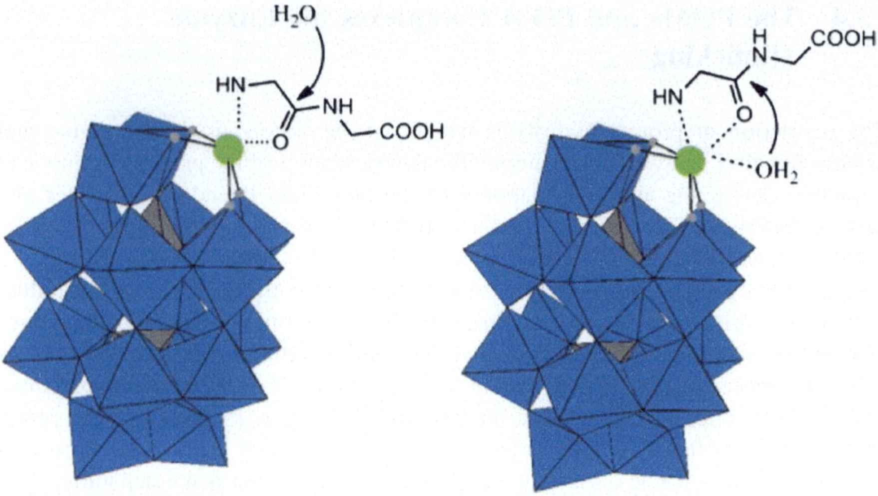

Fig. 13.10 Mechanism for the hydrolysis of GG: nucleophilic attack of solvent water (*left*) and coordinated water (*right*). Reprinted with the permission from Ref. [116]. Copyright 2012, American Chemical Society

cluster underwent a Lewis acid activation of the amide bond and that made it feasible for the nucleophilic attacking from water. The internal nucleophile water molecules are also regarded as the coordinated site and solvent, while the N-terminal amine group takes another coordinating entity for the effective attachment of the peptide. The hydrolysis rate can be simply modulated through the change of amino acid side chains. The increase of aliphatic side chain volume leads to the decrease of the hydrolysis rate, indicating that the steric hindrance from the side chain strongly affects the catalytic process [117]. The hydroxyl group in side chains of the amino acids can facilitate hydrolysis of amide bond by pushing N → O acyl rearrangement. The five-membered cyclic transition state formed from intramolecular attack by the side chain hydroxy group rearranges into an acylated serine intermediate that is more easily hydrolyzed. Amino acids with a carbonyl group side chain are favorable of hydrolysis as well due to the coordination of Zr ion and the activation of amide carbon toward nucleophilic attack by water molecules. In addition, peptides containing amino acids with positively charged side chains can be hydrolyzed at a higher rate due to the secondary interactions with the negatively charged POM surface. The type of POMs for hydrolysis was further extended. A Zr-substituted Lindqvist-type POM, $(Me_4N)_2[W_5O_{18}Zr(H_2O)_3]$, was demonstrated to have a high activity for hydrolysis of the X–Ser amino acid sequence, in which His–Ser is the fast one [118]. The efficient activity was confirmed to derive from the chelation of the Zr ion with His–Ser through the imidazole nitrogen, amine nitrogen, and amide carbonyl oxygen. A dimeric Zr-substituted Keggin-type POM, $(Et_2NH_2)_8[\{\alpha\text{-}PW_{11}O_{39}Zr(\mu\text{-}OH)(H_2O)\}_2]\cdot 7H_2O$ was studied on the hydrolysis of glycylserine (Gly-Ser) and glycylglycine (Gly-Gly). The highest catalytic activity of this POM for peptide bond hydrolysis was obtained at pD 5.5–6.0. The coordination of the POM cluster to dipeptides via the amine nitrogen and amide carbonyl oxygen results in the polarization of the peptide bond, making it more susceptible to hydrolysis [119]. The dimeric tetra-Zr-substituted Dawson-type POM $Na_{14}[Zr_4(P_2W_{16}O_{59})_2(\mu_3\text{-}O)_2(OH)_2(H_2O)_4]\cdot 57H_2O$ showed the hydrolyzed activity to hydrolyze dipeptides such as Gly-Gy, via its amide oxygen and amine nitrogen atoms [120].

Except the dipeptides, those oligopeptides, for example, triglycine, tetraglycine, glycylglycylhistidine, and glycylserylphenylalanine, can also be hydrolyzed by Zr-POM clusters, through selecting suitable conditions with high activity [121, 122]. The hydrolysis of polypeptide system also works well. Oxidized insulin chain B, as a 30-amino acid polypeptide, can be selectively cleaved by $K_{15}H[Zr(\alpha_2\text{-}P_2W_{17}O_{61})_2]\cdot 25H_2O$ POM cluster [123]. The hydrolytic reactivity of the POM cluster toward oxidized insulin chain B was examined by mixing the polypeptide and POM in equal amount in a pH 7 aqueous solution. The hydrolysis sites on the oxidized insulin chain B locate at Gly8–Ser9, Leu6–Cys(SO_3H)7, Gln4–His5, and Phe1–Val2 through the detection of HPLC-ESI-MS and MALDI-TOF at 37 °C. By increasing the temperature to 60 °C, the same cleavage sites were observed as those at 37 °C, with an additional cleavage site at Gly20–Glu21 (Fig. 13.11). The raise of the temperature does not have a major impact on the selectivity of hydrolysis, but leads to a faster appearance of the peptide fragments. The first cleavage site on oxidized insulin chain B was the Gly8–Ser9 peptide bond. The higher reactivity of the Gly–Ser bond is contributed by the hydroxyl group of serine accelerating N $\Rightarrow$ O acyl rearrangement, resulting in the hydrolysis of the Gly8–Ser9 bond. The second cleavage site was between the residues Leu6 and Cys(SO_3H)7. The negatively charged sulfonate group acting as a ligand for the positively charged Zr center generates an additional electrostatic interaction, which induces the fixation of Zr to the imidazole N of His5 closer to the carbonyl group of Leu6 and then causes

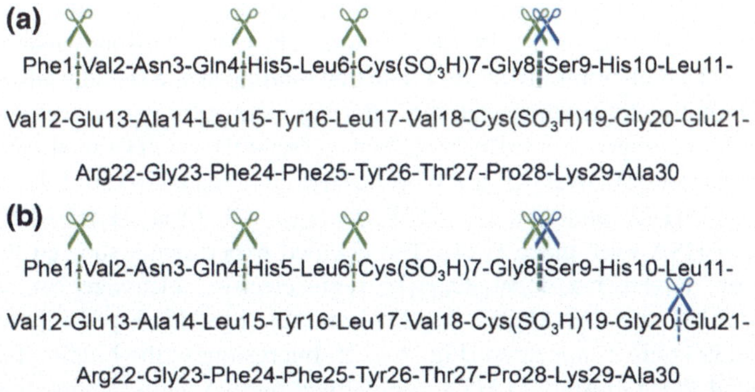

Fig. 13.11 Primary amino acid sequence of oxidized insulin chain B with the corresponding cleavage sites observed at pH 7.0 and **a** 37 °C and **b** 60 °C. Reproduced from Ref. [123] by permission of the Royal Society of Chemistry

its polarization for accelerated cleavage of the Leu6–Cys(SO_3H)7 bond. For the third cleavage site, Gln4–His5 involves two aspects. On one hand, the N in the imidazole of His5 provides binding site with POMs to cleave this peptide bond. On the other hand, deamidation to Gln4 can induce an electrostatic interaction between the positively charged Zr and the negative carboxylate side chain, resulting in the hydrolysis of Gln4–His5. The fourth cleavage site, hydrolyzed by the POM, appears at Phe1–Val2. The coordination of the Zr center to the N-terminal amino group of Phe1 leads to the hydrolysis of the peptide.

13.4.2 Selected Hydrolysis of POMs for Proteins

Selective hydrolysis of proteins is one of the most important procedures in analytical biochemistry and biotechnology. Currently, only limited studies have been reported on the hydrolysis of proteins by POMs. Parac-Vogt et al. also displayed the early example for protein hydrolysis regarding the catalysis of POM clusters [34]. Hen egg-white lysozyme (HEWL) was cleaved in the presence of a sandwiched [Ce(α-$PW_{11}O_{39})_2]^{10-}$ cluster at the condition of pH 7.4 and 37 °C. The cleavage sequences are Val-Xxx-Ala-Ala-Lys-Phe-Glu-Xxx-Asn-Phe-Xxx-Thr and Arg-Asn-Thr-Asp-Gly-Ser-Thr-Asp-Tyr-Ala-Gly, indicative of the hydrolysis of HEWL occurring between the residues tryptophan 28 and valine 29 (cleavage site I) and the residues asparagine 44 and arginine 45 (cleavage site II). In a contrast experiment, α-lactalbumin with almost identical 3D structure like HEWL but a very different surface potential (pI: HEWL = 11.3, α-lactalbumin = 4.2) exhibited no evidence of hydrolysis. Thus, the large difference in activity of POM toward HEWL and α-lactalbumin can be attributed to the differences in surface charge because the tertiary folding of these two proteins was almost identical. NMR studies demonstrated that the positively charged surface area surrounding cleavage site I strongly interacted with the negative POM framework. The binding near site II, cleavage of the Asn44–Arg45 bond appearing after 9 h incubation, was directed by the Ce ion binding mode rather than the POM skeleton. Thus, the Ce-sandwiched Keggin cluster showed a higher affinity at site I than that binding at site II and leading to site I hydrolyzed totally. As another example, after the incubation with a series of POMs with organic counterions, $(n\mathrm{Bu}_4\mathrm{N})_6[\{W_5O_{18}Zr(\mu\text{-}OH)\}_2]\cdot 2H_2O$, $(Et_2NH_2)_{10}[Zr(PW_{11}O_{39})_2]\cdot 7H_2O$, $(Et_2NH_2)_8[\{\alpha\text{-}PW_{11}O_{39}Zr(\mu\text{-}OH)(H_2O)\}_2]\cdot 7H_2O$, $K_{15}H[Zr(\mu_2\text{-}P_2W_{17}O_{61})_2]\cdot 25H_2O$, and $Na_{14}[Zr_4(P_2W_{16}O_{59})_2(\mu_3\text{-}O)_2(OH)_2(H_2O)_4]\cdot 10H_2O$, the hydrolysis of HSA took place [124]. The detailed experiments showed that in the presence of $K_{15}H[Zr(\alpha_2\text{-}P_2W_{17}O_{61})_2]$, regioselective cleavage of HSA at Arg114-Leu115 (site 1), Ala257-Asp258 (site 2), Lys313-Asp314 (site 3) and Cys392-Glu393 (site 4) appeared (Fig. 13.12). Interestingly, the binding from these Zr-substituted POM clusters to HSA is governed by two main factors: the electrostatic interaction between the POM framework and the binding sites on the protein, accompanying by other three weaker interactions resulting from the anchoring of Zr ion to HSA side chains. The site I situates at a positive patch of the protein, inducing

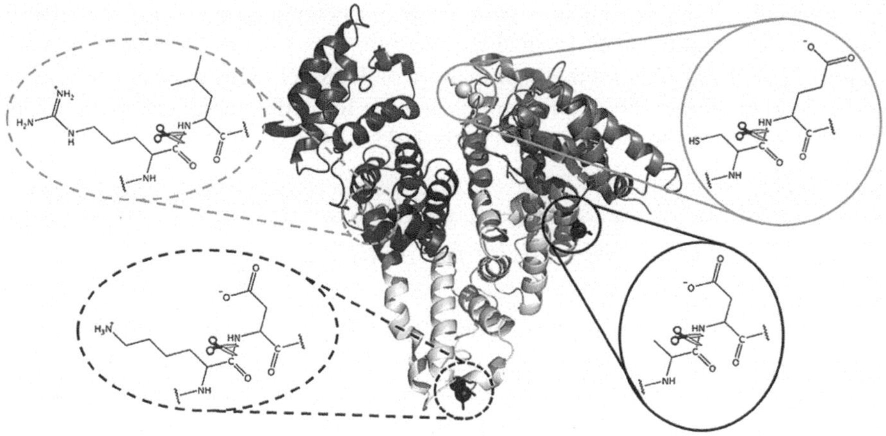

Fig. 13.12 Cartoon representation of HSA for the four cleavage sites. Reproduced from Ref. [125] by permission of John Wiley & Sons Ltd.

the electrostatic interaction to the negative charged POMs and then hydrolysis at this part. Meanwhile, the anchoring of Zr onto the acidic side chains is responsible for the hydrolysis at the remaining three sites [125]. The horse heart myoglobin (HHM) consisting of 153 amino acids including 8 Asp and 13 Glu residues was selected as a target to investigate the hydrolysis of proteins. Among series of Zr-substituted POMs, $(Et_2NH_2)_8[\{\alpha\text{-}PW_{11}O_{39}Zr\text{-}(\mu\text{-}OH)(H_2O)\}_2]\cdot 7H_2O$ has a remarkable selectivity toward the hydrolysis of Asp-X bonds in HHM. The HHM was hydrolyzed at six peptide bond sites, Asp4-Gly5, Asp20-Ile21, Asp44-Lys45, Asp60-Leu61, Asp126-Ala127, and Asp141-Ile142. The hydrolyzed peptide bonds in HHM were found to upstream from the aspartate residue, which demonstrated the site-selective hydrolysis by the POM used. The phenomena can be contributed to the protein surface charge distribution at the cleavage sites. As is shown in the results, each of the hydrolyzed peptide bonds situated near positively charged surface patch that can electrostatically interact with the negatively charged POM. Moreover, the carboxyl group in the Asp side chain helped for the fixation of the used POM catalyst to the protein through Zr coordination, which assisted the hydrolysis by nucleophilic attack to form a tetrahedral intermediate. All these interactions accelerated the hydrolysis of proteins [126].

13.5 The Healing Effects of POMs

Currently, more than two hundred kinds of POMs have been reported with medical activity and the fact encourages researchers to make more effort for applying the inorganic clusters in medical chemistry. Though there are still some intrinsic disadvantages, the POM's biological macromolecular targeting recognition, including their molecular properties such as polarity, redox potentials, surface charge distribution, shape, acidity, and designable structure, was regarded as the important features [127–132]. The fantastic properties of POM clusters induce covalent or electrostatic binding with organic groups and the formed complex can match with physiological environment of human body [133, 134].

13.5.1 The POMs' Bioactivity for Antivirus

One of the most possible medical properties of POMs can be attributed to their antiviral, antibacterial, and antitumor activities. The antiviral activity of POMs has been studied nearly about fifty years. However, the mechanism of viral inhibition of POMs is still unclear [135]. The inhibition for viruses is affected by the degree of cellular penetration and localization of a drug [84]. After that, RNA virus was studied, whose inhibitory mode was attributed to the interference of virus adsorption to target cells and hindrance of virus penetration to cells mainly affected by the main feature of the POM. The POM was considered to perhaps interrupt the enzymes of virus. For the moloney's leukemogenic virus, the therapeutic effect by POMs is believed to result from the inhibition of DNA polymerase though the inhibitory effect is found to be reversible [136]. Thus, most studies involve the inhibition of viral enzymes such as reverse transcriptase, protease in retroviruses, surface viral proteins, and so forth.

Recently, several POMs have been widely employed on the antiviral investigations. Among them, $K_7[Ti_2PW_{10}O_{40}]\cdot 3H_2O$ is confirmed to have the antiviral activity on human immunodeficiency virus (HIV1), herpesvirus hominis (HSV-1 and HSV-2), respiratory syncytial virus (RSV), human cytomegalovirus (HCMV), and so forth [137, 138]. The $K_6[SiNi(H_2O)W_{11}O_{39}]$ and $[(NH_4)_{12}H_2Eu_2(MoO_4)(H_2O)_{16}(Mo_7O_{24})_4]\cdot 13H_2O$ clusters with more complicated architecture and chemical composition exhibit antiviral activity on HIV-1, HSV-1, and HSV-2 [137, 139–141]. In addition, niobium ion-substituted clusters, $K_7[SiW_9Nb_3O_{40}]$ and $(Me_3NH)_7[SiW_9Nb_3O_{40}]$ display antiviral abilities on the influenza virus (Influenza A/B), RSV, murine leukemia sarcoma virus, and HIV [134, 142].

HIV is a lentivirus that causes virus infection and makes body acquire immunodeficiency syndrome (AIDS) over time. The $K_{13}[Ce(SiW_{11}O_{39})_2]\cdot 26H_2O$ and $K_6[BGa(H_2O)W_{11}O_{39}]\cdot 15H_2O$ clusters were used as potent anti-HIV agents by evaluating them in inhibiting HIV-1 and simian immunodeficiency viruses at a low concentration. The mechanism of anti-HIV activity for the POM clusters could be attributed to the inhibition of linkage for virus to cell [143]. After that, a series of POM clusters such as $K_7[P_2W_{17}NbO_{62}]$ was applied to evaluate their inhibitory effect on HIV-1 protease. The main interactions between POMs and HIV-1 protease include hydrogen bonding, including a terminal oxygen positioned between the Ile50 and Ile50′ residues of the enzyme flap, a central bridging oxygen close to protonated catalytic aspartyl residues, and a bridging oxygen in the Ni cap located near residues Arg 8 or Arg 8′. The inhibitory activity of POMs was ascribed not to

Table 13.3 Antiviral activity and cytotoxicity of $Cs_2K_4Na[SiW_9Nb_3O_{40}]$ in vitro

Virus	Cell line	EC_{50} (μg/mL)	CC_{50} (μg/mL)
Influenza *A*[a]	MDCK	7.4 ± 1.1	426.2 ± 8.7
Influenza B	MDCK	11.2 ± 3.1	475.2 ± 6.9
HSV-1	Vero	2.5 ± 1.3	1060.5 ± 9.3
HSV-2	Vero	7.3 ± 1.6	1358.5 ± 8.5
HIV-1	MT-4	3.2 ± 0.8	325.7 ± 5.4
HBV[b]	HepG2	11.4 ± 0.6	1784.0 ± 3.1

the binding to the active site of HIV-1 protease, but the linkage to a cationic pocket on the hinge region of the flaps covering the active site [89].

The property of POM clusters for broad-spectrum antiviral was also noticed by Yamamoto et al. [143] initially. Later, Li and Kang et al. reported that $Cs_2K_4Na[SiW_9Nb_3O_{40}]$ exhibited broad and high antivirus activities for influenza virus (Influenza A and Influenza B), HSV-1, HSV-2, and HIV-1 [144]. The antiviral activity and cytotoxicity showed in Table 13.3 revealed that the cluster possesses low EC50 value (50% virus inhibitory effective concentration) on Influenza A, Influenza B, HSV-1, HSV-2, and HIV-1 and low cytotoxicity with high CC50 value (50% cell inhibitory concentration) on MDCK, Vero, and MT-4 cells. A series of experiments indicated that this POM locates on the surface instead of the interior of the cell used. Further, the broad-spectrum antiviral features of this POM were mainly attributed to its localization on the cell surface. In addition, this type of POMs performs good anti-hepatitis B virus (HBV) properties [145].

13.5.2 The POMs with Antibacterial Property

Gram-positive bacterium and gram-negative bacterium are the common pathogenic bacteria. POMs in various morphologic structures such as Keggin, lacunary Keggin, Dawson, double-Keggin, Keggin-sandwiched, and so forth, were investigated in enhancing antibacterial activity of β-lactam antibiotics on methicillin-resistant staphylococcus aureus (MRSA), which is a gram-positive bacterium, and Escherichia coli (*E. coli*), which belongs to a gram-negative bacterium [146, 147]. Yamase et al. observed that the antibacterial activity of β-lactam antibiotics through the utilization of $K_6[P_2W_{18}O_{62}]\cdot 14H_2O$, $K_4[SiMo_{12}O_{40}]\cdot 3H_2O$ and $K_7[PTi_2W_{10}O_{40}]\cdot 6H_2O$ against methicillin-resistant MRSA and vancomycin-resistant *S. aureus* (VRSA) could be enhanced [148]. The biological reduction of the first two clusters was available within both MRSA and VRSA and the cells also kept alive, indicating that the POMs can penetrate through the cell wall consisting of peptidoglycan layers and reach cytoplasmic membrane. The inhibitory effect can be attributed to the suppression of transcription processes from mecA and pbp genes to mRNAs, which associates with the modified metabolisms exerted by the POMs' capture in the cytoplasmic membrane through the cell wall peptidoglycan layer.

The antibacterial features on gram-negative bacterium usually make expression on *E. coli.* Li et al. exhibited that the nanofibers consisting of cationic peptides covered polyanions displayed excellent antibacterial for *E. coli* [149]. The $H_4SiW_{12}O_{40}$ cluster has the ability to drive the self-assembly of short peptides consisting of an alternating sequence of hydrophilic (lysine) and hydrophobic (azobenzene) residues, into well-separated nanofibers with highly concentrated positive charges on the surface. The nanofibers were uniform with a diameter of *ca.* 13 nm in width and several micrometers in lengths. The compared experiments showed that the individual peptide and the POM cluster exhibit poor inhibitory ability. In contrast, the nanofibers displayed significantly enhancement for antibacterial activity (Fig. 13.13), in which the minimal inhibitory concentration of the peptide nanofibers is 60 μM. The dead assay for cell viability gave the mechanism for inhibition, which involved cell membrane disruption upon contact with the nanofibers. The enhanced antibacterial activity can be attributed to the large peptide aggregates deriving from the accumulation of the peptides on the surface of the cell membrane, which not only strengthen the binding affinity, but also avoid the slow accumulation process of individual peptide. Silver nanoparticles capped with $H_3PW_{12}O_{40}$, $H_3PMo_{12}O_{40}$, $(NH_4)_3[PMo_{12}O_{40}]{\cdot}nH_2O$ rhombic dodecahedral nanocrystals, complex with $H_5PMo_{10}V_2O_{40}$ and bamboo charcoal, nanocomposites formed by $H_5PMo_{10}V_2O_{40}$ and chitosan, also show obvious activity against *E. coli* though the nanoparticle itself has the antibacterial feature [150–153]. The multilayers formed by POM clusters and positive polyelectrolytes exhibit excellent antibacterial efficiency as well, indicative of the potentials as antibacterial, protective materials, and self-detoxifying material [154–157].

Antimony potassium tartrate and stibophen are favorable as antiprotozoal and anthelmintic agents. Consequently, incorporation of antimony into the POM frameworks for the antimicrobial activities has been widely investigated [158, 159].

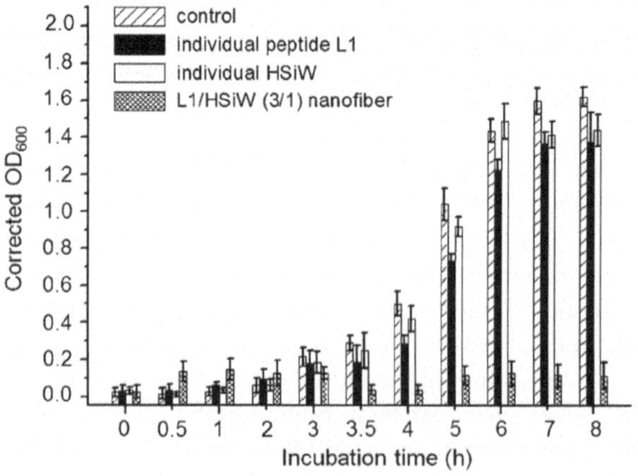

Fig. 13.13 Optical density of *E. coli* with incubation time in the presence of the individual peptide, individual POM, and nanofibers. Reproduced from Ref. [149] by permission of John Wiley & Sons Ltd.

Table 13.4 MIC determination of various polyanions against the growth of gram-positive and gram-negative bacteria

	MIC Determination (μg/mL)				
	1	2	$[As^{III}_2W_{19}O_{67}(H_2O)]^{14-}$	PhSb-1	PhSb-2
Gram-positive					
Paenibacillus sp.	500	250	No inhibition	250	125
Bacillus subtilis	250	250	No inhibition	125	62.5
Clavibacter michiganensis	500	250	No inhibition	250	250
Gram-negative					
Vibrio sp. *Gal 12*	250	250	No inhibition	125	62.5
Pseudomonas putida DSM 291	500	1000	No inhibition	125	62.5
Escherichia coli DH5α	1000	500	No inhibition	500	250

Reprinted with the permission from Ref. [161]. Copyright 2016, American Chemical Society

Kortz et al. reported a series of organoantimony(III)-containing heteropoly-tungstates to investigate their antibacterial effect. The POM clusters $[(PhSb^{III})_4(A\text{-}\alpha\text{-}PVW_9O_{34})_2]^{10-}$ and $[\{2\text{-}(Me_2NCH_2C_6H_4)Sb^{III}\}_3(B\text{-}\alpha\text{-}As^{III}W_9O_{33})]^{3-}$ have a slightly stronger activity against the gram-positive *B. subtilis*, compared to the gram-negative organism, *E. coli* [160]. The effective inhibition against six types of bacteria was studied for organoantimony(III)-functionalized POMs, $Cs_3KNa_6[Na\{2\text{-}(Me_2HN^+CH_2)C_6H_4Sb^{III}\}As_2^{III}W_{19}O_{67}(H_2O)]\cdot 43H_2O$ and $Rb_{2.5}K_{5.5}[\{2\text{-}(Me_2HN^+CH_2)C_6H_4Sb^{III}\}_2As_2^{III}W_{19}O_{67}(H_2O)]\cdot 18H_2O\cdot Me_2NCH_2C_6H_5$. Both of them successfully inhibit the growth of selected gram-positive and gram-negative bacterial strains [161]. In contrast, the reference $K_{14}[As_2^{III}W_{19}O_{67}(H_2O)]$ cluster exhibited no impact on the same bacterial species (Table 13.4). The inhibitory effect is concluded from the organoantimony(III)-containing POMs interfering with peptidoglycan production inside the cell wall and thus ultimately causing death of bacteria.

13.5.3 The POMs for Potential Antitumor Drugs

Anticancer POMs have been widely investigated for medical applications [162–165]. The antitumoral activity of POMs was found to be comparable with that of commercial drugs. The mechanism usually involves several aspects including the induction of cell apoptosis, nonspecific weak interactions with DNA, and the inhibition of ATP generation and angiogenesis-promoting factors, such as basic fibroblast growth factor [84, 162]. The early investigations dealt with the significant antitumoral effect of POMs, particularly $[NH_3Pr^i]_6[Mo_7O_{24}]\cdot 3H_2O$ against MX-1 murine mammary cancer cell line, Meth A sarcoma, and MM46 adenocarcinoma [166]. Yanagie et al. reported the anticancer activity of this POM for pancreatic carcinoma and further discussed the mechanism for cell apoptosis [167]. Wang et al. carried out the antitumoral investigation on which the POM clusters inhibit liver cancer SMMC-77721 cell line, human stomach SGC-7910 cell line, cervical cancer Hela cell line, and so forth. Recently, they reported that a POM macroanion $\{CoSb_6O_4(H_2O)_3[Co(hmta)SbW_8O_{31}]_3\}^{15-}$ (hmta: hexamethylenetetramine) exhibited a high cytotoxicity against various cancer cells, especially ovarian cancer cells. The prevention of cell cycle and binding with proteins in high affinity by POMs resulted in the cell apoptosis and cell proliferation inhibition [168].

However, in the present stage, no POM drugs have been practically applied after initial development for clinical trials, due to the two main crucial issues. One involves the toxicity of all POMs at higher dosages while the other refers to the negatively charged features of POM clusters which induce weak and unspecific interactions with biomolecules. The surface modification of POMs with organic molecules or biomolecules might be an efficient method to overcome the disadvantages because the organic moiety could recognize the specific units and guide POMs interacting with target biomolecules. Li et al. reported flowerlike hierarchical nanostructures by the co-assembly of dopamine and phosphotungstic acid [169]. The dimension and morphology of the nanostructures could be modulated through changing the ratio and concentration of dopamine and POMs, and pH value. The nanostructure was able to load doxorubicin hydrochloride, which was widely applied as an anticancer drug in chemotherapy. The release of doxorubicin was pH-dependent, in which nearly all the drugs were released from the nanostructure at pH 7.4 within 12 h. This POM assembly can be considered as a promising improvement for the oral delivery of doxorubicin in the treatment of certain cancers.

Liu et al. studied the organic–inorganic hybrid prepared by grafting a long-chain organoalkoxysilane lipid onto a POM cluster, with the formula of $[(C_{16}H_{33})_2NCONH(CH_2)_3SiNaP_5W_{29}O_{110}]^{n-}$ (Fig. 13.14). The hybrid with amphiphilic nature induced the spontaneous organization into micelles in aqueous solution. The treatment of human colorectal cancer cell lines HT29 with the hybrid demonstrated that increasing the concentrations of the hybrids resulted in a decrease of alive cells, and the cell viability reduced with the prolonged incubation time. The inhibition of the hybrid to normal human umbilical vein endothelial cells was much smaller than that toward cancer cell lines, indicative of the lower toxicity to normal cells. The cluster was captured by cancer cells and observed to locate in the cytoplasm of cells in 1 h incubation. Thus, the lipid groups on the hybrids boosted the penetration of cluster into the cells by inserting into the interior or binding to the surface of the cell membrane with high affinity [170]. After that, they extended the inhibition to other hybrids [171]. The nanoparticles in several dozens of nanometers, which are composed of Pt grafted-POM, $[PW_{11}O_{40}(SiC_3H_6NH_2)_2Pt(NH_3)_2C_{12}]^{3-}$ and 1,2-distearoyl-sn-glycero-3-phosphoethanolamine-N-[methoxy (polyethylene glycol)-2000], exhibited highly efficiency superior to classic cisplatin

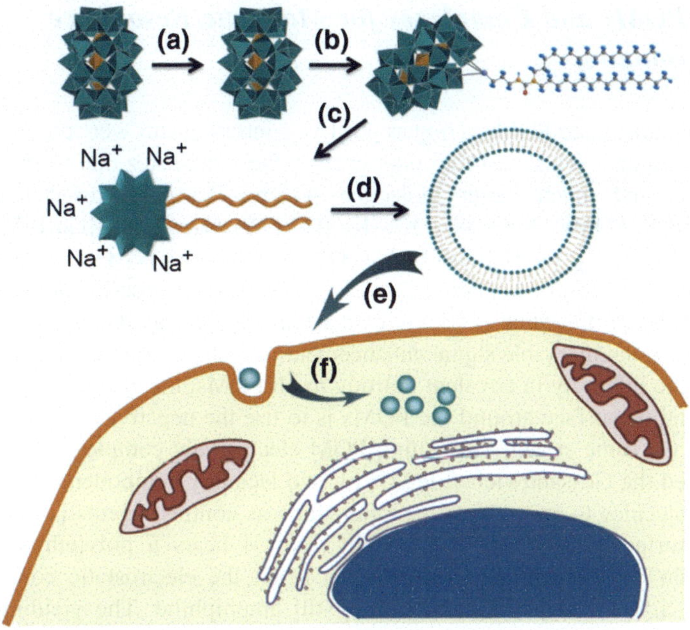

Fig. 13.14 Illustration of synthesis process of POM-based organic–inorganic hybrids and their vesicle formation: **a** synthesis of POM, **b** covalent reaction between POM and organoalkoxysilane lipid, **c** cation exchange, **d** formation of micelles, **e** entering the tumor cells, and **f** intracellular localization. Reproduced from Ref. [170] by permission of John Wiley & Sons Ltd.

in inhibiting cellular growth of HT29 cells and treating human colorectal cancer in mice. The mechanism involves platinum (IV)-to-platinum (II) reduction, DNA binding of platinum (II)-substituted POM, and subsequent apoptosis.

13.6 The POMs and Complexes for Bio-Imaging

Magnetic resonance imaging (MRI) is one of the most impressive medical imaging techniques owing to its noninvasive feature and high spatial resolution. The MRI signal is generally dependent on the longitudinal (T_1) and transverse (T_2) relaxation times of abundant hydrogen nuclei in a biological organism. To improve the MRI sensitivity, various kinds of paramagnetic materials are applied as contrast agents to enhance the signal contrast. Gadolinium ion complexes (Gd^{3+}) are generally selected as positive T_1 contrast agents because of the large paramagnetic moment and long electronic relaxation time, which can effectively shorten the longitudinal relaxation time of water protons for enhancement of T_1 relaxivity (r_1) and a brighter contrast [172, 173].

13.6.1 POMs and Complexes for Magnetic Resonance Imaging

The Gd ion-substituted POMs acting as positive contrast agents were reported to get a higher longitudinal relaxation rate than general Gd complexes due to the chemical composition and much larger molecular weight. Pei et al. used a series of Gd-substituted POMs, $K_9GdW_{10}O_{36}$, $K_{11}[Gd(PW_{11}O_{39})_2]$, $K_{17}[Gd(P_2W_{17}O_{61})_2]$, $K_{13}[Gd(SiW_{11}O_{39})_2]$, $K_{11}H_6[Gd_3O_3(SiW_9O_{34})_2]$, $K_{15}[Gd(BW_{11}O_{39})_2]$, and $K_{17}[Gd(CuW_{11}O_{39})_2]$, as contrast agents [174–177]. The in vivo experiment revealed that these POM clusters have higher relaxivity than that of widely used commercial product (Gd-DTPA) and remarkable signal enhancement was observed in liver and kidney.

Due to the difficulty in covalent grafting to all POMs, the general route to create a biocompatible surface around the POMs is to use the negatively charged feature. Following the same strategy in getting POM electrostatic complexes, Wu's group encapsulated the Gd-sandwiched POMs by two lacunary components with cationic organic molecules to investigate their functions as contrast agents [178]. Because the used surfactant molecule ($EO_{12}BphC_{10}NC_{12}$) bears a poly(ethylene oxide) (PEO) chain and a quaternary ammonium head, the electrostatic complex with $K_{13}[Gd(\beta_2\text{-}SiW_{11}O_{39})_2]\cdot 27H_2O$ cluster is still amphiphilic. The yielding hybrids exhibited water solubility and formed a regular vesicular structure with POM units locating at the center. The relaxivity (r_1) that is defined by the equation below describes the change of relaxation rate ($\Delta(1/T_1) = \Delta R_1$) of water protons normalized to the concentration [CA] of contrast agent. The value of r_1 can be modulated in a large scale by various factors, for example, external field, temperature, the electronic properties of the paramagnetic center, water residence time (τ_m), rotational correlation time (τ_R), first and second coordination sphere hydration (q), and the ion-to-water proton distance [179]. The obtained relaxation rate ($\Delta(1/T_1) = \Delta R_1$) values of Gd-POM complex below critical aggregation concentration (cac) was $r_1 = 61\ \text{mM}^{-1}\text{s}^{-1}$ and above cac was $r_1 = 2.3\ \text{mM}^{-1}\text{s}^{-1}$. The relevant characterizations proved that the relaxivity (r_1) of monodispersed Gd-POM complex was enhanced greatly while the aggregated hybrids were reduced oppositely, indicating the excellent features at fully dispersed state. After that, the organic–inorganic hybrid assembly of $[GdW_{10}O_{36}]^{9-}$ covered with a cationic polymer poly(hexyl-spermine)acrylamide was prepared and the MRI performance was evaluated [180]. The T_1-weighted MRI performance was found to be enhanced about three times. Two important issues were concluded to dominate the contrast effect. One involves the molecular weight that is proportional to the longer rotational correlation time of Gd-POM, whereas more importantly, the hydrophobic layer surrounding the POMs restricts the free penetration of external water molecules for the exchange with the coordinated water on the clusters:

$$r_1 = \frac{\Delta\left(\frac{1}{T_1}\right)}{[\text{CA}]}.$$

13.6.2 POM Complex for Bimodal Imaging

To improve the imaging performance for the POM CAs, dendritic molecules were selected to cover the Gd-POM by the same procedure [181]. Interestingly, for first- and second-generation dendritic cations covered $K_{13}[Gd(\beta_2\text{-}SiW_{11}O_{39})_2]$ complexes, self-assemblies were found always existing in the solutions. However, mono-micellar structure was obtained through the encapsulation of POM cluster with third-generation dendritic cation bearing triethylene glycol monomethyl ether terminal groups locating toward outside (Fig. 13.15). The measured relaxation rate r_1 value is calculated to be *ca.* 16.83 $mM^{-1}\ s^{-1}$, and the value is enhanced by increasing the sample concentration within certain range (from 0.01 to 0.09 mM), demonstrating the potential of this micelle as an effective contrast agent. In addition, the complex showed the ability to accommodate fluorescent dyes such as rhodamine B for both MRI and fluorescence imaging (FI). By incubation with

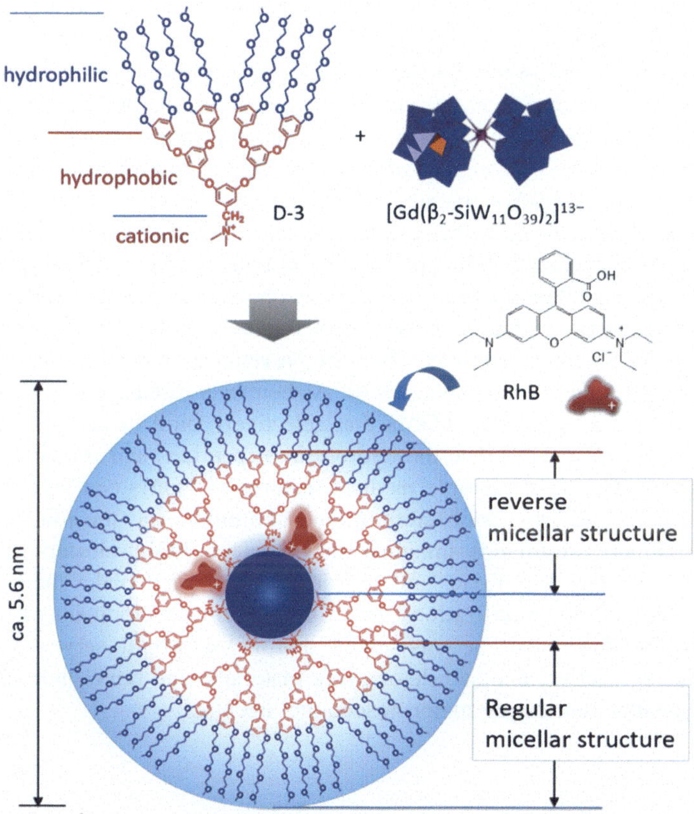

Fig. 13.15 Structural illustration of the complex micelle with dual structure acting as a carrier loading dye molecules for in vivo imaging. Reproduced from Ref. [181] by permission of John Wiley & Sons Ltd.

MCF-7 cells, the fluorescent imaging property of the prepared complex was demonstrated. Further, the in vivo experiments showed that the complex distributed in viscus organs but mainly accumulated in liver and almost fully metabolism of the complex was observed after 24 h, indicating excellent fluorescent property of the complex as an optical probe in practical clinic applications. Thus, the in vivo MRI and FI results confirmed the advantages of this hybrid as a bimodal contrast agent, especially as positive blood pool and liver specific MRI contrast agent. Other applications of the Gd-POM complex based on BSA-coated Gd-POM was reported, which was able to realize dual-modal MR/CT imaging and photothermal therapy/radiotherapy of cancer [182].

13.7 Summary and Outlook

In summary, POMs represent a class of nano-sized inorganic polyanionic clusters with which the framework structure and composition can be modulated widely through simple chemical synthesis. The biological activity of POMs displays great interests in comparison to the diverse potential applications in other fields. Due to the topology, active oxygen atoms, and negative charges, the POM clusters provide various binding modes with metal ions and organic groups closing to biological nanostructure surfaces through hydrogen bonding, coordination, electrostatic interaction, etc. The negative charges and nano-sized structure make POMs become promising candidates to raise the crystallization ability of proteins through a suitable combination or matched dimension with proteins or the domain of proteins. In addition, POMs have the functions of inhibitors on several enzyme families, such as kinase, phosphatases, sulfotransferases, nucleotidases, and so forth. Another significant inhibitory expression of POMs is the prevention of aggregation of amyloid β-peptides (Aβ), which is associated with Alzheimer's disease (AD). Further, the capability of metal-substituted POMs that selectively bind to certain residues or regions of proteins can promote the hydrolysis of peptides and proteins, like artificial enzymes. The most possible medical features of POMs can be attributed to their activity on antivirus, antibacterium, and antitumor, and the role of contrast agent in magnetic resonance imaging. In brief, the POM clusters are stable, effective, designable, low toxic, and biocompatible, which make them vigorous candidates for biological utilizations. From the current development of biological activity on POMs, it can be envisioned that POM would serve as effective antibacterial, protective, and self-detoxifying materials, and novel inorganic drugs as well as against worldwide major nosocomial agents.

References

1. Wang EB, Li YG, Lu Y, Wang XL (2009) Introduction in polyoxometalates. Northeast Normal University Press, Changchun
2. Poblet JM, López X, Bo C (2003) An *initio* and DFT modelling of complex materials: towards the understanding of electronic and magnetic properties of polyoxometalates. Chem Soc Rev 32(5):297–308
3. Yamase T (1998) Photo- and electrochromism of polyoxometalates and related materials. Chem Rev 98(1):307–326
4. Kaur J, Griffin K, Harrison B, Kozhevnikov IV (2002) Friedel-crafts acylation catalysed by heteropoly acids. J Catal 208(2):448–455
5. Alsalme AM, Wiper PV, Khimyak YZ, Kozhevnikova EF, Kozhevnikov IV (2010) Solid acid catalysts based on $H_3PW_{12}O_{40}$ heteropoly acid: acid and catalytic properties at a gas-solid interface. J Catal 276(1):181–189
6. Santos ICMS, Paz FAA, Simões MMQ, Neves MGPMS, Cavaleiro JAS, Klinowski J, Cavaleiro AMV (2008) Catalytic homogeneous oxyfunctionalization with hydrogen peroxide in the presence of a peroxotungstate. Appl Catal A: Gen 351(2):166–173
7. Lan Y, Wang EB, Song YH, Song YL, Kang ZH, Xu L, Li Z (2006) An effective layer-by-layer adsorption and polymerization method to the fabrication of polyoxometalate-polypyrrole nanoparticle ultrathin films. Polymer 47(4):1480–1485
8. Aoshima A, Tonomura S, Yamamatsu S (1990) New synthetic route of polyoxytetramethyleneglycol by use of heteropolyacids as catalyst. Polym Adv Technol 1(2):127–132
9. Murakami H, Kozeki T, Suzuki Y, Ono S, Ohtake H, Sarukura N, Ishikawa E, Yamase T (2001) Nanocluster crystals of lacunary polyoxometalates as structure-design-flexible, inorganic nonlinear materials. Appl Phys Lett 79(22):3564–3566
10. Yamase T, Ishikawa E, Fukaya K, Nojiri H, Taniguchi T, Atake T (2004) Spin-frustrated $(VO)_3^{6+}$-triangle-sandwiching octadecatungstates as a new class of molecular magnets. Inorg Chem 43(25):8150–8157
11. Liang J, Ma YY, Sun H, Li W, Wu LX (2013) Polyanion cluster patterning on polymer surface through microemulsion approach for selective adsorption of proteins. J Colloid Interface Sci 409:80–87
12. Sun H, Li HL, Wu LX (2009) Micro-patterned polystyrene surfaces directed by surfactant-encapsulated polyoxometalate complex via breath figures. Polymer 50(9):2113–2122
13. Berzelius JJ (1826) The preparation of the phosphomolybdate ion $[PMo_{12}O_{40}]^{3-}$. Pogg Ann 6:369–372
14. Keggin JF (1934) The structure and formula of 12-phosphotungstic acid. Proc R Soc A 144 (851):75–100
15. Zhao JW, Jia HP, Zhang J, Zheng ST, Yang GY (2007) A combination of lacunary polyoxometalates and high-nuclear transition-metal clusters under hydrothermal conditions. Part II: from double cluster, dimer, and tetramer to three-dimensional frameworks. Chem Eur J 13(36):10030–10045
16. Al-Oweini R, Sartorel A, Bassil BS, Natali M, Berardi S, Scandola F, Kortz U, Bonchio M (2014) Photocatalytic water oxidation by a mixed-valent $Mn_3^{III}Mn^{IV}O_3$ manganese oxo core that mimics the natural oxygen-evolving center. Angew Chem Int Ed 53(42):11182–11185
17. Hungerford G, Hussain F, Patzke GR, Green M (2010) The photophysics of europium and terbium polyoxometalates and their interaction with serum albumin: a time-resolved luminescence study. Phys Chem Chem Phys 12(26):7266–7275
18. Zheng L, Gu ZJ, Ma Y, Zhang GJ, Yao JN, Keita B, Nadjo L (2010) Molecular interaction between europium decatungstate and histone H1 and its application as a novel biological labeling agent. J Biol Inorg Chem 15(7):1079–1085

19. Crans DC, Mahroof-Tahir M, Anderson OP, Miller MM (1994) X-ray structure of $(NH_4)_6(Gly\text{-}Gly)_2V_{10}O_{28}\cdot 4H_2O$: model studies for polyoxometalate-protein interactions. Inorg Chem 33(24):5586–5590
20. Wang RY, Jia DZ, Zhang L, Liu L, Guo ZP, Li BQ, Wang JX (2006) Rapid synthesis of amino acid polyoxometalate nanotubes by one-step solid-state chemical reaction at room temperature. Adv Funct Mater 16(5):687–692
21. An HY, Wang EB, Xiao DR, Li YG, Su ZM, Xu L (2006) Chiral 3D architectures with helical channels constructed from polyoxometalate clusters and copper-amino acid complexes. Angew Chem Int Ed 45(6):904–908
22. Crans DC, Holst H, Keramidas AD, Rehder D (1995) A slow exchanging vanadium(V) peptide complex: vanadium(V)-glycine-tyrosine. Inorg Chem 34(10):2524–2534
23. Sadakane M, Dickman MH, Pope MT (2001) Chiral polyoxotungstates. 1. Stereoselective interaction of amino acids with enantiomers of $[Ce^{III}(\alpha_1\text{-}P_2W_{17}O_{61})(H_2O)_x]^{7-}$. The structure of DL-$[Ce_2(H_2O)_8(P_2W_{17}O_{61})_2]^{14-}$. Inorg Chem 40(12):2715–2719
24. Li HW, Wang YZ, Zhang T, Wu YQ, Wu LX (2014) Selective binding of amino acids on europium-substituted polyoxometalates and the interaction-induced luminescent enhancement effect. ChemPlusChem 79(8):1208–1213
25. Crans DC, Chen H, Anderson OP, Miller MM (1993) Vanadium(V)-protein model studies: solid-state and solution structure. J Am Chem Soc 115(15):6769–6776
26. Yan XH, Zhu PL, Fei JB, Li JB (2010) Self-assembly of peptide-inorganic hybrid spheres for adaptive encapsulation of guests. Adv Mater 22(11):1283–1287
27. Zhang T, Li HW, Wu YQ, Wang YZ, Wu LX (2015) Self-assembly of an europium-containing polyoxometalate and the arginine/lysine-rich peptides from human papillomavirus capsid protein L1 in forming luminescence-enhanced hybrid nanoparticles. J Phy Chem C 119(15):8321–8328
28. Zhang T, Li HW, Wu YQ, Wang YZ, Wu LX (2015) The two-step assemblies of basic-amino-ccid-rich peptide with a highly charged polyoxometalate. Chem Eur J 21 (25):9028–9033
29. Bijelic A, Rompel A (2015) The use of polyoxometalates in protein crystallography—an attempt to widen a well-known bottleneck. Coord Chem Rev 299:22–38
30. Zhang G, Keita B, Brochon JC, de Oliveira P, Nadjo L, Craescu CT, Miron S (2007) Molecular interaction and energy transfer between human serum albumin and polyoxometalates. J Phy Chem B 111(7):1809–1814
31. Zhang G, Keita B, Craescu CT, Miron S, de Oliveira P, Nadjo L (2008) Molecular interactions between Wells-Dawson type polyoxometalates and human serum albumin. Biomacromol 9(3):812–817
32. Zheng L, Ma Y, Zhang G, Yao J, Bassil BS, Kortz U, Keita B, de Oliveira P, Nadjo L, Craescu CT, Miron S (2009) Molecular interaction between a gadolinium-polyoxometalate and human serum albumin. Eur J Inorg Chem 34:5189–5193
33. Hungerford G, Suhling K, Green M (2008) Luminescence enhancement of a europium containing polyoxometalate on interaction with bovine serum albumin. Photochem Photobiol Sci 7(6):734–737
34. Stroobants K, Moelants E, Ly HG, Proost P, Bartik K, Parac-Vogt TN (2013) Polyoxometalates as a novel class of artificial proteases: selective hydrolysis of lysozyme under physiological pH and temperature promoted by a cerium(IV) Keggin-type polyoxometalate. Chem Eur J 19(8):2848–2858
35. Zhang G, Keita B, Craescu CT, Miron S, de Oliveira P, Nadjo L (2007) Polyoxometalate binding to human serum albumin: a thermodynamic and spectroscopic approach. J Phys Chem B 111(38):11253–11259
36. Schemberg J, Schneider K, Demmer U, Warkentin E, Müller A, Ermler U (2007) Towards biological supramolecular chemistry: a variety of pocket-templated, individual metal oxide cluster nucleations in the cavity of a Mo/W-storage protein. Angew Chem Int Ed 46 (14):2408–2413

37. Kowalewski B, Poppe J, Demmer U, Warkentin E, Dierks T, Ermler U, Schneider K (2012) Nature's polyoxometalate chemistry: X-ray structure of the Mo storage protein loaded with discrete polynuclear Mo–O clusters. J Am Chem Soc 134(23):9768–9774
38. Fenske D, Gnida M, Schneider K, Meyer-Klaucke W, Schemberg J, Henschel V, Meyer AK, Knöchel A, Müller A (2005) A new type of metalloprotein: the Mo storage protein from azotobacter vinelandii contains a polynuclear molybdenum-oxide cluster. ChemBioChem 6 (2):405–413
39. Zebisch M, Krauss M, Schafer P, Strater N (2014) Structures of *Legionella pneumophila* NTPDase1 in complex with polyoxometallates. Acta Cryst D 70(4):1147–1154
40. Poppe J, Warkentin E, Demmer U, Kowalewski B, Dierks T, Schneider K, Ermler U (2014) Structural diversity of polyoxomolybdate clusters along the three-fold axis of the molybdenum storage protein. J Inorg Biochem 138:122–128
41. Backvall JE (2011) Modern oxidation methods, 2nd edn. Wiley-VCH, Weinheim
42. Crans DC, Rithner CD, Theisen LA (1990) Application of time-resolved vanadium-51 2D NMR for quantitation of kinetic exchange pathways between vanadate monomer, dimer, tetramer, and pentamer. J Am Chem Soc 112(8):2901–2908
43. Kholdeeva OA, Maksimov GM, Maksimovskaya RI, Vanina MP, Trubitsina TA, Naumov DY, Kolesov BA, Antonova NS, Carbo JJ, Poblet JM (2006) Zr^{IV}-monosubstituted Keggin-type dimeric polyoxometalates: synthesis, characterization, catalysis of H_2O_2-based oxidations, and theoretical study. Inorg Chem 45(18):7224–7234
44. Ramakrishnan V (2010) Unraveling the structure of the ribosome (Nobel Lecture). Angew Chem Int Ed 49(26):4355–4380
45. Mauracher SG, Molitor C, Al-Oweini R, Kortz U, Rompel A (2014) Crystallization and preliminary X-ray crystallographic analysis of latent isoform PPO4 mushroom (Agaricus bisporus) tyrosinase. Acta Cryst F 70(2):263–266
46. Mauracher SG, Molitor C, Al-Oweini R, Kortz U, Rompel A (2014) Latent and active *ab*PPO4 mushroom tyrosinase cocrystallized with hexatungstotellurate(VI) in a single crystal. Acta Cryst D 70(9):2301–2315
47. Aureliano M, Crans DC (2009) Decavanadate ($V_{10}O_{28}^{6-}$) and oxovanadates: oxometalates with many biological activities. J Inorg Biochem 103(4):536–546
48. Aureliano M, Fraqueza G, Ohlin CA (2013) Ion pumps as biological targets for decavanadate. Dalton Trans 42(33):11770–11777
49. Rudenko G, Henry L, Henderson K, Ichtchenko K, Brown MS, Goldstein JL, Deisenhofer J (2002) Structure of the LDL receptor extracellular domain at endosomal pH. Science 298 (5602):2353–2358
50. Davies DR, Hol WG (2004) The power of vanadate in crystallographic investigations of phosphoryl transfer enzymes. FEBS Lett 577(3):315–321
51. Reinelt S, Hofmann E, Gerharz T, Bott M, Madden DR (2003) The structure of the periplasmic ligand-binding domain of the sensor kinase CitA reveals the first extracellular PAS domain. J Biol Chem 278(40):39189–39196
52. Almo SC, Bonanno JB, Sauder JM, Emtage S, Dilorenzo TP, Malashkevich V, Wasserman SR, Swaminathan S, Eswaramoorthy S, Agarwal R, Kumaran D, Madegowda M, Ragumani S, Patskovsky Y, Alvarado J, Ramagopal UA, Faber-Barata J, Chance MR, Sali A, Fiser A, Zhang ZY, Lawrence DS, Burley SK (2007) Structural genomics of protein phosphatases. J Struct Funct Genomics 8(2–3):121–140
53. Zebisch M, Krauss M, Schafer P, Lauble P, Strater N (2013) Crystallographic snapshots along the reaction pathway of nucleoside triphosphate diphosphohydrolases. Structure 21 (8):1460–1475
54. Bae JH, Lew ED, Yuzawa S, Tomé F, Lax I, Schlessinger J (2009) The selectivity of receptor tyrosine kinase signaling is controlled by a secondary SH2 domain binding site. Cell 138(3):514–524
55. Felts RL, Reilly TJ, Tanner JJ (2006) Structure of francisella tularensis AcpA: prototype of a unique superfamily of acid phosphatases and phospholipases C. J Biol Chem 281 (40):30289–30298

56. Schluenzen F, Tocilj A, Zarivach R, Harms J, Gluehmann M, Janell D, Bashan A, Bartels H, Agmon I, Franceschi F, Yonath A (2000) Structure of functionally activated small ribosomal subunit at 3.3 Å resolution. Cell 102(5):615–623
57. Bashan A, Yonath A (2008) The linkage between ribosomal crystallography, metal ions, heteropolytungstates and functional flexibility. J Mol Struct 890(1–3):289–294
58. Bijelic A, Molitor C, Mauracher SG, Al-Oweini R, Kortz U, Rompel A (2015) Hen egg-white lysozyme crystallisation: protein stacking and structure stability enhanced by a Tellurium(VI)-centred polyoxotungstate. ChemBioChem 16(2):233–241
59. Evans HR, Holloway DE, Sutton JM, Ayriss J, Shone CC, Acharya KR (2004) C3 exoenzyme from *Clostridium botulinum*: structure of a tetragonal crystal form and a reassessment of NAD-induced flexure. Acta Cryst D 60(8):1502–1505
60. Ladenstein R, Bacher A, Huber R (1987) Some observations of a correlation between the symmetry of large heavy-atom complexes and their binding sites on proteins. J Mol Biol 195 (3):751–753
61. Müller CE, Iqbal J, Baqi Y, Zimmermann H, Rollich A, Stephan H (2006) Polyoxometalates —a new class of potent ecto-nucleoside triphosphate diphosphohydrolase (NTPDase) inhibitors. Bioorg Med Chem Lett 16(23):5943–5947
62. Kraus W, Stephan H, Röllich A, Matějka Z, Reck G (2005) $K_6H_2[TiW_{11}CoO_{40}]\cdot 13H_2O$, with a monotitanoundecatungstocobaltate(II) anion. Acta Cryst E 61(3):i35–i37
63. Meißner T, Bergmann R, Oswald J, Rode K, Stephan H, Richter W, Zänker H, Kraus W, Emmerling F, Reck G (2006) Chitosan-encapsulated Keggin Anion $[Ti_2W_{10}PO_{40}]^{7-}$. Synthesis, characterization and cellular uptake studies. Trans Met Chem 31(5):603–610
64. Boglio C, Micoine K, Derat É, Thouvenot R, Hasenknopf B, Thorimbert S, Lacôte E, Malacria M (2008) Regioselective activation of oxo ligands in functionalized dawson polyoxotungstates. J Am Chem Soc 130(13):4553–4561
65. Kortz U, Vaissermann J, Thouvenot R, Gouzerh P (2003) Heteropolymolybdates of phosphate, phosphonate, and phosphite functionalized by glycine. Inorg Chem 42(4):1135–1139
66. Li J, Qi YF, Li J, Wang HF, Wu XY, Duan LY, Wang EB (2004) Heteropolymolybdate–amino acid complexes: synthesis, characterization and biological activity. J Coord Chem 57 (15):1309–1319
67. Micoine K, Hasenknopf B, Thorimbert S, Lacôte E, Malacria M (2007) A general strategy for ligation of organic and biological molecules to Dawson and Keggin polyoxotungstates. Org Lett 9(20):3981–3984
68. Kurth DG, Lehmann P, Volkmer D, Cölfen H, Koop MJ, Müller A, Du Chesne A (2000) Surfactant-encapsulated clusters (SECs): $(DODA)_{20}(NH_4)[H_3Mo_{57}V_6(NO)_6O_{183}(H_2O)_{18}]$, a case study. Chem Eur J 6(2):385–393
69. Volkmer D, Bredenkötter B, Tellenbröker J, Kögerler P, Kurth DG, Lehmann P, Schnablegger H, Schwahn D, Piepenbrink M, Krebs B (2002) Structure and properties of the dendron-encapsulated polyoxometalate $(C_{52}H_{60}NO_{12})_{12}[(Mn(H_2O))_3(SbW_9O_{33})_2]$, a first generation dendrizyme. J Am Chem Soc 124(35):10489–10496
70. Volkmer D, Du Chesne A, Kurth DG, Schnablegger H, Lehmann P, Koop MJ, Müller A (2000) Toward nanodevices: synthesis and characterization of the nanoporous surfactant-encapsulated Keplerate $(DODA)_{40}(NH_4)_2[(H_2O)_n \subset Mo_{132}O_{372}(CH_3COO)_{30}(H_2O)_{72}]$. J Am Chem Soc 122(9):1995–1998
71. Wang XH, Liu JF, Pope MT (2003) New polyoxometalate/starch nanomaterial: synthesis, characterization and antitumoral activity. Dalton Trans 5:957–960
72. Wang XH, Li F, Liu SX, Pope MT (2005) New liposome-encapsulated-polyoxometalates: synthesis and antitumoral activity. J Inorg Biochem 99(2):452–457
73. Yang Y, He JH, Wang XH, Li B, Liu JF (2004) Preparation, characterization and in vitro antitumoral activity of a nanosize liposome complex encapsulated polyoxotungstate $K_6H_2[CoW_{11}TiO_{40}]$. Transit Met Chem 29(1):96–99

74. Zhai FY, Li DL, Zhang CL, Wang XH, Li R (2008) Synthesis and characterization of polyoxometalates loaded starch nanocomplex and its antitumoral activity. Eur J Med Chem 43(9):1911–1917
75. Geisberger G, Paulus S, Gyenge EB, Maake C, Patzke GR (2011) Targeted delivery of polyoxometalate nanocomposites. Small 7(19):2808–2814
76. Menon D, Thomas RT, Narayanan S, Maya S, Jayakumar R, Hussain F, Lakshmanan VK, Nair SV (2011) A novel chitosan/polyoxometalate nano-complex for anti-cancer applications. Carbohydr Polym 84(3):887–893
77. Stephan H, Kubeil M, Emmerling F (2013) Müller CE (2013) Polyoxometalates as versatile enzyme inhibitors. Eur J Inorg Chem 10–11:1585–1594
78. Nakae T, Ishii J, Tokunaga M (1979) Subunit structure of functional porin oligomers that form permeability channels in the outer membrane of escherichia coil. J Biol Chem 254 (5):1457–1461
79. Boyd DW, Kustin K, Niwa M (1985) Do vanadate polyanions inhibit phosphotransferase enzymes? BBA-Protein Strct M 827(3):472–475
80. Prudent R, Moucadel V, Laudet B, Barette C, Lafanechère L, Hasenknopf B, Li J, Bareyt S, Lacôte E, Thorimbert S, Malacria M, Gouzerh P, Cochet C (2008) Identification of polyoxometalates as nanomolar noncompetitive inhibitors of protein kinase CK2. Chem Biol 15(7):683–692
81. van Huijsduijnen RH, Sauer WHB, Bombrun A, Swinnen D (2004) Prospects for inhibitors of protein tyrosine phosphatase 1B as antidiabetic drugs. J Med Chem 47(17):4142–4146
82. Turner TL, Nguyen VH, McLauchlan CC, Dymon Z, Dorsey BM, Hooker JD, Jones MA (2012) Inhibitory effects of decavanadate on several enzymes and leishmania tarentolae in vitro. J Inorg Biochem 108:96–104
83. Hulley P, Davison A (2003) Regulation of tyrosine phosphorylation cascades by phosphatases: what the actions of vanadium teach us. J Trace Elem Med Bio 16(4):281–290
84. Yamase T (2005) Anti-tumor, -viral, and -bacterial activities of polyoxometalates for realizing an inorganic drug. J Mater Chem 15(45):4773–4782
85. Seko A, Yamase T, Yamashita K (2009) Polyoxometalates as effective inhibitors for sialyl- and sulfotransferases. J Inorg Biochem 103(7):1061–1066
86. Melani A, Corti F, Stephan H, Müller CE, Donati C, Bruni P, Vannucchi MG, Pedata F (2012) Ecto-ATPase inhibition: ATP and adenosine release under physiological and ischemic in vivo conditions in the rat striatum. Exp Neurol 233(1):193–204
87. Dong ZX, Tan RK, Cao J, Yang Y, Kong CF, Du J, Zhu S, Zhang Y, Lu J, Huang BQ, Liu SX (2011) Discovery of polyoxometalate-based HDAC inhibitors with profound anticancer activity in vitro and in vivo. Eur J Med Chem 46(6):2477–2484
88. Iqbal J, Barsukova-Stuckart M, Ibrahim M, Ali SU, Khan AA, Kortz U (2012) Polyoxometalates as potent inhibitors for acetyl and butyrylcholinesterases and as potential drugs for the treatment of Alzheimer's disease. Med Chem Res 22(3):1224–1228
89. Judd DA, Nettles JH, Nevins N, Snyder JP, Liotta DC, Tang J, Ermolieff J, Schinazi RF, Hill CL (2001) Polyoxometalate HIV-1 protease inhibitors. A new mode of protease inhibition. J Am Chem Soc 123(5):886–897
90. Flutsch A, Schroeder T, Grutter MG, Patzke GR (2011) HIV-1 protease inhibition potential of functionalized polyoxometalates. Bioorg Med Chem Lett 21(4):1162–1166
91. Messmore JM, Raines RT (2000) Decavanadate inhibits catalysis by ribonuclease A. Arch Biochem Biophys 381(1):25–30
92. Kosik K (1992) Alzheimer's disease: a cell biological perspective. Science 256(5058): 780–783
93. Yamin G, Ono K, Inayathullah M, Teplow D (2008) Amyloid β-protein assembly as a therapeutic target of Alzheimers disease. Curr Pharm Design 14(30):3231–3246
94. Geng J, Li M, Ren JS, Wang EB, Qu XG (2011) Polyoxometalates as inhibitors of the aggregation of amyloid beta peptides associated with Alzheimer's disease. Angew Chem Int Ed 50(18):4184–4188

95. Li M, Xu C, Wu L, Ren JS, Wang EB, Qu XG (2013) Self-sssembled peptide-polyoxometalate hybrid nanospheres: two in one enhances targeted inhibition of amyloid β-peptide aggregation associated with Alzheimer's disease. Small 9(20):3455–3461
96. Gao N, Sun HJ, Dong K, Ren JS, Duan TC, Xu C, Qu XG (2014) Transition-metal-substituted polyoxometalate derivatives as functional anti-amyloid agents for Alzheimer's disease. Nat Commun 5:3422
97. Gotz J, Ittner LM (2008) Animal models of Alzheimer's disease and frontotemporal dementia. Nat Rev Neurosci 9(7):532–544
98. Zhang M, Mao X, Yu Y, Wang CX, Yang YL, Wang C (2013) Nanomaterials for reducing amyloid cytotoxicity. Adv Mater 25(28):3780–3801
99. Qing GY, Zhao SL, Xiong YT, Lv ZY, Jiang FL, Liu Y, Chen H, Zhang MX, Sun TL (2014) Chiral effect at protein/graphene interface: a bioinspired perspective to understand amyloid formation. J Am Chem Soc 136(30):10736–10742
100. Faller P, Hureau C, La Penna G (2014) Metal ions and intrinsically disordered proteins and peptides: from Cu/Zn amyloid-beta to general principles. Acc Chem Res 47(8):2252–2259
101. Geng J, Li M, Wu L, Ren JS, Qu XG (2012) Liberation of copper from amyloid plaques: making a risk factor useful for Alzheimer's disease treatment. J Med Chem 55(21): 9146–9155
102. Gao N, Dong K, Zhao AD, Sun HJ, Wang Y, Ren JS, Qu XG (2016) Polyoxometalate-based nanozyme: design of a multifunctional enzyme for multi-faceted treatment of Alzheimer's disease. Nano Res 9(4):1079–1090
103. Gao N, Sun HJ, Dong K, Ren JS, Qu XG (2015) Gold-nanoparticle-based multifunctional amyloid-β inhibitor against Alzheimer's disease. Chem Eur J 21(2):829–835
104. Guan YJ, Li M, Dong K, Gao N, Ren JS, Zheng YC, Qu XG (2016) Ceria/POMs hybrid nanoparticles as a mimicking metallopeptidase for treatment of neurotoxicity of amyloid-β peptide. Biomaterials 98:92–102
105. Grant K, Kassai M (2006) Major advances in the hydrolysis of peptides and proteins by metal ions and complexes. Curr Org Chem 10(9):1035–1049
106. Hlavaty JJ, Benner JS, Hornstra LJ, Schildkraut I (2000) Identification of the metal-binding sites of restriction endonucleases by Fe^{2+}-mediated oxidative cleavage. Biochemistry 39 (11):3097–3105
107. Grodsky NB, Soundar S, Colman RF (2000) Evaluation by site-directed mutagenesis of aspartic ccid residues in the metal site of pig heart NADP-dependent isocitrate dehydrogenase. Biochemistry 39(9):2193–2200
108. Fournié-Zaluski MC, Soleilhac JM, Turcaud S, Laï-Kuen R, Crine P, Beaumont A, Roques BP (1992) Development of [^{125}I]RB104, a potent inhibitor of neutral endopeptidase 24.11, and its use in detecting nanogram quantities of the enzyme by "inhibitor gel electrophoresis". Proc Natl Acad Sci USA 89(14):6388–6392
109. Cheng X, Shaltiel S, Taylor SS (1998) Mapping substrate-induced conformational changes in cAMP-dependent protein kinase by protein footprinting. Biochemistry 37(40):14005–14013
110. Humphreys DP, King LM, West SM, Chapman AP, Sehdev M, Redden MW, Glover DJ, Smith BJ, Stephens PE (2000) Improved efficiency of site-specific copper(II) ion-catalysed protein cleavage effected by mutagenesis of cleavage site. Protein Eng Des Sel 13(3):201–206
111. Molina H, Horn DM, Tang N, Mathivanan S, Pandey A (2007) Global proteomic profiling of phosphopeptides using electron transfer dissociation tandem mass spectrometry. Proc Natl Acad Sci USA 104(7):2199–2204
112. Meyer B, Papasotiriou DG, Karas M (2011) 100% Protein sequence coverage: a modern form of surrealism in proteomics. Amino Acids 41(2):291–310
113. Swaney DL, Wenger CD, Coon JJ (2010) Value of using multiple proteases for large-scale mass spectrometry-based proteomics. J Proteome Res 9(3):1323–1329
114. Bakhtiar R, Thomas JJ, Siuzdak G (2000) Mass spectrometry in viral proteomics. Acc Chem Res 33(3):179–187

115. Suh J, Chei WS (2008) Metal complexes as artificial proteases: toward catalytic drugs. Curr Opin Chem Biol 12(2):207–213
116. Absillis G, Parac-Vogt TN (2012) Peptide bond hydrolysis catalyzed by the Wells-Dawson $Zr(\alpha_2\text{-}P_2W_{17}O_{61})_2$ polyoxometalate. Inorg Chem 51(18):9902–9910
117. Vanhaecht S, Absillis G, Parac-Vogt TN (2013) Amino acid side chain induced selectivity in the hydrolysis of peptides catalyzed by a Zr(IV)-substituted Wells-Dawson type polyoxometalate. Dalton Trans 42(43):15437–15446
118. Ly HGT, Absillis G, Bajpe SR, Martens JA, Parac-Vogt TN (2013) Hydrolysis of dipeptides catalyzed by a airconium(IV)-substituted Lindqvist type polyoxometalate. Eur J Inorg Chem 26:4601–4611
119. Ly HGT, Absillis G, Parac-Vogt TN (2013) Amide bond hydrolysis in peptides and cyclic peptides catalyzed by a dimeric Zr(IV)-substituted Keggin type polyoxometalate. Dalton Trans 42(30):10929–10938
120. Ly HGT, Mihaylov T, Absillis G, Pierloot K, Parac-Vogt TN (2015) Reactivity of dimeric tetrazirconium(IV) Wells-Dawson polyoxometalate toward dipeptide hydrolysis studied by a combined experimental and density functional theory approach. Inorg Chem 54(23):11477–11492
121. Ly HGT, Absillis G, Parac-Vogt TN (2015) Comparative study of the reactivity of zirconium(IV)-substituted polyoxometalates towards the hydrolysis of oligopeptides. Eur J Inorg Chem 13:2206–2215
122. Stroobants K, Absillis G, Shestakova PS, Willem R, Parac-Vogt TN (2013) Hydrolysis of tetraglycine by a Zr(IV)-substituted Wells-Dawson polyoxotungstate studied by diffusion ordered NMR spectroscopy. J Cluster Sci 25(3):855–866
123. Sap A, Absillis G, Parac-Vogt TN (2015) Selective hydrolysis of oxidized insulin chain B by a Zr(IV)-substituted Wells-Dawson polyoxometalate. Dalton Trans 44(4):1539–1548
124. Stroobants K, Absillis G, Moelants E, Proost P, Parac-Vogt TN (2014) Regioselective hydrolysis of human serum albumin by Zr(IV)-substituted polyoxotungstates at the interface of positively charged protein surface patches and negatively charged amino acid residues. Chem Eur J 20(14):3894–3897
125. Stroobants K, Goovaerts V, Absillis G, Bruylants G, Moelants E, Proost P, Parac-Vogt TN (2014) Molecular origin of the hydrolytic activity and fixed regioselectivity of a Zr(IV)-substituted polyoxotungstate as artificial protease. Chem Eur J 20(31):9567–9577
126. Ly HGT, Absillis G, Janssens R, Proost P, Parac-Vogt TN (2015) Highly amino acid selective hydrolysis of myoglobin at aspartate residues as promoted by zirconium(IV)-substituted polyoxometalates. Angew Chem Int Ed 54(25):7391–7394
127. Zhang J, Song YF, Cronin L, Liu TB (2008) Self-assembly of organic-inorganic hybrid amphiphilic surfactants with large polyoxometalates as polar head groups. J Am Chem Soc 130(44):14408–14409
128. Fernandez JA, Lopez X, Bo C, de Graaf C, Baerends EJ, Poblet JM (2007) Polyoxometalates with internal cavities: redox activity, basicity, and cation encapsulation in $[X^{n+}P_5W_{30}O_{110}]^{(15-n)-}$ Preyssler complexes, with X = Na^+, Ca^{2+}, Y^{3+}, La^{3+}, Ce^{3+}, and Th^{4+}. J Am Chem Soc 129(40):12244–12253
129. Schmitz KS (2009) Macroion clustering in solutions and suspensions: the roles of microions and solvent. J Phy Chem B 113(9):2624–2638
130. Kong XJ, Long LS, Zheng Z, Huang RB, Zheng LS (2010) Keeping the ball rolling: fullerene-like molecular clusters. Acc Chem Res 43(2):201–209
131. Lunkenbein T, Kamperman M, Li ZH, Bojer C, Drechsler M, Förster S, Wiesner U, Müller AHE, Breu J (2012) Direct synthesis of inverse hexagonally ordered diblock copolymer/polyoxometalate nanocomposite films. J Am Chem Soc 134(30):12685–12692
132. Long DL, Burkholder E, Cronin L (2007) Polyoxometalate clusters, nanostructures and materials: from self assembly to designer materials and devices. Chem Soc Rev 36(1):105–121
133. Pope MT, Müller A (1991) Polyoxometalate chemistry: an old Field with new dimensions in several disciplines. Angew Chem Int Ed Engl 30(1):34–48

134. Rhule JT, Hill CL, Judd DA, Schinazi RF (1998) Polyoxometalates in medicine. Chem Rev 98(1):327–358
135. Shigeta S, Mori S, Kodama E, Kodama J, Takahashi K, Yamase T (2003) Broad spectrum anti-RNA virus activities of titanium and vanadium substituted polyoxotungstates. Antiviral Res 58(3):265–271
136. Fischer J, Ricard L, Weiss R (1976) The structure of the heteropolytungstate $(NH_4)_{17}Na[NaW_{21}Sb_9O_{86}]\cdot 14H_2O$. An inorganic cryptate. J Am Chem Soc 98(10):3050–3052
137. Take Y, Tokutake Y, Inouye Y, Yoshida T, Yamamoto A, Yamase T, Nakamura S (1991) Inhibition of proliferation of human immunodeficiency virus type 1 by novel heteropoly-oxotungstates in vitro. Antiviral Res 15(2):113–124
138. Kim GS, Judd DA, Hill CL, Schinazi RF (1994) Synthesis, characterization, and biological activity of a new potent class of anti-HIV agents, the peroxoniobium-substituted heteropolytungstates. J Med Chem 37(6):816–820
139. Inouye Y, Take Y, Tokutake Y, Yoshida T, Yamamoto A, Yamase T, Nakamura S (1990) Inhibition of replication of human immunodeficiency virus by a heteropolyoxotungstate (PM-19). Chem Pharm Bull 38(1):285–287
140. Inouye Y, Tokutake Y, Yoshida T, Yamamoto A, Yamase T, Nakamura S (1991) Antiviral activity of polyoxomol ybdoeuropate PM-104 against human immunodeficiency virus type 1. Chem Pharm Bull 39(6):1638–1640
141. Hill CL, Weeks MS, Schinazi RF (1990) Anti-HIV-1 activity, toxicity, and stability studies of representative structural families of polyoxometalates. J Med Chem 33(10):2767–2772
142. Ni L, Greenspan P, Gutman R, Kelloes C, Farmer MA, Boudinot FD (1996) Cellular localization of antiviral polyoxometalates in J774 macrophages. Antiviral Res 32(3): 141–148
143. Yamamoto N, Schols D, Clercq ED, Debyser Z, Pauwels R, Balzarini J, Nakashima H, Baba M, Hosoya M, Snoeck R (1992) Mechanism of anti-human immunodeficiency virus action of polyoxometalates, a class of broad-spectrum antiviral agents. Mol Pharmacol 42 (6):1109–1117
144. Wang J, Liu Y, Xu K, Qi YF, Zhong J, Zhang K, Li J, Wang EB, Wu ZY, Kang ZH (2014) Broad-spectrum antiviral property of polyoxometalate localized on a cell surface. ACS Appl Mater Interfaces 6(12):9785–9789
145. Zhang H, Qi YF, Ding YH, Wang J, Li QM, Zhang JZ, Jiang YF, Chi XM, Li J, Niu JQ (2012) Synthesis, characterization and biological activity of a niobium-substituted-heteropolytungstate on hepatitis B virus. Bioorg Med Chem Lett 22 (4):1664–1669
146. Yamase T, Fukuda N, Tajima Y (1996) Synergistic effect of polyoxotungstates in; combination with β-lactam antibiotics on antibacterial activity against methicillin-resistant staphylococcus aureus. Biol Pharm Bull 19(3):459–465
147. Fukuda N, Yamase T, Tajima Y (1999) Inhibitory effect of polyoxotungstates on the production of penicillin-binding proteins and β-lactamase against methicillin-resistant staphylococcus aureus. Biol Pharm Bull 22(5):463–470
148. Inoue M, Suzuki T, Fujita Y, Oda M, Matsumoto N, Yamase T (2006) Enhancement of antibacterial activity of beta-lactam antibiotics by $[P_2W_{18}O_{62}]^{6-}$, $[SiMo_{12}O_{40}]^{4-}$, and $[PTi_2W_{10}O_{40}]^{7-}$ against methicillin-resistant and vancomycin-resistant Staphylococcus aureus. J Inorg Biochem 100(7):1225–1233
149. Li JF, Chen ZJ, Zhou MC, Jing JB, Li W, Wang Y, Wu LX, Wang LY, Wang YQ, Lee M (2016) Polyoxometalate-driven self-assembly of short peptides into multivalent nanofibers with enhanced antibacterial cctivity. Angew Chem Int Ed 55(7):2592–2595
150. Daima HK, Selvakannan PR, Kandjani AE, Shukla R, Bhargava SK, Bansal V (2014) Synergistic influence of polyoxometalate surface corona towards enhancing the antibacterial performance of tyrosine-capped Ag nanoparticles. Nanoscale 6(2):758–765
151. He JZ, Pang H, Wang WQ, Zhang Y, Yan B, Li XR, Li SJ, Chen J (2013) Uniform $M_3PMo_{12}O_{40}\cdot nH_2O$ ($M = NH_4^+, K^+, Cs^+$) rhombic dodecahedral nanocrystals for effective antibacterial agents. Dalton Trans 42(44):15637–15644

152. Yang FC, Wu KH, Lin WP, Hu MK (2009) Preparation and antibacterial efficacy of bamboo charcoal/polyoxometalate biological protective material. Micropor Mesopor Mat 118(1–3):467–472
153. Fiorani G, Saoncella O, Kaner P, Altinkaya SA, Figoli A, Bonchio M, Carraro M (2014) Chitosan-polyoxometalate nanocomposites: synthesis, characterization and application as antimicrobial agents. J Cluster Sci 25(3):839–854
154. Gao SY, Wu ZX, Pan DM, Lin Z, Cao R (2011) Preparation and characterization of polyoxometalate–Ag nanoparticles composite multilayer films. Thin Solid Films 519 (7):2317–2322
155. Wu KH, Yu PY, Yang CC, Wang GP, Chao CM (2009) Preparation and characterization of polyoxometalate-modified poly(vinyl alcohol)/polyethyleneimine hybrids as a chemical and biological self-detoxifying material. Polym Degrad Stab 94(9):1411–1418
156. Feng YH, Han ZG, Peng J, Lu J, Xue B, Li L, Ma HY, Wang EB (2006) Fabrication and characterization of multilayer films based on Keggin-type polyoxometalate and chitosan. Mater Lett 60(13–14):1588–1593
157. Yu X, Chen CY, Peng J, Shi ZY, Shen Y, Mei JL, Ren ZX (2014) Antibacterial-active multilayer films composed of polyoxometalate and methyl violet: fabrication, characterization and properties. Thin Solid Films 571:69–74
158. Yang P, Lin Z, Bassil BS, Alfaro-Espinoza G, Ullrich MS, Li MX, Silvestru C, Kortz U (2016) Tetra-antimony(III)-bridged 18-tungsto-2-arsenates(V), $[(LSb^{III})_4(A\text{-}\alpha\text{-}As(V)W_9O_{34})_2]^{10-}$ (L = Ph, OH): turning bioactivity on and off by ligand substitution. Inorg Chem 55(8):3718–3720
159. Yang P, Bassil BS, Lin Z, Haider A, Alfaro-Espinoza G, Ullrich MS, Silvestru C, Kortz U (2015) Organoantimony(III)-containing tungstoarsenates(III): from controlled assembly to biological activity. Chem Eur J 21(44):15600–15606
160. Barsukova-Stuckart M, Piedra-Garza LF, Gautam B, Alfaro-Espinoza G, Izarova NV, Banerjee A, Bassil BS, Ullrich MS, Breunig HJ, Silvestru C, Kortz U (2012) Synthesis and biological activity of organoantimony(III)-containing heteropolytungstates. Inorg Chem 51 (21):12015–12022
161. Yang P, Lin Z, Alfaro-Espinoza G, Ullrich MS, Rat CI, Silvestru C, Kortz U (2016) 19-Tungstodiarsenate(III) functionalized by organoantimony(III) groups: tuning the structure-bioactivity relationship. Inorg Chem 55(1):251–258
162. Yanagie H, Ogata A, Mitsui S, Hisa T, Yamase T, Eriguchi M (2006) Anticancer activity of polyoxomolybdate. Biomed Pharmacother 60(7):349–352
163. Zhou Z, Zhang DD, Yang L, Ma PT, Si YA, Kortz U, Niu JY, Wang JP (2013) Nona-copper (II)-containing 18-tungsto-8-arsenate(III) exhibits antitumor activity. Chem Commun 49 (45):5189–5191
164. Yang HK, Cheng YX, Su MM, Xiao Y, Hu MB, Wang W, Wang Q (2013) Polyoxometalate-biomolecule conjugates: a new approach to create hybrid drugs for cancer therapeutics. Bioorg Med Chem Lett 23(5):1462–1466
165. El Moll H, Zhu W, Oldfield E, Rodriguez-Albelo LM, Mialane P, Marrot J, Vila N, Mbomekalle IM, Riviere E, Duboc C, Dolbecq A (2012) Polyoxometalates functionalized by bisphosphonate ligands: synthesis, structural, magnetic, and spectroscopic characterizations and activity on tumor cell lines. Inorg Chem 51(14):7921–7931
166. Yamase T, Fujita H, Fukushima K (1988) Medical chemistry of polyoxometalates. Part 1. Potent antitumor activity of polyoxomolybdates on animal transplantable tumors and human cancer xenograft. Inorg Chim Acta 151(1):15–18
167. Ogata A, Mitsui S, Yanagie H, Kasano H, Hisa T, Yamase T, Eriguchi M (2005) A novel anti-tumor agent, polyoxomolybdate induces apoptotic cell death in AsPC-1 human pancreatic cancer cells. Biomed Pharmacother 59(5):240–244
168. Zhang ZM, Duan XP, Yao S, Wang ZS, Lin ZK, Li YG, Long LS, Wang EB, Lin WB (2016) Cation-mediated optical resolution and anticancer activity of chiral polyoxometalates built from entirely achiral building blocks. Chem Sci 7(7):4220–4229

169. Li H, Jia Y, Wang AH, Cui W, Ma HC, Feng XY, Li JB (2014) Self-assembly of hierarchical nanostructures from dopamine and polyoxometalate for oral drug delivery. Chem Eur J 20(2):499–504
170. Fu L, Gao HQ, Yan M, Li SZ, Li XY, Dai ZF, Liu SQ (2015) Polyoxometalate-based organic-inorganic hybrids as antitumor drugs. Small 11(24):2938–2945
171. Sun TD, Cui W, Yan M, Qin G, Guo W, Gu HX, Liu SQ, Wu Q (2016) Target delivery of a novel antitumor organoplatinum(IV)-substituted polyoxometalate complex for safer and more effective colorectal cancer therapy in vivo. Adv Mater 28(34):7397–7404
172. Na HB, Hyeon T (2009) Nanostructured T1 MRI contrast agents. J Mater Chem 19 (35):6267–6273
173. Accardo A, Tesauro D, Aloj L, Pedone C, Morelli G (2009) Supramolecular aggregates containing lipophilic Gd(III) complexes as contrast agents in MRI. Coord Chem Rev 253 (17–18):2193–2213
174. Feng JH, Li XJ, Pei FK, Sun GY, Zhang X, Liu ML (2002) An evaluation of gadolinium polyoxometalates as possible MRI contrast agent. Magn Reson Imaging 20(5):407–412
175. Sun GY, Feng JH, Wu HF, Pei FK, Fang K, Lei H (2004) Gadolinium heteropoly complex $K_{17}[Gd(P_2W_{17}O_{61})_2]$ as a potential MRI contrast agent. J Magn Magn Mater 281(2–3):405–409
176. Sun GY, Feng JH, Wu HF, Pei FK, Fang K, Lei H (2004) Investigation of sandwiched gadolinium (III) complexes with tungstosilicates as potential MRI contrast agents. Magn Reson Imaging 22(3):421–426
177. Li ZF, Li WS, Li XJ, Pei FK, Li YX, Lei H (2007) The gadolinium complexes with polyoxometalates as potential MRI contrast agents. Magn Reson Imaging 25(3):412–417
178. Wang YL, Zhou SY, Kong DL, Yang HS, Chai WQ, Kortz U, Wu LX (2012) Self-assembly and alterable relaxivity of an organic cation-encapsulated gadolinium-containing polyoxometalate. Dalton Trans 41(33):10052–10059
179. Caravan P (2006) Strategies for increasing the sensitivity of gadolinium based MRI contrast agents. Chem Soc Rev 35(6):512–523
180. Chai WQ, Wang S, Zhao H, Liu GF, Fischer K, Li HL, Wu LX, Schmidt M (2013) Hybrid assemblies based on a gadolinium-containing polyoxometalate and a cationic polymer with spermine side chains for enhanced MRI contrast agents. Chem Eur J 19(40):13317–13321
181. Zhang SM, Zheng YM, Yin SY, Sun JZ, Li B, Wu LX (2017) A dendritic supramolecular complex as uniform hybrid micelle with dual structure for bimodal living imaging. Chem Eur J 23(12):2802–2810
182. Yong Y, Zhou LJ, Zhang SS, Yan L, Gu ZJ, Zhang GJ, Zhao YL (2016) Gadolinium polytungstate nanoclusters: a new theranostic with ultrasmall size and versatile properties for dual-modal MR/CT imaging and photothermal therapy/radiotherapy of cancer. NPG Asia Mater 8(5):e273

Chapter 14
Inorganic-Organic Hybrid Materials Based on Nanopolyoxometalates

Yitong Wang and Jingcheng Hao

Abstract Various types of nano-scale polyoxometalates (POMs) with beautiful topologies has been synthesized successfully by destroying the hydration shell of the anions caused by the extremely hydrophilic surface. Their magnetic, electronic, and photoluminescent properties and valuable applications in catalysis, medicine, and material science are discussed. Meanwhile, the last ten years have witnessed a remarkable development in terms of preformed organic-inorganic POM-based hybrid systems for the rational design of functional architectures, assemblies and materials. Hydrophilic POMs of different sizes and shapes can interact with hydrophobic cationic surfactants, the resulting materials show amphiphilic properties with electrostatic interactions between the hydrophilic and hydrophobic components, called Surfactant-Encapsulated Clusters (SECs) or Surfactant-Encapsulated-POMs (SEPs). This hydrophobic surfactant-encapsulated clusters (HSECs) can fabricated through covalent or non-covalent interaction, which can construct ordered self-assembly, e.g. robust onionlike structures, honeycomb films or giant vesicle. Moreover, This ordered giant vesicle acts as building block to fabricate three dimensional structures. In addition, SECs can further self-assemble to give a variety of nanostructures on various surfaces/interfaces, among them, the most representative nanostructures discussed below is ordered honeycomb films, which is carried out by a simple solvent-evaporation method. It is reasonable to assume that the condensed water microdroplets induced by the quick evaporation of solvents play an important role as template for the formation of pores. Various factors are being investigated to construct thin films with different morphologies. We hope the inorganic-organic hybrid functional materials based on POMs will bridge polyoxometalate chemistry and material chemistry, which can be further explored application in many fields.

Keywords Polyoxometalates · Surfactant · Self-assembly · Inorganic-organic hybrids · Porous films

Y. Wang · J. Hao (✉)
Key Laboratory for Colloid and Interface Chemistry,
Ministry of Education, Shandong University, Jinan 250100, China
e-mail: jhao@sdu.edu.cn

J. Li (ed.), *Supramolecular Chemistry of Biomimetic Systems*,
DOI 10.1007/978-981-10-6059-5_14

14.1 Introduction to Developed POMs

Polyoxometalates (POMs) chemistry has been known for two centuries, which has received tremendous impetus from the intensive review by Pope and Müller in 1991 [1] and has accordingly achieved considerable progress in the last two decades. Metal-oxygen cluster-anions (polyoxometalates, or POMs) are discrete, anionic metal-oxide clusters of early transition metals [2, 3]. Composed principally of Mo, W, V and Nb, they represent a large and important class of inorganic molecules and materials, whose relevance is being highlighted in a number of review articles [4–6], with diverse range of dynamic molecular structures, a wide range of physical and chemical properties, and large domains of applications [7–11].

In the last three decades, various polyoxometalates have been synthesized successfully by destroying the hydration shell of the anions caused by the extremely hydrophilic surface [12–14]. Meanwhile, the last ten years have witnessed a remarkable development in terms of preformed organic-inorganic POM-based hybrid systems for the rational design of functional architectures, assemblies and materials [15–18].

14.1.1 Structures of POMs

Polyoxometalate structures are based on $\{MO_x\}$ polyhedra sharing vertices, edges, or more rarely, faces. In general, the typical structures of polyoxoanions are Keggin-type [19], Dawson-type [20], Waugh-type [21], Anderson-type [22], Silverton-type [23] and Lindquist-type [24], as shown in Fig. 14.1 and the Keggin-type and the Dawson-type are the most common type. In addition, POMs reveal a huge variety of shapes, sizes and compositions, such as, a "Keplerate-type" sphere $\{Mo_{132}\}$ $(NH_4)_{42}[\{(Mo^{VI})Mo_5^{VI}O^{21}(H_2O)_6\}_{12}\ \{Mo_2O_4(CH_3COOH)\}_{30}]\cdot ca\cdot 300H_2O\cdot ca\cdot 10CH_3COONH_4$ and $\{Mo_{72}Fe_{30}\}$ $[Mo_{72}^{VI}Fe_{30}^{III}O_{252}(CH_3COO)_{12}\{Mo_2O_7(H_2O)\}_2\{H_2Mo_2O_8(H_2O)\}(H_2O)_{91}]\cdot ca\cdot 150H_2O$, a "giant wheel-type" cluster $\{Mo_{154}\}$ $Na_{15}[Mo_{154}O_{462}H_{14}(H_2O)_{70}]_{0.5}[Mo_{152}O_{457}H_{14}(H_2O)_{68}]_{0.5}\cdot ca\cdot 400H_2O$, a "protein-sized blue lemon" $\{Mo_{368}\}$ $(Na_{48}\ [H_{496}Mo_{368}O_{1464}S_{48}]\ \cdot ca\cdot 1000\ H_2O)$ [25], and sandwich-type clusters, $\{Mn_2Bi_2W_{20}\}$ $(Na_6(NH_4[(Mn(H_2O)_3(WO)_2(BiW_9O_{33})_2])$ [26] and $\{EuW_{10}\}$ $(Na_9[EuW_{10}O_{36}]\cdot 32H_20)$ [27]. Compared with the traditional salts, POMs are formed by linking metal oxide polyhedra which are composed of many more atoms (even thousands), and become giant, with diameters of several nanometers. The different colors of the POMs in Table 14.1 represent different types of metal oxide polyhedral: light blue, MoO_7 pentagon bi-pyramid polyhedron; dark blue, MoO_6 octahedron; yellow, FeO_6 octahedron; light green, CrO_6 octahedron; dark green, VO_6 octahedron, which enable researchers to construct POMs with different topologies and sizes [11]. Due to the excess of oxo-ligands over metal ions, POMs are usually highly negatively charged, i.e., macro-polyoxoanions with high solubility in water and polar solvents exhibit unique solution behaviors. In dilute solutions,

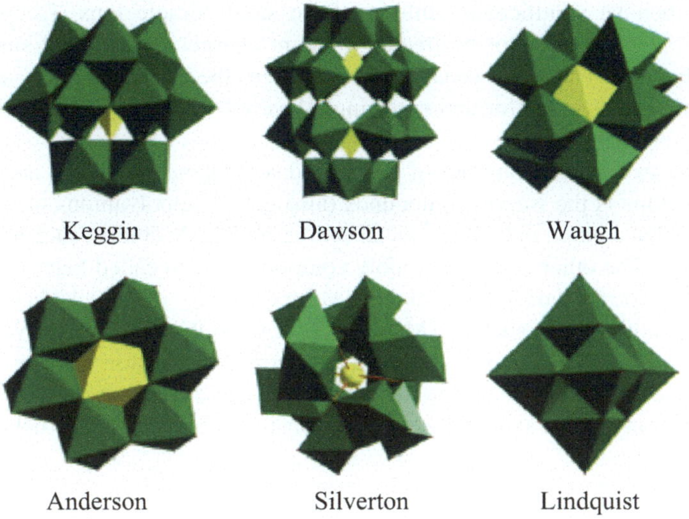

Fig. 14.1 Six typical structures of polyoxometalates

Table 14.1 Information on the structure, charge density, and self-assembly behavior of some typical giant polyoxometalate molecular clusters

Polyoxoanion		$\{Mo_{72}Fe_{30}\}$		$\{Mo_{72}Cr_{30}\}$	$\{Mo_{154}\}$	$\{Cu_{20}P_8W_{48}\}$	$\{P_4Y_9W_{43}\}$	$\{Mo_{132}\}$	$\{Mo_{72}V_{30}\}$
Molecular structure									
Negative charge	0-1	1-29		5	15	25	30	42	31
Size (nm)		2.5		2.5	3.6×1.2	2.2×1.1	4.0×2.0×2.0	2.9	2.5
Charge density (nm^{-2})	0-0.051 (pH < 2.9)	0.015-1.477 (2.9 < pH < 6.6)	Unstable (pH > 6.6)	0.255	0.543	1.291	1.08	1.590	1.579
Self-assembly in aqueous solution	No	Yes	No	Yes	Yes	Yes	Yes	Yes at high conc.	No
R_h of Blackberry (nm)		48-15		60	45	38	46		

Formula	References
$\{Mo_{72}Fe_{30}\}$ $[Mo^{VI}_{72}Fe^{III}_{30}O_{252}(CH_3COO)_{12}\{Mo_2O_7(H_2O)\}_2\{H_2Mo_2O_8(H_2O)\}(H_2O)_{91}]$ $\cdot ca\cdot 150H_2O$	[30]
$\{Mo_{72}Cr_{30}\}$ $[\{Na(H_2O)_{12}\}\{Mo^{VI}_{72}Cr^{III}_{30}O_{252}(CH_3COO)_{19}(H_2O)_{94}\}]$ $\cdot ca\cdot 120H_2O$	[34]
$\{Mo_{154}\}$ $Na_{15}[Mo_{154}O_{462}H_{14}(H_2O)_{70}]_{0.5}[Mo_{152}O_{457}H_{14}(H_2O)_{68}]0.5\cdot ca\cdot 400H_2O$	[28]
$\{Cu_{20}P_8W_{48}\}$ $K_{12}Li_{13}[Cu_{20}Cl(OH)_{24}(H_2O)_{12}(P_8W_{48}O_{184})]22H_2O$	[33]
$\{P_4Y_9W_{43}\}$ $K_{15}Na_6(H_3O)_9[(PY_2W_{10}O_{38})_4(W_3O_{14})]39H_2O$	[32]
$\{Mo_{132}\}$ $(NH_4)_{42}[\{(Mo^{VI})Mo^{VI}_5O^{21}(H_2O)_6\}_{12}\{Mo_2O_4(CH_3COOH)\}_{30}]$ $\cdot ca\cdot 300H_2O\cdot ca\cdot 10CH_3COONH_4$	[29]
$\{Mo_{72}V_{30}\}Na_8K_{14}(VO)_2[\{(Mo^{VI})Mo^{VI}_5O_{21}(H_2O)_3\}_{10}\{(Mo^{VI})Mo^{VI}_5O_{21}(H_2O)_3(SO_4)\}_2$ $\{V^{IV}O(H_2O)\}_{20}\{V^{IV}O\}_{10}(\{KSO_4\}_5)_2]$ $\cdot ca\cdot 150H_2O$	[31]

macroions behave significantly different from small simple ions (Debye-Hückel theory) because they cannot be treated as point charges or colloidal suspensions (DLVO theory). The size discrepancy between the polyoxoanions and their counter-ions results in the formation of single-layered, hollow, spherical, blackberry structures.

There are two generic families of POMs and some general features are shown in Table 14.1. One is the isopolycompounds (also called isopolyanions or isopolyoxometalates) that contain only the d^0 metal cations and oxide anions, such as $\{Mo_{154}\}$ and $\{Mo_{132}\}$. The other is the heteropoly compounds (also called heteropolyanions or heteropolyoxometalates) that contain one or more p-, d-, or f-block "heteroatorns" in addition to other ions, such as $\{Mo_{72}Fe_{30}\}$, $\{Cu_{20}P_8W_{48}\}$ and $\{P_4Y_9W_{43}\}$. Over half of the elements in the periodic table are known to function as heteroatoms in heteropoly compounds. These heteroatoms can reside either buried in (not solvent accessible) or at surface (solvent accessible) positions in the POM structures.

14.1.2 *Properties of POMs*

Polyoxometalates are composed of traditional metals and coordinated oxygen atoms. Due to the active outer shell electron of traditional metals, polyoxometalates show excellent electronic versatility. Many researches have demonstrated nano-scale POM clusters are versatile, including electrochemical properties, magnetic properties and photoluminescent properities.

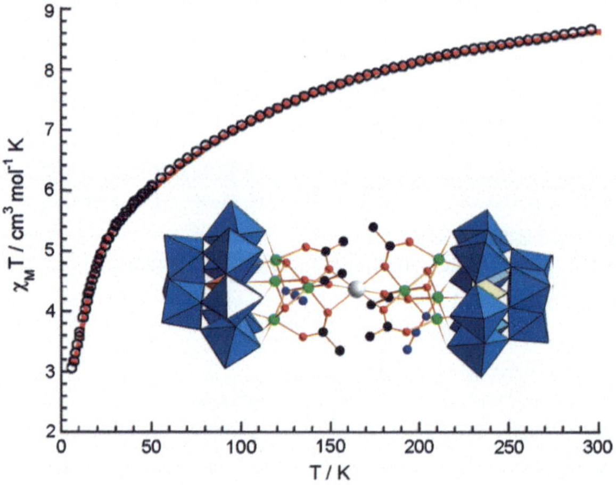

Fig. 14.2 Thermal dependence of $\chi_M T$ for $\{Ni_4SiW_9\}$. The *solid line* represents the best-fit parameters using Hamiltonian. The inset schematically shows polyhedral and ball-and-stick representation of $\{Ni_4SiW_9\}$. Color code: *blue octahedra*, $\{WO_6\}$; *orange tetrahedra*, $\{SiO_4\}$; *green spheres*, Ni; *gray sphere*, Na; *red spheres*, O; *blue spheres*, N; *black sphere*, C. Reprinted with the permission from Ref. [35]. Copyright 2008, American Chemical Society

Pichon et al. discovered and characterized that the high-nuclearity Ni^{II}-substituted polyoxometalate {Ni_4SiW_9} ($Na_{15}[Na\{(A\text{-}\alpha\text{-}SiW_9O_{34})Ni_4(CH_3COO)_3(OH)_3\}_2\cdot4NaCl\cdot36H_2O$) possesses ferromagnetic and antiferromagnetic interactions among clusters concomitantly [35]. As shown in Fig. 14.2, the χ_MT product continuously decreases from 300 K (χ_MT = 8.65 cm^3 mol^{-1} K, with the χ_MT value calculated for eight non-interacting Ni^{II} centers being 8.82 cm^3 mol^{-1} K, assuming g = 2.1) to 2 K (χ_MT = 2.72 cm^3 mol^{-1} K), indicating that antiferromagnetic exchange interactions are predominant in the {Ni_4SiW_9}. In addition, it has been proved that the Keplerate-type {$Mo_{72}Fe_{30}$} possesses paramagnetism, which were also investigated by experimental investigation and theoretical simulation with the classical and quantum Heisenberg model [36].

The composition of the metal and the shape of the framework define the properties of a POM. Due to the traditional metals, POMs possess rich electrochemical properties [37–40]. For example, in order to study the electrochemical properties of POMs, the wheel-shaped Cu_{20}-tungstophosphate $K_{12}Li_{13}[Cu_{20}Cl(OH)_{24}(H_2O)_{12}(P_8W_{48}O_{184})]22H_2O$ was selected by measuring cyclic voltammetry and controlled potential coulometry in pH = 0 and pH = 5 media [38]. Figure 14.3a shows the cyclic voltammogram (CV) of $Cu_{20}P_8W_{48}$ in a pH = 0 medium. A large current intensity composite reduction wave is observed with its main peak located at 0.113 V, followed by small, hardly distinguishable waves. On potential reversal, small waves are also seen, followed by a large current reoxidation wave located at +0.034 V. This last wave displays the characteristic shape usually encountered for the reoxidation of deposited Cu^0 on electrode surface. Figure 14.3b shows the CV of $Cu_{20}P_8W_{48}$ in a pH = 5 medium. The pattern is composed of two closely spaced reduction waves and well-separated from another set of two closely spaced waves. The authors presumed the first two reduction waves were the two-step reduction of Cu^{2+} to Cu^0 through Cu^+. The two waves featuring the Cu^{2+}/Cu^+ and Cu^+/Cu^0 reduction steps are located at −0.162 and −0.252 V, respectively, and are far from the W^{VI} reduction waves which appear at −0.682 and −0.796 V. It can be conclude that the influence of pH is reflected in the overall negative potential shift of the CV when the medium acidity decreases and W^{VI} reduction waves have much better separation from Cu^{2+} waves at pH = 5 than at pH = 0. The activity in the reduction of nitrate and the intermediate or final products of the polyanion $Cu_{20}P_8W_{48}$ was tested at pH = 0 and pH = 5. However, the waves only can be observed at pH = 5, as shown in Fig. 14.3c, d, a medium in which heteropolyacid-based catalysts are more difficult to obtain than in more acidic solutions. It must be concluded that an efficient catalysis of the reduction of nitrite can be achieved during the reduction of $Cu_{20}P_8W_{48}$.

Moreover, some lanthanide-containing and transition metal-containing polyoxometalates are photoluminescent. For example, a Eu-containing $Na_9[EuW_{10}O_{36}]$ $32H_2O$ [27] polyoxometalates is strongly luminescent under UV light excitation (Fig. 14.4).

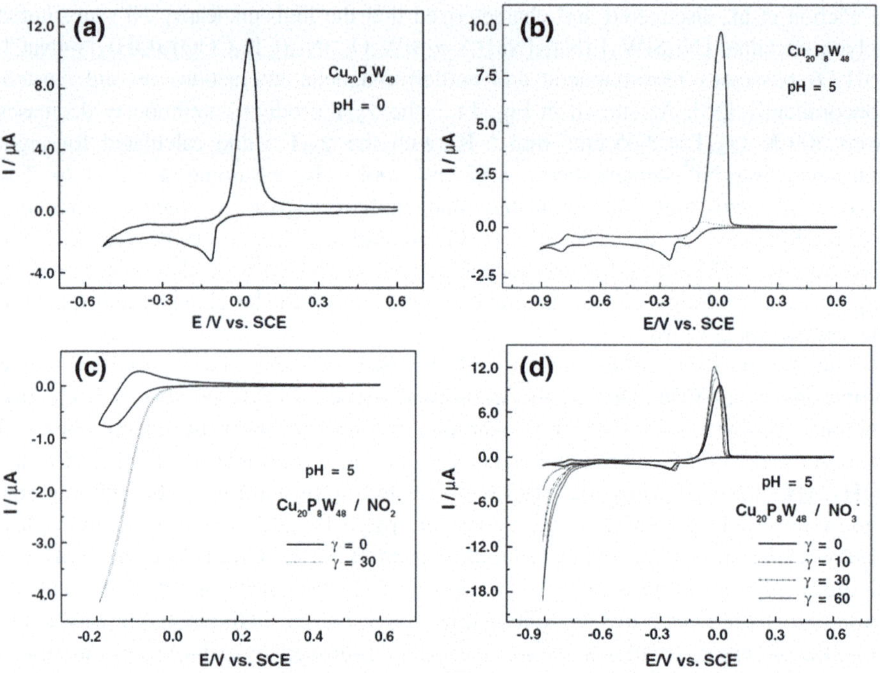

Fig. 14.3 **a** Cyclic voltammograms of 4×10^{-5} M solutions of $Cu_{20}P_8W_{48}$ in H_2SO_4 (pH = 0) medium. The scan rate was 2 mV s^{-1}, the working electrode was glassy carbon (3 mm diameter disc), and the reference electrode was SCE. **b** Cyclic voltammograms of 4×10^{-5} M solutions of $Cu_{20}P_8W_{48}$ in 1 M CH_3COOLi + CH_3COOH (pH = 5) medium. The scan rate was 2 mV s^{-1}, the working electrode was glassy carbon (3 mm diameter disc), and the reference electrode was SCE. **c** Cyclic voltammograms (scan rate: 2 mV s^{-1}) for the electrocatalytic reduction of nitrate with a 4×10^{-5} M solution of $Cu_{20}P_8W_{48}$ in a pH = 5 medium (1 M CH_3COOLi + CH_3COOH). **d** Cyclic voltammograms (scan rate: 2 mV s^{-1}) for the electrocatalytic reduction of nitrite with a 4×10^{-5} M solution of $Cu_{20}P_8W_{48}$ in a pH = 5 medium (1 M CH_3COOLi + CH_3COOH). The excess parameter defined as $\gamma = C^0 (NO_x)/C^0 (Cu_{20}P_8W_{48})$. Reprinted from Ref. [38], Copyright 2012, with permission from Elsevier

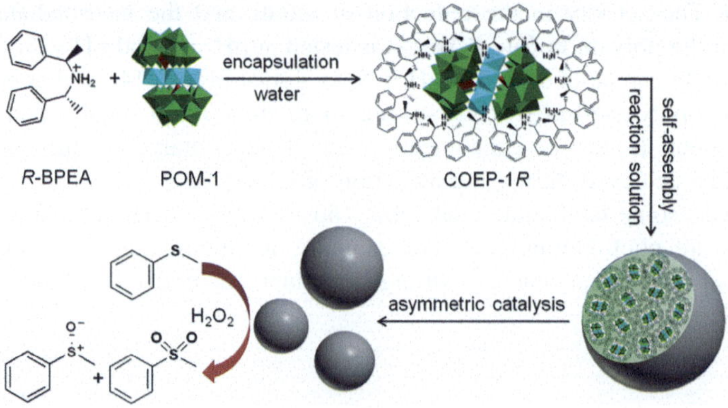

Fig. 14.4 Schematic illustration of the preparation procedure of COEP-1*R* assemblies and their asymmetric catalytic oxidation for methylphenyl sulfide with H_2O_2 (30%). Reproduced from Ref. [45] by permission of the Royal Society of Chemistry

14.1.3 Applications of POMs

Because of their remarkable properties, POMs have a great deal of potential to meet contemporary societal demands regarding health, environment, energy and information technologies, and broad applications as catalysts, photo-electronic/magnetic materials, and biologically active materials. Recently, the researches of polyoxometalates are focused on the development of novel POM-based molecular and composite materials for various applications. POMs hybrid materials can be classified into two categories, which are defined according to the interactions between the hybrid materials and POMs moieties, namely, non-covalent and covalent interactions, respectively. Non-covalently substituted POMs can be assembled through electrostatic interactions, hydrogen bonding, and/or van der Waals interactions [41–44]. Covalent functionalization by grafting organic moieties onto POMs result in the formation of another hybrids. It is also worth noting that covalently substituted POMs still have the capacity for further self-assembly via a variety of interactions, providing an opportunity to construct various interesting supramolecular assemblies. Modifications of POM clusters can enhance the original functionalities, which may lead to hybrid materials with not only value-adding properties, but also synergistic effects.

Due to their specific redox and acidic properties, POMs, as a type of commonly effective catalyst, have been applied for numerous organic transformations. Wu et al. [45] developed chiral polyoxometalate complexes which can be used as microreactors, and exhibited an efficient asymmetric catalytic activity for the oxidation of sulfide with up to 72% enantiomeric excess. The authors used a chiral organic cation to encapsulate POM with high catalytic activity through electrostatic interaction, resulting in the formation of a chiral organic cation-encapsulated POM (COEP) complex. They successfully constructed COEP-based spherical self-assembly complexes in the reaction solutions (Fig. 14.4). The synergistic effect of the two components predominates the catalytic process, in which the POM contributes to the catalysis of sulfoxidation and the chiral organic part provides a chiral microenvironment.

The polyanionic feature of POMs allows them to form complexes with many proteins or peptides containing cationic residues through ionic bonds, thus resulting in potential biological applications [46–50]. Alzheimer's disease (AD) is the most common form of dementia, which is characterized by cerebral extracellular amyloid plaques and intracellular neurofibrillary tangles [51]. Although the molecular mechanisms of AD pathogenesis are not clearly understood owing to its complexity, many efforts have been made to development of Ab inhibitors [52]. Qu's group has demonstrated that the POMs with Wells-Dawson or a Keggin structure showed moderate to high inhibition of Ab aggregation [47, 53]. Furthermore, they developed some POMs/peptides hybrid platforms as the inhibitors of the aggregation of Amyloid β peptides associated with AD. Through the self-assembly of phosphotungstate, $K_8[P_2CoW_{17}O_{61}]$, and Aβ15-20 (Ac-QKLVFF-NH_2), Li et al. demonstrated the spontaneous formation of hybrid colloidal spheres in water, which was predominantly driven by electrostatic interactions. Furthermore, the enhanced inhibition efficiency and specific targeted-Aβ ability have been considered important and useful in clinical treatment with reduced side effects of POM [54]. In addition, Lee et al. generated multivalent nanofibers through electrostatic interactions between polyoxometalates and short peptides. Because of the supramolecular nature of the ensemble, POMs thus enabled the enhancement of the antimicrobial efficacy and biological stability of short peptides in situ [55]. It is expected that the extraordinary properties of POMs could be used to develop hierarchical assemblies for the fields of biological chemistry and materials science.

In addition to their conventional hotspots as catalysts, antimicrobial and virus inhibitors, polyoxometalates are also attractive in many other applications. Eu-containing POM of $Na_9[EuW_{10}O_{36}]\cdot 32H_2O$ (EuW_{10}) shows the highest luminescent quantum yield and strongest fluorescence. Song et al. [56]. fabricated the well-ordered, ultra-thin films based on the hybrid assembly of $Na_9[EuW_{10}O_{36}]\cdot 32H_2O$ (denoted as EuW_{10}) and exfoliated MgAl layered double hydroxide (LDH) monolayers by utilizing the layer-by-layer (LBL) technique, as shown in

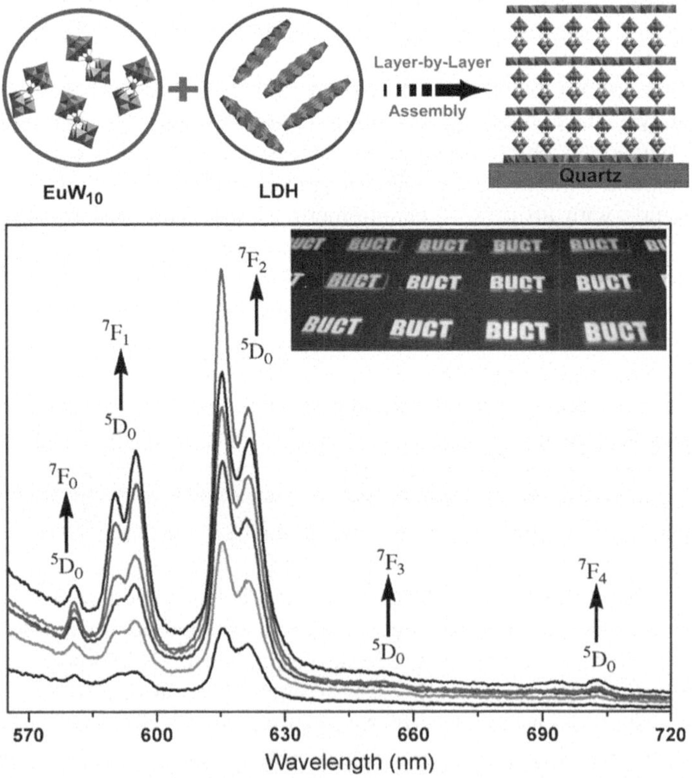

Fig. 14.5 **a** Schematic representation of the assembly process of the nanocomposite films of $(EuW_{10}/LDH)_n$. **b** Fluorescence spectra and photographs under UV irradiation at 254 nm with *n* increasing from 3 to 18. Reproduced from Ref. [56] by permission of John Wiley & Sons Ltd

Fig. 14.5. Anisotropic luminescence spectroscopy measurements show that the $(EuW_{10}/LDHs)_n$ ultra-thin films display well-defined red luminescence with an anisotropy value, r, of about 0.15. The cooperative and orientation effects between the LDH and polyanions play significant roles in the improved properties of the ultra-thin films. This work could benefit the design and fabrication of novel electro-optical devices based on the ultra-thin films of hybrid materials. The dream goal of energy research is to transpose incredibly efficient process and make an artificial device whereby the catalytic splitting of water is finalized to give a continuous production of oxygen and hydrogen.

This chapter is mainly devoted to a review of recent developments by our group in thin films formed by nanoengineered polyoxometalates and their various applications. Section 14.2 gives an overview of supramolecular architectures assembled from amphiphilic hybrid polyoxometalates, including the multilayer thin films, giant vesicles and onionlike structures. In Sect. 14.3, fabrication of honeycomb films of HSECs at air/water interface has been concluded, including the fabrication method, mechanism, morphology modulation and the promising applications of honeycomb films in detail.

14.2 Supramolecular Architectures Assembled from Amphiphilic Hybrid Polyoxometalates

One of the important aspects of POM clusters is their high solubility in a variety of solvents, which places POMs in a unique position as 'molecular' metal oxides. Mostly, POMs are ionized as giant anions (several nanometers) and countered by small cations, such as Na^+, NH_4^+, and H^+, in aqueous solutions, which is very different from the case for other metal-oxide materials. Once POMs are dissolved, the clusters interact electrostatically with cationic species, and this intrinsic driving force enables associations between POMs and cationic ions, molecules, complexes and polymers, as well as positively charged solid surfaces of materials. Consequently, POMs can be reliably modified through various techniques and methods to develop sophisticated materials and devices.

Hydrophilic POM macroions can interact with organic cations or cationic surfactants mainly through electrostatic interactions to construct inorganic-organic amphiphilic hybrids. One example is the surfactant encapsulated POM clusters (SECs). Kurth et al. created SECs by replacing the countercations of anionic molybdovanadato POMs by long alkyl chained cationic ammonium surfactants [57, 58]. A close packing of the alkyl chains was observed, and it was concluded that there is a compact shell around the POM. The surfactant shell improves the stability of the embedded POM as well as its solubility in nonpolar, aprotic organic solvents. Furthermore, it can change the surface chemical properties of POMs in order to realize the integration of multifunctional POMs into thin films and ordered three-dimensional aggregations.

14.2.1 Multilayer Films Containing POMs by Layer-by-Layer Technique on Planar Substrates

The origins of layer-by-layer assembly (LbL) can be traced back to the work of Iler in 1966, in which he fabricated multilayers by alternative deposition of positively and negatively charged colloid particles on planar surfaces [59]. It has been evolved as a powerful approach to produce different composite films on various types of substrates. Compared with other methods, the LBL technique has significant advantages in film stability and catalyst reusability. What is more, each nanometer-scaled POM molecule has a homogeneous diameter and surface charge when dissolved in a polar solvent. Our group fabricated the ultrathin films based on TiO_2 and tungstophosphate ($H_3PW_{12}O_{40}$, abbreviated as PW_{12}) by LbL self-assembly method on quartz, silicon, and ITO substrates and first applied them to photodegrade dye effluents [60]. Based on electrostatic interaction, $(TiO_2/PW_{12})_n$ multilayer films were fabricated and characterized on quartz slides, silicon wafers, and microscopic glass slides. These solid substrates were treated according to the literature [61]. The hydrophilic substrates were first immersed in poly(allylamine hydrochloride) (PAH) aqueous solution for 15 min, then rinsed by diluted $HClO_4$ solution at pH = 2.5 and dried by N_2. Subsequently, poly(styrenesulfonate) (PSS) was assembled in the same way as PAH. The procedure described above was repeated once to gain a surface with uniform charge. Then, the substrates with four precursor layers were alternately dipped in positively charged TiO_2 colloid (16 g L^{-1}, pH $\sim$ 2.0) for 15 min and 3 mmol L^{-1} negative charged PW_{12} solution for another 15 min until the desired number of bilayers was obtained. Rinsing and drying were conducted after each deposition cycle. The multilayer films with the desired number were described as $(TiO_2/PW_{12})_n$, where n was the number of bilayers. The $(TiO_2/PW_{12})_n$ composite films (with n = 2, 4, 6, 8, 10) are assembled on $(PAH/PSS)_2$ precursor film-coated quartz slides. The self-assembly process of multilayer films can be monitored by UV-vis spectroscopy, as shown in Fig. 14.6a. The $(PAH/PSS)_2$ precursor films show no absorbance in

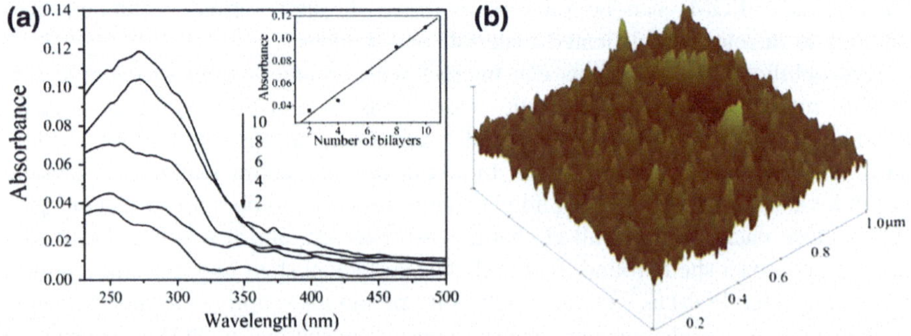

Fig. 14.6 a UV-vis spectra of the $(TiO_2/PW_{12})_n$ composite films assembled on quartz slides. Inset shows the relationship of absorbance at 253 nm versus the number of bilayers of $(TiO_2/PW_{12})_n$ films. b $(TiO_2/PW_{12})_4$ multilayer film assembled on silicon slides. Reprinted with the permission from Ref. [60]. Copyright 2011 American Chemical Society

230–500 nm UV-vis region. The maximum absorbance of composite films is 253 nm, which is attributed to charge transfer from O to W atom in the Keggin unit, increasing with TiO_2 and PW_{12} deposition. The inset shows the absorbance at 253 nm as a function of bilayer number and an approximately linear relationship is observed. The detailed information about the surface morphology and homogeneity of multilayer films are investigated by AFM (Fig. 14.6b).

In past decades, due to the low cost, innoxiousness, chemical inertness, and high photocatalytic performance, TiO_2 has been extensively used as photocatalysts for the degradation of organic dyes [62]. However, the photocatalytic efficiency of TiO_2 is remarkably decreased owing to the fast recombination of photogenerated electron hole pairs. POMs, considered efficient electron acceptors, can successively transfer the photogenerated electrons from the TiO_2 conduction band to the empty d orbital of theirs, which can improve the photocatalytic efficiency owing to the synergistic effect of POMs and TiO_2. In order to evaluate the photocatalytic efficiency of the composite films, methyl orange (MO), was used as a probe for the photocatalytic degradation. The maximum absorption peak of MO shifts from 463 nm under nearly neutral condition to 487 nm at pH 4.0 and to 507 nm at pH 2.0, which is due to the delocalization of lone pair electrons on the azo group in two different structure of MO. We carried out the photocatalytic tests at pH 2.0. Figure 14.7a shows the UV-vis absorption spectra changes of aqueous MO (10 mg mL^{-1}) solution in the presence of $(TiO_2/PW_{12})_n$ composite films. The intensity of MO absorption peaks gradually decreases as irradiation time increases, which suggests that MO is almost completely decomposed. The kinetics of MO photodecomposition by composite films with different number of bilayers was investigated. The plots of $\ln c_0/c$ (where c_0 stands for the initial concentration and c is the residual concentration at time t) versus time for MO photodegradation are shown in Fig. 14.7b, which indicates that photodecomposition of diluted MO solutions agrees well with apparent first-order kinetics.

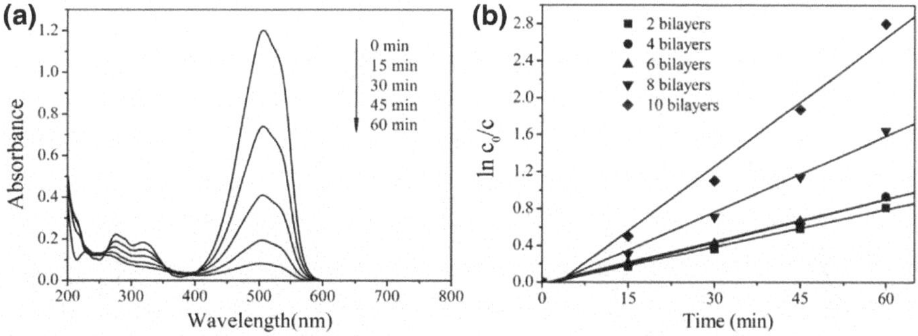

Fig. 14.7 **a** UV-vis absorption spectra changes of MO (10 mg L^{-1}, pH = 2.0) solution in the presence of $(TiO_2/PW_{12})_{10}$ composite films. **b** Reaction kinetic study of photodegradation of MO (the initial concentration: 10 mg L^{-1}, pH = 2.0) by $(TiO_2/PW_{12})_n$ films. Reprinted with the permission from Ref. [60]. Copyright 2011 American Chemical Society

Coincidentally, we demonstrated polyoxometalates which have different compositions, shapes, and sizes can interact with TiO_2 to self-assemble into multilayer film on silicon wafers and glass slides for photocatalytic degradation of methyl orange (MO). We choose nano-scaled POMs with different shapes, sizes, and compositions, including $H_4SiW_{12}O_{40}$ (SiW_{12}) [62, 63] and $Na_4W_{10}O_{32}$ ($W_{10}O_{32}$) [64], to be the typical examples. The main role of surfactants in dyeing processes is to increase the solubility of dyes in water and improve the dye-uptake and dye fastness, to reduce the dyeing temperature. In fact, actual dye wastewaters contain not only dyes, but also other auxiliaries such as salts and surfactants. Therefore, to make sure the effect of surfactants and salts in the MO degradation process is a burning question. In our study, photocatalytic processes were performed in the presence of two inorganic salts and two different surfactants, NaCl, Na_2SO_4, sodium dodecylsulfate (SDS, cmc = 8.6 mmol L^{-1}), and polyoxyethylene dodecyl ether ($C_{12}E_{23}$, cmc = 0.043 mmol L^{-1}). Figures 14.8a, b turn out that the presence of inorganic salts, especially Na_2SO_4, exhibits an obvious inhibition effect on the MO degradation. The removal efficiency of dye is deceased when the concentration of inorganic salts is increased. The difference between two inorganic salts may be mostly due to the aggregation or association of ionic dye in water. As shown in Figs. 14.8c, d, the surfactants with different types exhibit two completely different results. It can be observed that the nonionic surfactant shows the excellent

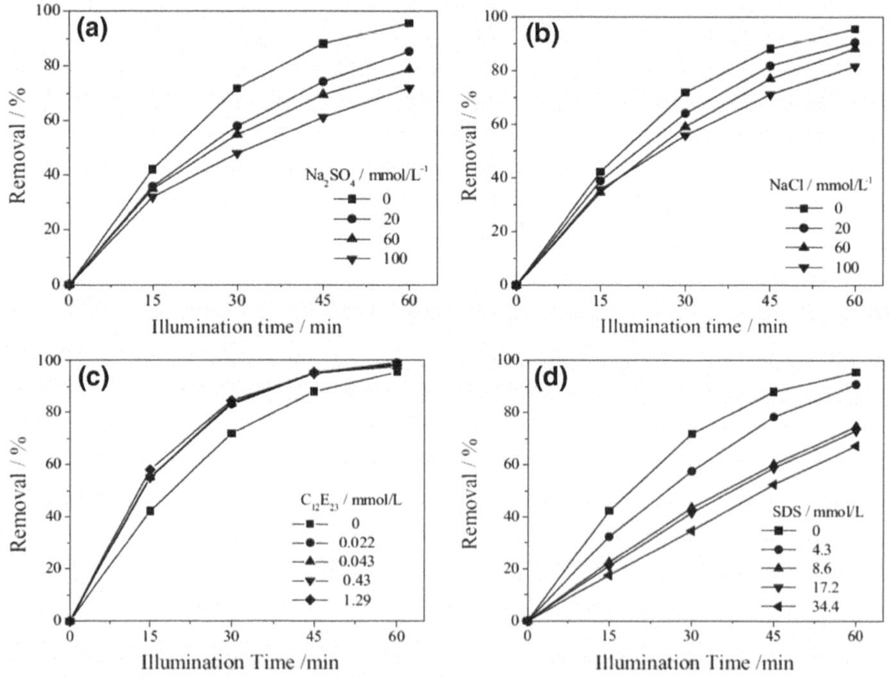

Fig. 14.8 Removal efficiency of MO at different concentration of inorganic salts: **a** Na_2SO_4 and **b** NaCl and **c** $C_{12}E_{23}$ and **d** SDS. Reprinted from Ref. [63], Copyright 2014, with permission from Elsevier

acceleration effect on the photocatalytic degradation of MO, while for the anionic surfactant SDS, a markedly suppressing effect is observed. For $C_{12}E_{23}$, oxygen in the EO groups of $C_{12}E_{23}$ has extra lone pairs of electrons,while W atom of SiW_{12} polyanions has empty d orbitals. Hence, SiW_{12} polyanion and nonionic surfactant $C_{12}E_{23}$ can form a weak complex through electrostatic interaction, which is responsible for the enhancement of MO removal efficiency. Contrary to $C_{12}E_{23}$, in premicellar or micellar regions formed by anionic surfactant SDS, ionic pairs formed due to the weak interaction between dye and SDS first adhere to the micelles surfaces, and then incorporate into the micelles, in which MO may become inert to react with TiO_2/SiW_{12} composite films. This work will provide valuable information for the researchers who engage in treatment of textile effluents.

14.2.2 Self-patterning Porous Films Composed of POMs-Based Giant Vesicles

Cationic surfactants with long enough alkyl chains can interact with POMs stoichiometrically through electrostatic interaction, which can improve the stability of POMs in nonpolar solvent and change the interface property of POMs. Therefore, cationic surfactants encapsulated POMs macroanions can transfer from water to organic solution for forming POMs/surfactants complexes. These organic-inorganic nanoscale hybrids can change the surface property of POMs to be hydrophobic in behavior. These complexes still retain amphiphilic property to dissolve in organic solvents and self-assemble to form various ordered aggregates such as thin hybrid films [63], vesicles [66], liquid crystals [67, 68], gels with the structures of fibers [69], and so on. Many groups have reported single- or multi-layer films fabricated by POMs. However, to the best of our knowledge, very few self-assembled three dimensional structures based on POMs and organic components were used to fabricate films as building blocks. Conceptually, we reported a new fabrication of the complexes by a Keplerate-type polyoxomolybdate, $\{Mo_{72}Fe_{30}\}$, and a double-chain cationic surfactant DODMABr to produce the giant inverse vesicles in organic solvent [70]. The 2.5-nm-diameter, "Keplerate" molecule $\{Mo_{72}Fe_{30}\}$ exists as almost neutral molecules in crystals, but in solution they behave like a weak acid: the water ligands attached to the Fe^{III} centers tend to partially deprotonate, thus making $\{Mo_{72}Fe_{30}\}$ clusters slightly negatively charged (carrying several localized charges). The degree of deprotonation depends on the solution pH, from almost 0 at pH 3.0–22 at pH 4.9. When the $\{Mo_{72}Fe_{30}\}$ aqueous solution mixed with enough amount of cation surfactant dioctadecyldimethylammonium chloride (DODMABr), $\{Mo_{72}Fe_{30}\}$ can be transfer into the organic phase (chloroform phase) by forming the hydrophobic DODMABr-$\{Mo_{72}Fe_{30}\}$ complexes through electrostatic interactions between the $\{Mo_{72}Fe_{30}\}$ anions and the $DODMA^+$ cations. The yellow color of the aqueous phase (from $\{Mo_{72}Fe_{30}\}$) continuously vanishes, while the color of the $CHCl_3$ phase turns gradually yellow.

In order to determine the composition of DODMABr-{$Mo_{72}Fe_{30}$} complexes, the negative charge number of {$Mo_{72}Fe_{30}$} polyanions can calculate by an exponential equation:

$$N_{\{Mo72Fe30\}} = M_{\{Mo72Fe30\}} 10^{-pH\{Mo72Fe30\}} / c_{\{Mo72Fe30\}}$$

where $N_{\{Mo72Fe30\}}$ is the negative charge number of {$Mo_{72}Fe_{30}$} polyanions; $M_{\{Mo72Fe30\}}$ is the molecular weight of {$Mo_{72}Fe_{30}$}; $pH_{\{Mo72Fe30\}}$ is the pH value of {$Mo_{72}Fe_{30}$} aqueous solution and $c_{\{Mo72Fe30\}}$ is the concentration of {$Mo_{72}Fe_{30}$} aqueous solution.

In this study, {$Mo_{72}Fe_{30}$} aqueous solution has a certain concentration is 8 mg mL^{-1} {$Mo_{72}Fe_{30}$} and the pH value is 2.91. So the composition of {$Mo_{72}Fe_{30}$}/DODMABr complexes could be calculated to be {$Mo_{72}Fe_{30}$} $(DODMA)_3$. In the $CHCl_3$-CH_3OH (3:1)mixed solution, {$Mo_{72}Fe_{30}$} $(DODMA)_3$ complex can form densely packed multilamellar inverse vesicles due to the larger inorganic polar head groups comparing to conventional surfactants.

Densely packed multi-lamellar inverse vesicles in organic solution via self-assembly formed by the {$Mo_{72}Fe_{30}$}$(DODMA)_3$ complexes were demonstrated by the SEM and TEM observations (Fig. 2.4). Differing from normal vesicles, reverse vesicles are those with hydrophilic groups inwards and hydrophobic countparts outwards arranged in non-aqueous solutions. Figures 14.9a, b show two different types of spherical vesicles and spherical vesicles with collapse parts. TEM images indicated the enlarged image of part of a multilayer vesicle, which were the hollow sphere composed of small particles with the average 2.47 $\pm$ 0.07 nm in diameter, which exactly corresponds to the size of a {$Mo_{72}Fe_{30}$}. Furthermore, the bilayer thickness of the {$Mo_{72}Fe_{30}$}$(DODMA)_3$ vesicles can be calculated to be about 7.3 nm according to the extended chain length of double chain cationic DODMABr (2.4 nm) [71] and the diameter of {$Mo_{72}Fe_{30}$} (2.5 nm). The average wall thickness of the vesicle is about (50.6 $\pm$ 13.0) nm from Fig. 14.9, thus the vesicles of the {$Mo_{72}Fe_{30}$}$(DODMA)_3$ complexes are about (7 $\pm$ 2) bilayers.

The giant vesicles can self-assemble to form porous films on solid surfaces via the breath figure method. When the {$Mo_{72}Fe_{30}$}$(DODMA)_3$ complexes were prepared by 0.3 mL {$Mo_{72}Fe_{30}$} aqueous solution (8 mg mL^{-1}) mixed with DODMABr organic solution, the relatively excess hydrophobic DODMABr molecules exist on the single crystal silicon slides, water permeation on substrate surfaces was hampered and the humid nitrogen stream could not drive the {$Mo_{72}Fe_{30}$}$(DODMA)_3$ complexes to form porous films (Fig. 14.10a). Increasing amount of {$Mo_{72}Fe_{30}$} to 0.45 mL in DODMABr organic solution, as shown in Fig. 14.10b, the porous structures became clearer and the porous films can easily be obtained, but the pore size is polydispersed from 200 nm to more than 2.0 μm. With a further increase amount of {$Mo_{72}Fe_{30}$} to 0.6 mL, highly ordered porous films with large size holes of 7–10 μm (Fig. 14.10d) surrounding pores with uniform sizes of 1.0 μm (Fig. 14.10e) can be successfully obtained by large areas. In order to observe the interior structure of the porous films more distinctly,

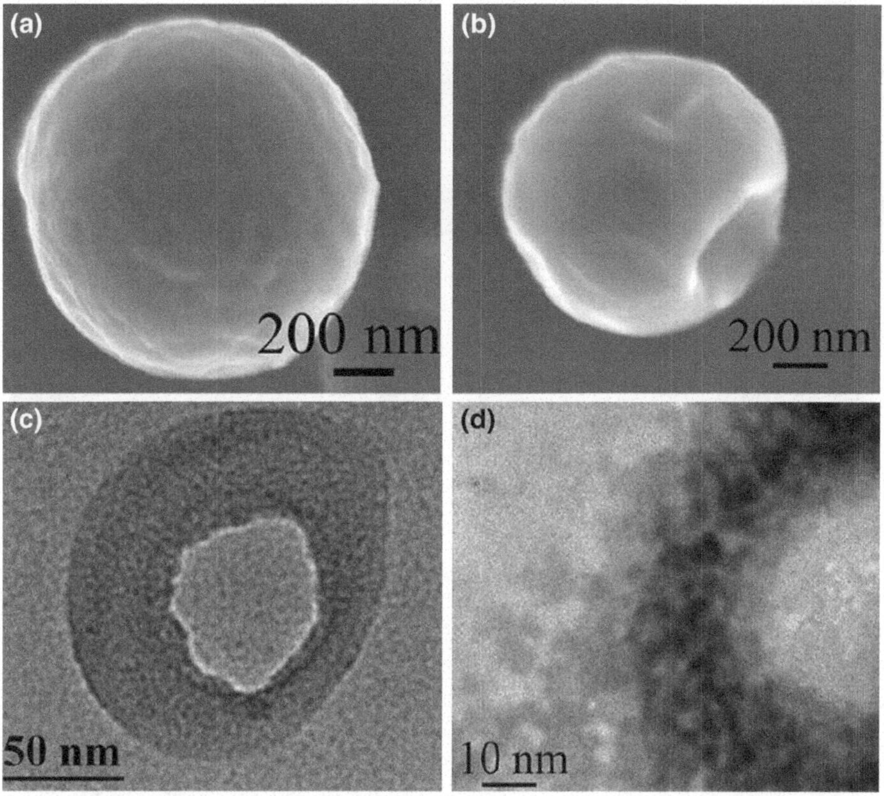

Fig. 14.9 **a**, **b** SEM enlarged images. **c**, **d** TEM enlarged images of giant inverse vesicles of $\{Mo_{72}Fe_{30}\}(DODMA)_3$ complexes in organic phase. Reprinted from Ref. [70], Copyright 2016, with permission from Elsevier

a cross-sectional view of the porous films was shown in Fig. 14.10f. The interior of porous films are filled with giant vesicles in a regular arrangement, indicating the giant vesicles acted as building blocks of the porous films. The structure of giant vesicles is reserved completely during the process of organic solvent evaporation because of higher structural strength. In order to further analyze the porous film structures formed by the vesicles of the $\{Mo_{72}Fe_{30}\}(DODMA)_3$ complexes, the XPS analysis of the porous film surface was characterized. The high resolution XPS spectra of C1 s were devided into two peaks centered at 284.7 and 286.4 eV, which are attributed to C–H/C–C and C–N, respectively. The high resolution XPS spectra of N1 s were deconvoluted into three peaks centered at 398.7, 399.9 and 403 eV (Fig. 14.11b). The major peak at 399.9 eV was attributed to C–N. The peak at 402–403 eV was selectively attributed to $N^{+}OH^{-}$ inter- and intra-molecular hydrogen bond, and the weak peak at 403 eV resulted from the interaction between small amount of $\{Mo_{72}Fe_{30}\}$ anions and $DODMA^{+}$. Meanwhile, N1 s binding energy of the C–N bond is decreased because of the interaction of $\{Mo_{72}Fe_{30}\}$ macroanions

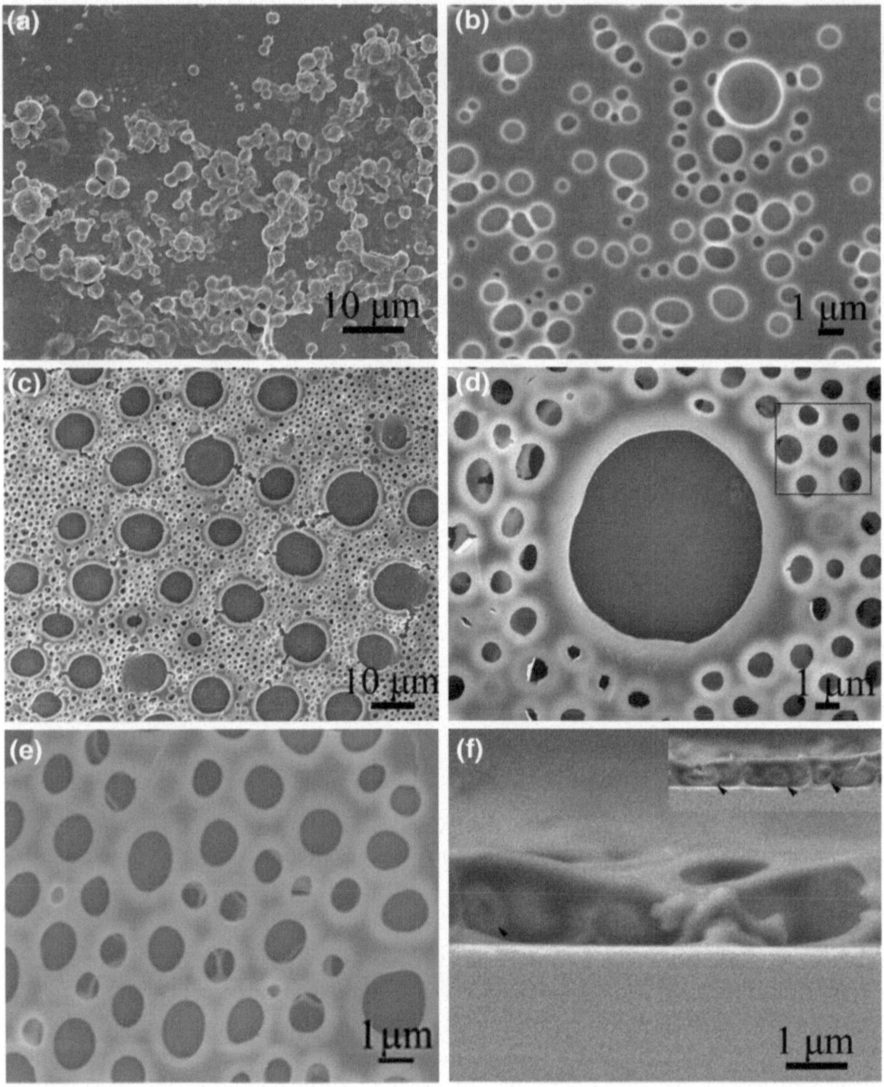

Fig. 14.10 FE-SEM images of porous films of vesicles formed by $\{Mo_{72}Fe_{30}\}(DODMA)_3$. The organic solution of $\{Mo_{72}Fe_{30}\}(DODMA)_3$ vesicles were prepared by different volumes of 8 mg mL^{-1} $\{Mo_{72}Fe_{30}\}$ aqueous solution equilibrated with 4 mL 1 mg mL^{-1} DODMABr in $CHCl_3$–CH_3OH (3:1) solution for 2 weeks. $V_{\{Mo72Fe30\}}$ = 0.30 mL (**a**), 0.45 mL (**b**), 0.6 mL (**c**, **d**). An enlarged image of the wall of the porous films (**e**) and a cross-sectional survey of the chosen area arbitrarily marked with red rectangle in Fig. 14.10d, one can see the vesicles with collapsed parts (*arrows marked*, **f**). Reprinted from Ref. [70], Copyright 2016, with permission from Elsevier

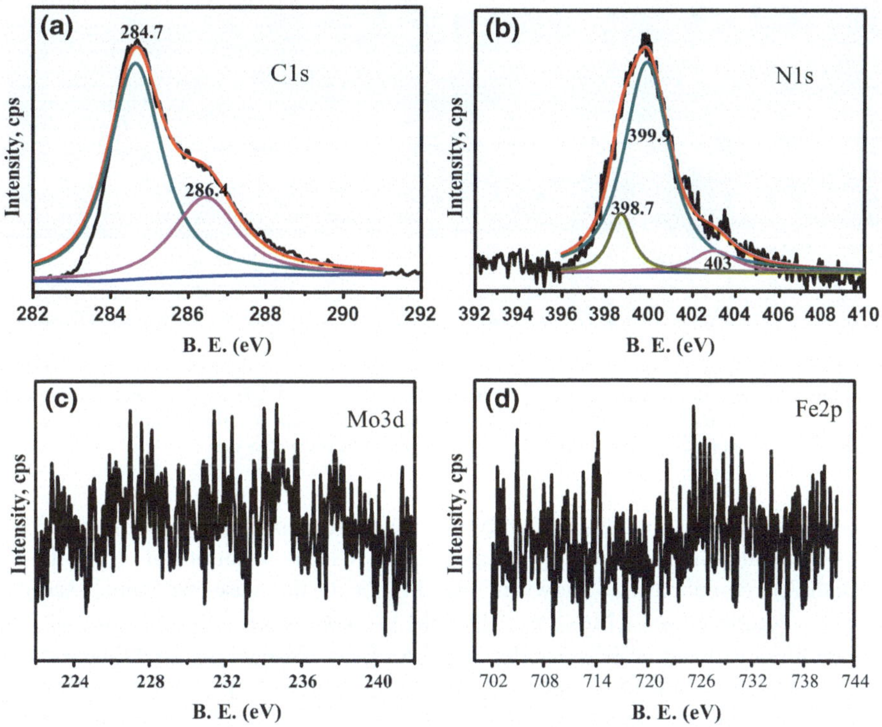

Fig. 14.11 High resolution XPS spectra of C1s (**a**), N1s (**b**), Mo3d (**c**) and Fe2p (**d**) of porous films of vesicles formed by $\{Mo_{72}Fe_{30}\}(DODMA)_3$. The organic solution of the $\{Mo_{72}Fe_{30}\}(DODMA)_3$ vesicles were prepared by 0.6 mL 8 mg mL^{-1} $\{Mo_{72}Fe_{30}\}$ aqueous solution equilibrated with 4 mL 1 mg mL^{-1} DODMABr in $CHCl_3$–CH_3OH (3:1) solution. Reprinted from Ref. [70], Copyright 2016, with permission from Elsevier

and $DODMA^+$, result in a weak peak appeared at 398.7 eV. However, as shown in Fig. 14.11c and d, no obvious peaks appeared in the high resolution XPS spectra of Mo3d and Fe2p, indicating that the surface of the porous films were mainly formed by excess DODMABr, and probably had few $\{Mo_{72}Fe_{30}\}(DODMA)_3$ complexes.

In order to investigate whether the honeycomb films can form by changing the composition of $\{Mo_{72}Fe_{30}\}$/DODMABr complexes. Another $\{Mo_{72}Fe_{30}\}$/(DODMA) complexes in organic phase formed by adjusting the concentration of $\{Mo_{72}Fe_{30}\}$ aqueous solution to 1 mg mL^{-1}. In this case, seven H^+ ions can be released from one $\{Mo_{72}Fe_{30}\}$ molecule in aqueous solution, in which the $\{Mo_{72}Fe_{30}\}(DODMA)_7$ complexes were obtained. The vesicles formed by $\{Mo_{72}Fe_{30}\}(DODMA)_7$ complexes are much smaller than those formed by $\{Mo_{72}Fe_{30}\}(DODMA)_3$ complexes. When the same volume of aqueous solution was mixed with the $CHCl_3$–CH_3OH (3:1) organic solution, almost all CH_3OH in organic phase was transferred into aqueous solution, $\{Mo_{72}Fe_{30}\}(DODMA)_7$ vesicles retained in $CHCl_3$ phase. Giant vesicles cannot be formed because of weak

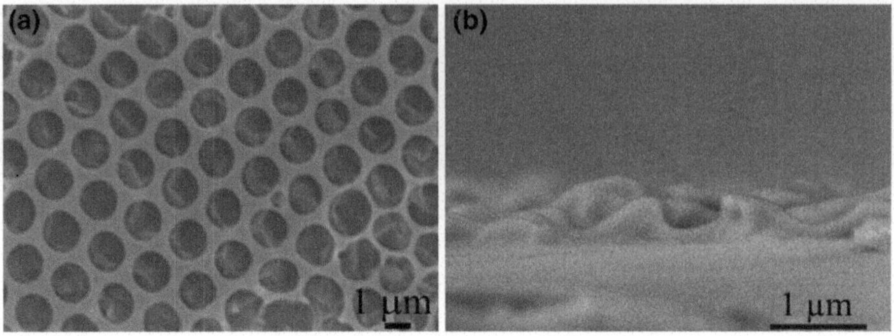

Fig. 14.12 FE-SEM images of highly ordered honeycomb films of the $\{Mo_{72}Fe_{30}\}(DODMA)_7$ complexes formed by 4 mL 1 mg mL^{-1} DODMABr in $CHCl_3$-CH_3OH (3:1) solvent equilibrated with 4 mL 1 mg mL^{-1} $\{Mo_{72}Fe_{30}\}$. Surface morphology (**a**) and a cross-sectional survey (**b**). Reprinted from Ref. [70], Copyright 2016, with permission from Elsevier

polarity of organic solvent. The highly ordered honeycomb films were prepared on single crystal silicon slides by dropping the $CHCl_3$ solution of $\{Mo_{72}Fe_{30}\}(DODMA)_7$ complexes, as shown in Fig. 14.12a. In this case, we cannot see the vesicles composed of $\{Mo_{72}Fe_{30}\}(DODMA)_7$ complexes situated interior the porous films. It must be because the vesicles of the $\{Mo_{72}Fe_{30}\}(DODMA)_7$ complexes were too small to be survived the drying process in preparation of honeycomb films via breath figure method (Fig. 14.12b). This phenomenon demonstrates that the concentrations of POMs and surfactants play an important role in the preparation of highly ordered honeycomb films.

14.2.3 Robust Onionlike Structures Formed by a Fullerene C_{60}-POM Hybrid

Polyoxometalates (POMs) have remarkable properties and a great deal of potential to meet contemporary societal demands regarding health, environment, energy and information technologies. However, implementation of POMs in various functional architectures, devices or materials requires a processing step. Most developments have considered the modification of POMs through non-covalent or covalent interaction. We have discussed the interaction of POMs and surfactants through non-covalent in Sects. 14.2 and 14.3. In this section, we will investigate the covalent modification. The covalently modified amphiphilic hybrid POMs are attractive. Some POM clusters possess multiple sites available for functionalization, which can be done by linking one or more hydrophobic organic functional groups to one POM. In addition, the amphiphilic nature of these hybrids extends the functionality of POM clusters in organic media, which used as multifunctional oxidation or acidification catalysts with good selective recognition of substrates.

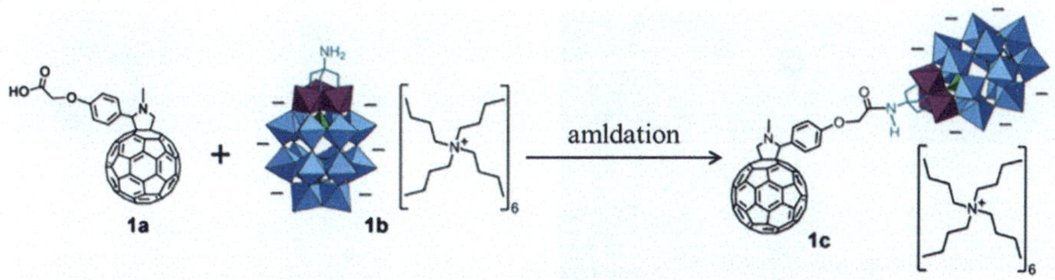

Fig. 14.13 Synthetic route to fullerene C_{60}-Dawson POM hybrid compound **1**. Reproduced from Ref. [72] by permission of the Royal Society of Chemistry

In view of the outstanding optical, electrochemical and photodynamic activities of C_{60} and well-known magnetic and catalytic properties of POMs, the idea of covalently linking them to create a new class of amphiphiles becomes quite attractive. However, C_{60}-POM hybrids as structural models are rarely reported. Recently our group reported the synthesis and aggregation behavior of a fullerene C_{60}-Dawson POM hybrid (**1**) for the first time [72]. Our strategy (Fig. 14.13) to create **1** includes amide coupling between a monosubstituted C_{60} carboxylic acid (**1a**) and a tris-modified, Dawson trivanadium substituted heteropolytungstate (**1b**) in a THF/CH_3CN mixture solvent. C_{60} exhibits poor solubility in polar solvents including DMSO, DMF and CH_3CN while Dawson POMs are not. After covalent attachment, **1** shows quite a similar solubility to the POM moiety, may due to the larger size of Dawson POM compared to C_{60}. The compound **1** possesses unique amphiphile molecular structure, resulting in dissolving in these POM-friendly solvents with the C_{60} unit as the solvophobic part. We first checked the aggregation behavior of **1** in DMSO and the typical results are summarized in Fig. 14.14. Observations of two typical samples showed the formation of vesicles with diameters of 30–110 nm (image Fig. 14.14a, b). From the high resolution TEM (HRTEM, image c,) we can see the thickness of an individual layer is $\sim$3.5 nm (between the arrow heads) and a distance of 5.0 nm between two adjacent layers (averaged from four repeating units indicated between the two arrows). In scanning electron microscopy (SEM) observations (image d), contrast variations have also been noticed at the edges of the onionlike aggregates, which further confirms a core-shell structure.

Multiple intermolecular forces including solvent-phobic interaction, van der Waals attraction and p–p stacking among C_{60} spheres as well as electrostatic interaction and the like-charge attraction among POM units driven the formation of the onionlike structures demonstrated here. We presumed a structural model of molecular bilayers with the POM units staying outside and C_{60} moieties hiding inside, which can be proofed from, as shown in Fig. 14.15, ^{13}C NMR where signals of the carbon atoms from C_{60} appear as a bulge between 140 and 150 ppm, which is in sharp contrast to those of C_{60} monoadducts in a good solvent, which exist as a variety of sharp peaks in the same region. It can be speculated that a structural model of molecular bilayers can be envisaged with the POM units staying outside

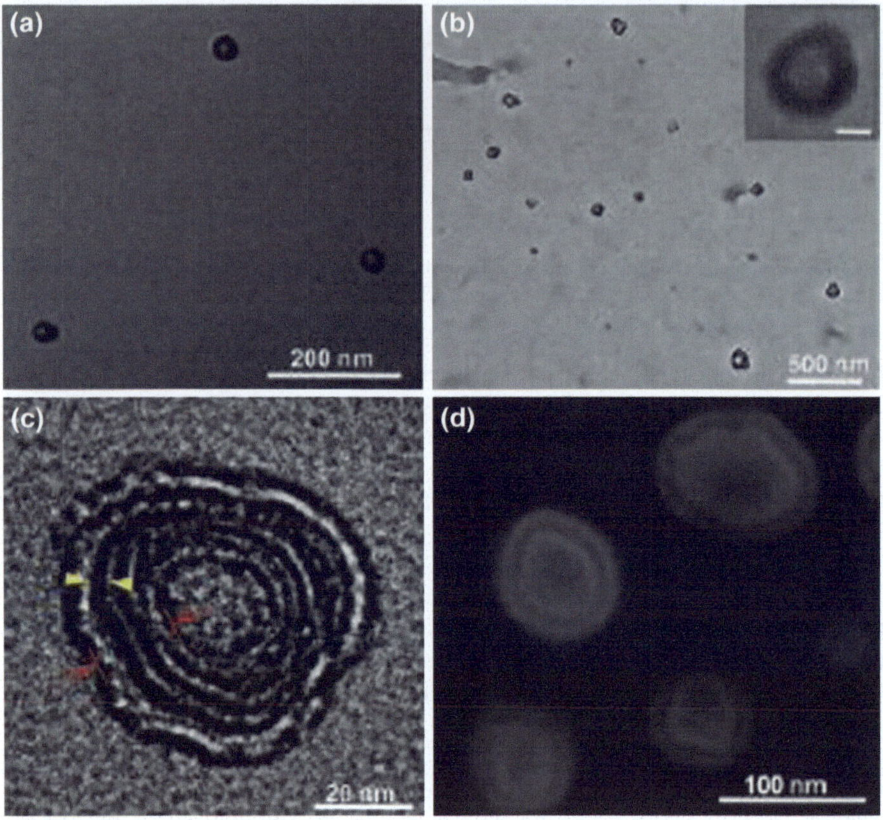

Fig. 14.14 Self-assembly of compound **1** in DMSO. TEM (**a**, **b**), HRTEM (**c**) and SEM (**d**) images. The concentration of **1** is 5.0 (**a**, **d**) and 20.0 (**b**, **c**) mg mL^{-1}, respectively. The scale bar for the magnified image inside image **b** is 20 nm. Reproduced from Ref. [72] by permission of the Royal Society of Chemistry

and C_{60} moieties hiding inside. Therefore, the C_{60} spheres are internally encapsulated in a quasi-solid state with obvious inter-sphere interactions upon aggregate formation. In addition, the aggregate formation also leads to peak broadening and baseline roughening in ^{1}H NMR (Fig. 14.15), which is consistent with similar phenomenon observed for alkyl chains embedded in molecular aggregates in water, such as vesicles. When solvents replaced by other solvents with similar polarities, such as CH_3CN and DMF, onionlike structures with similar sizes and number of layers also can be obtained. Interestingly, the onionlike structures exhibit high resistance towards the addition of water. The solutions remain stable up to the addition of 99% (v/v) water for the two series of samples with 5.0 and 0.6 mg mL^{-1} **1** in DMSO. The relatively strong interactions among the C_{60} units in water as well as the decreased water solubility of the POM moieties are contributed to the robustness of the onionlike structures towards the addition of water.

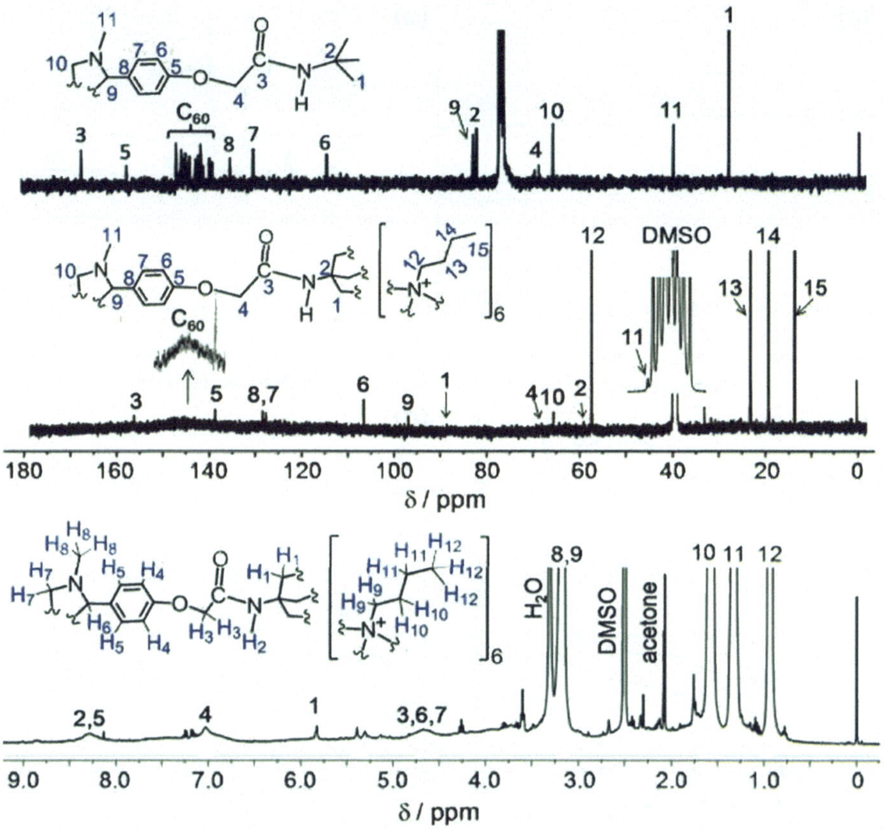

Fig. 14.15 *Top* ^{13}C NMR of the monosubstituted C_{60} carboxylic ester (1a″, in $CDCl_3$) and **1** (in d6-DMSO). Bottom: ^{1}H NMR of **1** (in d6-DMSO). The C_{60} unit and POM moiety has been omitted in the molecular structures for better clarity. Reproduced from Ref. [72] by permission of the Royal Society of Chemistry

The onionlike structures formed by compound **1** have rich physicochemical properties originating from both C_{60} and POM moieties, which makes them thoroughly different from their counterparts formed by surfactant mixtures [73, 74]. The magnetic properties of **1** both in the solid state and bulk solutions were investigated by electron spin resonance (ESR) measurements, as shown in Figs. 14.16a, b. Compared to that in the solid state as evidenced by the well-defined anisotropic eight-line splittings of the $^{23}V(IV)$ nuclei derived from the POM unit (Fig. 14.16b), an un-splitting single line was noticed within the whole investigated temperature range, which can be ascribed to the faster spin-lattice interaction in bulk solution due to the formation of aggregates. Furthermore, the electrochemical properties of **1** along with the two precursor molecules 1a and 1b have been investigated by cyclic voltammetry (Fig. 14.16c). C_{60} and their derivatives are well-known for their capability of the production of singlet oxygen ($^1O_2$) under light. Figure 14.16d

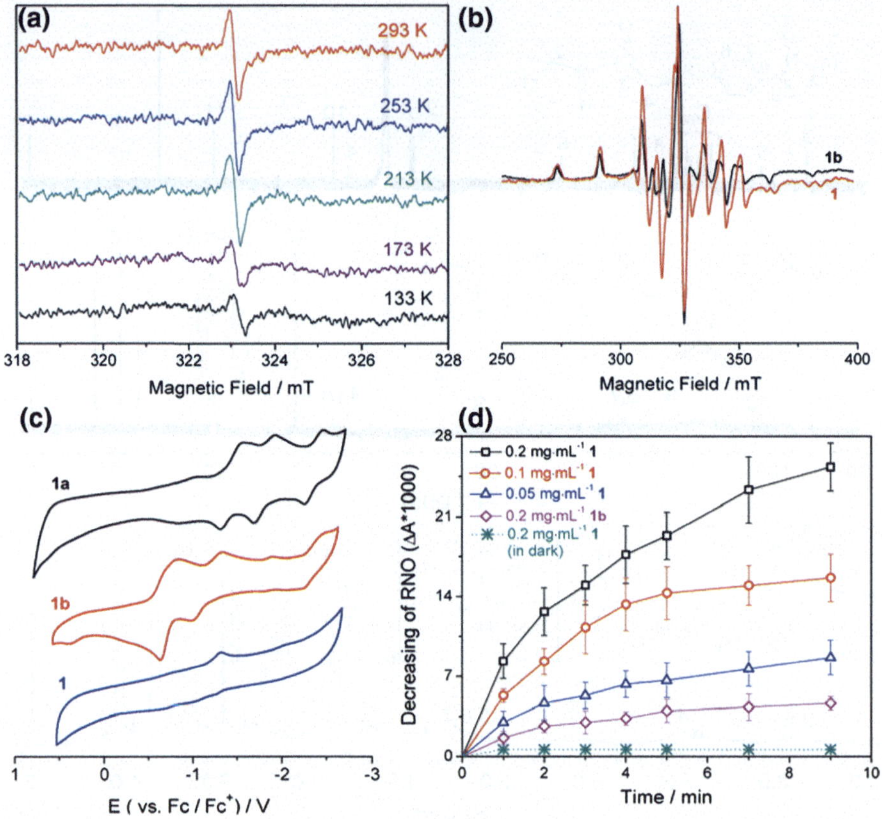

Fig. 14.16 **a** Temperature-dependent ESR spectra of 5.0 mg mL^{-1} **1** in CH_3CN. **b** ESR spectra of 1b and 1 in the solid state at room temperature. **c** Cyclic voltammetry of 1a (in dichlorobenzene), 1b (in DMF) and 1 (in DMF) with 0.1 mol L^{-1} tertbutylammonium hexafluorophosphate as the supporting electrolyte. **d** Generation of $^1O_2$ for **1** in aqueous solutions containing 1% (by volume) DMSO under 532 nm green light with a power density of 0.1 W cm^{-2}. For comparison, the activity in the dark is also given. Reproduced from Ref. [72] by permission of the Royal Society of Chemistry

shows the capability of $^1O_2$ production observed in the onionlike structures can be unambiguously attributed to the presence of the C_{60} unit. Moreover, a dose-dependent manner has been revealed under the irradiation of green light. The photodynamic properties of the onionlike structures indicate that the C60 moieties well preserve the capability of $^1O_2$ production even trapped in the molecular bilayers. This is exciting, considering the great potential of the biological activities exhibited by onionlike structures formed by compound **1** in biomedicine.

14.3 Self-assembled Honeycomb Films of Hydrophobic Surfactant-Encapsulated Clusters (HSECs) at Air/Water Interface

Nature provides inexhaustible inspiration to mankind, which continues to encourage us to develop new methods and approaches to the construction of artificial advanced materials. Bees, the artificers in nature, always build their hives in a hexagonal arrangement to minimize work and enhance the stability of the hive. In 1994, Francois et al. [75]. developed the polymeric porous honeycomb film with hexagonally arranged pores 0.2–10 μm in diameter, walls 0.1–0.2 μm thick, and 10–30 μm high prepared by exposing a drop of polystyrene-b-polyparaphenylene solution in carbon disulfide (CS_2) to a flow of moist air. As shown in Fig. 14.17, the micrometer-sized pores exhibited a highly regular hexagonal arrangement, which are similar to the natural hive. From then on, many researchers made efforts to fabricate the honeycomb porous film by various materials and methods. Colloidal sphere templates [76, 77], emulsion templates [78], and other "lithograph" [79] have been used to construct ordered porous arrays with controlled pores in a long range by replicating periodic structures from inorganic oxides, colloidal metal particles, and polymers. In 2005, Wu et al. found that the surfactant-encapsulated POMs can self-assemble into vesicles in chloroform solution and can be further transferred into three-dimensional microporous architectures under moist air [80], which created opportunities for combining inorganic chemistry and colloidal surface chemistry, allowing us to fabricate microsized patterns of inorganic functional units through stepwise self-assembly of preorganized building blocks.

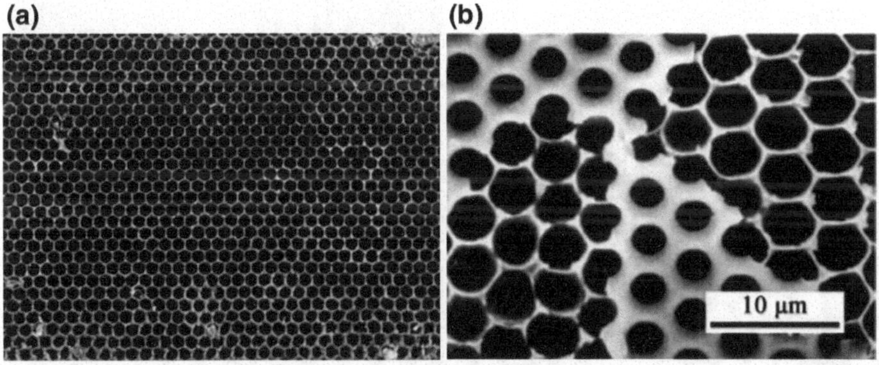

Fig. 14.17 **a** Photograph of natural hive built by worker-bees. **b** SEM image of PS-b-PPP honeycomb film. Reprinted by permission from Macmillan Publishers Ltd: Ref. [75], copyright 1993

14.3.1 Fabricating Honeycomb Films of HSECs at Air-Water Interface

Hydrophilic POMs of different sizes and shapes can interact with hydrophobic cationic surfactants, the resulting materials show amphiphilic properties with electrostatic interactions between the hydrophilic and hydrophobic components, called Surfactant-Encapsulated Clusters (SECs) or Surfactant-Encapsulated-POMs (SEPs), which can further self-assemble to give a variety of nanostructures on various surfaces/interfaces.

Our research group presented a unique system in 2007 [81]. We report the unprecedented self-assembly of hydrophobic double-chain cationic surfactant hybrid materials DODMACl-{$Mo_{72}Fe_{30}$} at the air/water interface without any extra airflow, which leads to the formation of highly ordered honeycomb films.

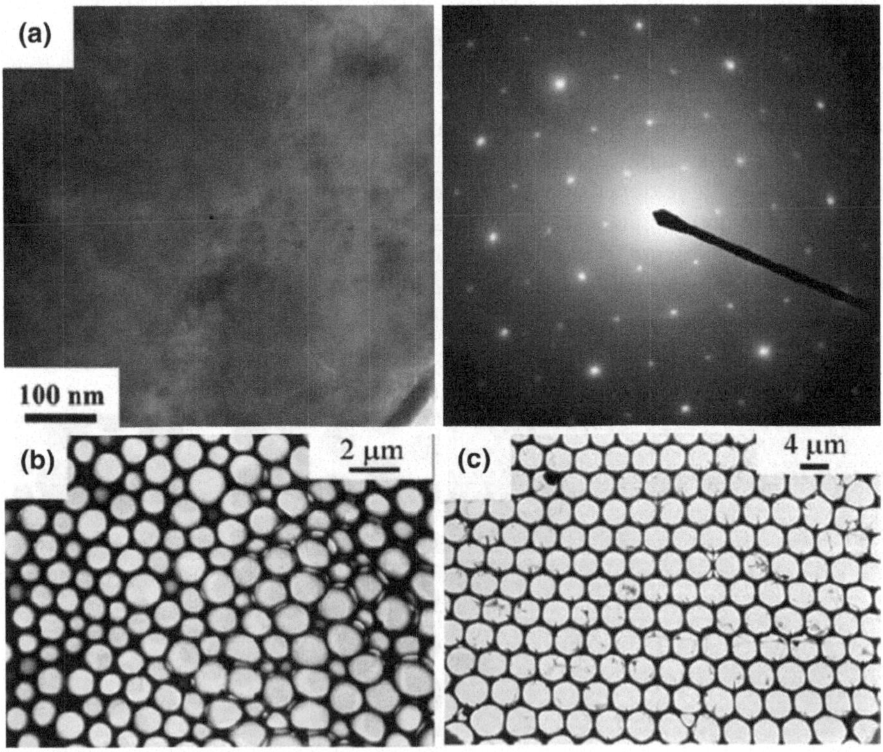

Fig. 14.18 **a** TEM images (*left column*) and the corresponding electron diffraction patterns (*right column*) of organized crystalline-structured monolayer films. TEM images of {$Mo_{72}Fe_{30}$}-DODMA complexes at **b** $c_{DODMACl}$ = 1.2 mg mL^{-1} and **c** $c_{DODMACl}$ = 2.0 mg mL^{-1} in $CHCl_3$. Reproduced from Ref. [80] by permission of John Wiley & Sons Ltd

In this study, we use the 2.5-nm-diameter, "Keplerate" molecule $\{Mo_{72}Fe_{30}\}$ as our model system, and they can interact with cation surfactant dioctadecyldimethylammonium chloride (DODMACl) to form the DODMACl -$\{Mo_{72}Fe_{30}\}$ complexes in organic solvent, which has discussed in Sect. 14.2.2 in detail. Then, an amount of $CHCl_3$-phase solution of DODMACl-$\{Mo_{72}Fe_{30}\}$ complexes was dropped onto a pure water surface, the honeycomb nanostructures of $\{Mo_{72}Fe_{30}\}$-DODMA complexes formed after evaporating $CHCl_3$, as shown in Fig. 14.18.

We demonstrated that the regularity of honeycomb architectures heavily depends on the surfactant concentration, given the fixed $\{Mo_{72}Fe_{30}\}$ concentration in aqueous solution. When the DODMACl concentration is lower than 1.2 mg mL^{-1} in $CHCl_3$, $\{Mo_{72}Fe_{30}\}$-DODMA complexes formed organized crystalline structures at the air/water interface by the packing of the complexes. When the concentration of DODMACl reaches 1.2 mg mL^{-1} in $CHCl_3$ ($c_{DODMACl}$ = 2.02 $mmolL^{-1}$), unique self-assembly of $\{Mo_{72}Fe_{30}\}$-DODMA complexes into honeycomb nanostructures was observed, as shown in Fig. 14.18b. The thickness of the wall between pores is in general smaller than 0.5 μm and the pore sizes are not

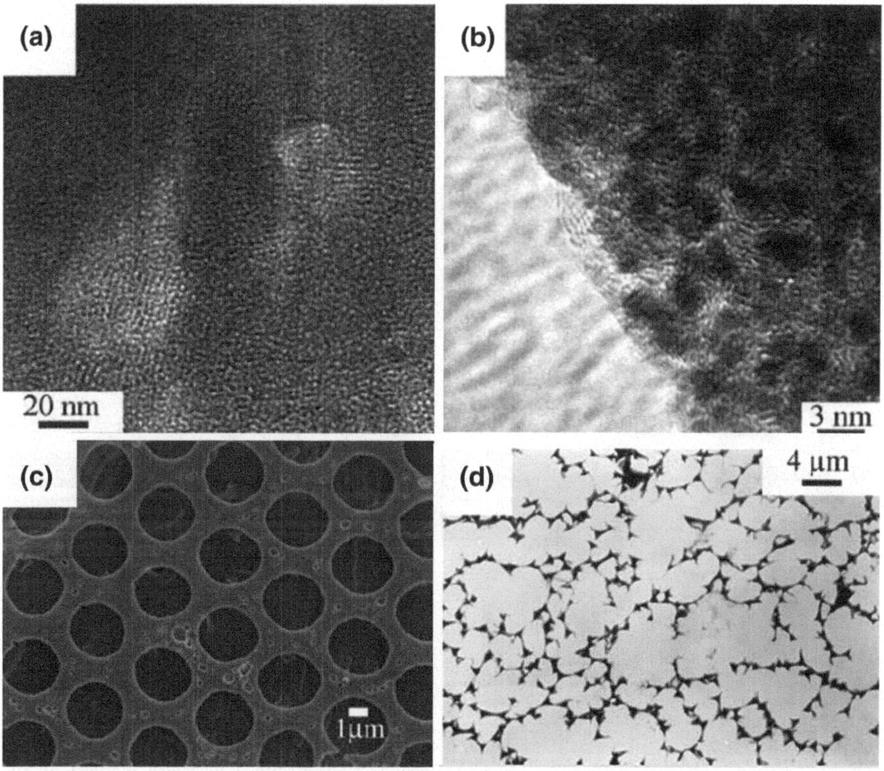

Fig. 14.19 **a**, **b** HRTEM images of the highly ordered honeycombs, which reveal the ordered texture with $\{Mo_{72}Fe_{30}\}$ at $c_{DODMACl}$ = 2.0 mg mL^{-1} in $CHCl_3$. **c** SEM image of highly ordered honeycomb architectures of $\{Mo_{72}Fe_{30}\}$-DODMA complexes. **d** TEM image of $\{Mo_{72}Fe_{30}\}$-DODMA complexes at $c_{DODMACl}$ = 4.0 mg mL^{-1} in $CHCl_3$. Reproduced from Ref. [80] by permission of John Wiley & Sons Ltd

homogeneous, ranging from 0.5 to 1.5 μm. At $c_{DODMACl} = 2.0$ mmolL^{-1}, the nanostructure formed by {$Mo_{72}Fe_{30}$}-DODMA complexes (Fig. 14.18c) becomes more orderly, with honeycomb holes of uniform pore size (3.5 μm). The average thickness of the walls between the pores remains about 0.6 μm. HRTEM images revealed that the walls are composed of many 2.5-nm-diameter dark objects (Figs. 14.19a, b), which are contributed by single {$Mo_{72}Fe_{30}$} clusters. This result confirms that the {$Mo_{72}Fe_{30}$} clusters play a critical role in constructing the uniform honeycomb structures, together with the surfactant. SEM images indicated small holes of highly ordered honeycomb films exist inside the 500-nm thick walls. When the concentration of DODMACl is higher than 4.0 mg mL^{-1} in $CHCl_3$ ($c_{DODMACl} = 6.73$ mmolL^{-1}), the unique honeycomb structures are not observed any more, as shown in Fig. 14.19d.

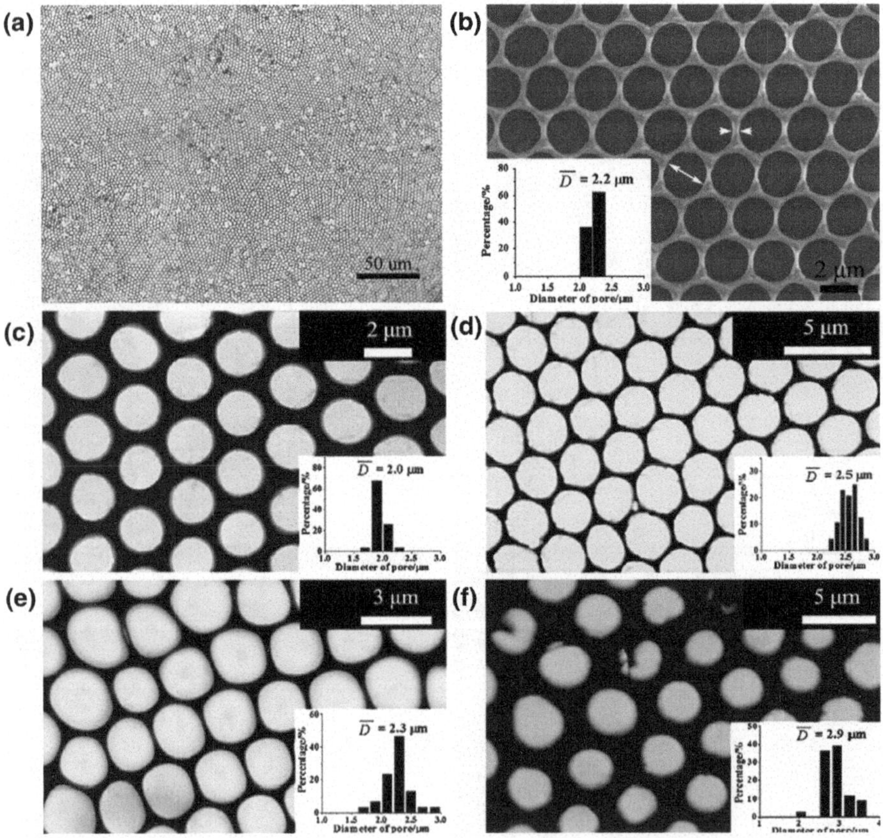

Fig. 14.20 **a** Optical microscope image (magnification ×400) of the porous honeycomb films of DODMA^{+}-encapsulated {$Mn_2Bi_2W_{20}$}. **b** SEM image. A pore and a wall are marked by the arrowheads. TEM images of the honeycomb films based on different DODMA^{+}-encapsulated polyoxometalates, including **c** ring-shaped {Mo_{154}} at $c_{DODMACl} = 2.6$ mg mL^{-1}, **d** Keplerate spherical {Mo_{132}} at $c_{DODMACl} = 2.5$ mg mL^{-1}, **e** hedgehog-shaped {Mo_{368}} at $c_{DODMACl} =$ 0.7 mg mL^{-1}, and **f** Keggin {PW_{12}} at $c_{(DODMA)3\{PW12\}} = 1.0$ mg mL^{-1}. The insets show their pore size distribution. Reprinted from Ref. [81], Copyright 2011, with permission from Elsevier

Moreover, we fabricated highly ordered honeycomb-structured macroporous films by self-assembly of various surfactant-encapsulated polyoxometalates which have different compositions, shapes, and sizes, at the air/water interface without any extra moist airflow across the solution surface (Fig. 14.20) [81]. We choose a series of nanoscaled POMs with different shapes, sizes, and compositions, including $\{Mn_2Bi_2W_{20}\}$, $\{Mo_{154}\}$, $\{Mo_{132}\}$, $\{Mo_{368}\}$, and $\{PW_{12}\}$, to be the typical examples for inorganic cores. Based on the investigations of above POMs by our group, it can be concluded that, the self-assembly of surfactant-encapsulated POMs to the ordered macroporous honeycomb films at the air/water interface is a universal phenomenon, no matter with the compositions, shapes, and sizes of POMs.

14.3.2 Mechanism of Self-assembly of SECs into Honeycomb Films

How do polymers and/or HSECs self-assemble into ordered honeycomb films on solid substrates with a moist airflow and/or directly at the air/water interface? Although it should be very complicated, we are still trying to propose a resonable mechanism of the self-assembly at the air/water interface [82]. During the self-assembly, the rapid evaporation of organic solvent will debase the temperature around the solution surface. The water vapor around the solution surface will be condensed to become micrometer water droplets. These micrometer water droplets act as important templates for macropores, and the organic solvent concentrate and deposit around these water droplets [83, 84]. Meanwhile, the induced Marangoni convection [85] and local capillary forces mainly drive the water droplets to hexagonally arrange in a minimum energy state. After the complete evaporation of solvent, the surface will warm up. Thus the saturated water vapor pressure is increased, and the enveloped water droplet will be evaporated again. Consequently, pores are formed and materials are solidified in walls. The porous honeycomb films are finally obtained. In fact, the stability of the water droplets which are the

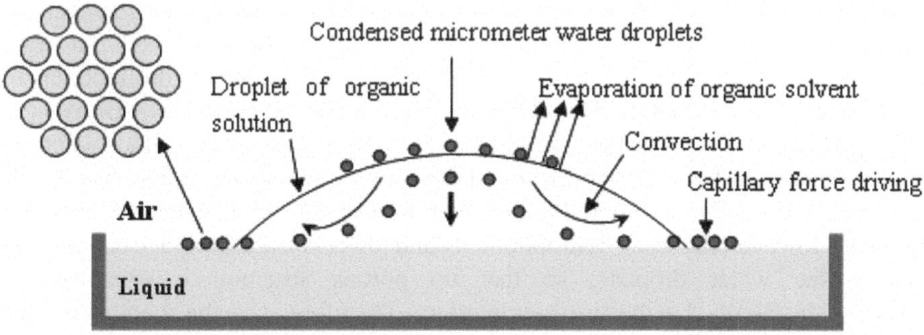

Fig. 14.21 The formation mechanism of periodic hexagonal honeycomb films at the air/water interface. Reprinted from Ref. [82], Copyright 2009, with permission from Elsevier

templates for honeycomb morphology is the crucial step (Fig. 14.21). Therefore, influences can be used to control the energy state of water droplets, and consequently, the morphology of the hybrid films can be modulated.

14.3.3 Morphology Modulation of Honeycomb Films of SECs

It can be speculated that the stabilization of micrometer water droplets is the crucial step during the self-assembly to porous honeycomb films. To further this hypothesis, we verified the honeycomb film morphology becomes changeable morphologies by modulating water droplet state.

Originally, we speculated that the fully hydrophobic complexes would spread homogeneously on the surface of water, similar to the case of spreading a layer of oil at the air/water interface. However, in reality we found that the arrangement of complexes was very complicated and dependent on surfactant concentration. As discussed in Sect. 14.3.1, an optimum surfactant concentration is fit for self-assembly into a honeycomb film. At lower or higher concentrations, no ordered honeycomb film is observed.

The surface properties of SECs, such as the hydrophobicity and the surface coverage, are also very importance in forming ordered honeycomb nanostructures. We chose three cation surfactants with different hydrophobic tails to encapsulate the $\{Mn_2Bi_2W_{20}\}$, resulting in the formation of $(DODMA)_{10}\{Mn_2Bi_2W_{20}\}$, $(DDDMA)_{10}\{Mn_2Bi_2W_{20}\}$ and $(CTA)_{10}\{Mn_2Bi_2W_{20}\}$ complex. Both $(DODMA)_{10}\{Mn_2Bi_2W_{20}\}$ and $(DDDMA)_{10}\{Mn_2Bi_2W_{20}\}$ can self-assemble into relatively ordered porous films at the air/water interface at the same amount of $\{Mn_2Bi_2W_{20}\}$, while for $(CTA)_{10}\{Mn_2Bi_2W_{20}\}$, there only formed less ordered fragment of complexes (Fig. 14.22). Furthermore, the hydrophobicity of SECs mainly lies in the surfactant chains and the surface coverage degree of polyoxometalates by encapsulating surfactants. Their hydrophilic lipophilic balance (HLB) values of surfactants and the surface coverage degree of polyoxometalates by surfactants in SECs which is equal to the percentage of surfactant volume to SECs volume are approximately calculated (Table 14.2). As we know, the smaller HLB value is, the stronger the hydrophobic of surfactants is. So the DODMA is more hydrophobic and has a larger surface coverage, DDDMA comes second. Therefore, we infer that hydrophobicity of the three complex is $(DODMA)_{10}\{Mn_2Bi_2W_{20}\} > (DDDMA)_{10}\{Mn_2Bi_2W_{20}\} > (CTA)_{10}\{Mn_2Bi_2W_{20}\}$. For the $(DODMA)_{10}\{Mn_2Bi_2W_{20}\}$ complex, the condensed water droplets can be well stabilized and form two dimensional microcosmic w/o structures, and finally, an ordered honeycomb film is self-organized at the air/water interface. However, the weak hydrophobicity and lower surface coverage cannot well stabilize the water droplets, so that no porous structure is obtained for $(CTA)_{10}\{Mn_2Bi_2W_{20}\}$ at the air/water interface. Therefore, it can be concluded that a higher hydrophobicity and lager surface coverage are beneficial to the self-assembly

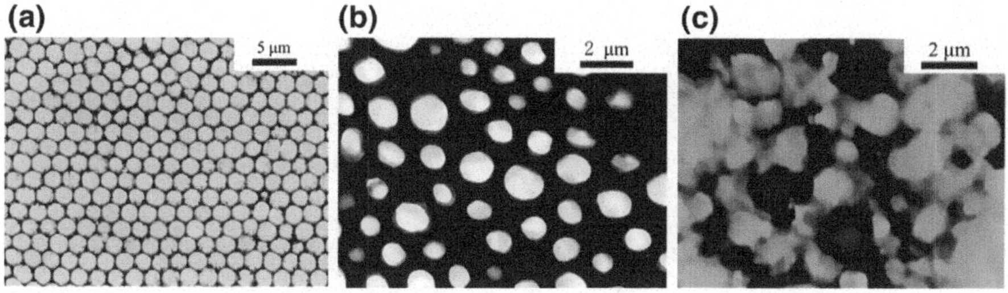

Fig. 14.22 TEM images of honeybomb films by casting three chloroform solutions at the air/water interface: **a** 1.7 mg mL^{-1} $(DODMA)_{10}\{Mn_2Bi_2W_{20}\}$; **b** 1.5 mg mL^{-1} (DDDMA) $_{10}\{Mn_2Bi_2W_{20}\}$; and **c** 1.3 mg mL^{-1} (CTA) $_{10}\{Mn_2Bi_2W_{20}\}$. Reprinted from Ref. [82], Copyright 2009, with permission from Elsevier

Table 14.2 The properties of the three cationic surfactants and the surface coverage of $\{Mn_2Bi_2W_{20}\}$ by surfactants (a is a constant)

Cationic surfactants	$DODMA^+$	$DDDMA^+$	CTA^+
HLB values	a-17.1	a-11.4	a-7.6
Volume of hydrophobic chains (nm^3)	1.02	0.70	0.46
Surface coverage of POM in SECs (%)	79	72	63

of honeycomb film at the air/water interface. Moreover, the air/solution interface not only provides a flat medium for preparing honeycomb films, but also the surface current of the supporting solutions can influence the morphologies of the obtained films.

In addition, the evaporation rate of solvents is another important factor during the process of self-assembly [82]. It is found that both carbon disulfide and chloroform solution of $(DODMA)_{10}\{Mn_2Bi_2W_{20}\}$ can form highly ordered honeycomb films at the air/water interface (Figs. 14.23a, b), with a pore size of ~ 1 μm for carbon disulfide as the solvent and ~ 2 μm taking chloroform as the solvent. Taking the boiling points of carbon disulfide and chloroform being 46 and 61 °C, respectively into account, it is easy to conclude that the higher the boiling point of the solvents is, the larger the pore sizes of the honeycomb films are. We presume that the condensed water droplets have more time to coalesce and grow during the self-organization if the boiling point of solvent is high. Consequently, the pores templated by water droplets are much larger. However, if the boiling points of solvents are too high, such as cyclohexane and n-heptane with the boiling points of 81 and 98 °C, respectively, it is also shown that no porous architectures are observed (Figs. 14.23c, d). It is because that solvent volatility is too slow to produce a sufficient temperature gradient around the organic solution surface; thus there are not enough condensed micrometer water droplets to template the formation of pores.

We also found that supporting solutions can influence the morphologies of the obtained films. The honeycomb porous film with a few disorderly arranged pores at

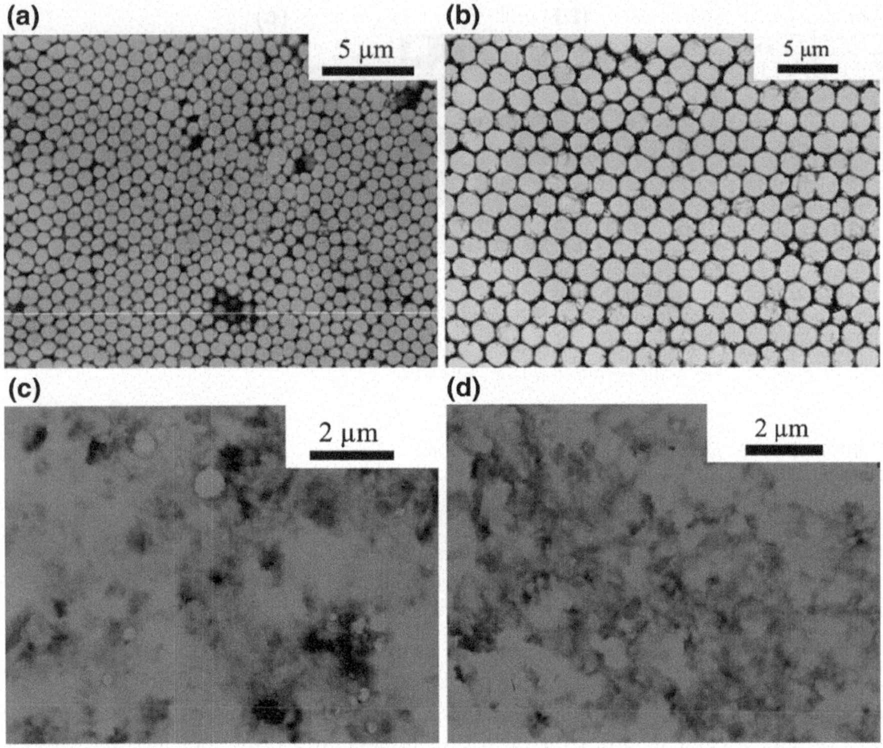

Fig. 14.23 TEM images of the honeycomb films prepared by dropping 1.7 mg mL^{-1} $(DODMA)_{10}\{Mn_2Bi_2W_{20}\}$ solution at the air/water interface by using different solvents: **a** carbon disulfide, **b** chloroform, **c** cyclohexane, and **d** *n*-heptane, respectively. Reprinted from Ref. [82], Copyright 2009, with permission from Elsevier

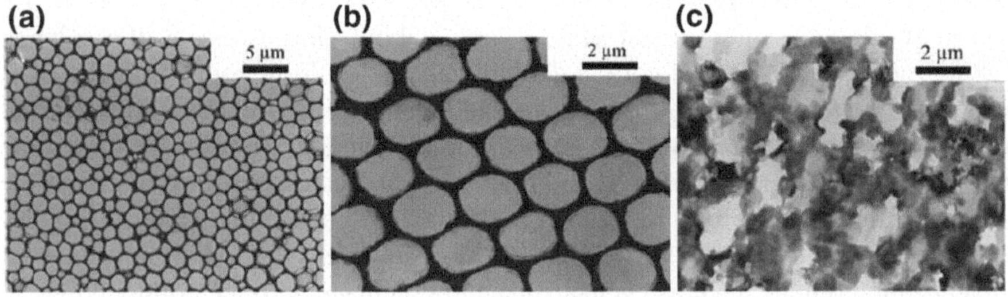

Fig. 14.24 TEM images of the thin films of $DODMA^+$-encapsulated $\{Mn_2Bi_2W_{20}\}$ at c_{DODMA+} = 1.0 mg mL^{-1} fabricated at the **a** air/water, **b** air/1 mol L^{-1} NaCl solution, and **c** air/2 mmol L^{-1} TTAB solution interfaces. Reprinted from Ref. [82], Copyright 2009, with permission from Elsevier

the air/water interface is fabricated by the $DODMA^+$ at $c_{DODMA+} = 1.0$ mg mL^{-1} and $\{Mn_2Bi_2W_{20}\}^{10-}$ at $c_{\{Mn2Bi2W20\}} = 1.0$ mg mL^{-1}, which are shown in Fig. 14.24a. The same chloroform solution can self-assemble into a more close-packed hexagonal porous film by changing the supporting solution to the air/1 mol L^{-1} NaCl solution interface, shown in Fig. 14.24b. However, if 2 mmol L^{-1} TTAB solution acts as the supporting solution, because of their amphiphilicity, the surface tension can be reduced and cause a strong interfacial turbulence. Consequently, thin films with very disordered fragments without pores are observed (Fig. 14.24c). It can be concluded that the air/solution interface not only provides a flat medium for preparing honeycomb films, but also can influence the morphologies of the obtained films.

14.3.4 Excellent Properties of Honeycomb Films of SECs

As discussed above in Sect. 14.2, the nano-scale POM clusters are versatile, including magnetic properties, electrochem87ical properties and photoluminescent.

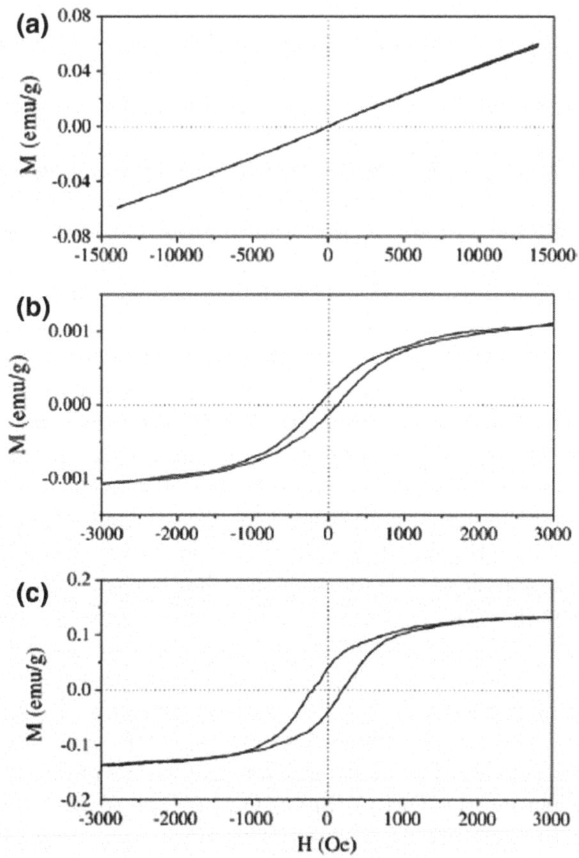

Fig. 14.25 Magnetic hysteresis loops of **a** $\{Mn_2Bi_2W_{20}\}$ crystals, **b** $(DODMA)_{10}\{Mn_2Bi_2W_{20}\}$ complexes, and **c** the ordered porous honeycomb film of $DODMA^+$-encapsulated $\{Mn_2Bi_2W_{20}\}$. Reprinted from Ref. [81], Copyright 2011, with permission from Elsevier

Combining the properties of POMs and the specificities of surfactants, hence, the self-assembled honeycomb films of surfactant encapsulated polyoxometalates could literally be described as a novel class of materials. Based on our investigations, it is found that $\{Mn_2Bi_2W_{20}\}$ crystals possess paramagnetism from the magnetic hysteresis loop, shown in Fig. 14.25 [81]. When the $\{Mn_2Bi_2W_{20}\}$ clusters are encapsulated by DODMA through electrostatic interaction, the formed $(DODMA)_{10}\{Mn_2Bi_2W_{20}\}$ complexes exhibit a little ferromagnetism (Fig. 14.25a). It can be calculated the coercive force (Hc) of 127 Oe and the residual magnetization/saturationmagnetization (Mr/Ms) of 0.12 (1.3×10^{-4} emu g^{-1}/ 1.1×10^{-3} emu g^{-1}) from the magnetic hysteresis loop (Fig. 14.25b), Further, the ordered honeycomb film of $DODMA^+$-encapsulated $\{Mn_2Bi_2W_2\}$ self-assembled at the air/water interface by $(DODMA)_{10}\{Mn_2Bi_2W_{20}\}$ complexes magically possesses a stronger ferromagnetism with Hc of 193 Oe and Mr/Ms of 0.27 (0.037 emu g^{-1}/ 0.137 emu g^{-1}) which may due to the synergistic effect of the POMs and surfactants.

In addition, ordered honeycomb films of SECs exhibit excellent electrochemical properties. As an example, the electrochemical properties of the ordered honeycomb film of $DODMA^+$-encapsulated $\{Mo_{368}\}$ is studied by using a cyclic voltammetry (CV) method, and the reference electrode, counter electrode and electrolyte solution are saturated calomel electrode, platinum sheet and 0.5 mol L^{-1} H_2SO_4 soultion, respectively. The honeycomb film of $DODMA^+$-encapsulated $\{Mo_{368}\}$ was transferred onto the ITO-glass which used as the working electrode, and then the bare region was carefully covered by the insulating nail polish in order to precisely detect the honeycomb film-coated electrode surface. Compared with the dotted lines which indicate that no active species are behaved for bare ITO and DODMACl in 0.5 mol L^{-1} H_2SO_4 solution. Comparing the cyclic voltammetry of $\{Mo_{368}\}$ and the honeycomb film (as shown in Fig. 14.26, solid line), one can speculate that the electrochemical properties of the honeycomb film of $DODMA^+$-encapsulated $\{Mo_{368}\}$ are derived from the inorganic $\{Mo_{368}\}$. However, the changes may be ascribed to the electrostatic interaction of POMs and surfactants.

Due to the intrinsic fluorescence of the europium polyoxometalate anion, $[EuW_{10}O_{36}]^{9-}$, the porous honeycomb films of $(DODMA)_9[EuW_{10}O_{36}]$ complexes can emit fluorescence when they are excited by UV light, presenting excelent photoluminescent property [86]. In the polyoxometalate structure of $Na_9[EuW_{10}O_{36}]\cdot 32H_2O$, Eu^{3+} is located in the center of $[EuW_{10}O_{36}]$, which achieves 8-fold coordination by the attachment of twoW_5O_{18} ligands, forming a distorted square antiprism. The photo excitation of the $O \rightarrow W$ charge transfer (CT) of $[EuW_{10}O_{36}]$ induces the emission of Eu^{3+}. When the UV light at a wavelength of 270 nm irradiated the honeycomb film, the luminescence derived from the ${}^5D_0 \rightarrow {}^7F_J$ ($J \approx 1, 2, 3,$ and 4) transition can be observed. As shown in Fig. 14.27c (black curve), the emission band of $[EuW_{10}O_{36}]$ crystals at 593 nm corresponds to the ${}^5D_0 \rightarrow {}^7F_1$ emission transition. A strong band at 619 nm is attributed to the ${}^5D_0 \rightarrow {}^7F_2$ emission transition. The much less intense bands at 650 and 698 nm are assigned to the${}^5D_0 \rightarrow {}^7F_3$ and ${}^5D_0 \rightarrow {}^7F_4$ emission transitions, respectively. The weak emission peak at 580 nm in $[EuW_{10}O_{36}]$ crystals disappears when the $(DODMA)_9[EuW_{10}O_{36}]$ complexes formed, and the relative intensities

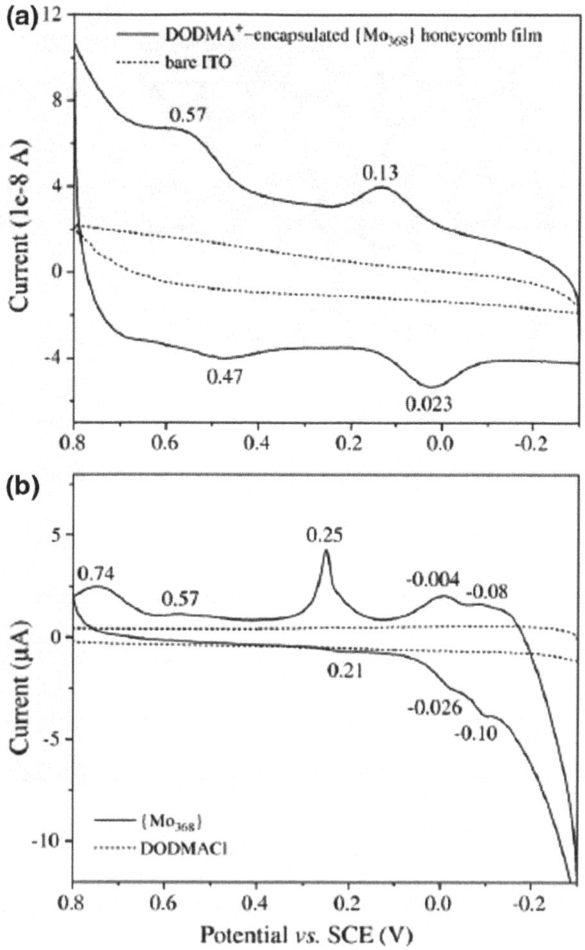

Fig. 14.26 Cyclic voltammograms of **a** the honeycomb film of DODMA$^+$-encapsulated {Mo_{368}} (*solid line*) and bare ITO (*dotted line*), **b** {Mo_{368}} (*solid line*) and DODMACl (*dotted line*). All the curves were measured in 0.5 mol L^{-1} H_2SO_4 solution. Scan rate 50 mV s^{-1}. Reprinted from Ref. [81], Copyright 2011, with permission from Elsevier

and the band widths are also a little changed, which may be attributed to the change of the Eu^{3+} microenvironment in $(DODMA)_9[EuW_{10}O_{36}]$ complexes.

In order to broaden applications of the honeycomb porous films, the surfactant encapsulated POMs honeycomb films used as templates to directionally electrodeposit gold nanoparticles into their macropores also have been investigated [87]. As shown in Figs. 14.28a, Au nanoparticles are well-deposited into the honeycomb pores but not on the walls. It is reasonable according to the formation mechanism of honeycomb films discussed in the Sect. 14.3.2. After transferring the honeycomb films of $(DODMA)_{10}\{Mn_2Bi_2W_{20}\}$ onto ITO-coated glass, the holes are the bare ITO conductive surface, which can provide channels for electron transfer onto the working electrode. However, the electro-deposition is prevented owing to the insulation of DODMA. Consequently, Au nanoparticles prefer to electro-deposit into the macropores rather than the walls. Rhodamine 6G (R6G) is a

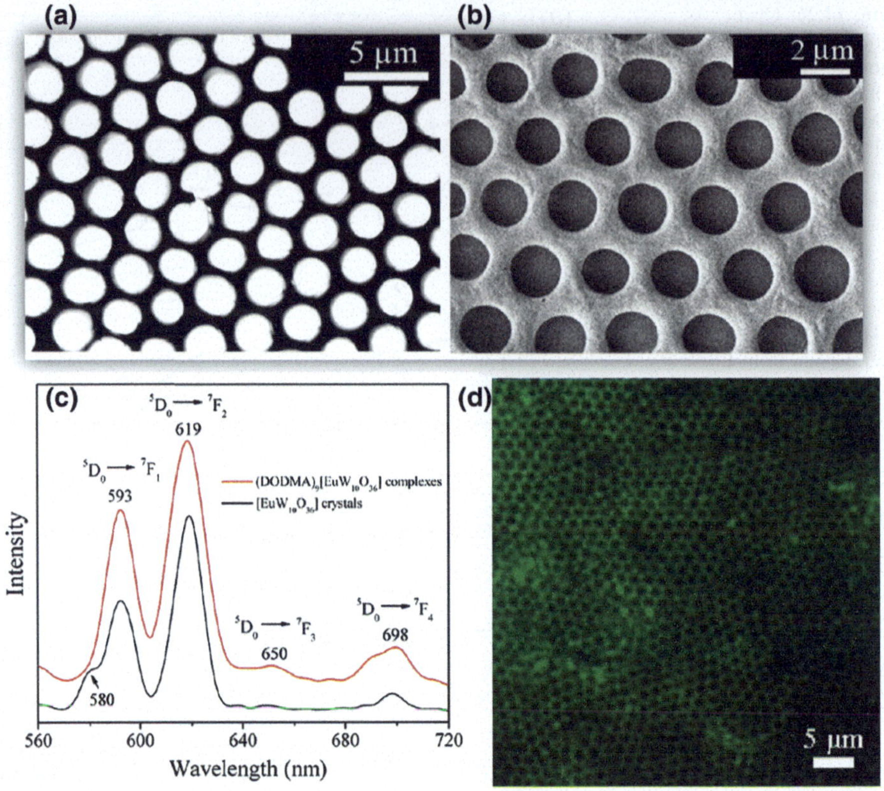

Fig. 14.27 Micrograph images of self-assembled honeycomb films of $(DODMA)_9[EuW_{10}O_{36}]$ complexes at the air/water interface. **a** TEM image. **b** SEM image. **c** Fluorescence spectra of an aqueous solution of $[EuW_{10}O_{36}]$ crystals (*black curve*) and a chloroform solution of $(DODMA)_9[EuW_{10}O_{36}]$ complexes (*red curve*), excited at 270 nm. The positions of $^5D_0 \rightarrow {}^7F_J$ ($J \approx 1, 2, 3,$ and 4) emission bands are indicated. **d** Typical fluorescence micrograph of the porous honeycomb film of $(DODMA)_9[EuW_{10}O_{36}]$ complexes, which is irradiated by a mercury lamp. Reprinted with permission from Ref. [86]. Copyright 2010 by American Physical Society

strongly fluorescent xanthene derivative, which can be used as a model to evaluate the enhanced Raman scattering signal of organic molecules by honeycomb films. In Fig. 14.28b, no Raman signal for R6G was detected for bare ITO-coated glass (curve 1). Coincidentally, R6G was not detected for the honeycomb film-covered ITO-coated glass because of the coverage of the honeycomb films, it was also difficult to detect the glass vibrations. Interestingly, R6G signals were obviously detected for directly electrodeposited Au nanoparticles (curve 3) and Au-filled honeycomb films (curve 4). In particular, curve 4 presents a much stronger SERS effect, which may be due to the hierarchical structure of the microscaled array and nanoscaled particles. The results indicate that the hierarchical Au-filled honeycomb film is a good detector of R6G molecules.

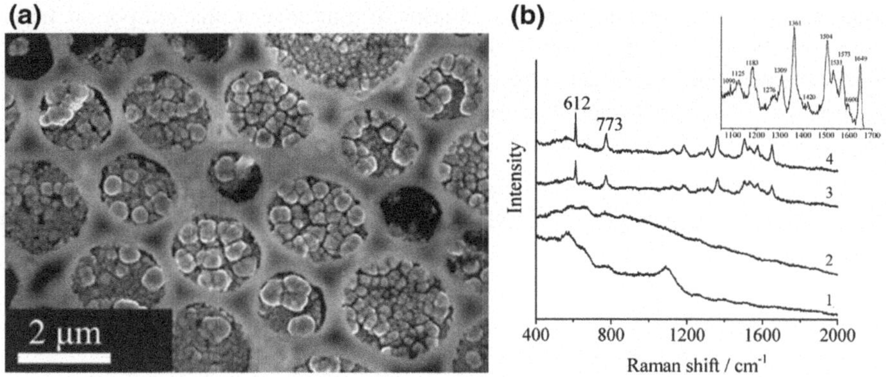

Fig. 14.28 **a** SEM image of the hierarchical array with Au nanoparticles directionally electrodeposited into the honeycomb macropores. **b** Raman scattering spectra of R6G on (1) the bare ITO-coated glass, (2) the honeycomb film-covered ITO-coated glass and (3, 4) the Au-deposited ITO-coated glass. The Au nanoparticles were electrodeposited without the honeycomb film template in (3) and with the honeycomb film template in (4). The honeycomb film template was dissolved in chloroform before Raman preparation for sample D. The inset shows the magnification of curve d from 1050 to 1700 cm^{-1}. The Raman measurements were all conducted using an excitation wavelength of 532 nm and with a 20 s accumulation. Reproduced from Ref. [87] by permission of The Royal Society of Chemistry (RSC) on behalf of the Centre National de la Recherche Scientifique (CNRS) and the RSC

14.3.5 Conclusion

Nano-scale polyoxometalates (POMs) are a class of discrete anionic metal oxides that can be viewed as transferable building blocks, which can be applied in the preparation of functional materials. The past decades have witnessed the significant advancement in the synthesis and application of POMs-based materials. More amazingly, based on their excellent water solubility, according to electrostatic interactions, these inorganic macroions demonstrate some features that usually are believed to belong only to complex biological molecules, such as the self-recognition in dilute solutions. Meanwhile, polyoxometalates-based organic-inorganic hybrid materials possess amphiphilic properties by self-assembling into supramolecular architectures, e.g. vesicles, reverse vesicles or onionlike structures in polar and non-polar solvents. Moreover, self-assembly of surfactant-encapsulated POM clusters into honeycomb films at the air/water interface, which is templated by condensed water micro-droplets are functionally attractive. So far, only a few interesting properties of amphiphilic hybrid POMs have been studied. There are still challenges and much more exploration is necessary in the development of POMs. Many efforts have been made to comprehensively understand fascinate chemical and physical properties of POMs and to design rational synthetic strategies and synthesize novel amphiphilic hybrid POMs. We believe that a better understanding of mechanism will no doubt help in judicious design and improve the function of highly efficient POM-based materials. In addition, the basic investigations of the structure and assembly of POMs, organic-inorganic hybrids of multilayer films and porous films containing POMs are also worthy of thorough investigation for applications in catalysis, medicine, smart materials, and beyond will open new vistas in materials research and build a bridge between polyoxometalate chemistry and material chemistry.

References

1. Pope MT, Müller A (1991) Polyoxometalate chemistry: an old field with new dimensions in several disciplines. Angew Chem Int Ed Engl 30:34–48
2. Pope MT (1983) Heteropoly and isopoly oxometalates. Springer, Berlin
3. Rao CNR, Müller A, Cheetham AK (Eds) (2006) The chemistry of nanomaterials: synthesis, properties and applications. Wiley, New York
4. Gouzerh P, Proust A (1998) Main-group element, organic, and organometallic derivatives of polyoxometalates. Chem Rev 98:77–112
5. Weinstock IA (1998) Homogeneous-phase electron-transfer reactions of polyoxometalates. Chem Rev 98:113–170
6. Müller A, Peters F, Pope MT, Gatteschi D (1998) Polyoxometalates: very large clusters nanoscale magnets. Chem Rev 98:239–272
7. Mizuno N, Kamata K (2011) Catalytic oxidation of hydrocarbons with hydrogen peroxide by vanadium-based polyoxometalates. Coord Chem Rev 255:2358–2370
8. Hill CL (2007) Special issue: polyoxometalates in catalysis-foreword. J Mol Catal A: Chem 262:1–1
9. Rhule JT, Hill CL, Judd DA, Schinazi RF (1998) Polyoxometalates in medicine. Chem Rev 98:327–358
10. Hasenknopf B (2005) Polyoxometalates: introduction to a class of inorganic compounds and their biomedical applications. Front Biosci 10:275
11. Long DL, Burkholder E, Cronin L (2007) Polyoxometalate clusters, nanostructures and materials: from self-assembly to designer materials and devices. Chem Soc Rev 36:105–121
12. Kim KC, Pope MT (1999) Cation-directed structure changes in polyoxometalate chemistry. equilibria between isomers of bis (9-tungstophosphatodioxouranate (VI)) complexes. J Am Chem Soc 121:8512–8517
13. Briot E, Piquemal JY, Vennat M, Brégeault JM, Chottard G, Manoli JM (2000) Aqueous acidic hydrogen peroxide as an efficient medium for tungsten insertion into MCM-41 mesoporous molecular sieves with high metal dispersion. J Mater Chem 10:953–958
14. Zhai QG, Wu XY, Chen SM, Zhao ZG, Lu CG (2007) Construction of Ag/1, 2, 4-triazole/polyoxometalates hybrid family varying from diverse supramolecular assemblies to 3-D rod-packing framework. Inorg Chem 46:5046–5058
15. Landsmann S, Luka M, Polarz S (2012) Bolaform surfactants with polyoxometalate head groups and their assembly into ultra-small monolayer membrane vesicles. Nat Commun 3:1299
16. Izzet G, Abécassis B, Brouri D, Piot M, Matt B, Serapian SA, Bo C, Proust A (2016) Hierarchical self-assembly of polyoxometalate-based hybrids driven by metal coordination and electrostatic interactions: from discrete supramolecular species to dense monodisperse nanoparticles. J Am Chem Soc 138:5093–5099
17. Liu H, Hsu CH, Lin ZW, Shan WP, Wang J, Jiang J, Huang MJ, Lotz B, Yu XF, Zhang WB, Yue K, Cheng SZD (2014) Two-dimensional nanocrystals of molecular janus particles. J Am Chem Soc 136:10691–10699
18. Sun HG, Tu YF, Wang CL, Van Horn RM, Tsai CC, Graham MJ, Sun B, Lotz B, Zhang WB, Cheng SZD (2011) Hierarchical structure and polymorphism of a sphere-cubic shape

amphiphile based on a polyhedral oligomeric silsesquioxane-[60] fullerene conjugate. J Mater Chem 21:14240–14247
19. Keggin JF (1934) Structure and formula of 12-phosphotungstic acid. Proc R Soc A 144:75–79
20. Dawson B (1953) The structure of the 9 (18)-heteropoly anion in potassium 9 (18)-tungstophosphate, K_6 $(P_2W_{18}O_{62})\cdot 14H_2O$. Acta Cryst 6:113–126
21. Waugh JCT, Schoemaker DP, Pauling L (1954) On the structure of the heteropoly anion in ammonium 9-molybdomanganate, $(NH_4)_6MnMo_9O_{32}\cdot 8H_2O$. Acta Cryst 7:438–441
22. Anderson JS (1937) Constitution of the poly-acids. Nature 140:850–851
23. Tsay YH, Silverton JVZ (1973) Krist 137:256
24. Lindquist I (1950) A crystal structure investigation of the paramolybdate ion. Ark Kemi 2:325–341
25. Müller A, Beckmann E, Bögge H, Schmidtmann M, Dress A (2002) Inorganic chemistry goes protein size: a Mo368 nano-hedgehog initiating nanochemistry by symmetry breaking. Angew Chem Int Ed 41:1162–1167
26. Bösing M, Nöh A, Loose I, Krebs B (1998) Highly efficient catalysts in directed oxygen-transfer processes: synthesis, structures of novel manganese-containing heteropolyanions, and applications in regioselective epoxidation of dienes with hydrogen peroxide. J Am Chem Soc 120:7252–7259
27. Sugeta M, Yamase T (1993) Crystal structure and luminescence site of Na_9 $(EuW_{10}O_{36})\cdot$ $32H_2O$. Bull Chem Soc Jpn 66:444–449
28. Müller A, Das SK, Fedin VP, Krickemeyer E, Beugholt C, Bögge H, Schmidtmann M, Hauptfleisch B (1999) Rapid and simple isolation of the crystalline molybdenum-blue compounds with discrete and linked nanosized ring-shaped anions: Na_{15} $[Mo_{\{126\}}\{VI\}$ $Mo_{\{28\}}VO_{462}H_{14}$ $(H_2O)_{70}]$ $_{0.5}[Mo_{\{124\}}\{VI\}Mo_{\{28\}}VO_{457}H_{14}(H_2O)_{68}]$ $0.5\cdot$ca.$400H_2O$ and Na_{22} $[Mo_{\{118\}}\{VI\}Mo_{\{28\}}$ $VO_{442}H_{14}$ $(H_2O)_{58}]\cdot$ca.$250H_2O$. Z Anorg Allg Chem 625:1187–1192
29. Müler A, Krickemeyer E, Bögge H, Schmidtmann M, Peters F (1998) Organizational forms of matter: an inorganic super fullerene and keplerate based on molybdenum oxide. Angew Chem Int Ed 37:3359–3363
30. Müller A, Sarkar S, Shah SQN, Bögge H, Schmidtmann M, Sarkar S, Kögerler P, Hauptfleisch B, Trautwein AX, Schünemann V (1999) Archimedean synthesis and magicnumbers: "sizing" giant molybdenum-oxide-based molecular spheres of the keplerate type. Angew Chem Int Ed 38:3238–3241
31. Botar B, Kogerler P, Hill CL (2005) $[\{(Mo)Mo_5O_{21}(H_2O)_3(SO_4)\}_{12}(VO)_{30}(H_2O)_{20}]^{36-}$: a molecular quantum spin icosidodecahedron. Chem Commun 25:3138–3140
32. Howell RC, Perez FG, Jain S, Horrocks JWD, Rheingold AL, Francesconi LC (2001) A new type of heteropolyoxometalates formed from lacunary polyoxotungstate ions and europium or yttrium cations. Angew Chem Int Ed 113:4155–4158
33. Mal SS, Kortz U (2005) The wheel-shaped Cu_{20} Tungstophosphate $[Cu_{20}Cl(OH)_{24}(H_2O)_{12}$ $(P_8W_{48}O_{184})]_{25}$-Ion. Angew Chem Int Ed 44:3777–3780
34. Todea AM, Merca A, Bögge H, Slageren J, Dressel M, Engelhardt L, Luban M, Glaser T, Henry M, Müller A (2007) Extending the (Mo)Mo5}12M30 capsule keplerate sequence: A {Cr30 cluster of s = 3/2 metal centers with a Na(H2O)12 encapsulate. Angew Chem Int Ed 46:6106–6110
35. Pichon C, Mialane P, Dolbecq A, Marrot J, Rivière E, Bassil BS, Kortz U, Keita B, Nadjo L, Sécheresse F (2008) Octa-and nonanuclear nickel (II) polyoxometalate clusters: synthesis and electrochemical and magnetic characterizations. Inorg Chem 47:11120–11128
36. Müller A, Luban M, Schröder C, Modler R, Kögerler P, Axenovich M, Schnack J, Canfield P, Bud'ko S, N N (2001) Classical and quantum magnetism in giant keplerate magnetic molecules. Chem Phys Chem 2:517–521
37. Lopez X, Maestre JM, Bo C, Poblet JM (2001) Electronic properties of polyoxometalates: A DFT study of α/β-$[XM_{12}O_{40}]$ n-relative stability (M = W, Mo and X a main group element). J Am Chem Soc 123:9571–9576

38. Jabbour D, Keita B, Nadjo L, Kortz U, Mal SS (2005) The wheel-shaped Cu_{20}-tungstophosphate $[Cu_{20}Cl(OH_{24}(H_2O)_{12}(P_8W_{48}O_{184})^{25-}$, redox and electrocatalytic properties. Electrochem Commun 7:841–847
39. Keita B, Zhang G, Dolbecq A, Mialane P, Sécheresse F, Miserque F, Nadjo L (2007) MoV-MoVI Mixed valence polyoxometalates for facile synthesis of stabilized metal nanoparticles: Electrocatalytic oxidation of alcohols. J Phys Chem C 111:8145–8148
40. Mbomekalle IM, Keita B, Lu YW, Nadjo L, Contant R, Belai N, Pope MT (2004) Synthesis, characterization and electrochemistry of the novel Dawson-type tungstophosphate $[H_4PW_{18}O_{62}]^{7-}$ and first transition metal ions derivatives. Eur J Inorg Chem 2004:276–285
41. Liu TB, Imber B, Diemann E, Liu G, Cokleski K, Li HL, Chen ZQ, Müller A (2006) Deprotonations and charges of well-defined Mo72Fe30 nanoacids simply stepwise tuned by pH allow control/variation of related self-assembly processes. J Am Chem Soc 128:15914–15920
42. Kistler ML, Liu TB, Gouzerh P, Todea AM, Müller AM (2009) Molybdenum-oxide based unique polyprotic nanoacids showing different deprotonations and related assembly processes in solution. Dalton Trans 26:5094–5100
43. Todea AM, Merca A, Bögge H, Glaser T, Pigga JM, Langston MLK, Liu T, Prozorov R, Luban M, Schröder C, Casey WH, Müller A (2010) Porous capsules $\{(M)\ M_5\}_{12}FeIII_{30}$ ($M = Mo^{VI}$, W^{VI}): sphere surface supramolecular chemistry with 20 ammonium ions, related solution properties, and tuning of magnetic exchange interactions. Angew Chem Int Ed 49:514–519
44. Kistler ML, Patel KG, Liu TB (2009) Accurately tuning the charge on giant polyoxometalate type keplerates through stoichiometric interaction with cationic surfactants. Langmuir 25:7328–7334
45. Wang Y, Li H, Qi W, Yang Y, Yan Y, Li B, Wu LX (2012) Supramolecular assembly of chiral polyoxometalate complexes for asymmetric catalytic oxidation of thioethers. J Mater Chem 22:9181–9188
46. Lee IS, Long JR, Prusiner SB, Safar JG (2005) Selective precipitation of prions by polyoxometalate complexes. J Am Chem Soc 127:13802–13803
47. Geng J, Li M, Ren JS, Wang EB, Qu XG (2011) Polyoxometalates as inhibitors of the aggregation of amyloid β peptides associated with Alzheimer's disease. Angew Chem Int Ed 50:4184–4188
48. Geng J, Li M, Ren JS, Wang EB, Qu XG (2011) Polyoxometalates as inhibitors of the aggregation of amyloid β peptides associated with Alzheimer's disease. Angew Chem 123:4270–4274
49. Yan XH, Zhu PL, Fei JB, Li JB (2010) Self-assembly of peptide-inorganic hybrid spheres for adaptive encapsulation of guests. Adv Mater 22:1283–1287
50. Goovaerts V, Stroobants K, Absillis G, Parac-Vogt TN (2013) Molecular interactions between serum albumin proteins and Keggin type polyoxometalates studied using luminescence spectroscopy. Phys Chem Chem Phys 15:18378–18387
51. Kosik KS (1992) Alzheimer's disease: a cell biological perspective. Science 256:780–783
52. Yamin G, Ono K, Inayathullah M, Teplow DB (2008) Amyloid β-protein assembly as a therapeutic target of Alzheimer's disease. Curr Pharm Des 14:3231–3246
53. Gao N, Sun H, Dong K, Ren J, Duan T, Xu C, Qu XG (2014) Transition-metal-substituted polyoxometalate derivatives as functional anti-amyloid agents for Alzheimer's disease. Nat Commun 5:3422
54. Li M, Xu C, Wu L, Ren J, Wang E, Qu XG (2013) Self-assembled peptide-polyoxometalate hybrid nanospheres: two in one enhances targeted inhibition of amyloid β-peptide aggregation associated with Alzheimer's disease. Small 9:3455–3461
55. Li J, Chen Z, Zhou M, Jing J, Li W, Wang Y, Wu L, Wang L, Wang Y, Lee M (2016) Polyoxometalate-driven self-assembly of short peptides into multivalent nanofibers with enhanced antibacterial activity. Angew Chem Int Ed 55:2592–2595

56. Xu J, Zhao S, Han Z, Wang X, Song YF (2011) Layer-by-layer assembly of Na_9 [$EuW_{10}O_{36}$]·$32H_2O$ and layered double hydroxides leading to ordered ultra-thin films: cooperative effect and orientation effect. Chem Eur J 17:10365–10371
57. Kurth DG, Lehmann P, Volkmer D, Cölfen H, Koop MJ, Müller A, Chesne AD (2000) Surfactant-encapsulated clusters (SECs): $(DODA)_{20}(NH_4)[H_3Mo_{57}V_6(NO)_6O_{183}(H_2O)_{18}]$, a case study. Chem Eur J 6:385–393
58. Caruso F, Kurth DG, Volkmer D, Koop MJ, Müller A (1998) Ultrathin molybdenum polyoxometalate-polyelectrolyte multilayer films. Langmuir 14:3462–3465
59. Iler RK (1966) Multilayers of colloidal particles. J Colloid Interface Sci 21:569–594
60. Niu P, Hao J (2011) Fabrication of titanium dioxide and tungstophosphate nanocomposite films and their photocatalytic degradation for methyl orange. Langmuir 27:13590–13597
61. Fan D, Hao J (2009) Fabrication and electrocatalytic properties of chitosan and Keplerate-type polyoxometalate $\{Mo_{72}Fe_{30}\}$ hybrid films. J Phys Chem B 113: 7513-7516
62. Han F, Kambala V, Srinivasan M, Rajarathnam D, Naidu R (2009) Tailored titanium dioxide photocatalysts for the degradation of organic dyes in wastewater treatment: a review. Appl Catal A 359:25–40
63. Niu P, Hao J (2012) Efficient degradation of methyl orange via multilayer films of titanium dioxide and silicotungstic acid. Sci China Chem 55:2366–2372
64. Niu P, Hao J (2014) Efficient degradation of organic dyes by titaniumdioxide–silicotungstic acid nanocomposite films: Influenceof inorganic salts and surfactants. Colloids Surf A 443:501–507
65. Niu P, Hao J (2013) Photocatalytic degradation of methyl orange by titanium dioxide-decatungstate nanocomposite films supported on glass slides. Colloids Surf A 431:127–132
66. Volkmer D, Chesne AD, Kurth DG, Schnablegger H, Lehmann P, Koop J, Müller A (2000) Toward nanodevices: synthesis and characterization of the nanoporous surfactant-encapsulated Keplerate $(DODA)_{40}(NH_4)_2[(H_2O)_n \subset Mo_{132}O_{372}(CH_3COO)_{30}(H_2O)_{72}]$. J Am Chem Soc 122:1995–1998
67. Zhang J, Song Y, Cronin L, Liu T (2008) Self-assembly of organic–inorganic hybrid amphiphilic surfactants with large polyoxometalates as polar head groups. J Am Chem Soc 130:14408–14409
68. Jia Y, Zhang J, Zhang Z, Li Q, Wang E (2014) Metal-centered polyoxometalates encapsulated by surfactant resulting in the thermotropic liquid crystal materials. Inorg Chem Commun 43:5–9
69. Li W, Wu L (2014) Liquid crystals from star-like clusto-supramolecular macromolecules. Polym Int 63:1750–1764
70. He Z, Ai H, Li B, Wu L (2012) A supramolecular gel based on an adenine symmetrically grafted Anderson-type polyoxometalate complex. Chin Sci Bull 57:4304–4309
71. Sun H, Yang Q, Hao J (2016) Self-patterning porous films of giant vesicles of $\{Mo_{72}Fe_{30}\}$ $(DODMA)_3$ complexes as frameworks. Adv Colloid Interface Sci 235:14–22
72. Tanford C (1972) Micelle shape and size. J Phys Chem 76:3020–3024
73. Zhou SJ, Feng YQ, Chen MJ, Li Q, Liu BY, Cao JM, Sun XF, Li HG, Hao J (2016) Robust onionlike structures with magnetic and photodynamic properties formed by a fullerene C_{60}-POM hybrid. Chem Commun 52:12171–12174
74. Song A, Dong S, Jia X, Hao J, Liu W, Liu T (2005) An onion phase in salt-free zero-charged catanionic surfactant solutions. Angew Chem Int Ed 44:4018–4021
75. Dong R, Zhong Z, Hao J (2012) Self-assembly of onion-like vesicles induced by charge and rheological properties in anionic-nonionic surfactant solutions. Soft Matter 8:7812–7821
76. Widawski G, Rawiso M, Francois B (1994) Self-organized honeycomb morphology of star-polymer polystyrene films. Nature 369:387–389
77. Sun F, Cai W, Li Y, Cao B, Lei Y, Zhang L (2004) Morphology-controlled growth of large-area two-dimensional ordered pore arrays. Adv Funct Mater 14:283–288
78. Kulinowski KM, Jiang P, Vaswani H, Colvin VL (2000) Porous metals from colloidal templates. Adv Mater 12:833–838

79. Nishikawa T, Nishida J, Ookura R, Nishimura SI, Scheumann V, Zizlsperger M, Lawall R, Knoll W, Shimomura M (2000) Web-structured films of an amphiphilic polymer from water in oil emulsion: fabrication and characterization. Langmuir 16:1337–1342
80. Shimomura M, Sawadaishi T (2001) Bottom-up strategy of materials fabrication: a new trend in nanotechnology of soft materials. Curr Opin Colloid Interface Sci 6:11–16
81. Fan D, Jia X, Tang P, Hao J, Liu T (2007) Self-patterning of hydrophobic materials into highly ordered honeycomb nanostructures at the air/water interface. Angew Chem Int Ed 119:3406–3409
82. Tang P, Hao J (2011) Macroporous honeycomb films of surfactant-encapsulated polyoxometalates at air/water interface and their electrochemical properties. Adv Colloid Interface Sci 161:163–170
83. Tang P, Hao J (2009) Formation mechanism and morphology modulation of honeycomb hybrid films made of polyoxometalates/surfactants at the air/water interface. J Colloid Interface Sci 333:1–5
84. Pitois O, Francois B (1999) Crystallization of condensation droplets on a liquid surface. Colloid Polym Sci 277:574–578
85. Stenzel MH (2002) Formation of regular honeycomb-patterned porous film by self-organization. Aust J Chem 55:239–243
86. Block MJ (1956) Surface tension as the cause of Bénard cells and surface deformation in a liquid film. Nature 178:650–651
87. Tang P, Hao J (2010) Photoluminescent honeycomb films templated by microwater droplets. Langmuir 26:3843–3847
88. Tang P, Hao J (2010) Directionally electrodeposited gold nanoparticles into honeycomb macropores and their surface-enhanced Raman scattering. New J Chem 34:1059–1062

The manufacturer's authorised representative in the EU is Springer Nature Customer Service Centre GmbH, Europaplatz 3, 69115 Heidelberg, Germany. If you have any concerns regarding our products, please contact ProductSafety@springernature.com

Printed and bound by CPI Group (UK) Ltd, Croydon, CR0 4YY
30/09/2026
02225446-0015